PHASE CONJUGATE LASER OPTICS

WILEY SERIES IN LASERS AND APPLICATIONS

D. R. VIJ, Editor
Kurukshetra University

OPTICS OF NANOSTRUCTURED MATERIALS • Vadim Markel

LASER REMOTE SENSING OF THE OCEAN: METHODS AND APPLICATIONS • Alexey B. Bunkin

COHERENCE AND STATISTICS OF PHOTONICS AND ATOMS • Jan Perina

METHODS FOR COMPUTER DESIGN OF DIFFRACTIVE OPTICAL ELEMENTS • Victor A. Soifer

PHASE CONJUGATE LASER OPTICS • Arnaud Brignon and Jean-Pierre Huignard (eds.)

PHASE CONJUGATE LASER OPTICS

Arnaud Brignon
Jean-Pierre Huignard
Editors

A WILEY-INTERSCIENCE PUBLICATION

JOHN WILEY & SONS, INC.

Copyright © 2004 by John Wiley & Sons, Inc. All rights reserved.

Published by John Wiley & Sons, Inc., Hoboken, New Jersey.
Published simultaneously in Canada.

No part of this publication may be reproduced, stored in a retrieval system, or transmitted in any form or by any means, electronic, mechanical, photocopying, recording, scanning, or otherwise, except as permitted under Section 107 or 108 of the 1976 United States Copyright Act, without either the prior written permission of the Publisher, or authorization through payment of the appropriate per-copy fee to the Copyright Clearance Center, Inc., 222 Rosewood Drive, Danvers, MA 01923, 978-750-8400, fax 978-646-8600, or on the web at www.copyright.com. Requests to the Publisher for permission should be addressed to the Permissions Department, John Wiley & Sons, Inc., 111 River Street, Hoboken, NJ 07030, (201) 748-6011, fax (201) 748-6008.

Limit of Liability/Disclaimer of Warranty: While the publisher and author have used their best efforts in preparing this book, they make no representations or warranties with respect to the accuracy or completeness of the contents of this book and specifically disclaim any implied warranties of merchantability or fitness for a particular purpose. No warranty may be created or extended by sales representatives or written sales materials. The advice and strategies contained herein may not be suitable for your situation. You should consult with a professional where appropriate. Neither the publisher nor author shall be liable for any loss of profit or any other commercial damages, including but not limited to special, incidental, consequential, or other damages.

For general information on our other products and services please contact our Customer Care Department within the U.S. at 877-762-2974, outside the U.S. at 317-572-3993 or fax 317-572-4002.

Wiley also publishes its books in a variety of electronic formats. Some content that appears in print, however, may not be available in electronic format.

Library of Congress Cataloging-in-Publication Data:

Brignon, Arnaud.
  Phase conjugate laser optics / Arnaud Brignon, Jean-Pierre Huignard.
    p. cm.
  ISBN 0-471-43957-6 (Cloth)
  1. Lasers. 2. Electrooptics. 3. Optical phase conjugation. I. Huignard, J.-P. (Jean-Pierre), 1944– II. Title.
TA1675 .B75 2003
621.36′6–dc22

                                                                                             200321226

Printed in the United States of America

10 9 8 7 6 5 4 3 2 1

CONTENTS

Foreword	xiii
Contributors	xv
Preface	xvii

Chapter 1. Overview of Phase Conjugation   1

Jean-Pierre Huignard and Arnaud Brignon

1.1	General Introduction	1
1.2	Phase Conjugation Through Four-Wave Mixing	5
	1.2.1 Phase Conjugation and Holography	5
	1.2.2 The Basic Formalism of Four-Wave Mixing	6
	1.2.3 Self-Pumped Phase Conjugation	8
1.3	The Nonlinear Materials	10
	1.3.1 Optical Kerr Effects	10
	1.3.2 Stimulated Brillouin Scattering	10
	1.3.3 Photorefraction	11
	1.3.4 Free Carriers in Semiconductors	11
	1.3.5 Saturable Amplification	12
	1.3.6 Saturable Absorption	12
	1.3.7 Molecular Reorientation in Liquid Crystals	13
	1.3.8 Thermal Gratings	13
1.4	The Criteria for the Choice of Materials	14
1.5	Conclusion	15
	References	15

Chapter 2. Principles of Phase Conjugating Brillouin Mirrors   19

Axel Heuer and Ralf Menzel

2.1	Introduction	19
2.2	Theoretical Description of the SBS Process	21
	2.2.1 General Equations	22
	2.2.2 Optical Phase Conjugation by SBS	25

v

		2.2.3	SBS Threshold	30
		2.2.4	Numerical Calculations 2D Model (Focused Geometry)	31
		2.2.5	Numerical Calculations 3D Model (Focused Geometry)	35
		2.2.6	Numerical Calculations for Waveguides	38
	2.3	Realization of SBS Mirrors		43
		2.3.1	Bulk Media SBS Mirrors	45
		2.3.2	Optical Fibers	49
		2.3.3	Tapered Fibers	52
		2.3.4	Liquid Waveguides (Capillaries)	54
	2.4	Summary		56
		References		57

Chapter 3. Laser Resonators with Brillouin Mirrors — 63

Martin Ostermeyer and Ralf Menzel

3.1	Introduction	63
3.2	Survey of Different Resonator Concepts with Brillouin Mirrors (SBS-PCRs)	64
3.3	Stability and Transverse Modes of Phase Conjugating Laser Resonators with Brillouin Mirror	70
3.4	Q-Switch via Stimulated Brillouin Scattering	78
3.5	Resonance Effects by Interaction of Start Resonator Modes with the SBS Sound Wave	82
3.6	Longitudinal Modes of the Linear SBS Laser	84
	3.6.1 Transient Longitudinal Mode Spectrum	84
	3.6.2 Mode Locking	90
	3.6.3 Analytical Pulse Shape Description	93
	3.6.4 Impact of Acoustic Decay Time on Longitudinal Modes	96
	3.6.5 Summary	99
3.7	High Brightness Operation of the Linear-SBS Laser	99
	References	105

Chapter 4. Multi-Kilohertz Pulsed Laser Systems with High Beam Quality by Phase Conjugation in Liquids and Fibers — 109

Thomas Riesbeck, Enrico Risse, Oliver Mehl, and Hans J. Eichler

4.1	Introduction	109
4.2	Amplifier Ssetups	110
4.3	Active Laser Media Nd:YAG and Nd:YALO	112
4.4	Design Rules for MOPA Systems	114
4.5	Beam Quality Measurement	116

4.6	Characterization of Fiber Phase Conjugate Mirror	117
4.7	Flashlamp-pumped Nd:YALO MOPA Systems with Fiber Phase Conjugator	123
4.8	Actively Q-Switched Flashlamp-pumped Nd:YAG MOPA Systems with Fiber Phase Conjugator	129
4.9	Continously Pumped Nd:YAG MOPA Systems with Fiber Phase Conjugator	131
4.10	500-Watt Average Output Power MOPA System with CS_2 as SBS Medium	137
4.11	Conclusion and Outlook	142
	References	143

Chapter 5. High-Pulse-Energy Phase Conjugated Laser System   147

C. Brent Dane and Lloyd A. Hackel

5.1	Introduction	147
5.2	High-Energy SBS Phase Conjugation	149
	5.2.1 The Question of Fidelity Versus Input Energy	149
	5.2.2 The Experimental Measurement of SBS Wavefront Fidelity	150
	5.2.3 The Input Pulse Rise-Time Requirement	152
5.3	A 25-J, 15-ns Amplifier Using a Liquid SBS Cell	154
	5.3.1 Design Considerations for the 15-ns System	154
	5.3.2 Optical Architecture of the 15-ns System	154
	5.3.3 SBS Phase Conjugation with a Liquid Cell	158
	5.3.4 Operation of the 25-J/Pulse 15-ns Laser System	161
	5.3.5 Summary of the 15-ns High-Energy Laser System	168
5.4	A Long Pulse 500-ns, 30-J Laser System	168
	5.4.1 Design Considerations for the 500-ns System	168
	5.4.2 Optical Architecture of the 500-ns System	169
	5.4.3 Long-Pulse SBS Phase Conjugation	172
	5.4.4 Output Characteristics of the 500-ns Laser System	176
	5.4.5 Summary of the 500-ns High-Energy Laser System	182
5.5	A 100-J Laser System Using Four Phase-Locked Amplifiers	184
	5.5.1 Design Considerations for the Phase-Locked System	185
	5.5.2 Optical Architecture of the Phase-Locked System	186
	5.5.3 Output Characteristics of the 100-J Laser System	191
	5.5.4 Summary of the 100-J Phase-Locked Laser System	197
5.6	Summary and Conclusions	198
	References	201

Chapter 6. Advanced Stimulated Brillouin Scattering for Phase Conjugate Mirror Using LAP, DLAP Crystals and Silica Glass — 205

Hidetsugu Yoshida and Masahiro Nakatsuka

6.1	Introduction	205
6.2	Crystal Structure of LAP and DLAP	206
6.3	Basic Characteristics for Stimulated Brillouin Scattering	207
	6.3.1 Damage Threshold	207
	6.3.2 Physical Properties of SBS	207
	6.3.3 SBS Reflectivity	209
6.4	Application of Solid-State SBS Mirrors to High-Power Lasers	213
	6.4.1 Correction of Aberrations	213
	6.4.2 High-Peak Power Laser System with LAP Phase Conjugate Mirror	215
	6.4.3 High-Energy Operation of Nd Lasers with Silica Glass Phase Conjugate Mirror	217
6.5	Conclusion	220
	References	220

Chapter 7. Stimulated Brillouin Scattering Pulse Compression and Its Application in Lasers — 223

G. A. Pasmanik, E. I. Shklovsky, and A. A. Shilov

7.1	Introduction	223
7.2	Phenomenological Description of Brillouin Compression	224
7.3	Theoretical Analysis of Brillouin Pulse Compression	227
7.4	Numerical Simulation	233
7.5	Characterization of Materials Used for SBS Compressors	237
7.6	Experimental Study of Brillouin Pulse Compression	239
7.7	Application of SBS Pulse Compression to Diode-Pumped Solid-State Lasers with High Pulse Repetition Rate	247
7.8	Conclusion	252
	References	253

Chapter 8. Principles and Optimization of $BaTiO_3$:Rh Phase Conjugators and their Application to MOPA Lasers at 1.06 µm — 257

Nicolas Huot, Gilles Pauliat, Jean-Michel Jonathan, Gérald Roosen, Arnaud Brignon, and Jean-Pierre Huignard

8.1	Introduction	257

8.2	Overview of Material Properties		258
	8.2.1	Characterization with CW Illumination	258
	8.2.2	Performances of Oxidized Crystals	264
	8.2.3	Characterization with Nanosecond Illumination	267
8.3	Self-Pumped Phase Conjugation		272
	8.3.1	Internal Loop Self-Pumped Phase Conjugate Mirror	272
	8.3.2	Ring Self-Pumped Phase Conjugation	272
8.4	Dynamic Wavefront Correction of MOPA Laser Sources		285
	8.4.1	Origin of Aberrations in Nd:YAG Amplifier Rods	285
	8.4.2	MOPA Laser Sources Including a Photorefractive Self-Pumped Phase Conjugate Mirror	286
	8.4.3	Comparison of Photorefractive Self-Pumped Phase Conjugation to Other Existing Techniques	291
8.5	Conclusion		293
	References		294

Chapter 9. Spatial and Spectral Control of High-Power Diode Lasers Using Phase Conjugate Mirrors — 301

Paul M. Petersen, Martin Løbel, and Sussie Juul Jensen

9.1	Introduction		302
9.2	Laser Diode Arrays with Phase Conjugate Feedback		303
9.3	Frequency-Selective Phase Conjugate Feedback with an Etalon in the External Cavity		306
	9.3.1	Experimental Setup	306
	9.3.2	Characteristics of the On-Axis Configuration	308
	9.3.3	Far-Field Spatial Characteristics in the Off-Axis Configuration	309
	9.3.4	The Improvement of the Spatial Brightness	311
	9.3.5	Spectral Characteristics of the Laser System	312
	9.3.6	The Improvement of the Temporal Coherence	313
9.4	Tunable Output of High-Power Diode Lasers Using a Grating in the External Cavity		314
9.5	Stability of the Output of Diode Lasers with External Phase Conjugate Feedback		318
	9.5.1	Long-Term Stability of the Phase Conjugate Laser System	320
	9.5.2	The Influence of External Reflections of the Output Beam	320
9.6	Frequency Doubling of High-Power Laser Diode Arrays		323
9.7	Conclusions and Perspectives		325
	References		325

Chapter 10. Self-Pumped Phase Conjugation by Joint Stimulated Scatterings in Nematic Liquid Crystals and Its Application for Self-Starting Lasers — 331

Oleg Antipov

10.1	Introduction	331
10.2	Self-Pumped Phase Conjugation by Joint Stimulated Scattering	333
	10.2.1 Geometrical Features of Joint Stimulated Scattering	333
	10.2.2 Theoretical Description of Phase Conjugation by Joint Stimulated Scattering in a Nonlinear Layer with Feedback Loop	334
	10.2.3 Experimental Investigations of Self-Pumped Phase Conjugation of Laser Beams in Nematic Liquid-Crystal Lasers	345
10.3	Self-Starting Lasers with a Nonlinear Mirror Based on Nematic Liquid Crystals	351
	10.3.1 Theoretical Description of the Principle of Self-Starting Lasers	351
	10.3.2 Numerical Computation of the Self-Starting Laser with an NLC Mirror	353
	10.3.3 Experimental Investigation of the Self-Starting Lasers	357
10.4	Conclusion	363
	References	364

Chapter 11. Self-Adaptive Loop Resonators with Gain Gratings — 367

Michael J. Damzen

11.1	Introduction	367
11.2	Theory of Multiwave Mixing in Gain Media	371
	11.2.1 Rate Equation for the Laser Gain Coefficient	371
	11.2.2 The Optical Field Equation	372
	11.2.3 The Intensity Interference Pattern	372
11.3	The Steady-State Regime	374
11.4	The Transient Regime	377
11.5	The General Time Regime	379
	11.5.1 Instantaneous Coupling Coefficients	380
	11.5.2 Time-Integrated Coupling Coefficients	381
11.6	Self-Pumped Phase Conjugation	382
	11.6.1 Experimental Setup	385
	11.6.2 Spatial and Phase Conjugation Behavior	386
	11.6.3 Energy and Temporal Behavior	386

11.7	Double Phase Conjugation	387
11.8	Self-Starting Adaptive Gain-Grating Lasers	390
11.9	Self-Adaptive Loop Resonators Using a Thermal Grating Hologram	392
11.10	Experimental Characterization of a Thermal Grating	396
	11.10.1 Time Dynamics and Diffraction Efficiency Results	397
	11.10.2 Spatial Issues and Phase Conjugation Results	398
11.11	Experimental Operation of a Self-Adaptive Loop Resonator Using a Thermal Grating "Hologram"	399
	11.11.1 Experimental Adaptive Laser System	400
	11.11.2 Experimental Results of Adaptive Resonator	401
	References	404

Index 407

■ FOREWORD

Research activities in laser physics and in photonics technologies over the last two decades have continuously produced a large diversity of new advances. Several examples illustrate the major impact of optics in the quantum sciences, engineering, metrology, communication fiber networks, or high-capacity data storage. Besides these established fields of research and development for industry or for the consumer markets, laser optics will certainly disseminate in the near future in new areas such as biology, chemistry, medicine, or nanotechnologies. The constant progress of new generations of solid-state lasers will support these objectives for the extension of the fields of applications of photonics.

The performances, reliability, and cost effectiveness of diode pumping has largely contributed to the current maturity of the laser technologies. It permits the realization of more efficient sources and the extraction of more energy from the amplifying media in the continuous or pulse operating modes. These requirements are challenging innovative approaches for the design of new laser architectures emitting high power and high brightness beams whose quality is close to the diffraction limit. This volume, edited by A. Brignon and J.-P. Huignard of Thales Research and Technology, contributes to these ambitious objectives by reviewing original nonlinear optical techniques that permit a dynamic correction of any beam distortion due to passive or active optical elements in the cavity. Optical phase conjugation possesses the fascinating ability to restore a perfect beam after it is reflected by a nonlinear mirror. The function has stimulated a great deal of research into the physics of the nonlinear phenomena and beam interactions which promise to have the best characteristics for realizing this unconventional optical component. The authors of the different chapters of this volume are major players in this field, and they clearly highlight the original concepts of nonlinear optics involved for the demonstration of novel laser architectures based on conjugate mirrors for delivering laser beams with a high spatial quality. Basic phenomena, laser structures, and experiments for beam characterization are treated in great detail in the different chapters of the volume. This collection of chapters provides the status of the current developments in the field. It represents a full complement of a long period of basic research efforts involving the multidisciplinary expertise of scientists and engineers encompassing optical material sciences, laser physics, and laser engineering. We hope that this book will stimulate further activities for the discovery of new nonlinear media since the concept of phase conjugation will undoubtedly apply for scaling future high-energy laser performances beyond traditional limits. We are confident that controlling all the key parameters of the sources such as power, spectral bandwidth, and brightness through self-adaptive nonlinear optical technics will contribute to the widespread development of lasers systems that satisfy the requirements of industrial and scientific applications.

Dominique Vernay
Technical and Scientific Manager
Thales—Paris

CONTRIBUTORS

Oleg Antipov, Institute of Applied Physics, Russian Academy of Sciences, 603950 Nizhny Novgorod, Russia

Arnaud Brignon, Thales Research and Technology—France, 91404 Orsay, France

Michael J. Damzen, The Blackett Laboratory, Imperial College, London SW7 2BW, United Kingdom

C. Brent Dane, Lawrence Livermore National Laboratory, Livermore, California 94550, USA

Hans J. Eichler, Technische Universität Berlin, Optisches Institut, 10623 Berlin, Germany

Lloyd A. Hackel, Lawrence Livermore National Laboratory, Livermore, California 94550, USA

Axel Heuer, University of Potsdam, Institute of Physics, Chair of Photonics, 14469 Postdam, Germany

Jean-Pierre Huignard, Thales Research and Technology—France, 91404 Orsay, France

Nicolas Huot, Laboratoire Charles Fabry, Institut d'Optique, 91403 Orsay, France

Sussie Juul Jensen, Optics and Fluid Dynamics Department, Risø National Laboratory, DK-4000 Roskilde, Denmark

Jean-Michel Jonathan, Laboratoire Charles Fabry, Institut d'Optique, 91403 Orsay, France

Martin Løbel, Optics and Fluid Dynamics Department, Risø National Laboratory, DK-4000 Roskilde, Denmark

Oliver Mehl, Technische Universität Berlin, Optisches Institut, 10623 Berlin, Germany

Ralf Menzel, University of Potsdam, Institute of Physics, Chair of Photonics, 14469 Postdam, Germany

Masahiro Nakatsuka, Institute of Laser Engineering, Osaka University, Osaka 565-0871, Japan

Martin Ostermeyer, University of Potsdam, Institute of Physics, Chair of Photonics, 14469 Postdam, Germany

G. A. Pasmanik, Passat, Toronto, Ontario, Canada M3J 3H9

Gilles Pauliat, Laboratoire Charles Fabry, Institut d'Optique, 91403 Orsay, France

Paul M. Petersen, Optics and Fluid Dynamics Department, Risø National Laboratory, DK-4000 Roskilde, Denmark

Thomas Riesbeck, Technische Universität Berlin, Optisches Institut, 10623 Berlin, Germany

Enrico Risse, Technische Universität Berlin, Optisches Institut, 10623 Berlin, Germany

Gérald Roosen, Laboratoire Charles Fabry, Institut d'Optique, 91403 Orsay, France

A. A. Shilov, Passat, Toronto, Ontario, Canada M3J 3H9

E. I. Shklovsky, Passat, Toronto, Ontario, Canada M3J 3H9

Hidetsugu Yoshida, Institute of Laser Engineering, Osaka University, Osaka 565-0871, Japan

PREFACE

Since the discovery of the laser in the 1960s, a great amount of research activity has led to an impressive increase of the overall performances of the sources emitting in the visible or in the infrared spectral regions. The most significant achievements for solid-state lasers in the last 10 years are the increase in laser output power or pulse energy by orders of magnitude due to the introduction of the diode pumping of the gain media. This technology also led to a remarkable improvement of the electrical to optical efficiency as well as compactness and reliability of the sources. All these recent technological breakthroughs have contributed to the fast evolution of the field of photonics and a growing interest in solid-state lasers for many different industrial and scientific applications. For example, in manufacturing, material processing, or the medical areas, lasers are now routinely used to focus high-energy densities on a surface. This ability also opens new opportunities in basic science interactions for plasma physics or X-ray generation with sources delivering ultrashort pulses. Also due to the directivity of optical antennas, lasers will undoubtly be applied in LIDAR imaging systems, for ground or space communications or for monitoring of the atmosphere. All these applications clearly require sources delivering high-quality optical beams whose divergence must not exceed the diffraction limit during beam propagation. In other terms, the wavefront emitted by a high-power laser must be free of any aberrations or distortions which would degrade the brightness of the source and thus would lead to a decrease of the system performances. Attaining these operating conditions is an important challenge, since thermal loading due to strong pumping of the gain media induces aberrated thermal lenses, which severely affects the beam quality. It thus results in wavefront aberrations that reduce the brightness of the source and that evolve when changing the operating conditions of the source. Adaptive correction of phase aberrations in a laser cavity or in a master-oscillator power-amplifier structure is thus a crucial problem that must be taken into account in solid-state laser sources. An elegant approach offering a great potential to solve this question involves nonlinear optical phase conjugation. This technique permits the generation of a complex phase conjugate replica of a wavefront after beam reflection on a nonlinear mirror, thus leading to a compensation of any wavefront distorsions. This nonlinear reflection can be interpreted as the conjugate wavefront generation due to a dynamic hologram in a material that exhibits a third-order nonlinearity. Since the discovery of the effect in the early 1970s, optical phase conjugation is now an established field of nonlinear optics, and it has opened very important scientific and technological advances in laser physics over the last decades. This book is devoted to the current development in the field of Phase

Conjugate Lasers with the objective of showing the impact of these innovative concepts on the architectures and performances of a new class of solid-state lasers. Phase conjugate lasers exhibit adaptive correction of their own aberrations whatever their operating conditions, and they provide maximum brightness to the user for a large diversity of scientific or industrial applications. The critical issue of this very attractive approach is to identify the most efficient media and nonlinear mechanisms that operate at the required wavelengths. In this perspective, the book presents the basic physical phenomena and materials involved for efficient generation of the conjugate waves for specific examples of laser sources. The book also develops in detail an analysis of the laser architectures and nonlinear mirrors that are best suited to operate in continuous-wave or pulsed regimes, respectively. The ability of phase conjugate lasers to deliver beams with a high spatial and spectral quality is clearly outlined in the different chapters.

After a brief overview of the basic principles of nonlinear optical phase conjugation in Chapter 1, a large part of the book is devoted to lasers, including a Brillouin phase conjugating mirror. In Chapter 2 the principles, the basic properties, the materials (bulk and fiber geometry), and performances of stimulated Brillouin scattering (SBS) mirrors are presented. Such nonlinear mirrors can be implemented inside a laser resonator as shown in Chapter 3. Besides the demonstration of high brightness operation, the authors analyze in detail the stability and the mode structures of these unconventional nonlinear resonators. To achieve high power with a near-diffraction-limited beam, master-oscillator power-amplifier (MOPA) configurations are demonstrated in Chapter 4 in which both liquid and glass fiber Brillouin conjugators are used. The fiber presents the advantages of compactness and lower energy threshold due to the long interaction length of the fiber medium. However, to achieve very high energy the use of SBS liquid cells is required as presented in Chapter 5. Using the capability of phase conjugation to phase-lock several beams issued from different amplifiers, the authors demonstrate up to 100 J of output energy while keeping the beam quality close the diffraction limit. Some applications may require solid-state SBS mirrors instead of liquid cells. For that purpose, the authors of Chapter 6 investigate and characterize SBS properties of bulk solid-state materials like organic crystals and glasses. The previous chapters have concerned the ability of SBS mirrors to compensate for phase aberrations of gain media. It is also important to highlight (as done in Chapter 7) that an SBS nonlinear mirror can perform pulse compression in the time domain. This brings the opportunity of controlling both spatial and temporal characteristics of laser pulses with the same nonlinear mechanism. In the following chapters, alternative nonlinear mechanisms are presented. In particular, infrared-sensitive photorefractive crystals are used in Chapter 8. The authors detail the specific properties of this type of nonlinear material and demonstrate dynamic correction of MOPA laser sources. It is also shown in Chapter 9 that photorefractive crystals can be used to realize a semiconductor laser diode cavity with phase conjugate feedback for spatial and spectral filtering of the modes. In Chapter 10, a nematic liquid crystal cell is implemented in a laser resonator to perform phase conjugation and correction of intracavity distortions. This relies on the large anisotropy and nonlinear effects in liquid crystals. Thermal gratings can also be used to build a self-adaptive phase conjugate loop resonator as demonstrated in Chapter 11. In all these studies, two distinct materials are employed for the gain medium and the phase conjugate mirror. It is finally shown in Chapter 11 that laser gain media can perform phase conjugation by using gain saturation as the nonlinear mechanism. Self-adaptive holographic loop resonators are demonstrated using this interaction.

This book gives a complete review of the state of the art of phase conjugate lasers, including laser demonstrators, performance, technology, and selection of the most important and promising classes of nonlinear media.

We express our warm thanks to all our co-authors for their very valuable contributions and for their fruitful discussions and cooperation during the preparation of this book.

Jean-Pierre Huignard
Arnaud Brignon
Paris, 2003

CHAPTER 1

Overview of Phase Conjugation

JEAN-PIERRE HUIGNARD and ARNAUD BRIGNON

Thales Research and Technology—France, 91404 Orsay, France

1.1 GENERAL INTRODUCTION

The discovery in the early 1970 by Zel'dovich et al. [1] that a nonlinear process could generate a phase conjugate replica of a complex incident wavefront has opened a wide interest in the laser and optics community. Since the first experiments done with a ruby laser and Brillouin scattering in a gas cell, the field of optical phase conjugation has stimulated a lot of research and development activities that cover both the fundamental and applied parts of the field of laser optics. The important new aspects of optical phase conjugation which are of prime interest are the following: First, phase conjugation is a nonlinear mechanism that reverses both the direction of propagation and the phase of an aberrated wavefront; second, the generation of the conjugate beam can be viewed as a dynamic holographic recording process in a medium that exhibits a third-order nonlinearity. Such an unconventional optical device is now known as a phase conjugator or a nonlinear phase conjugate mirror. The major applications of phase conjugation will rely on these remarkable physical properties, which are illustrated in Fig. 1.1. It shows the now well-known comparison between a classical mirror on Fig. 1.1a which satisfies the conventional reflection law for the incident wavefront, while Fig. 1.1b shows the function of a nonlinear mirror which reverses the sign of the incident wave vector $\vec{k}_i$ at any point of the incident wavefront propagating in the $+z$ direction. In other words, if $\mathbf{E}_i = E_i \exp(i\omega_0 t - ik_i z)$ is the incident scalar optical field expression, the returned conjugate field $\mathbf{E}_c$ due to the nonlinear mirror is expressed by $\mathbf{E}_c = E_i^* \exp(i\omega_0 t + ik_i z)$. This field propagates in the $-z$ direction with complex amplitude E_i^* and at frequency ω_0. We will show later that the intensity of the conjugate field is affected in the general case by a nonlinear reflection coefficient R (R can be larger than one) and in some interactions by a slight frequency shift $\delta \ll \omega_0$. Figure 1.2 illustrates the situation where an incident wavefront is disturbed by an aberrating medium (atmospheric turbulence, passive or active optical components, etc.). Due to phase reversal, a diffraction-

---

Phase Conjugate Laser Optics, edited by Arnaud Brignon and Jean-Pierre Huignard
ISBN 0-471-43957-6   Copyright © 2004 John Wiley & Sons, Inc.

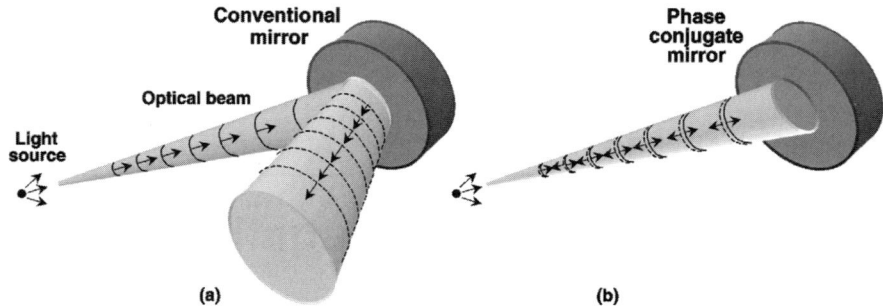

Figure 1.1. Comparison of beam reflection by (a) a conventional mirror and (b) a nonlinear phase conjugate mirror.

limited wave can be recovered after double passing through severely aberrated optical components and beam reflection on the nonlinear mirror. In particular—and this is the main subject treated in this volume—a phase conjugate mirror permits the compensation of any static or dynamic aberrations due to high gain medium in a laser cavity or in a master oscillator power amplifier architecture. These important properties are described in Fig. 1.3. Figure 1.3a shows a laser oscillator whose cavity consists of a classical and a conjugate mirror: A stable oscillation can occur because of the compensation of the thermal lensing effects and aberrations due to the highly pumped gain media. In such conditions a diffraction-limited beam can be extracted from the cavity. The alternative approach is presented in Fig. 1.3b. The oscillator emits a low-energy beam with a diffraction-limited quality. It is then amplified by the gain medium operating in a double-pass configuration. Due to the conjugate mirror, the returned beam is compensated for any aberrations due to the high-gain laser amplifier. A diffraction-limited beam is extracted by 90° polarization rotation. So, according to these remarkable properties, it is expected that we can realize a new class of high-power and high-brightness phase conjugate lasers delivering a beam quality that fits the requirements for scientific and industrial applications. This

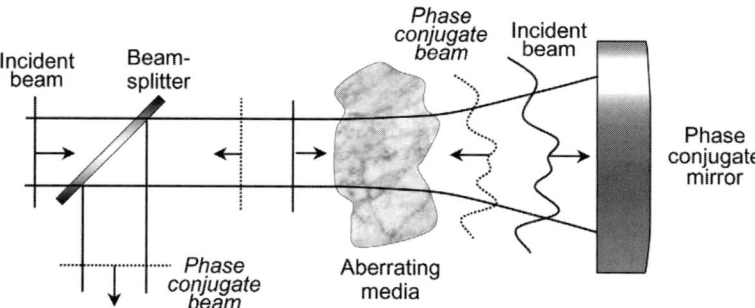

Figure 1.2. Compensation of the aberrations due to a phase distorting media by wavefront reflection on a phase conjugate mirror.

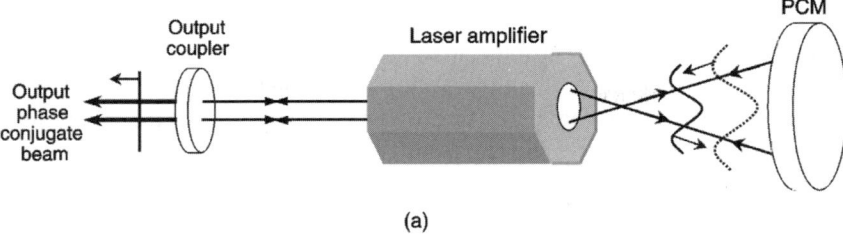

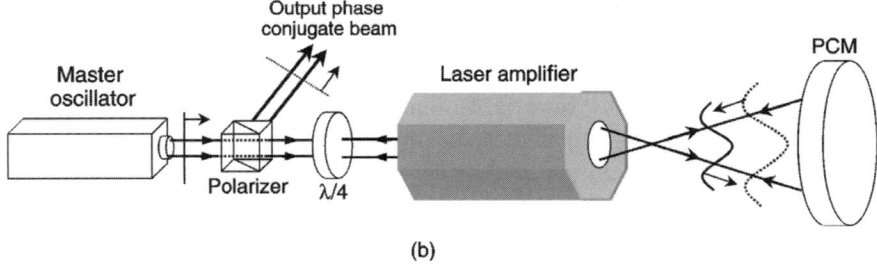

Figure 1.3. The two main laser architectures involving a phase conjugate mirror for correction of the aberrations due to thermal effects in the gain medium. (a) Laser oscillator with intracavity phase conjugate mirror and (b) master-oscillator power amplifier with a phase conjugate mirror.

capability of aberration compensation was also shown in the earliest research works on Fourier optics and holography. Kogelnik [2] had already demonstrated that static aberrations can be compensated by using conventional holographic recording. After processing of the photographic media and proper readout of the hologram, it generated the backward conjugate wave for a clear image restoration through a distorting media. The analogy of phase conjugation with dynamic holography was then outlined by Yariv [3] and in early experiments with photorefractive crystals [4], and it contributed to extend the field of applications, thus including parallel image processing, optical correlation for pattern recognition, holographic interferometry for non destructive testing, incoherent to coherent image conversion, novelty filters for moving object detection [3].

It was then recognized that when doing simultaneous recording and readout of a volume hologram with beams having the same or nearly the same wavelength, the incident or conjugate waves can be amplified [5–8]. This is due to the energy transfer from the pump beams which interfere with the probe beam in the volume of the nonlinear media. These phenomena of mutual coupling of waves interfering in the nonlinear media are of great importance in view of applications and have led to remarkable unified treatments of the fields of nonlinear optics and dynamic holography. These interactions have led to outstanding applications to coherent beam amplification and to amplified phase conjugation. The possibility of amplifying the amplitude of a complex wavefront through a third-order nonlinearity permits (a) the demonstration of novel types of optical resonators, including a nonlinear mirror with gain, or (b) the attainment of high-gain and low-noise image amplification. We must also outline that pioneering experiments on dynamic holography were done in the 1970s in different media allowing the recording of elementary gratings by interfering two beams at the same wavelength. The gratings was due to a change of the absorption or of the index of refraction of the media. In these early experiments, the holographic technique was mainly used to probe the spatiotemporal evolution of the physical mechanisms responsible for grating formation in materials like semiconductors or saturable absorbers [9, 10].

The mechanisms of dynamic holography and phase conjugation have stimulated a great interest in the research laboratories for nearly two decades. First, it was important to analyze and to characterize with details the physics of the third-order nonlinearities $\chi^{(3)}$ when the material is illuminated by interference of a pump and probe beams. It induces a spatial modulation of the material complex dielectric constant which generates an amplitude or a phase volume grating. Third-order effects occur in isotropic transparent media, and there is no restriction to material that exhibits an inversion symmetry center as for the second-order nonlinearities. In particular, the most established $\chi^{(3)}$ nonlinear mechanisms for phase conjugation are Kerr effects, Brillouin scattering, and Raman scattering. However, other effects that may also provide an efficient index or amplitude modulation through mechanisms with a response time ranging several nanoseconds to seconds are of great interest for phase conjugation [11]. This is the situation encountered with photorefractive effects, free carrier generation in semiconductors, laser gain media, and so on; several chapters of this book will present laser architectures and system performances based on these nonlinearities used to realize an efficient phase conjugate mirror. Another aspect of the field which has been intensively covered in the research labs is the capability of the dynamic hologram to exchange energy between the incident and conjugate waves that interfere in the dynamic media. First observed and analyzed by holographic self-diffraction phenomena in photorefractive crystals like $LiNbO_3$, [12, 13], these effects were subjects to intense research activities either using the formalism of wave propagation in the nonlinear media, or based on the point of view of holography and self-diffraction that can reinforce (or reduce) the intensity of one of the interfering probe (or pump) beams. Both approaches complement each other with regard to the understanding of the physical phenomena involved for the generation of a phase conjugate wavefront. They permit the prediction of novel beam interactions and applications and also enable us to compare materials and mechanisms through the calculation and experimental measurements of the two-wave or four-wave mixing gain coefficients per unit of length of the nonlinear media. This parameter, as well as the required laser characteristics such as wavelength, continuous or pulsed operation, and incident energy or power levels, will contribute to our making the right choice of the nonlinear mechanism and materials for a reliable operation of the laser including a phase conjugator [14, 15]. Hereafter, we will detail the main optical configurations as well as the physical mechanisms and materials that have been proposed and studied during the early stages of the field [16–18]. They are now at the origin of the current developments of optical phase conjugation for novel laser cavities and architectures which are treated in this book.

1.2 PHASE CONJUGATION THROUGH FOUR-WAVE MIXING

1.2.1 Phase conjugation and holography

It is now well established from previous works that the basic geometry for phase conjugation consists of two counterpropagating plane waves that are the pump beams of amplitudes E_1 and E_2 and that interfere with a probe beam of amplitude E_p. In general, the beams have the same polarization state, and the probe is incident at an arbitrary angle on the recording medium having a third-order nonlinearity. A clear interpretation of conjugate beam generation shown in Fig. 1.4 is the following: The probe and signal beam interfere in the nonlinear material and create an interference periodic pattern that modulates the properties of the media. The resulting grating wave vector amplitude is $k = 2\pi/\Lambda$. The fringe period Λ is given by the formula $\Lambda = \lambda/2 \sin \theta$, where λ is the laser wavelength, and $\pm \theta$ is the pump and probe beam incident angles with respect to the normal to the nonlinear media; it results in a complex volume hologram due to a spatial distribution of the refractive index (Kerr-like media), of the absorption (saturable absorbers), or of the gain when the conjugator is the laser medium itself. The second antiparallel pump beam is then diffracted under Bragg conditions by the dynamic volume hologram; following the classical formalism of holography, it generates a backward conjugate wavefront whose complex amplitude can written as $E_c = E_p^* E_1 E_2$. It is named the conjugate image beam in holography, and its amplitude is proportional to E_p^*. Since this geometry involves waves E_1, E_2, E_p, and E_c which are simultaneously present in the material, it is known as four-wave mixing (4WM). The reflectivity of the nonlinear mirror is thus defined as $R = |E_c(0)/E_p(0)|^2$. Considering that ω_1, ω_2, and ω_p are the respective frequencies of the pump and probe beams, the conjugate beam exhibits a frequency $\omega_c = \omega_1 + \omega_2 - \omega_p$. However, starting from this general formula, two particular situations are of interest: the degenerate four-wave mixing where all frequencies are equal to ω_0 (Fig. 1.4a) and the nearly degenerate four-wave mixing case (Fig. 1.4b) where $\omega_c = \omega_0 - \delta$, $\delta \ll \omega_0$ when the probe and pump frequencies are respectively $\omega_0 + \delta$ and ω_0. This last situation is com-monly encountered in several types of efficient nonlinearities (Brillouin scattering, photorefractives, etc.) where the frequency detuning δ range respectively from several gigahertz to few hertz. It is viewed as a nonlinear interaction involving a moving holographic grating. The formalism describing the coupling of the interfering optical fields via a given material nonlinearity is another important aspect of the work done on phase conjugation. The analysis must identify the conditions for efficient interactions between the beams for energy transfer from the pump to the incident and conjugate beams. In particular, it will be shown that interactions involving nearly degenerate four-wave mixing or a nonlocal spatial response of the material (phase-shifted volume grating) permit the attainment of high gain coefficients. It will result in

6 OVERVIEW OF PHASE CONJUGATION

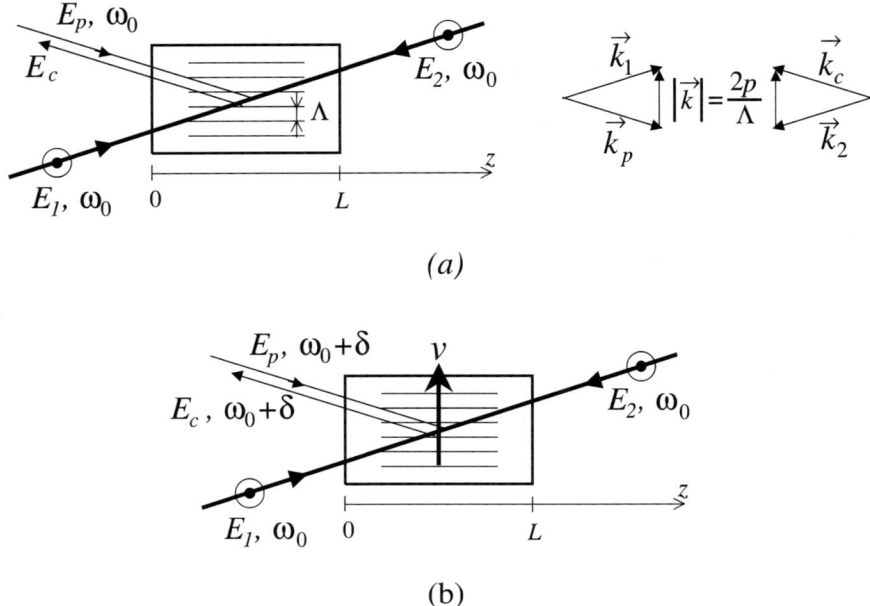

Figure 1.4. Generation of a phase conjugate wavefront by four-wave mixing interaction. The pump and probe beams interfere to create a dynamic hologram in the nonlinear medium. (a) Degenerate four-wave mixing and (b) nearly degenerate four-wave mixing.

amplification (or depletion) of the probe or amplified conjugate beam reflectivity. These unique properties make the 4WM interactions extremely useful for the applications developed in this book. Hereafter, we give some of the major relations that govern the amplitudes of the transmitted E_p ($z = L$) and conjugate E_c ($z = 0$) waves, thus allowing us to calculate the reflectivity of the nonlinear mirror.

1.2.2 The basic formalism of four-wave mixing

The original equations of degenerate (or nearly degenerate) 4WM were first derived independently by Yariv and Pepper [5] and Bloom and Bjorklund [8] by considering a set of coupling wave equations which will describe the space and time evolution of the waves that interact all together in the nonlinear media. The equations relate the fact that the hologram is dynamic: It adapts in real time (with an inertia due to a finite response time) to the interference pattern due to mutual interference of the beams in the gratings that they are writing and reading. Note that with all beams having the same or nearly the same frequency ($\delta \ll \omega_0$), we consider that there is a perfect phase matching of the $\vec{k}$ vectors $\vec{k}_c = \vec{k}_1 + \vec{k}_2 - \vec{k}_p$, where $\vec{k}_i$ is the optical wave vector. In the hypothesis of nondepleted pump beam and equal frequencies, we have from Refs. [5–9] the following set of coupled equations for the steady-state amplitudes of the optical fields E_c and E_p:

$$\frac{dE_c}{dz} = i\frac{\omega_0}{2}\sqrt{\frac{\mu}{\varepsilon}}\chi^{(3)}E_p^*E_1E_2 \tag{1}$$

$$\frac{dE_p}{dz} = -i\frac{\omega_0}{2}\sqrt{\frac{\mu}{\varepsilon}}\chi^{(3)}E_c^*E_1E_2 \tag{2}$$

where ω_0 the optical frequency of the beams, $\chi^{(3)}$ is the third-order nonlinearity in SI units, and μ and ε are the susceptibility and permittivity of the media. From these general expressions we can derive more simple formulas when introducing a coupling constant K that will be proportional to the pump beam amplitudes E_1 and E_2 and to the third-order nonlinear coefficient of the media χ^3:

$$\frac{dE_p}{dz} = -iK^*E_c^* \tag{3}$$

$$\frac{dE_c}{dz} = -iK^*E_p^* \tag{4}$$

where $K^* = i(\omega_0/2)\sqrt{\mu/\varepsilon}\chi^{(3)}E_1E_2$. Solving these equations leads to the following expressions for the transmitted and reflected fields for a material interaction length L:

$$E_c(0) = -iE_p^*(0)\frac{K^*}{|K|}\tan|K|L \tag{5}$$

$$E_p(L) = E_p(0)\frac{1}{\cos|K|L} \tag{6}$$

Two parameters will be of interest in view of applications of degenerate four-wave mixing: first, the reflectivity $R = |E_c(0)/E_p(0)|^2$ of the nonlinear phase conjugate mirror; and second, the gain $G = |E_p(L)/E_p(0)|^2$, which characterizes probe beam amplification. According to the above relations, the parameters R and G which are characteristic of the degenerate 4WM interaction are expressed by the following expressions:

$$R = \tan^2|K|L \tag{7}$$

$$G = \frac{1}{\cos^2|K|L} \tag{8}$$

These formula clearly show that amplified reflection and transmission can occur for $|KL|$ satisfying $\pi/4 < |K|L < 3\pi/4$. In these conditions, the nonlinear interaction can be seen as a parametric amplifier for both the reflected and transmitted waves due to efficient energy transfer from the pump beams. Also an important consequence is the existence of self-oscillation when $|K|L = \pi/2$. It physically corresponds to an optical oscillation without mirror feedback for zero intensity probe beam.

Since amplification appears to be independent of the pump-probe beam angle in the above formula, 4WM interaction could permit to amplify complex incident and conjugate wavefront carrying spatial informations. However, it will be shown that with real materials there often exist an optimum range of pump-probe beam angles due to the physical mechanisms involved for the generation of the photoinduced grating index modulation. This angular bandpass will limit the number of transverse modes that can be amplified or conjugated, and it determines the space bandwidth product of this new type of image amplifier.

As an example, stimulated Brillouin scattering and photorefractive phenomena at zero applied field are more efficient for high spatial frequency gratings arising from interference of contrapropagating beams. On the other hand, other effects will require large grating periods due a low beam angle between probe and pump.

An important consequence of the 4WM interaction is its capability of realizing a laser cavity including a phase conjugate mirror having a continuous gain as demonstrated first by Feinberg and Hellwarth [19] with $BaTiO_3$ crystal and shown in Fig. 1.5. A classical mirror of amplitude reflectivity r is introduced on the path of the incident probe; and due to the phase conjugate properties of the nonlinear mirror whose reflectivity can be higher than unity, there can be a stable oscillation buildup in the cavity. The beam oscillating in the cavity starts from the coherent noise due to the pump beams E_1 and E_2. Its frequency is equal to the pump frequencies ω_0 in the case of a degenerate interaction and is independent of the distance between the mirror and the nonlinear medium. This experiment is generally realized with interfering pump and probe beams having the same state of polarization. If this condition is not satisfied due to the optical components that introduce both spatial aberrations and depolarizations, a special experimental arrangement is required to process the two polarization states for obtaining vectorial phase conjugation.

1.2.3 Self-pumped phase conjugation

Another important configuration for conjugation shown in Fig. 1.6 consists only of an aberrated signal beam that is reflected back by the nonlinear medium as a conjugate wavefront. This particular interaction, called self-pumped phase conjugation [20], does not require external contrapropagative pump beams, which

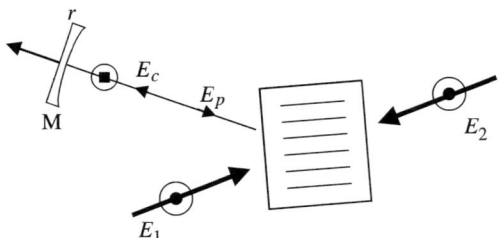

Figure 1.5. Phase conjugate oscillator. The oscillation is due to the nonlinear phase conjugate mirror that exhibits a reflectivity higher than unity.

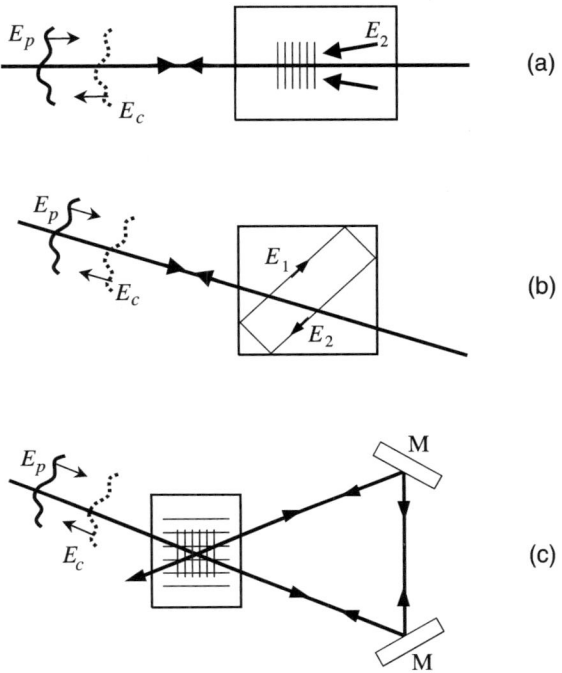

Figure 1.6. Self-pumped phase conjugate (SPPC) configurations. (a) SPPC by nonlinear backscattering, (b) SPPC due to self-induced internal feedback loop inside the nonlinear material, and (c) SPPC due to an external feedback loop.

interfere with the signal wave as shown in the Fig. 1.4a. In view of applications, it thus appears very convenient to use this interaction whose main characteristics are thus the following: The maximum conjugate beam reflectivity is equal to unity since there is no gain is due to energy transfer from a pump beam as is achieved in the conventional 4WM geometry; the phase conjugate beam originates from the coherent noise due to the signal beam. As shown in the Fig. 1.6a, it generates in the nonlinear medium complex reflection (or transmission) types of holographic volume gratings which are due to the interference of the signal with scattered plane waves components that propagate in the same or in opposite directions. It thus results in regions in the nonlinear medium where there is the equivalent of a 4WM interaction that self-generates the conjugate wave of the incident one. Also, the beams inside the volume of the media may form a loop (Fig. 1.6b), but the interaction can also be reinforced by using and external ring cavity geometry [21] to perform self-pumped phase conjugation with higher reflectivity (Fig. 1.6c). This geometry is very simple to use in experiments, and it has a great potential for improving the brightness of laser sources. However, this self-pumped interaction is not encountered in all nonlinear mechanisms. Self-pumped phase conjugation is very well suited in interactions involving stimulated Brillouin scattering or photorefractive backscattering phenomena and in particular several chapters of this book will illustrate applications showing the great potential of these mechanisms for beam quality control in laser cavities or in master oscillator power amplifier laser architectures.

1.3 THE NONLINEAR MATERIALS

There is potentially a wide variety of nonlinear physical mechanisms that can be used for optical phase conjugation of coherent laser beams. The generation of a phase conjugate wavefront arises from the third-order material nonlinearity; moreover, analogy with holography introduced above affords the opportunity of using materials allowing the recording of phase or amplitude dynamic volume of the holograms. Therefore, the mechanisms presented hereafter have been the subject of extensive research activities during the last two decades—both on the basic material properties and in depth analysis of interactions between beams that interfere in the volume of the nonlinear media.

1.3.1 Optical Kerr effects

This mechanism is a basic third-order nonlinearity that appears in isotropic materials such as gases, liquids, or solids in the presence of a strong electric field due to an intense optical beam or to an external applied voltage. On the microscopic scale it corresponds to a reorientation of dipoles in the presence of the high field. In such conditions the material refractive index n becomes a function of the light intensity according to the relation of the form $n = n_0 + n_2 I$, where I is the intensity of the optical beam and n_2 is the nonlinear refractive index, which is related to the electro-optic Kerr coefficient of the material. Kerr nonlinearity may exhibit an extremely short response time (picosecond range), but it generally requires high peak intensities ($>10 \text{ GW} \cdot \text{cm}^{-2}$) to induce efficient index nonlinearity for phase conjugation or ultrafast optical switching. Main materials used for experiments in wave mixing are liquids such as carbone disulfide (CS_2), nitrobenzene, acetone, and benzene or solids like quartz, silica, and glass (bulk or fibers).

1.3.2 Stimulated Brillouin scattering

The stimulated Brillouin scattering (SBS) effect originates from the electrostrictive effect in the transparent dielectric media such as liquids, gases or solids. The interference patterns due to the incident and spontaneous scattered optical fields generate a traveling acoustic wave that modulates the material refractive index through the elasto-optic effect. It thus results from the interaction the equivalent of a coherent moving phase grating that propagates at the sound velocity and that diffracts in the backward direction a phase conjugate replica of the incident high-intensity laser beam. SBS is a phenomena that exhibits a threshold, and its typical response time ranges in the nanosecond due to the phonon lifetime. Also, because of the moving grating the retro-reflected Stokes wave is frequency-shifted by the

Doppler effect corresponding to the sound velocity by several gigahertz. As will be shown in several chapters of this book, SBS is a very well suited interaction for self-pumped phase conjugation and most efficient materials are high-pressure gases (SF_6, N_2, Xe, etc.), liquids (CS_2, CCl_4, $GeCl_4$, $SiCl_4$, $TiCl_4$, freon, acetone, etc.), or solids (silica fibers or bulk, quartz, organic crystals such as LAP, DLAP, etc.).

1.3.3 Photorefraction

Photorefraction is a particular type of nonlinearity which arises in materials that exhibit linear (or quadratic) electro-optic (EO) effects. The illumination of the material by a two-beam interference pattern generates a photoinduced charge distribution. The photo-generated carriers (electrons or holes) are trapped, and thus it results in a space-charge field in the volume of the material which modulates the refractive index through the electro-optics coefficient. Microscopic phenomena for space-charge buildup involves electrons charge diffusion or drift under an external applied electric field. There are several specific characteristics of the photorefractive effect which differ from other known nonlinear mechanisms. First, there is no threshold effect and the material responds to the incident energy. In such conditions the material response time can be easily controlled from a fraction of a microsecond to several seconds; second, the amplitude of the photoinduced index modulation is mainly determined by the value of the electro-optic coefficient and by the trapping center density. Also, photorefractive materials have a dark storage time constant, equivalent to memory effect. Photorefractive materials have already demonstrated their great importance in experiments based on the recording and erasure of holograms for optical information processing, high-gain wave mixing, and phase conjugation with low-power visible or near-infrared laser beams. Most of the experiments were performed with different types of EO materials such as $LiNbO_3$, $BaTiO_3$, and $Bi_{12}(Si,Ge,Ti)O_{20}$; semiconductors such as GaAs, InP, and CdTe; PLZT ceramics; and, more recently, doped EO polymers or liquid crystals.

1.3.4 Free carriers in semiconductors

Semiconductors are excellent candidates for nonlinear optics experiments since they possess a large variety of physical mechanisms that can provide efficient and fast spatial index or absorption modulation under continuous-wave (CW) or pulsed laser illumination. The first 4WM experiments were based on electron–hole generation under pulsed laser for transient gratings recording at a wavelength for which photon energy is larger than the bandgap energy. The generated plasma induces a large modification of the semiconductor dielectric constant, thus leading to the recording of both phase and amplitude gratings. However, the index modulation will, in general, be the dominant contribution in efficient 4WM interactions involving large period gratings ($\Lambda > 10$ μm). Due to spatial plasma diffusion, small period gratings are washed out and their contribution to the diffraction is very much reduced. Temporal dynamics in semiconductors ranges from picoseconds to several 10 ns, depending of the material and laser wavelength used in the experiments.

Microcrystallites of a semiconductor can also be included in a glass matrix. It results in a bulk and isotropic material that may exhibit relatively efficient 4WM even for small period gratings as was demonstrated in CdSe-doped glasses in the green–red spectral region. Also, major advances in the semiconductor technology now permit the confinement of electron–hole pairs in potential walls called multiquantum wells. It permits an increase in the electronic density of states, thus leading to enhanced optical nonlinearities as already demonstrated in wave mixing experiments done on multiquantum well structures with GaAlAs–GaInAs on GaAs substrates. These materials may operate efficiently at semiconductor laser wavelength with low incident power.

1.3.5 Saturable amplification

The laser gain media itself is also of great interest for its capabilities of performing real-time holography and phase conjugation through saturable amplification. When two beams interfere in a flash lamp of diode pumped gain media, they create a spatial modulation of the beam amplitudes due to gain saturation effect. In other words, in bright region of the fringes, the amplification is reduced due to the gain saturation, while dark regions of the fringes receive a higher gain. It thus results in a spatial gain grating that diffracts the incident waves and generates a phase conjugate of the incident probe wave through a 4WM configuration. We note that in this interaction the amplitude grating is shifted by π with respect to the incident interference fringes. Also, the beams that interfere in the gain media will be amplified through the laser gain, and energy transfer will also occur through self-diffraction phenomena. Therefore, four-wave mixing in the laser media offers very attractive features: Highly efficient diffraction can be obtained for transmission- or reflection-type gratings, fast response time, and self-matching nonlinearity with respect to the laser wavelength. This interaction can operate in different temporal regimes and does not require high optical powers. Conjugate beams with high reflectivities can be obtained after optimizing the pump beam fluences with respect to (a) the saturation fluence of the laser media and (b) pump-to-probe beam ratios. With such conditions, self-pumped phase conjugate loop resonators using four-wave mixing in the gain media are demonstrated; they can be injected or self-starting. It will be shown that this interaction is well suited in laser media having a high gain length product, and experiments have been performed in solid-state saturable amplifiers such as ruby, Nd:YAG, Nd:YVO$_4$, Nd:glass, Nd:YLF, Ti:sapphire, liquid rhodamine dyes, or CO$_2$ and Cu vapor gas lasers.

1.3.6 Saturable absorption

The saturable absorption was recognized early as an efficient mechanism for efficient phase conjugation. It is based on the dependence of the absorption and refractive index of a two-level system as a function of the incident average intensity due to pump and probe beams. It is expected that transitions with a large cross section and a long relaxation time are more easily saturable. Also another important parameter is the frequency detuning off the transition line center. The effect of detuning the laser permits to have a dominant contribution of the phase grating, thus leading to a significant increase of the conjugate beam reflectivity. After optimizing the conditions of interactions—in particular, using large grating periods—amplified reflections by four-wave mixing was observed in atomic sodium vapor. However, for applications, it may be more attractive to use solid-state saturable absorbers such as Cr^{3+}-doped, Nd^{3+}-doped, and color-center crystals. More recently it was also shown that Cr^{4+}-doped GSGG and Cr^{4+}:YAG possess a broad absorption band around 1 μm and can be used for phase conjugation at 1.06 μm. Such characteristics are well suited for further applications on laser beam control, dynamic holography performed at this important wavelength.

1.3.7 Molecular reorientation in liquid crystals

Organics may offer very attractive properties when used in wave-mixing and dynamic holographic experiments: Large-size materials can be fabricated; and due to the large diversity of chemical compounds for material synthesis, it is expected that their nonlinearities can be optimized for a given application. The most promising media are based on polymers, liquid crystal, or a mixture of polymer-dispersed liquid crystal. In these materials the photoinduced index anisotropic called the Fredericks transition arises from molecular reorientation due to the electric field. It results in an index spatial modulation that gives rise to a dynamic grating or to self-phase modulation. Due to the large birefringence of liquid crystal, quite efficient photoinduced nonlinearities can be obtained in the visible or near-IR wavelengths. Typically, a nonlinear response in liquid crystal requires an equivalent electric field of 1 to 10 V · μm^{-1}, while other polymer materials require more than 100 V · μm^{-1}. Another remarkable property is that the nonlinear response can be further increased by doping with a dye such as dispersed red or orange. Although large gain coefficients have been measured in wave mixing experiments, the detailed mechanisms for molecular reorientation at low power levels in dye-doped liquid crystals are not yet fully explained. Several mechanisms with different time constants may be present: space-charge field at the interfaces, photomolecular alignment, and photoisomerization, thus leading to the recording of dynamic or nearly permanent holograms. Also, a different structure can be made: It consists of a liquid crystal layer deposited on a photoconductive film. In wave-mixing experiments, this hybrid structure behaves as a nonlinear media for dynamic phase holography. Liquid crystals thus provide the opportunity of new types of third-order nonlinearities; moreover, their physical and chemical properties continue to progress due to the development of the display market.

1.3.8 Thermal gratings

The energy or incident power absorbed by a media at a given wavelength in wave-mixing interactions also induces spatial index modulation through the contribution of the thermal coefficient $\partial n/\partial T$. This intensity-dependent change of the refractive index in liquids or solids may result in very significant self-diffraction effects for phase conjugate beam generation. From early works on these mechanisms, it is well established from heat equations that the steady-state index modulation varies as the square of the grating period and is inversely proportional to the thermal conductivity of the material. Therefore, for small grating periods, thermal diffusion tends to reduce both the photoinduced index modulation amplitude and the diffraction efficiency of thermal holograms. For that reason, wave-mixing experiments based on thermal nonlinearities will benefit from using a small pump-probe beam angle or a longer wavelength up to 10 μm. Also, in the pulse regime, if the pulse duration is short compared to the thermal relaxation time, the induced index nonlinearity is reduced. To perform efficient wave mixing, phase conjugation materials that display high thermal coefficient must be used, and this condition can be achieved in semiconductors like HgCdTe where $\partial n/\partial T = 10^{-3}$ at $\lambda = 10.6$ μm, or in liquid crystals where $\partial n/\partial T = 10^{-3}$ at room temperature and $\partial n/\partial T = 10^{-2}$ near the transition temperature.

1.4 THE CRITERIA FOR THE CHOICE OF MATERIALS

This review (which summarizes the main physical mechanisms) and nonlinear media highlight the great diversity and the intense research activities that have been pursued in the field of phase conjugation. Both the fundamental and applied aspects of the field have stimulated remarkable innovative concepts and subsequent new applications. To identify the most convenient mechanism is now an important task since each material exhibits very specific properties and the choice will result from a compromise with respect to the requirements due to the applica-tions. To illustrate this complex situation, we outline in the following list the main parameters which will contribute to identify the most suitable nonlinearity and material:

- Operating mode of the source: pulsed, CW, wavelength
- Required conjugate beam reflectivity and beam interactions (4WM or self-pumped)
- Required response time of the nonlinearity (can range from seconds to nanoseconds or even shorter).
- Good optical quality and stability, high damage threshold and compactness.

Beside these most important parameters, other characteristics such as low speckle noise wavefront generation, conjugate beam fidelity, and material reliability may also contribute to the selection of the good material for an optimum operation of the nonlinear mirror in a full-scale laser system. Nonlinear optical phase conjugation undoubtedly offers a great potential for energy scaling of laser sources, and several comprehensive books or review papers exist on this subject, on both the basic and applied aspects of this field [22–33].

1.5 CONCLUSION

Optical phase conjugation is now established as a domain of nonlinear optics, and further noticeable advances are expected due to the interest in the development of high-energy or high average power laser sources. The concept of phase conjugation permits the restoration of a beam whose quality is close to the diffraction limit whatever the phase aberrations present on the optical path of the laser beam. Moreover, this property is maintained even when changing the laser pulse energy or pulse repetition rate. This permits the attainment of a great flexibility in the operating conditions of the source for adjusting its characteristics to the requirements of the applications. Also, another very important property of phase conjugation is the capability of combining and phase locking of several laser sources, thus leading to an improvement of power and brightness of the emitted beam. For that purpose the concept of phase conjugation permits to overpass the limitations of classical laser architectures for power or energy scaling. To demonstrate and to integrate these concepts in laser systems require efficient third-order nonlinear materials for conjugate beam generation through a dynamic hologram. The main objective of this book is therefore to highlight the most suitable class of nonlinear mechanisms such as Brillouin scattering, photorefraction, saturable amplification, or thermal index modulation. Beside these established effects, there is still a great potential for optimizing the properties of existing media or for the discovery of new fundamental mechanisms or laser architectures based on the remarkable properties of nonlinear optical phase conjugation.

REFERENCES

1. B. Ya. Zel'dovich, V. I. Popovichev, V. V. Ragul'skiy, and F. S. Faizullov, Connection between the wave fronts of the reflected and exciting light in stimulated Mandel'shtam–Brillouin scattering, *Zh. Eksp. Teor. Fiz. Pis'ma Red.* **15**, 160 (1972) [English translation: *Sov. Phys. JETP* **15**, 109 (1972)].
2. H. Kogelnik, Holographic image projection through inhomogeneous media, *Bell Syst. Tech. J.* **44**, 2451 (1965).
3. A. Yariv, Phase conjugate optics for real time holography, *IEEE J. Quantum Electron.* **QE-14**, 650 (1978).
4. J. P. Huignard, J. P. Herriau, Ph. Aubourg, and E. Spitz, Phase conjugate wavefront generation via real time holography in photorefractive $Bi_{12}SiO_{20}$ crystals, *Opt. Lett.* **5**, 519 (1979).
5. A. Yariv and D. M. Pepper, Amplified reflection, phase conjugation and oscillation in degenerate four wave mixing, *Opt. Lett.* **1**, 16 (1977).
6. B. Fischer, M. Cronin Golomb, J. O. White, and A. Yariv, Amplified reflection, transmission and self oscillation in real time holography, *Opt. Lett.* **6**, 519 (1981).
7. R. W. Hellwarth, Generation of time reversed wavefronts by nonlinear refraction, *J. Opt. Soc. Am.* **67**, 1 (1977).

8. D. M. Bloom and G. C. Bjorklund, Conjugate wavefront generation and image reconstruction by four wave mixing, *Appl. Phys. Lett.* **31**, 592 (1977).
9. J. P. Woerdman, Formation of a transient free carrier hologram in Si, *Opt. Commun.* **2**, 212 (1971).
10. B. I. Stepanov, E. V. Ivakin, and A. S. Rubanov, Recording two-dimensional and three-dimensional dynamic holograms in bleachable substances, *Dokl. Aka. Nauk SSSR* **196**, 567 [English translation: *Sov. Phys. Dokl.-Tech. Phys.* **16**, 46 (1971)].
11. R. A. Fisher, *Optical Phase Conjugation*, Series in Quantum Electronics, Principles and Applications, Academic Press, New York (1983).
12. D. L. Staebler and A. J. Amodei, Coupled wave analysis of holographic storage in $LiNbO_3$, *J. Appl. Phys.* **43**, 1042 (1972).
13. P. Günter, Holography, coherent light amplification and optical phase conjugation with photorefractive materials, *Phys. Rep.* **93**, 201 (1982).
14. C. R. Guliano, Applications of optical phase conjugation, *Phys. Today* **34**, 27 (1981).
15. S. G. Odoulov, E. N. Sal'kova, M. S. Soskin, and L. G. Sukhoverkhova, Removal of distortions induced in laser beam amplifiers by methods of dynamic holography, *Ukr. Fiz. Zh.* **23**, 562 (1978).
16. S. Odoulov, M. Soskin, and M. Vasnetsov, Compensation for time-dependent phase inhomogeneity via degenerate four-wave mixing in $LiTaO_3$, *Opt. Commun.* **32**, 355 (1980).
17. D. M. Pepper, Nonlinear optical phase conjugation, Special issue of *Opt. Eng.* **21**, (1982).
18. J.-P. Huignard and A. Marrackchi, Coherent signal beam amplification in two wave mixing experiments with $Bi_{12}SiO_{20}$ crystals, *Opt. Commun.* **38**, 249 (1981).
19. J. Feinberg and R. W. Hellwarth, Phase conjugating mirror with continuous wave gain, *Opt. Lett.* **5**, 519 (1980).
20. J. Feinberg, Self-pumped continuous wave phase conjugator using internal reflection, *Opt. Lett.* **7**, 486 (1982).
21. M. Cronin-Golomb, B. Fischer, J. O. White, and A. Yariv, Passive self pumped phase conjugate mirror: Theoretical and experimental investigation, *Appl. Phys. Lett.* **41**, 689 (1982).
22. V. L. Vinetsky, N. V. Kukhtarev, V. B. Markov, S. G. Odulov, and M. S. Soskin, Amplification of coherent beams by dynamic holograms in ferroelectric crystals, *Izv. Akad. Nauk SSSR* **41**, 811 (1977).
23. D. M. Pepper, (ed.), Nonlinear optical phase conjugation, Special issue of *Opt. Eng.* **21**, 155 (1982).
24. B. Ya. Zeldovich, N. F. Pilipetsky, and V. U. Shkunov, *Principles of Phase Conjugation*, Springer-Verlag, Berlin (1985).
25. G. C. Valley, A review of stimulated Brillouin scattering excited with a broad band pump laser, *IEEE J. Quantum Electron.* **22**, 704, (1986).
26. D. A. Rockwell, A review of phase conjugate solid state lasers, *IEEE J. Quantum Electron.* **24**, 1124 (1988).
27. P. Yeh, Two wave mixing in nonlinear media, *IEEE J. Quantum Electron.* **25**, 484 (1989).
28. P. Günter and J.-P. Huignard (eds.), *Photorefractive Materials and Their Applications*, Vol. 61–62, Topics in Applied Physics, Springer-Verlag, Berlin (1989).

29. S. Odoulov, M. Soskin, and A. Khyzhniak, *Optical Oscillators with Degenerate Four-Wave Mixing (Dynamic Grating Lasers)*, Harwood Academic Publishers, Chur (1991).
30. P. Yeh, Photorefractive phase conjugators, *IEEE* **80**, 3, 486 (1992).
31. L. Solymar, D. J. Webb, and A. Grunnet-Jepsen, *The Physics and Applications of Photorefractive Materials*, Clarendon, Oxford (1996).
32. H. J. Eichler and O. Mehl, Phase conjugate mirrors, *J. Nonlinear Optical Physics Mater.* **10**, 43 (2001).
33. R. Menzel, *Photonics—Linear and Nonlinear Interactions of Laser Light and Matter*, Springer-Verlag, Berlin (2001).

CHAPTER 2

Principles of Phase Conjugating Brillouin Mirrors

AXEL HEUER and RALF MENZEL
University of Potsdam, Institute of Physics, Chair of Photonics,
14469 Potsdam, Germany

2.1 INTRODUCTION

Stimulated Brillouin scattering (SBS) is one of the most common nonlinear optical processes for achieving phase conjugating mirrors [1]. Although SBS is a third-order nonlinear process, it can be driven by one pump beam, only, which will finally be phase conjugate reflected in the SBS volume mirror [2]. As a result, phase conjugating SBS mirrors are simple self-pumped devices that can easily be applied in master oscillator double-pass amplifier (MOPA) configurations [3–5] or as high-reflecting mirrors in laser oscillators [6]. The wavefront inversion with respect to the propagation direction of the light, the phase conjugation, allows for the compensation of phase distortions from, for example, the active material in solid-state lasers in the second pass. As a result, almost ideal wavefronts can be obtained at the output of these lasers. Thus, the beam quality of such laser systems containing phase conjugating SBS mirrors can be diffraction-limited, although the phase distortions from the highly pumped and thus strongly thermally stressed active material would not allow beam propagation factors better than 10. Therefore, these phase conjugation SBS mirrors find applications in high brightness solid-state laser systems [3–6].

Although the nonlinear optical process of the SBS phase conjugation does not have a threshold in the sense of (for example) the laser threshold, an onset of, reflection can be observed as a function of the incident pump power or pump pulse energy. Therefore, an SBS threshold can be defined for a reflectivity of, for example, 1% or 2% [2]. At high incident pump powers or pulse energies the nonlinear reflectivity of the SBS mirror saturates in the best cases very close to 1. The highest observed reflectivities were above 95%. In between, a strong increase of reflectivity can be obtained, and usually pump powers or pulse energies 10–20 times above the threshold value are sufficient for practical applications.

Phase Conjugate Laser Optics, edited by Arnaud Brignon and Jean-Pierre Huignard
ISBN 0-471-43957-6   Copyright © 2004 John Wiley & Sons, Inc.

Finally, the phase conjugating SBS mirror as nonlinear optical device can be characterized by this threshold value, maximum reflectivity, the dynamic range, and the fidelity of the phase conjugation. In addition, the useful wavelength range, size, and toxicity may be important for practical use. Because the buildup of the phase conjugating mirror in the SBS materials is based on interference effects in the volume, the coherence demands of the pump light are another important feature. While for very narrow bandwidth lasers with very long coherence lengths in the range of kilometers the SBS threshold can be very low, for practical purposes SBS mirrors with coherence demands in the range of less than 1 m are required.

Therefore, phase conjugating SBS mirrors are usually designed for a special purpose—that is, for the special laser system they should be applied to. The following theoretical descriptions are helpful in developing the design criteria for these SBS mirrors and thus in optimizing the features of the optical phase conjugation based on SBS.

The SBS as a self-pumped process is normally achieved by focusing a laser beam into the SBS medium. In this medium the light is scattered at a self-induced acoustic wave grating. The stimulated scattering is based on the spontaneous Brillouin scattering of the incident light at the hypersound waves in the matter. The name of the process was taken from Léon Brillouin, who studied first, around 1920 [7], the scattering of light by acoustic waves (phonons). The noise of the acoustic waves is stochastic and has its origin in thermal fluctuations; this scattering is spontaneous and nondirectional.

The general aspects of the Brillouin scattering are shown in Fig. 2.1. An incident light beam with a wavelength λ_P, a frequency ν_P, and a wave vector $\vec{k}_P$ travels in the material through an acoustic wave of the wavelength Λ_B, frequency ν_B, and wave vector $\vec{k}_B$. This acoustic wave provides a propagating refractive index

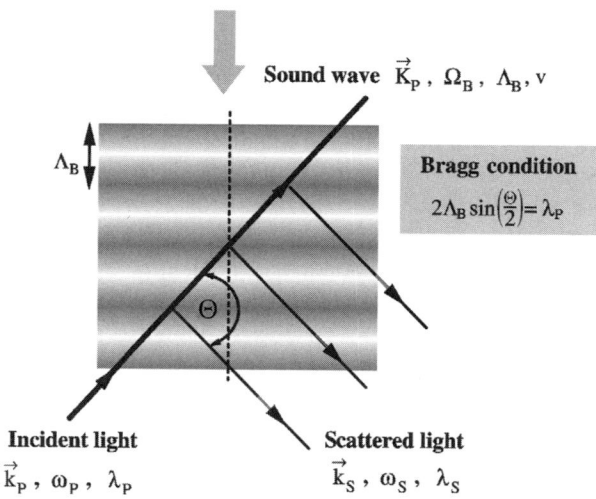

Figure 2.1. Schematic illustration of the Brillouin scattering.

modulation in the material. Due to this refractive grating, part of the incident light is scattered, fulfilling the Bragg condition:

$$2\Lambda_B \sin\left(\frac{\Theta}{2}\right) = \lambda_P \tag{1}$$

Since the acoustic wave is moving with the velocity v, the frequency ν_S of the scattered light is shifted by the Doppler effect:

$$\nu_S = \left(1 \pm 2\frac{vn}{c}\sin\left(\frac{\Theta}{2}\right)\right) \cdot \nu_P = \nu_P \pm \nu_B \tag{2}$$

The frequency shift of the scattered light $|\nu_P - \nu_S|$ is identical with the frequency of the acoustic wave ν_B. If the acoustic wave and the incident light wave have the same direction, the shift has a negative sign (Stokes), and for the case of counter propagating waves the sign is positive (anti-Stokes). In the particle picture, this is equivalent to a generation ($-$ sign) and an annihilation ($+$ sign) of an acoustic phonon. Equation (2) also shows that the frequency shift and therefore the frequency of the matching acoustic wave depends on the scattering angle and reaches its maximum for the backward direction $\theta = 180°$.

$$\nu_B = \frac{2vn}{\lambda_P} \tag{3}$$

Typical ranges of the Brillouin shift for backward scattering are 100 MHz for gaseous materials, 1 GHz for liquids, and 10 GHz for solid-state matter. Phase conjugating SBS mirrors are based on backward scattering. In this case, only the Stokes component exists in the Brillouin spectrum. SBS demands sufficiently high light powers, as they are supplied by lasers only. Therefore the first experimental examinations of SBS were made in 1964 [8–11] after the invention of the laser. A comprehensive summary of these early experiments is given in Ref. 12.

For a better understanding the SBS process can be divided into four steps (see Fig. 2.2). The process starts with spontaneous scattering at thermal density fluctuations. The spontaneous Brillouin scattering is unidirectional, and only a very small fraction will be scattered exactly in the backward direction. This small amount of the scattered light will interfere with the incident light and build up an interference pattern with a beat frequency equal to the Brillouin shift ν_B. This light beat amplifies the acoustic wave via electrostriction. Since this acoustic wave has the right frequency ν_B and the right direction, the scattering is increased and a positive feedback loop is completed. Thus the back-reflected part of the incident light increases rapidly and will reach saturation value of possibly more than 90%.

2.2 THEORETICAL DESCRIPTION OF THE SBS PROCESS

The basic equations describing the SBS process are known in detail, and therefore the theoretical description of SBS phase conjugation is more or less a technical problem, only. But it turns out that the system of partial differential equations

1. Spontaneous scattering from noise:

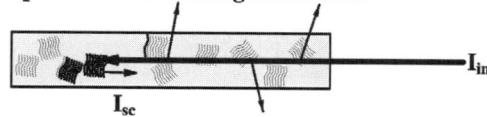

2. Interference between incident and scattered light:

3. Amplification of the sound wave by electrostriction:

4. Increased scattering from the intensified sound wave:

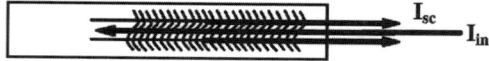

Figure 2.2. Schematic illustration of the SBS process.

cannot be solved analytically, and even numerical approaches demand further approximations. Therefore, different levels of simplifications were realized, and the resulting solutions are useful for special situations only. Nevertheless, important consequences can be derived from these models with respect to the expected nonlinear behavior on one side and the fidelity of the phase conjugation on the other. Especially the latest modeling based on mixtures of transversal modes allows the description of SBS phase conjugation with very high accuracy.

2.2.1 General equations

The general description of the SBS process has to take into account the growing of the reflected Stokes light from the SBS process, the depletion of the incident pump light, and the growing of the hypersound wave inside the SBS material. The interaction is based on the electrostriction in the SBS material. The thermally generated hypersound noise causes spontaneous Brillouin scattering, which is necessary for starting the stimulated scattering.

For the description it is assumed that the incident pump light propagates along the z axis into the SBS material. The total electric field amplitude $\vec{E}$ can then be represented as the sum of the two counterpropagating fields:

$$\vec{E} = \vec{E}_P(\vec{r}, t) + \vec{E}_S(\vec{r}, t) \qquad (4)$$

with the component of the pump light

$$\vec{E}_P = \left(\frac{1}{2} E_P(\vec{r}, t) \exp[i(k_P z - \omega_P t)] + c.c.\right) \vec{e}_P \qquad (5)$$

and the reflected Stokes light

$$\vec{E}_S = \left(\frac{1}{2} E_S(\vec{r}, t) \exp[i(-k_S z - \omega_S t)] + c.c.\right) \vec{e}_S \qquad (6)$$

with $k_{P/S} = 2\pi/\lambda_{P/S}$ and $\omega_{P/S} = 2\pi\nu_{P/S}$. The incident field propagates in the positive z direction toward the right side in Fig. 2.3, and the reflected or Stokes component propagates in the negative z direction to the left. $E_P(\vec{r}, t)$, $E_S(\vec{r}, t)$ are the complex amplitudes of the electric fields and $\vec{e}_P$, $\vec{e}_S$ are normalized polarization vectors. The frequencies of the two fields ω_P, ω_S and the wave numbers k_P, k_S are related to frequency and the wave vector of the acoustic wave (ω_B, k_B) by Eq. (2) ($\omega_B = \omega_P - \omega_S$; $\vec{k}_B = \vec{k}_P - \vec{k}_S$).

The propagation of an electromagnetic wave in a Brillouin medium with neglectable absorption ($\alpha \approx 0$) is described by the wave equation

$$\Delta \vec{E} - \frac{n^2}{c^2}\frac{\partial^2 \vec{E}}{\partial t^2} = \frac{4\pi}{c^2}\frac{\partial^2 \vec{P}^{NL}}{\partial t^2}, \qquad \vec{P}^{NL} = \frac{1}{4\pi}\frac{\gamma^e}{\rho_0}\vec{E}\rho \qquad (7)$$

The nonlinear polarization $\vec{P}^{NL}$ describes the coupling between the electric field $\vec{E}$ and the acoustic wave, which is represented by the modulation density ρ. γ^e is the electrostrictive coupling coefficient, and ρ_0 is the stationary density of the medium. In the coordinates of Fig. 2.3 the density modulation ρ also propagates along the positive z direction.

$$\rho = \frac{1}{2}\rho(\vec{r}, t) \exp[i(k_B z - \omega_B t)] + c.c. \qquad (8)$$

The combination of the equation of continuity and the Navier–Stokes equation leads to an acoustic wave equation [12]:

$$-\frac{\partial^2}{\partial t^2}\rho + v^2 \Delta \rho + \frac{\Gamma_B}{k_B^2}\frac{\partial}{\partial t}\Delta \rho = \frac{\gamma^e}{8\pi}\Delta E^2 \qquad (9)$$

The term on the right side describes the already-mentioned electrostrictive force. It moves the matter in the electric field via dipole interaction. The third term on the left

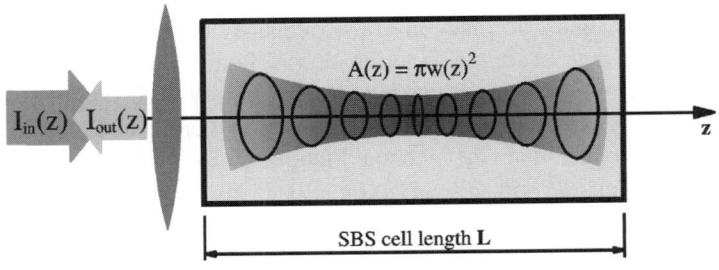

Figure 2.3. Definitions for modeling the SBS process.

side describes the damping of the acoustic wave by the phonon decay rate Γ_B.

$$\Gamma_B = \frac{1}{\tau_B} = \frac{\eta k_B^2}{\rho_0} \propto \frac{1}{\lambda_P^2} \tag{10}$$

The phonon decay rate as the reciprocal of the phonon lifetime τ_B is a function of the material viscosity η and density ρ_0. It is roughly proportional to the square of the wavelength of the incident light.

$$\tau_B \propto \lambda_P^2 \tag{11}$$

To obtain a set of coupled differential equations for the complex electric field amplitudes $E_P(\vec{r}, t)$, $E_S(\vec{r}, t)$ and for the normalized acoustic wave amplitude

$$S(\vec{r}, t) = \frac{\omega_P \gamma^e}{2cn\rho_0} \rho(\vec{r}, t) \tag{12}$$

we introduce Eq. (4) into Eq. (7) and introduce Eq. (8) into (9). Furthermore, the approximation of the slowly varying amplitudes (SVA) is used [2]. We also neglect the propagation of the acoustic wave, and all the optical frequencies are set to be equal ($\omega_S \approx \omega_P$). With these assumptions we achieve the following system of coupled equations:

$$\frac{\partial E_P(\vec{r}, t)}{\partial z} + \frac{n}{c}\frac{\partial E_P(\vec{r}, t)}{\partial t} + \frac{i}{2k_P}\nabla_T^2 E_P(\vec{r}, t) = i\frac{S(\vec{r}, t)}{2} E_S(\vec{r}, t)$$

$$\frac{\partial E_S(\vec{r}, t)}{\partial z} - \frac{n}{c}\frac{\partial E_S(\vec{r}, t)}{\partial t} + \frac{i}{2k_S}\nabla_T^2 E_S(\vec{r}, t) = -i\frac{S^*(\vec{r}, t)}{2} E_P(\vec{r}, t) \tag{13}$$

$$\frac{\partial S(\vec{r}, t)}{\partial t} + \frac{1}{2}\Gamma_B S(\vec{r}, t) = i\frac{g_B}{2}\Gamma_B E_P(\vec{r}, t) E_S^*(\vec{r}, t)$$

The Laplace operator $\nabla_T^2 = \partial^2/\partial x^2 + \partial^2/\partial y^2$ refers to the transversal distributions to the field, and g_B is the so-called Brillouin gain coefficient:

$$g_B = \frac{\gamma^{e^2} \omega_P^2}{c^3 n \rho_0 v \Gamma_B} \tag{14}$$

This coefficient is almost no function of the frequency or wavelength of the pump light because Γ_B is a quadratic function of the frequency, too [see Eq. (10)]. Typical values are of the order of magnitude 10^{-8} cm/W. The significance of this coefficient becomes obvious in case of stationary scattering of light pulses with durations much longer than the phonon lifetime, resulting in time-independent differential equations $\partial E_P/\partial t$, $\partial E_S/\partial t$, $\partial S/\partial t = 0$. If also pump intensity depletion is neglected ($I_P \gg I_S$), with $I_P = |E_P|^2$ and $I_S = |E_S|^2$ the following solution is obtained:

$$I_S(z) = I_S(L)\exp(G(z)) \quad \text{with} \quad G(z) = g_B I_P(L-z) \tag{15}$$

The Stokes intensity I_S grows exponentially in the backward direction. The amplification starts at the point $z = L$ with an initial value $I_S(L)$, and L indicates the interaction length inside the Brillouin medium. This initial value of the backscattered intensity is given by the amount of spontaneous scattered light. If the intensity $I_S(L)$ is a from outside into the Brillouin cell-injected Stokes beam, it operates as an SBS amplifier.

A general analytic solution of the differential Eq. (13) does not exist. In Sections 2.2.4–2.2.6, different numerical methods for solving this system of differential equations are discussed.

2.2.2 Optical phase conjugation by SBS

Optical phase conjugation means the reversing of the phase front of the propagating light wave with respect to the propagation direction. This means that the phase front of the reflected light is perfectly identical with the phase front of the incident light if the propagation direction is changed by 180° in the phase conjugating mirror. It is obvious that this phase conjugate reflection can be realized in volume mirrors only. In comparison to conventional mirrors, where only the perpendicular component of the wave vector is reflected at the mirror, in phase conjugation all components of the wave vector of the incident light wave will be inverted. For the theoretical description we assume an incident wave $\vec{E}(\vec{r}, t)$:

$$\vec{E}(\vec{r}, t) = \frac{1}{2}\vec{A}(\vec{r})\exp(i(\vec{k}\vec{r} - \omega t)) + c.c. = \frac{1}{2}\vec{E}(\vec{r})\exp(-i\omega t) + c.c. \quad (16)$$

If the spatial part of the incident wave $\vec{E}_{PC}(\vec{r}, t)$ is complex conjugated, the resulting reflected and then phase conjugated wave with the electric field $\vec{E}_{PC}(\vec{r}, t)$ is described by

$$\vec{E}_{PC}(\vec{r}, t) = \frac{1}{2}\vec{E}^*(\vec{r})\exp(-i\omega t) + c.c. = \frac{1}{2}\vec{A}^*(\vec{r})\exp(-i(\vec{k}\vec{r} + \omega t)) + c.c. \quad (17)$$

This process is formally equivalent with the transformation $t \rightarrow -t$ in the term describing the phase of the light wave. Therefore in the literature [13] sometimes optical phase conjugation was assigned to time reversal. This point of view is indeed true if the optical path of the light propagation is described. The phase conjugate reflected light will indeed follow exactly the same path as the incident light, but in the opposite direction. Also, all phase conditions stay constant. Therefore, the phase conjugate reflected light can combine at beam splitters or scattering elements to the original single beam. Thus, no loss occurs at the beam splitters, and a perfect incident Gaussian beam will be reconstructed after passing the scattering elements the second time [16, 17] (see Fig. 2.4). But the time reversal of the phase term of the optical wave does not mean also time reversal in the amplitude of, for example, a reflected light pulse. Therefore, the time reversal does not include, for example, the pulse shape reversal of an incident pulse. So far, the "time reversal" in connection with optical phase conjugation has to be used careful.

The nonlinear optical element realizing the phase conjugation is therefore called a phase conjugating mirror (PCM). If this mirror is realized with stimulated

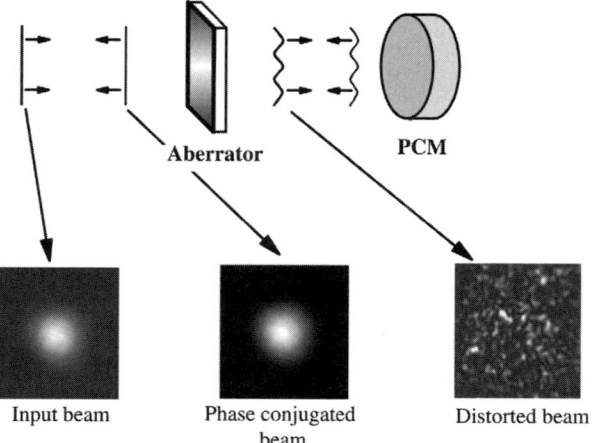

Figure 2.4. Schematic of the compensation of phase distortions by a phase conjugating mirror (PCM).

Brillouin scattering, the resulting mirror is called SBS-PCM or SBS mirror. These PCMs are most useful for the compensation of phase distortions, for example, in high-power solid-state lasers [1].

In contrast to generated four-wave mixing [13], in stimulated Brillouin scattering the phase conjugate wave $E_S(\vec{r}) \propto E_P^*(\vec{r})$ cannot directly be derived as a solution of the system of differential equations [Eq. (11)]. And indeed, the phase conjugation based on the SBS does not completely fulfill Eq. (15). Although since 1960 experiments investigated SBS, the first realization of optical phase conjugation based on SBS was obtained in 1972 [14, 15].

The mechanism of optical phase conjugation based on stimulated Brillouin scattering can be understood with Fig. 2.2. The interference field of the incident and the scattered light will be stronger because the reflected Stokes light will have the same wavefront as the incident pump light. Because of the positive feedback, this interference mechanism will select the wavefront match of the reflected light. So far, the phase conjugate wave has the highest gain in the reflection in stimulated Brillouin scattering. Therefore, the phase conjugated wave will grow from noise. It will finally be the dominant wave in SBS-PCMs.

Mathematically, this process can be modeled analytically in the stationary case [18]. Therefore, we assume that both electric fields of the pump light E_P and the reflected light E_S consist of a large number of different modes

$$E_P(\vec{r}) = \sum_i a_i(z) A_i(\vec{r})$$

$$E_S(\vec{r}) = \sum_i b_i(z) B_i(\vec{r})$$

(18)

These modes are part of a complete orthonormal system of solutions of differential equations which can be described by

$$\frac{\partial A_i(\vec{r})}{\partial z} + \nabla_T^2 A_i(\vec{r}) = 0, \qquad \frac{\partial B_i(\vec{r})}{\partial z} + \nabla_T^2 B_i(\vec{r}) = 0 \qquad (19)$$

$$\int_{-\infty}^{\infty} A_i(\vec{r}) A_j^*(\vec{r})\, dxdy = \delta_{ij}, \qquad \int_{-\infty}^{\infty} B_i(\vec{r}) B_j^*(\vec{r})\, dxdy = \delta_{ij} \qquad (20)$$

Using these fields of Eq. (17) in the system of differential equations [Eq. (12)], the stationary case can be written as

$$\sum_i A_i(\vec{r}) \frac{\partial a_i(z)}{\partial z} = -\frac{g_B}{2} \Gamma_B \sum_{jkl} b_j^*(z) B_j^*(\vec{r}) b_k(z) B_k(\vec{r}) a_l(z) A_l(\vec{r})$$

$$\sum_i B_i(\vec{r}) \frac{\partial b_i(z)}{\partial z} = -\frac{g_B}{2} \Gamma_B \sum_{jkl} a_j^*(z) A_j^*(\vec{r}) a_k(z) A_k(\vec{r}) b_l(z) B_l(\vec{r})$$

(21)

Equation (20) is then multiplied with the complex conjugate of one component (i.e., transversal fundamental mode) of orthonormal system (A_n^*, B_n^*) and then integrated over the whole space using the orthonormal condition of Eq. (19), resulting in the following equations:

$$\frac{\partial a_n(z)}{\partial z} = -\frac{g_B}{2} \Gamma_B \sum_{jkl} b_j^*(z) b_k(z) a_l(z) \cdot \int_{-\infty}^{\infty} B_j^*(\vec{r}) B_k(\vec{r}) A_l(\vec{r}) A_n^*(\vec{r})\, dxdy$$

$$\frac{\partial b_n(z)}{\partial z} = -\frac{g_B}{2} \Gamma_B \sum_{jkl} a_j^*(z) a_k(z) b_l(z) \cdot \int_{-\infty}^{\infty} A_j^*(\vec{r}) A_k(\vec{r}) B_l(\vec{r}) B_n^*(\vec{r})\, dxdy$$

(22)

From these equations it can be seen that only these modes are important because they have large coefficients A_n and B_n, meaning that they have a large overlap integral between the incident and the reflected modes. If it is further assumed that the incident light field consists of one mode, only

$$E_P(\vec{r}) = a_0(z) A_0(\vec{r}), \qquad \text{with} \quad A_0(\vec{r}) = B_0^*(\vec{r}) \qquad (23)$$

and this field of Eq. (22) is used in Eq. (21) results in

$$\frac{\partial b_n(z)}{\partial z} = -|a_0(z)|^2 \sum_l b_l(z) \cdot g_{ln}(z) \qquad (24)$$

with

$$g_{ln}(z) = \frac{g_B}{2} \Gamma_B \int_{-\infty}^{\infty} |B_0(\vec{r})|^2 B_l(\vec{r}) B_n^*(\vec{r})\, dxdy \qquad (25)$$

with the gain coefficient $g_{in}(z)$ for the Stokes components. It can be seen that the coefficient g_∞ with B_l, $B_n = B_0$ is the largest. In Ref. 19 a factor of 2 higher gain coefficient compared to the gain coefficients of the other modes was given. With this high gain coefficient, this phase conjugate mode will be amplified exponentially much more than all other modes. If all parameters of the SBS phase conjugating mirror are designed in the way that this gain coefficient relationship will be established, the almost perfect phase conjugation can be obtained.

$$E_S(\vec{r}) = \sum_i b_i(z)B_i(\vec{r}) \propto b_0(z)B_0(\vec{r}) \propto E_P^*(\vec{r}) \qquad (26)$$

So far, SBS phase conjugation is realized by the selection of the well-suited Stokes components via higher amplification. In case of saturation of the SBS process the difference between the different overlap integrals of Eq. (24) will decrease [20]. In this case, besides the phase conjugate component in the Stokes signal, also other components can be amplified from noise. Numerical simulations resulted for focused Gauss beam as pump field in this case in 95% phase conjugated reflected signal. For practical purposes, this value usually will be sufficient for the applications of phase conjugating SBS mirrors.

As a measure for the quality of the phase conjugation, the fidelity F is introduced. In theoretical studies, this fidelity is defined as a normalized correlation function [20].

$$F(z) = \frac{|\int E_S(\vec{r})E_P(\vec{r})\,dxdy|^2}{\int |E_S(\vec{r})|^2 dxdy \int |E_P(\vec{r})|^2 dxdy} \qquad (27)$$

The value of $F(z)$ is in the range from 0 to 1. In the case of perfect phase conjugation, the wavefronts of the pump and the Stokes light are proportional to each other, $E_S(\vec{r}) \propto E_P^*(\vec{r})$, and then $F(z) = 1$.

In agreement with the experiments, it can be shown theoretically [19–22] that the fidelity obtained with the focusing geometry is smaller than the fidelity that is observed by the phase conjugation in waveguides, especially if bell-shaped transversal beam profiles such as for example, the diffraction-limited Gaussian beams shall be phase conjugated with an SBS mirror because transversally spatial gain narrowing can be obtained in the focus range of the light. As result of this transversal gain profile of the SBS, the wings of the pump light beam will be reflected weaker than the central part. The reflected beam will then show a smaller beam waist radius w_0 compared to the pump beam. This beam waist reduction can be described by a factor β ($0 < \beta < 1$) [23, 24]. This factor β is a function of the intensity and will increase with higher pump intensities up to values close to 1. If the pump light shows aberrations, a spatial filtering can occur from this spatial gain narrowing [25–27]. As result, a higher spatial frequency that will occur in the Fourier plane in a larger distance from the beam axis can be depleted during the SBS reflection. This can decrease the fidelity of the SBS phase conjugation.

The fidelity of the phase conjugated beam as defined in Eq. (27) is experimentally hard to detect [21, 28]. Therefore, in many cases an "energy in the bucket" technique is used to detect the far-field fidelity [29, 30]. In this case the energy of the reflected light is measured behind a pinhole that is placed in the focal length of the lens (L3 in Fig. 2.5b). This pinhole is calibrated for the incident pump light as shown in Fig. 2.5a.

The fidelity results from these measurements by

$$F = C_{\text{cal},F} \frac{E_3}{E_2} \quad \text{with} \quad C_{\text{cal},F} = \frac{E_{1,\text{cal}}}{E_{3,\text{cal}}} \qquad (28)$$

Up to here the SBS was treated for polarized beams only. It was shown [31, 32] that for depolarized beams a full spatial and polarization phase conjugation could not be obtained with the conventional setup, focusing a beam into the SBS medium. To use SBS for the correction of both phase and depolarization distortions, because they occur, for example, in multimode optical fibers [33], a special scheme has to be used. This scheme was first described by Basov et al. [31]. It includes a polarizer, which splits the depolarized light into two beams with orthogonal polarization. Then the

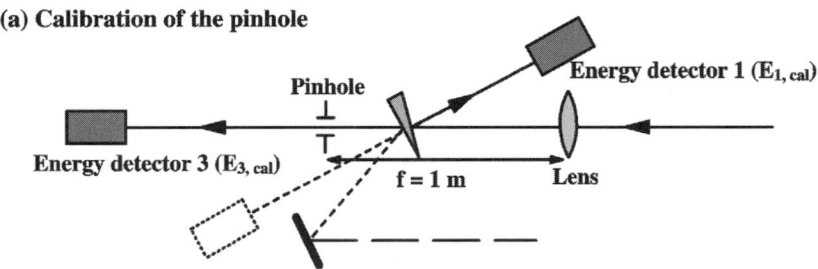

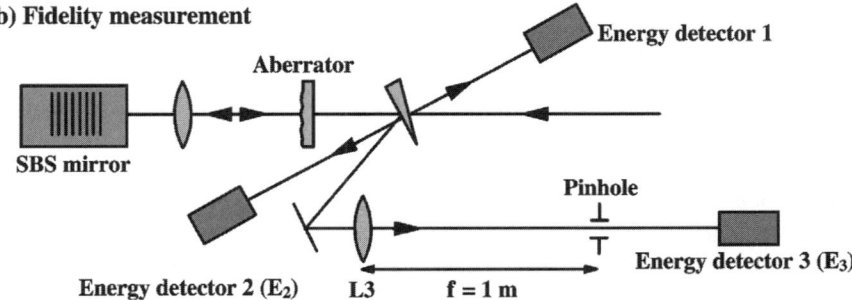

Figure 2.5. Measurement of the fidelity of phase conjugation of Gaussian beams with the "Energy in the Bucket" method. (a) The pinhole is calibrated by measuring $E_{1,\text{cal}}$ and $E_{3,\text{cal}}$. (b) Measurement of the energy of the incident E_1 and the reflected E_2 and the phase conjugated share E_3.

2.2.3 SBS threshold

As mentioned earlier, SBS mirrors do not have a threshold. Nevertheless, the onset of the reflectivity as a function of the incident pump power is very steep, and therefore it became common to define this onset as SBS threshold by the definition of a certain value of, for example, 1–2% reflectivity [34, 35, 91].

If, for example, the SBS threshold is defined as the input power of a pump light for 1% reflectivity, the necessary Brillouin gain can be calculated for stationary SBS from Eq. (15):

$$g_B P_{th} \frac{L_{eff}}{A_{eff}} = G_{th} \cong 21 \Rightarrow P_{th} = \frac{21 A_{eff}}{g_B L_{eff}} \quad (29)$$

The value of the Brillouin gain at the threshold follows from the intensity of the spontaneous scattered Stokes wave at thermally excited sound waves. The share of the spontaneous scattered intensity toward the phase conjugated Stokes wave was assumed by $I_S(L) = fI_P$ with $f = 10^{-11}$. The order of magnitude of f can be estimated [35] from

$$f = \left(\frac{1}{\exp(\hbar \omega_B / kT - 1)} + 1 \right) \cdot g_B \hbar \omega_S \frac{1}{\tau_B} \frac{L_{eff}}{4 A_{eff}} \quad (30)$$

The SBS threshold gain G_{th} is a function of the material parameters, especially the Brillouin gain coefficient g_B and the lifetime τ_B as well as of the pump geometry. Because the SBS is an optical nonlinear volume effect, the smaller the interaction cross section A_{eff} and the longer the effective interaction length L_{eff}, the lower the threshold. The effective cross section in Eq. (28) uses the approximation of the transversal flat top profile with an equivalent diameter which depends on the nonlinearity exponent of the process [36]. Long interaction lengths and small constant effective cross sections lead to a lower threshold which can be realized easily with waveguide geometries. In focusing geometry L_{eff} an A_{eff} are functions of the square of the focal length of the used lens and so no advantage can be realized by stronger focusing; the observed SBS threshold is almost independent of the focal length of the used lens as shown in Ref. 2.

But for a given pump light the interaction length cannot be increased too much because of the necessary coherence of the SBS process. If the linewidth of the pump light is smaller than the Brillouin linewidth ($\Delta \nu_P \ll \Delta \nu_B$), the effective interaction length L_{eff} is not a function of the coherence length L_{coh} of the pump light. In this case, L_{eff} is given by the geometrical length of the SBS waveguide structure or the Rayleigh length of the focusing geometry or by the absorption length in absorbing SBS material which roughly follows from $L_{eff} \approx 1/\alpha$ [37].

But if broadband pump light is used ($\Delta\nu_P \gg \Delta\nu_B$), the effective interaction length will be given by the above values as long as this length L is smaller than the coherence length $L < L_{coh}$ [38]. In this case the SBS threshold is not a function of the longitudinal mode structure of the pump light. The scattered light reproduces the spectrum of the pump light [39, 40]. Is L larger than L_{coh} the effective interaction length L_{eff} is then given by a coherence length $L_{eff} \approx L_{coh}$ [42]. All these discussions are related to stationary SBS scattering. Stationarity is realized with continuous-wave (CW) pump light or pump pulse durations much longer than the phonon life time τ_B; that is, $\tau_P > 100\, \tau_B$ [12]. With shorter pump pulses that do not fulfill this condition, the temporal evolution of the sound wave distribution has to be recognized, and transient effects have to be considered. As an approximation for transient SBS, Eq. (29) can be modified, too [41]:

$$E_{th} \cong \frac{G_{th} A_{eff}}{g_B L_{eff}} \left(\tau_P + G_{th} \frac{\tau_B}{2} \right) \tag{31}$$

The energy E_{th} of the pulse at the threshold follows approximately from the power at threshold P_{th} and the pulse duration τ_P by $E_{th} = P_{th}/\tau_P$. In case of transient SBS, the threshold energy E_{th} is a weak function of the pulse duration only and therefore a good measure. In this case the threshold power decreases almost reciprocally linearly with the pulse duration. For those short pulses, other nonlinear processes may become dominant over optical phase conjugation based on SBS [43]. Therefore, for example, for glass fibers the Brillouin threshold can be higher than the threshold for Raman scattering if the pulse duration is very short ($\tau_P \ll \tau_B$). In this case, instead of a backscattered SBS Stokes signal, a forward scattering Raman pulse can be obtained [37].

As with the discussed linewidth and pulse duration, also the polarization of the pump light can have a strong influence on the value of the SBS threshold [32].

Especially in waveguides the polarization of the pump and the Stokes light may vary along the interaction length. This can increase the threshold by a factor of 1 to 2. Complete depolarization resulted in a threshold increase by a factor of 1.5 [37]. Besides the already-mentioned threshold reduction using waveguides, other methods were developed. The threshold of a conventional SBS cell can be decreased by a factor of 10, if a feedback loop for the Stokes and pump pulse is applied [44, 45]. Using a Harriot cell to contain the SBS medium will increase the interaction length by the number of the realized foci [46]. A distinct reduction of the SBS threshold is achieved when an amplifying medium is used [47, 48].

2.2.4 Numerical calculations 2D model (focused geometry)

If focusing geometry is used in SBS phase conjugation, the buildup of the sound wave and thus the SBS reflectivity as a function of time and space is crucially dependent on the varying intensity distribution along the optical axis, as shown in Fig. 2.3. Therefore, at least the change in cross section of the pump and Stokes beam along the optical axes should be recognized in the numerical calculations of the SBS process. In the simplest case the change of the diameter of the focused pump light can be considered in a very simple way as it was shown in Ref. 2. The varying focus along the z axis can be recognized by the change of the cross section $A[z]$ leading to the following set of equations:

$$\frac{\partial I_P(z,t)}{\partial z} - \frac{n}{c_0}\frac{\partial I_P(z,t)}{\partial t} = -S(z,t)\sqrt{I_P I_S} - \frac{I_P}{A(z)}\frac{\partial A(z)}{\partial z} \quad (32)$$

$$\frac{\partial I_S(z,t)}{\partial z} - \frac{n}{c_0}\frac{\partial I_S(z,t)}{\partial t} = -S(z,t)\sqrt{I_P I_S} - \frac{I_S}{A(z)}\frac{\partial A(z)}{\partial z} \quad (33)$$

$$\frac{\partial S(z,t)}{\partial t} = \frac{1}{2\tau_B}\left\{g_B\sqrt{I_P I_S} - [S(z,t) - S_0]\right\} \quad (34)$$

In these equations the coherence length was assumed to be longer than the interaction length in the SBS material. These equations can be solved numerically with standard procedures. The parameters as refractive index, Brillouin gain coefficient g_B and phonon lifetime τ_B are given in Tables 2.1–2.3. The amplitude S_0 from thermal noise is the only fit parameter of this model. It has to be chosen for the onset of the SBS reflectivity—that is, the SBS threshold. Some values are given in Ref. 36. The results for one example are depicted in Fig. 2.6.

TABLE 2.1. Parameters of Some Useful SBS Gases[a]

Gas	λ (nm)	p (bar)	g_B (cm/GW)	τ_B (ns)	$\Delta\nu_B$ (MHz)	ν_B (MHz)
Xe	694	40	44	14.6	10.9	429
N_2	694	1	2.4×10^{-3}	0.10	1575	1017
		100	15	6.3	25	1127
CO_2	694	1	1×10^{-2}	0.2	723	768
		50	46	10	16.0	641
CH_4	694	1	7×10^{-3}	0.1	1591	1297
		100	69	7.4	21	1345
Ar	694	1	1.5×10^{-3}	0.1	1730	925
		100	13	7.2	2204	990
$CClF_3$	694	30	55	8.2	19	403
		20	20	7.5	21	458
C_2F_6	694	30	50	4.2	38	
		20	10	4.5	35.4	
SF_6	1064	20	14	17.3	9.2	240

[a] The refractive indices of these gases are very close to 1. The refractive index of SF_6 at 20 bar is, for example, 1.023.

TABLE 2.2. Parameters of Some Useful SBS Liquids[a]

Liquids	λ (nm)	n	ρ_0 (g/cm^3)	g_B (cm/GW)	τ_B (ns)	$\Delta\nu_B$ (MHz)	ν_B (GHz)
GeCl$_4$	1064	1.46	1.87	12	2.3	69.2	2.1
SnCl$_4$	1064	1.36	2.33	11	1.75	182	2.21
PCl$_3$	1064		1.57	8.6			2.76
SiCl$_4$	1064	1.41	1.48	10			2.16
TiCl$_4$	1064	1.61	1.73	14	1.47	108.3	3.0
CCl$_4$	1064	1.46	1.59	3.8	0.6	265.3	2.76
H$_2$O	694	1.33	1.0	4.8	0.5	317	5.91
D$_2$O	1064	1.33	1.1	3.1			3.66
CH$_4$O (methanol)	532	1.33	0.791	13.7	0.4	334	5.4
C$_2$H$_6$O (ethanol)	694	1.36	0.790	12	0.45	353	4.55
n-Pentane	532	1.36	0.626	18.0	0.69	230	5.31
n-Hexane	532	1.38	0.655	16.6	0.67	238	5.76
n-Heptane	532	1.39	0.684	12.6	0.69	230	6.10
n-Octane	532	1.40	0.699	12.6	0.51	312	6.36
Cyclohexane	694	1.43	0.779	6.8	0.21	774	5.55
Benzene	694	1.50	0.877	18	0.55	289	6.47
C$_2$Cl$_3$F$_3$ (Freon-113)	1064	1.36	1.58	6.2	0.84	189	1.74
CS$_2$	1064	1.60	1.266	130	6.4	24.8	3.76
Acetone	1064	1.36	0.790	20	2	79.6	2.67
FC-72	1064	1.2	1.68	6–6.5	1.2	270	1.1
FC-75	1064	1.3	1.77	4.5–5	0.9	350	1.34

[a] The data of these materials are collected from Refs. 53–67.

TABLE 2.3. Parameters of Some Useful SBS Solids[a]

Solids	λ (nm)	n	ρ_0 (g/cm^3)	g_B (cm/GW)	τ_B (ns)	$\Delta\nu_B$ (MHz)	ν_B (GHz)
SiO$_2$ (bulk)	532	1.46	2.202	2.9	0.98 ± 0.04	163.0	32.6
SiO$_2$ (fiber)	532	1.46		2.5	2.3	69.2	32
KD*P	532	1.507	2.355	5.1	1.5	107	30.5
KDP (THG)	532	1.481	2.332	6.5	2.1	72.9	28.5
d-LAP	532	1.584	1.6	29.85	1.9	82.3	19.6
LAA	532			24.9	1.6	100.4	21.0
LHG-8	532	1.531	2.83	2.74	0.73	219	27.8
GGG	532	1.9	7.09	1.02	0.6	12.5	26.3
BK7	532	1.520	2.510	2.15	0.96	165	34.6
CaF$_2$	532	1.435	3.179	4.11	3.49	45.6	37.1
Plexiglas	532	1.494	1.190		0.63	253.7	15.7

[a] The data of these materials are collected from Refs. 53–67.

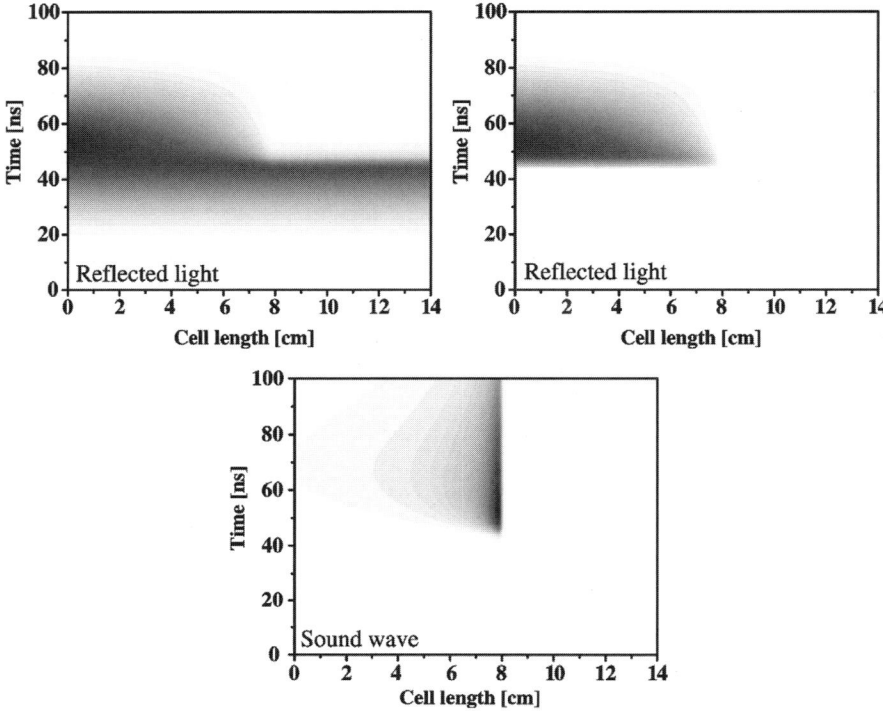

Figure 2.6. Gray scale plots of the intensity distributions of the pump light and the reflected light and the amplitude of the acoustic wave. For the calculation a Brillouin cell filled with SF_6 was assumed. The cell had a total length of 140 mm. The focal spot having a beam diameter of 40 μm was 80 mm behind the entrance window.

For these calculations a Gaussian pump light beam with a wavelength of 1064 nm was assumed. This beam had a beam diameter of 40 μm inside the cell 80 mm behind the entrance window. Such a beam would follow, for example, by focusing a 2-mm waist diameter with a focal length of 120 mm positioned 40 mm in front of the entrance window of the cell. As SBS material, SF_6 with a pressure of 20 bars was assumed. As can be seen from these calculations, all of the SBS reflection takes place in front of the waist of the focused pump light, and thus the reflected light occurs on the left side of the cell. In this case the sound wave was very highly concentrated around the focus of the beam. The calculations were done for a pump pulse energy of 41 mJ for a pulse duration of 30 ns (FWHM). As can be seen from the pump light graph, the leading edge of the pulse is needed to establish the sound wave inside the cell. This leading edge of the pump pulse transmits through the SBS cell virtually undepleted.

But from a certain time on the sound wave amplitude is strong enough to reflect the light, and therefore the following pump light is depleted strongly and a strong reflected Stokes light can be obtained. This temporally onset behavior of the SBS

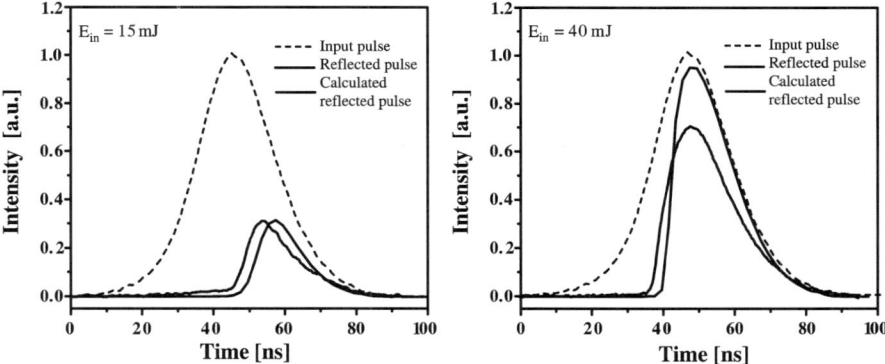

Figure 2.7. Pulse shapes of the incident pump light and the reflected Stokes light from SBS in SF$_6$ for the geometry as given by Fig. 2.3 for different energies of the pump pulse.

reflectivity is shown in Fig. 2.7. For two different input pulse energies the reflected pulses show different pulse shapes but in all cases the onset of the SBS reflectivity can be obtained.

These temporal profiles of the incident and the reflected pulses can be measured in the nanosecond range easily, and thus the calculations can be proven for the quality of the modeling. It turned out that the two-dimensional model can reach accuracies of about 20–30% in reproducing the pulse shape of the reflected light. Therefore, higher accuracies demand more sophisticated models as the 3D model described below.

2.2.5 Numerical calculations 3D model (focused geometry)

A very good approximation of the system of partial differential Eq. (12) would be reached considering the full transversal beam shape profile of the incident and the reflected light as well as for the sound wave amplitude distribution inside the SBS material as a function of time. But to our knowledge there is no successful trial to include the whole complexity of that system for successful calculation of a practical case as, for example, obtained in the previous chapter. Therefore, another approach was used to include a large part of this complexity in the calculations [49, 50]. In this model the calculations were based on a set of transversal modes for the incident and the reflected beam interacting inside the SBS cell. The change of this mode distribution for both beams was calculated then as a function of the propagation direction z and time. About 10 different modes were necessary to be included in these calculations for both the incident and the reflected light beams. Under these circumstances the computations take several minutes on main frame computers. Although only first trials were done with this complicated model, the potential seems quite promising. Especially the temporal shape for reflected pulses could be calculated with much higher accuracy as discussed in Section 2.2.4.

The partial differential equations (13) shall be solved with the mode mixture for the pump and the Stokes electric field:

$$E_P(r_\perp, z, t) = \sum_m a_m(z, t) A_m(r_\perp, z) \quad (35)$$

$$E_S(r_\perp, z, t) = \sum_m b_m(z, t) B_m(r_\perp, z) \quad (36)$$

This stick position method is based on the orthonormal basis of modes A_n and B_n. The position vector $r_\perp$ is directed perpendicular to the propagation direction of the two beams. The modes A_n and B_n satisfy the homogeneous Maxwell equations, and therefore the system (12) can be written as [49]

$$\left(\frac{n}{c}\frac{\partial}{\partial t} + \frac{\partial}{\partial z}\right) b_n = g_1 g_2 \sum_{i,j,k} C_{ij}^* a_k g_{knij} \quad (37)$$

$$\left(\frac{n}{c}\frac{\partial}{\partial t} - \frac{\partial}{\partial z}\right) a_n = -g_1 g_2 \sum_{i,j,k} C_{ij} b_k g_{knij}^* \quad (38)$$

with

$$C_{ij}(z, t) = \int_0^t \left[a_i(z, \tau) b_j^*(z, \tau) + f_{ij}(z, \tau)\right] e^{-\Gamma(t-\tau)} d\tau \quad (39)$$

$$g_{knij} = \int_{-\infty}^{+\infty} A_i^* A_k B_j B_n^* d^2 r \quad (40)$$

where f_{ij} describes the noise term that stimulates the SBS process. This noise term is chosen with spatial and temporal Gaussian distributions as mentioned above. These noise terms are delta-correlated as

$$\langle f_{ij}(z, t) f_{kl}^*(z', t') \rangle = Q_0 \delta_{ik} \delta_{jl} \delta(z - z') \delta(t - t') \quad (41)$$

with the constant Q_0 as it was given in [35]

$$Q_0 = \frac{2kT\rho_0 \Gamma}{g_B^2 v^2} \quad (42)$$

This Q_0 determines the size of the fluctuations of f. It is a function of the temperature T. In this equation, ρ_0 is the mean density and v the velocity of the sound wave in the material as used before. The variance of f is a free parameter that is used to fit the experimental results for a certain material and is kept constant for all experiments in the same material. As basis, A_n and B_n cylindrically symmetrical Laguerre–Gaussian modes were used:

$$A_n(r, z) = \left(\frac{2}{\pi}\right)^{1/2} \frac{1}{\omega(z)} e^{i(n+1/2)\psi(z)} L_n\left(\frac{2r^2}{\omega(z)^2}\right) \times \exp\left[-i\frac{kr^2}{2R(z)} - \frac{r^2}{\omega(z)^2}\right] \quad (43)$$

where the spot size $\omega^2(z)$, radius of curvature $R(z)$, and the phase angle $\psi(z)$ are given by the expressions

$$\omega^2(z) = \omega_0^2\left[1 + \left(\frac{z}{z_R}\right)^2\right], \qquad R(z) = z + \frac{z_R^2}{z}, \qquad \psi(z) \tan^{-1}\left(\frac{z}{z_R}\right) \qquad (44)$$

With this model it was possible to calculate the reflectivity of the SBS—for example, in Freon-113 as shown in Fig. 2.8. The agreement between experimental and theoretical data is even better than that of the two-dimensional model as described in Section 2.2.4. A similar result was reached with the nonstationary scattering in SF_6 as shown in Fig. 2.9. But the real potential of this three-dimensional model was obtained in modeling the temporal shape of the reflected pulses of the Stokes wave. With the two-dimensional model it was not possible to get good agreement for a whole set of different input energies. Therefore, the real proof was the modeling of these curves as can be seen from Fig. 2.10. The agreement of the calculation and the measured temporal profiles is astonishingly good even for different input energies of the pump pulse. The same good accuracy was reached in the nonstationary scattering in SF_6 as shown in Fig. 2.11.

Therefore, it can be concluded that also considering the transversal beam profile is important in the SBS reflection. This model is complete in the sense that it includes stationary and transient effects in SBS phase conjugation, it is started from noise, and the noise term is the only fit parameter that has to be determined only once for a certain material, and it includes the transverse effects of the SBS on the beam profile of the incident and the reflected pulses.

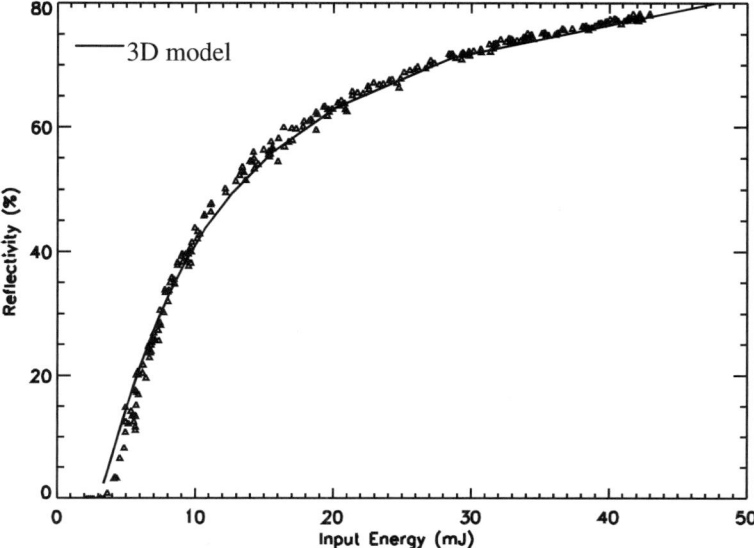

Figure 2.8. SBS reflectivity observed in Freon-113 as a function of the pulse energy.

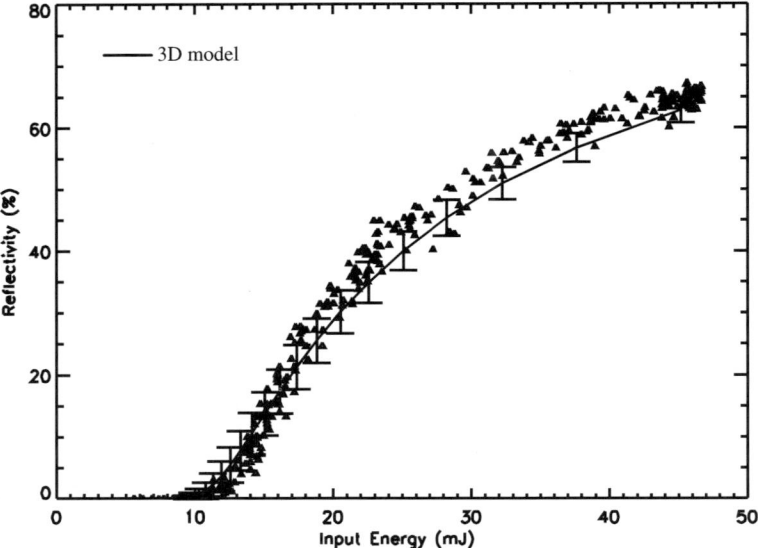

Figure 2.9. SBS reflectivity observed in SF_6 as a function of the pulse energy.

This model should allow for new and more complete predictions of the fidelity in SBS phase conjugation using focusing geometry in bulk materials. Also, the temporal profiles of the reflected light can be calculated with a very high accuracy including relaxation oscillations [51] at the beginning of the reflection process as well as fluctuations that are generated from the noise terms in the SBS material [52].

Therefore, this model seems to be up to now the most complete and useful description of the SBS process if high accuracy in all of these detailed parameters is wanted. The drawback is the high computation power that is necessary to calculate the scattering process including all these effects. So far, from the given models the best should be chosen for the particular application.

2.2.6 Numerical calculations for waveguides

The solution of the system of differential equations was given in Section 2.2.1 [Eq. (13)], while considering the transversal distribution of the electric field in the waveguide structure is not possible on personal computers, yet. A calculation of the complicated intensity distribution within a multimode fiber is difficult even if the nonlinear interactions are neglected.

If it is possible to assume a flat-top intensity profile inside the fiber, the computation effort is reduced drastically. The resulting system of differential equations is now one-dimensional in space and can therefore be integrated on a two-

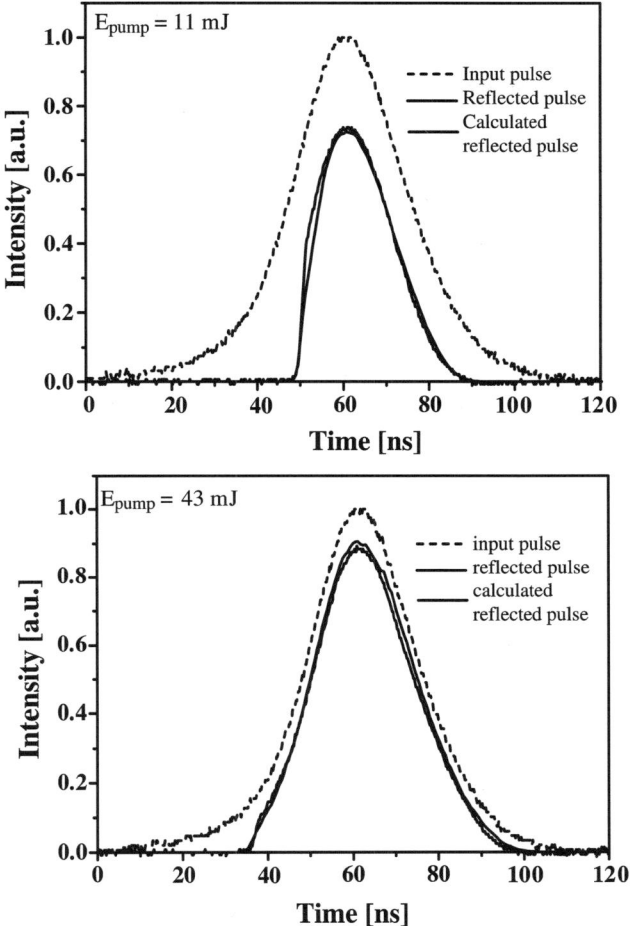

Figure 2.10. Calculated and measured pump and Stokes pulses in Freon-113.

dimensional space–time grid. The equations read as follows:

$$dE_P(z, t) = i\frac{S(z, t)}{2} E_S(z, t)\, dz$$

$$dE_S(z, t) = i\frac{S^*(z, t)}{2} E_P(z, t)\, dz \qquad (45)$$

$$dS(z, t) = \frac{\Gamma_B}{2} \left(ig_B E_P(z, t) E_S^*(z, t) - (S(z, t) - S_0(z, t)) \right) dt$$

The space steps dz follow from the temporal steps dt by the simple relation $dz = \pm(c/n)dt$. The plus sign relates to the pump pulse propagation in positive z direction, and the negative sign represents the reflected Stokes pulse. In this equation

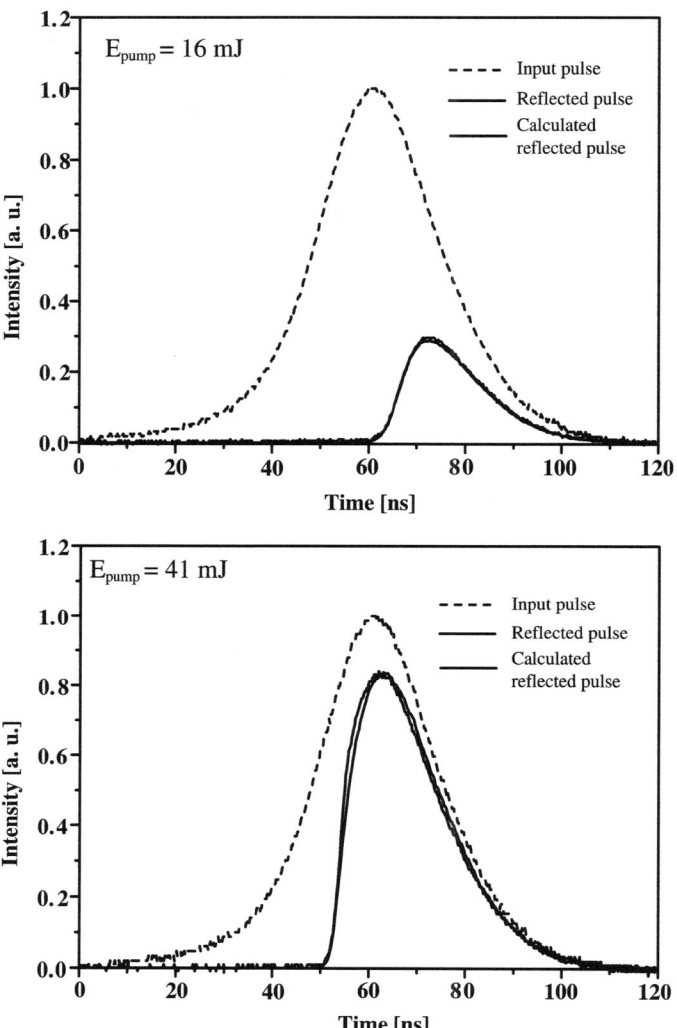

Figure 2.11. Calculated and measured pump and Stokes pulses in SF_6.

the amplitude of the sound wave $S(z, t)$ is started by a noise term of a size $S_0(z, t)$. As it was used in Eq. (40), it is assumed that this noise term has a Gaussian distribution.

With this model the intensity distribution of a pump light and the reflected Stokes light inside a fiber with a core diameter of 25 μm and a length of 2 m was calculated as shown in Fig. 2.12. Intensities coded as gray scale were calculated in this example for an input pulse energy of 0.05 mJ. As can be seen from this figure, almost no reflection can be obtained as long as the power of the rising pump pulse is below threshold. Thus, this leading part of the pump wave travels through the fiber almost undepleted and can be obtained at the end of the fiber experimentally. Nevertheless,

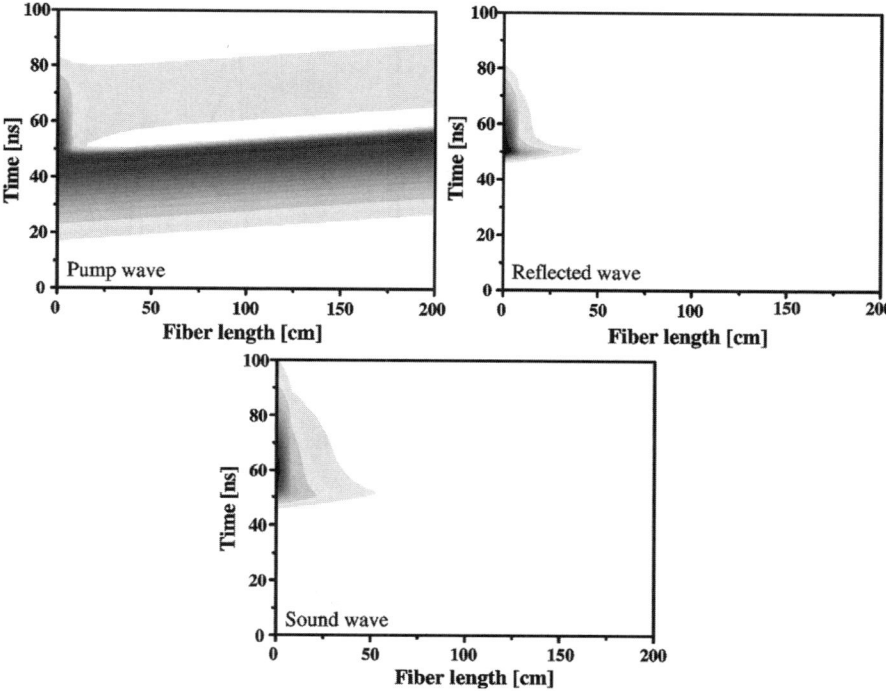

Figure 2.12. Gray scale plot of the intensity distributions of the pump and the Stokes light waves as well as the sound wave as a function of the fiber length for a pump pulse energy of 50 µJ.

some sound wave is generated and, after reaching the threshold, the sound wave and the Stokes wave increase rapidly. In this case the pump wave is almost depleted completely.

Although the fiber in this computation has a length of 2 m, and this length is necessary for starting the SBS reflectivity and lowering the threshold of the SBS process, Fig. 2.12 shows that the main reflection takes place in the very front part of the fiber. Only the first 10 to 20 cm are responsible for the main reflection of the pump light. This has important consequences for the quality and the parameters of the phase conjugation via SBS fiber mirrors. Most important, almost no depolarization takes place for the main part of the reflected light as can be obtained experimentally, too [85]. The reflected light is polarized 98% in the same way as the incident light was polarized. Therefore, these fibers can be used in master oscillator double-pass amplifier (MOPA) schemes with polarization outcoupling of the phase conjugated reflected and second time-amplified light, easily.

Nevertheless, the starting process of the SBS reflectivity occurs over the whole length of the fiber and therefore the threshold is mostly determined by this fiber length and of course the coherence length of the laser which was assumed to be long enough. As another consequence of this distributed starting process, sometimes

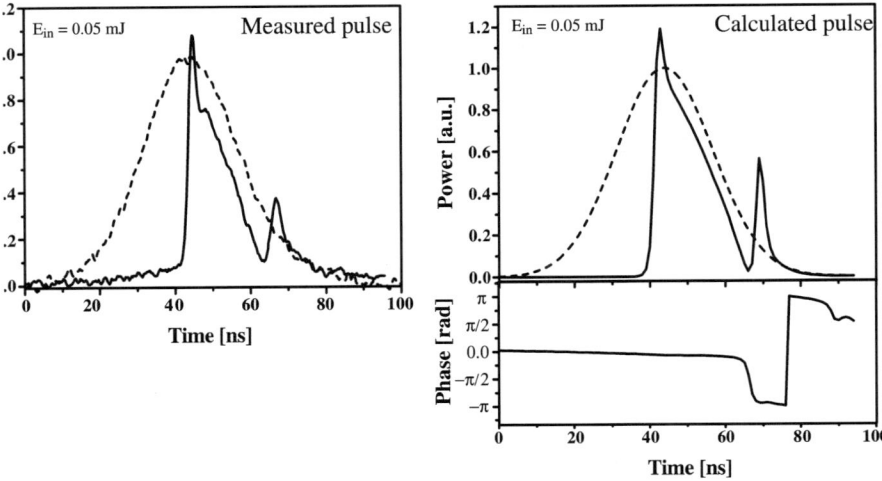

Figure 2.13. Measured and calculated Stokes pulse with strong modulations of phase and amplitude as a consequence of the starting process in the fiber.

strong modulations can be observed in the reflected pulses. A typical result is shown in Fig. 2.13. Although this behavior is stochastic, Fig. 2.13 shows that the measured pulses can also be obtained by the computation, and in this case the phase fluctuations can be given easily, too, as shown in the figure. Experimentally and theoretically, it was determined that roughly 30% of the measurements show this type of modulation. The real structure of the modulations is very different from pulse to pulse, although the overall reflectivity is almost the same within a few percent. Experimentally, no indications for the influence of the pulse energy or the type of fiber could be obtained; that is, in all cases the fluctuations were observed with roughly the same type of structure and probability.

The only common feature of these modulations is that after a first strong increase of the reflectivity, a strong decrease can be observed which is followed by a very sharp pulse spike that can be of much higher peak power than the incident pump pulse. From Fig. 2.13 it follows also that synchronously with the strong decrease of the reflection of the SBS a phase shift of roughly π can be observed. Similar behavior was obtained in monomode fibers [35, 82], but was also obtained earlier with some bulk SBS materials such as Freon [52]. The reason for these fluctuations is to our knowledge the stochastic generation of the sound wave based on fluctuations in the noise structure inside the fiber. This noise is then amplified by the nonlinear process up to the remarkable values as shown in Fig. 2.13. Therefore, this amplitude fluctuations occur more dramatically if the transit time through the SBS medium is much longer than the phonon lifetime [35].

Analyzing this process in more detail, we can assume that these phase jumps can happen at different positions along the fiber. The reflected light shows in this case a phase shift of π, and so far destructive interference between the different light waves can be obtained. As a result, the SBS reflection cancels out as observed in Fig. 2.13. The zone of depleted reflectivity will move through the fiber and is sometimes called phase wave. As a result of these mutually canceling sound waves, the pump wave is not that much depleted any more, and at another point the SBS process can start again. Then the pump light is reflected almost all over the whole fiber, and SBS pulse compression can take place [70].

As mentioned before, the fluctuations in the SBS signal did not have any impact on the energy reflectivity of the mirror. Thus, the 70% smooth pulse shapes and the 30% modulated pulses showed the about same reflectivity. In addition, the fidelity was roughly the same for modulated or nonmodulated pulses. More detailed knowledge should be available from fidelity measurements with high temporal resolution. The three-dimensional model described in Section 2.2.5 predicts a decrease of the fidelity while the fluctuation occurs [49].

2.3 REALIZATION OF SBS MIRRORS

The easiest method to realize phase conjugating SBS mirrors involves the utilization of suitable bulk material. In this case the incident pump beam which should be phase conjugated is focused into the material. This type of phase conjugating mirrors is easy to handle, very reliable, and simple to build. The only precondition in this case is sufficient beam power or pulse energy to overcome the threshold of these self-pumped devices. Therefore they are usually used for pulsed lasers providing peak powers in the range of several 10 or 100 kW.

High reflectivity and good fidelity of these phase conjugating SBS mirrors were realized by choosing the material and the focusing conditions as a function of light parameters. The parameters of several materials are given in Table 2.1 to Table 2.3. If low thresholds are needed, materials with high Brillouin gain have to be selected. For gases the gain coefficient is roughly proportional to the square of the material density. Therefore, usually the threshold is decreased by applying higher pressures with gaseous materials. The Brillouin lifetime is proportional to the density of a gas itself.

Lower lifetimes are usually also favorable for decreasing threshold of the SBS material because of the more stationary behavior (see Section 2.2.3). The lifetime of a sound wave scales for all materials roughly with the square of the wavelength of the pump light λ_{pump}^2 [see Eq. (11)] for wavelengths from the UV up to the IR. Therefore it is very difficult to realize optical phase conjugation based on SBS bulk materials in the near or far infrared between 3 and 10 μm [41, 54].

Another important feature of the SBS material with respect to the application in high-power laser systems is the possible absorption at the wavelengths the SBS mirror shall be operated. For instance many of the listed liquids in Table 2.2 show some absorption in the infrared. This absorption usually causes heating and can even produce optical damage. In any case the resulting refractive index distortions by heated material will disturb the optical quality of the SBS phase conjugation [68]. As a rule of thumb, the absorption coefficient should be smaller than $10^{-6}\,\text{cm}^{-1}$. Therefore, the absorption of a material has to be checked usually experimentally.

High chemical purity is one of the preconditions for using these materials in practical devices. But even Uvasol quality can be insufficient, because an additional problem occurs from impurities in form of little particles. They are usually not specified in the chemical purity table. These particles are often brought in the SBS material from the chemical purification process. These are, for example, little glass or ceramic particles. If necessary, these particles can be removed by very good filters or by pump-and-freeze techniques [61]. The distortions from impurities may especially occur using organic liquids for SBS phase conjugation. As a result, these liquids may show little gas bubbles during operation and may even turn finally from transparent to a yellow or brown color over time.

As discussed in the theoretical part, if the coherence length is longer than the interaction length in the material the reflectivity does not depend on the focusing in the bulk material. Long or short Rayleigh lengths from focusing will show roughly the same reflectivity as discussed, for example, in Refs. 2, 59, and 69. But usually the fidelity of SBS phase conjugation is better for a stronger focusing. This also makes these mirrors more handy because of the shorter geometrical length of the device. But very strong focusing is limited by optical damage or other nonlinear processes in the material which are both dependent on higher orders of the power of the intensity. Therefore, the focusing has to be adapted to the damage threshold of the material and the onset of the other nonlinear processes.

Very long Rayleigh lengths (i.e., very weak focusing) within the range of meters can cause additional pulse shaping effects from the propagation time of the light inside the SBS material. Most prominent in this is pulse compression which can be reached with optical lengths of the SBS material of the order of half the pulse length of the used light. In this case, maximum compression values of more than 1:10 can be reached. The shortest observed pulses by pulse compression via SBS reflection are in the range of 1 ns or even below [70–75].

Competing nonlinear optical processes are self-focusing, self-phase modulation, and stimulated Raman scattering. If these processes are activated to a large extent, the fidelity and the reflectivity of these SBS mirrors can be decreased drastically.

Another important feature of the phase conjugating SBS mirrors is the frequency shift of the reflected light. This frequency shift is characteristic for the material and also a function of the used wavelength of the applied light [Eq. (3)]. It is usually in the range of a few 100 MHz for liquids and gigahertz for solid materials. This frequency shift can cause problems if the reflected light shall be amplified in very narrow bandwidth amplifiers which do not cover this frequency shift from the phase conjugating SBS mirror. So far, this frequency shift has to be considered in designing the application.

Last but not least, the possible application of a certain SBS material may be restricted by toxicity or other safety reasons caused by the properties of the material.

Therefore, from the large list of possible and known SBS materials, only a small number are prominently used in laser applications. One is Freon-113 and other derivatives of this class of organic liquids. SF_6, nitrogen, and xenon are frequently used as gases. Newer and very promising materials are the heavy fluorocarbon liquids FC-72 and FC-75 [63] and the organic crystal LAP [66].

2.3.1 Bulk media SBS mirrors

In the simplest case the SBS liquid or gases bulk material will be filled in suitable cells with transparent windows for the incident pump light. The gas cells as shown in Fig. 2.14 have to contain the material with pressures of 20–100 bar (2–10×10^6 Pa). Therefore, the transparent windows have to be thick enough and the ceiling should keep the pressure inside the cell over years.

For pressures up to about 20 bars, simple ceiling based on O-rings is usually sufficient. The window thickness should be in the range of 3–5 mm. For higher pressures it is necessary to professionally design these cells or use some commercially available Raman cells. Window thicknesses of 10 mm or more may be necessary. In this case, also possible optical birefringence in the crystalline cell windows (e.g., of quartz) has to be taken into account. The high pressure on the window may cause some tension which can produce birefringence distortions in the cell. These can disturb crucially the fidelity of the SBS phase conjugation and the polarization outcoupling. The focusing lens can be used as an entrance window, and therefore the whole device becomes very compact and reliable. If the construction is well-designed, the gas cells can be used over years without any maintenance. Gas cells are very reliable because optical breakdown or other distortions in the cell usually will not prevent their use. The damaged material will just move very soon out of the focus region which is used in the SBS process.

Using liquids as SBS material is the easiest way to design such self-pumped nonlinear optical devices. The liquid can be filled in a suitable glass container which can be easily home-made or bought (see Fig. 2.15). These cells show typical lengths

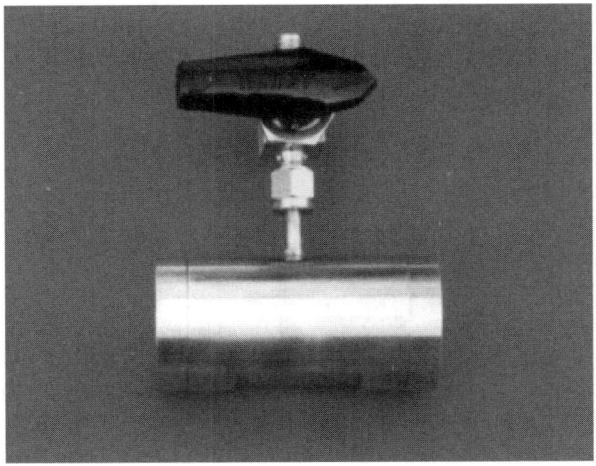

Figure 2.14. Gas cell as used as phase conjugating SBS mirror.

Figure 2.15. SBS mirror based on liquid material.

in the range of a few centimeters up to several 10 cm as a function of the application. For compression purposes, even several meters cell length is sometimes applied. The windows have to be sealed against the possibly aggressive liquid, and so far glue is usually not useful for aggressive SBS material. A disadvantage of liquids as SBS mirrors can be the possible bubble production or photochemical destruction of the matter. So far, the liquid may be exchanged in a month's or year's period as a function of the used powers in the setup.

Because of the damage problems, solids are used in bulk material SBS phase conjugation very seldom [66, 67]. But optical glasses are very useful in waveguide structures (optical fibers), as will be discussed below.

The energy reflectivity for some bulk materials as a function of the pulse energy of the incident pump light is shown in Figs. 2.16 and 2.17.

In Fig. 2.16 the reflectivity behavior for some liquids is given. The curves were normalized for the threshold energy of the incident pump light. This threshold energy is given in the figure, too. The used pulses had a wavelength of 532 nm and a pulse duration of 17 ns, and they were focused with a lens of 100-mm focal length into the cell. The threshold energies vary by almost a factor of 3 for the used materials from n-hexane and iso-octane starting with 0.30 mJ up to 1.1 mJ for water. But the shape of the reflectivity curve was almost independent of the used material as was discussed earlier [2].

Similar results were reached with gases as SBS materials. The results of the reflectivity as a function of the input pulse energy are given in Fig. 2.17. As can be seen, the reflectivity of these gases shows again almost the same shape if the input pulse energy is normalized for the threshold pulse energy. Nevertheless, in comparison to Fig. 2.16, now the spreading of the curves is slightly higher than for

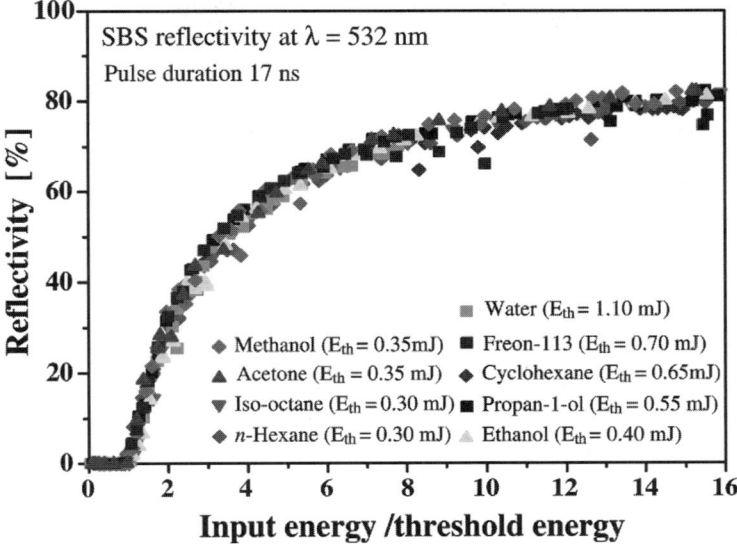

Figure 2.16. Measured SBS reflectivity of different materials at a laser wavelength of 532 nm. The reflectivity is plotted as a function of the normalized input pulse energy.

the more stationary scattering in the liquids. The reason here is the pulse shaping of the Stokes light as a consequence of the nonstationary scattering because of the long lifetimes in the gases (see Table 2.1). The threshold energy for these materials is almost one order of magnitude higher than that for the liquids measured at the

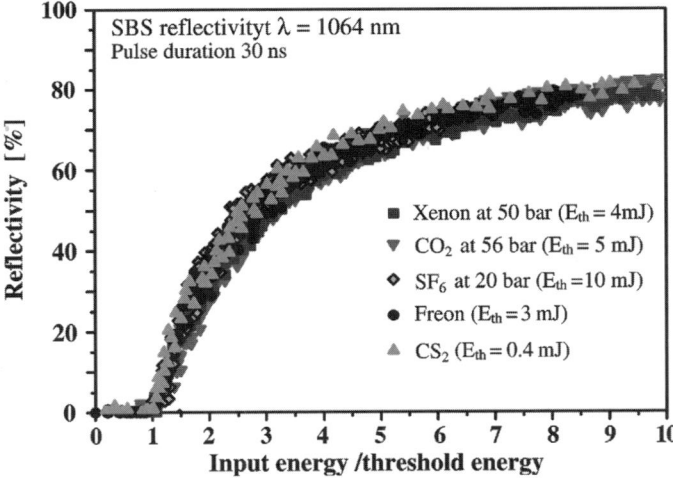

Figure 2.17. Measured SBS reflectivity of different materials at a laser wavelength of 1064 nm and 30-ns pulse duration. The reflectivity is plotted as a function of the normalized input pulse energy.

wavelength 532 nm. CS_2 as reference material with the highest known Brillouin gain shows a much smaller threshold.

The difference between stationary and nonstationary SBS reflection can be much more easily detected by the temporary profiles of the reflected light in comparison with the incident light as shown in Fig. 2.18. As can be seen from this figure showing two reflectivity situations with roughly the same energy reflectivity value in case of more stationary scattering as obtained for Freon-113 ($\tau_B = 0.84$ ns), the pulse of the reflected light is much more symmetric than nonstationary scattering in SF_6 ($\tau_B = 17.3$ ns). The onset at the beginning of the reflectivity is just a small part of the pulse, and overall the reflected light is almost not delayed to the incident light. In the case of SF_6, much more delay occurs. The onset of the reflected light occurs much later. Here the long lifetime of the SBS material is the prominent value causing this effect. Whereas the pulse seems to be delayed compared to the input pulse, the trailing edge of the reflected light is almost identical with the incident pulse.

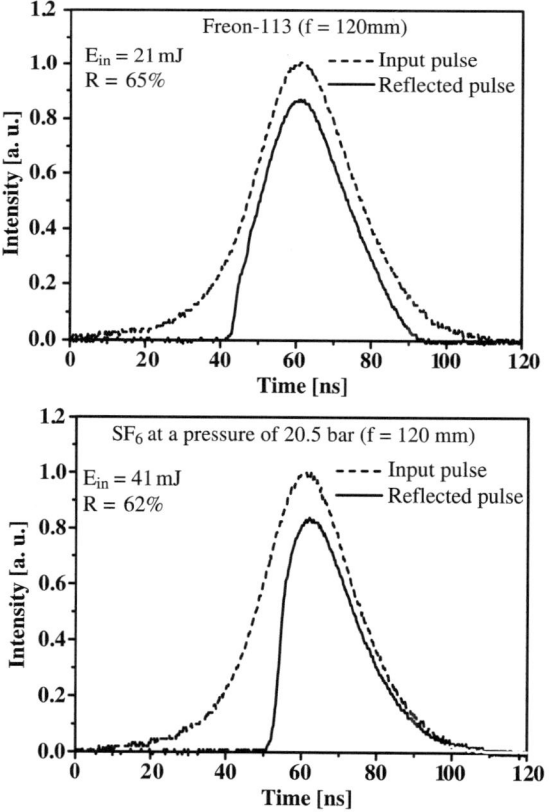

Figure 2.18. Measured pulse shapes of the input and the reflected pulse for two different media Freon-116 and SF_6.

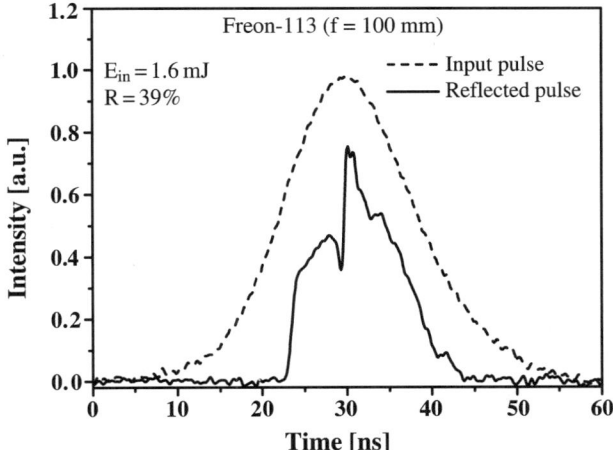

Figure 2.19. Measured shape of the reflected pulse by SBS in Freon-113. The pulse shows the typical spiked structure. The wavelength of the incident light was 532 nm.

If the lifetime of the used SBS material is very short compared to the input pulse duration, also instabilities like amplitude and phase fluctuation may occur more frequently as a result of the high nonlinearity of this optical process. One example is shown in Fig. 2.19. In this case the focusing and the whole setup were selected in a way that these instabilities can be obtained. As can be seen, the reflected light shows some fluctuations that even may not be fully resolved from the temporal resolution of the measuring system, which was roughly 100 ps in this case. Although this type of pulse can show also quite good reflectivity values, they may not be useful for practical applications. Therefore a well-designed focusing and the selection of the best material for the given pulse parameters may be important for save operation of the phase conjugating SBS mirror.

2.3.2 Optical fibers

Optical fibers typically allow very long interaction lengths of several kilometers. Therefore, in such fibers the SBS threshold can be decreased to very low values. So far, SBS reflectivity and even phase conjugation was obtained with continuous-wave (cw) lasers with an output power of 500 mW only [76–78]. An important precondition for these very low thresholds is a high coherent laser beam with a bandwidth below 100 MHz [84]. For highly coherent light the SBS has the lowest threshold of all nonlinearities inside a long optical fiber. The SBS is the main limiting factor for the power transmission in such fibers. In many applications the light which has to be phase conjugated does not fulfill the coherent requirements to use optical fibers of more than several meters length.

Nevertheless, fibers can also be used to reflect pulsed laser light with comparably small pulse energies if the fiber length and thus the coherence length is in the range

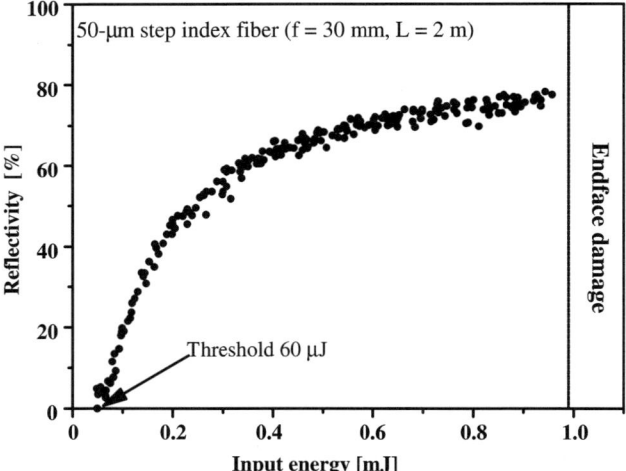

Figure 2.20. The reflectivity of a step index fiber with a 50-μm core diameter and a length of 2 m. The SBS threshold was 60 μJ, and damage of the fiber endface occurs at about 1 mJ.

of a few tens of centimeters up to meters only [79–83]. One example is given in Fig. 2.20. The applied step index fiber with a core diameter of 50 μm showed the threshold pulse energy of 60 μJ for a laser wavelength of 1064 nm. The incident beam was focused with a 30-mm lens into the fiber. Comparing this result with the above-obtained bulk reveals that the threshold is almost a factor 5 smaller than that in the observed bulk situation. Therefore, fibers seem to be very useful for low-threshold optical phase conjugation.

But the main problem using fibers is the endface damage occurring at low pulse energies. In the case of Fig. 2.20 the damage occurred at about 1 mJ input pulse energy. So far the useful dynamic range using fibers as SBS phase conjugators is usually quite small. The dynamic range can be estimated from the following formula:

$$\frac{P_{\text{damage}}}{P_{\text{threshold}}} = \frac{I_{\text{damage}} A_{\text{eff}} g_B L_{\text{eff}}}{21 A_{\text{eff}}} = 25 \cdot \frac{L_{\text{eff}}}{[m]} \qquad (46)$$

The maximum value from this formula is about a factor of $25/m$. Nevertheless, for certain applications (e.g., in MOPA systems), fibers may be useful for SBS phase conjugation. They are even of more interest because they can be bought from stock as telecommunication fibers. Because of the high demands in telecommunications, these fibers have very high quality and so far all problems of possible absorption and damping are negligible as used in optical phase conjugation.

Besides the shown step index fibers, gradient index fibers cannot be used for SBS. It turned out that the thresholds in gradient index fibers seem to be even a little bit smaller than in step index fibers. But the fidelity of the optical phase conjugation in gradient index fibers was usually quite low. Therefore, gradient index fibers seem not to be very useful for optical phase conjugating SBS mirrors in laser devices.

Another typical effect using optical fibers with long pulses from AOM Q-switching of cw lasers are the strong amplitude fluctuations. Two examples are shown in Fig. 2.21. These experiments are all single shots using the step index fiber of Fig. 2.20 under constant conditions. As can be seen from this figure, the pulse shape of the reflected light can be really different. High modulations can be obtained, and therefore the temporal reflectivity of the light will be different from shot to shot. So far, these fibers in this simple configuration may cause problems in a real laser system. Nevertheless, it should be mentioned that by designing the fiber length and the fiber core diameter in the right way they may be very useful, especially for applications in MOPA systems.

The reflectivity in a 47-m-long fiber as a function of the average input power of this long pulse is shown in Fig. 2.22. As can be seen from this figure, the average threshold power is in this case 112 mW and the peak power of the used AOM pulse is 102 W. These values are obtained with a single-mode Yb:YAG laser with a

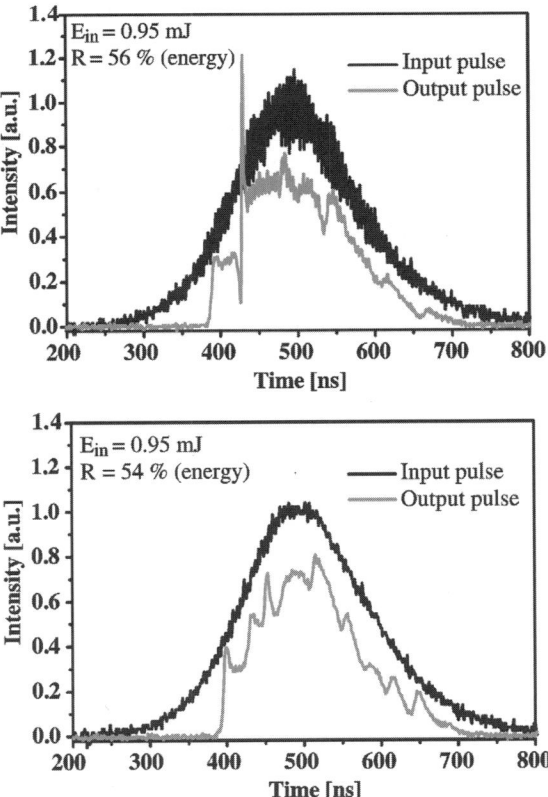

Figure 2.21. Two typical temporal shapes of the reflected pulse at the same input energy for long pulses of 220-ns duration.

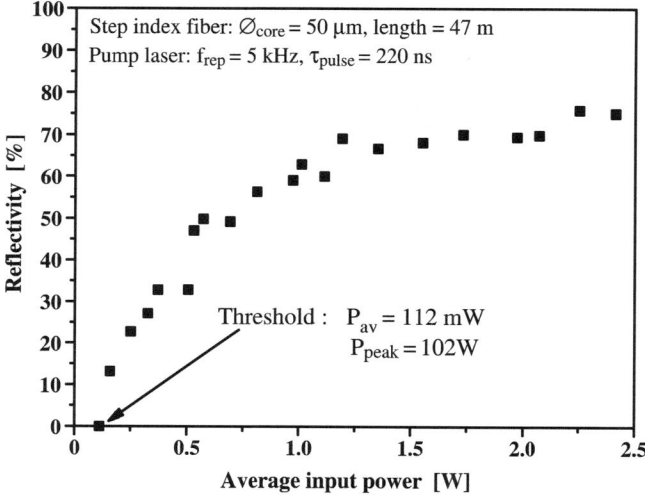

Figure 2.22. Reflectivity of a step index fiber as a function of the average input power using a Yb:YAG laser with a wavelength of 1030 nm and a pulse duration of 220 ns.

wavelength of 1030 nm, a repetition rate of 5 kHz, and a pulse duration of 220 ns. The step index fiber of 50-μm core diameter was 47 m long.

If the pulse duration of this type of laser is increased, the effective interaction length is increased, too, and for very long pulses the steady-state limit can be reached as shown in Fig. 2.23. So far, these experiments show that not always kilometer lengths are necessary to realize optical phase conjugation for realistic cw or quasi-cw lasers. The applied core diameter of 50 μm is still much above a monomode fiber diameter and allows about 500 modes inside the fiber. Therefore it is useful for optical phase conjugation. Nevertheless, also in this case the dynamic problem with this type of fiber is not solved and also the highest obtained reflectivities in the range below 80% may be too small for practical applications. On the other hand, using this type of fibers in MOPA systems seems to be reasonable because there the losses in the fiber do not significantly decrease the overall efficiency of the whole double-pass amplifier.

2.3.3 Tapered fibers

A very well operating solution for these above-described problems using fibers for optical phase conjugation was given in Ref. 86. In this case the SBS oscillator–SBS amplifier idea developed in Refs. 87–89 was combined with the advantages of optical fibers. The resulting tapered fiber structure resulted in very high reflectivity, large dynamic range, and very good fidelity with one device (see Fig. 2.24). The only limitation of this concept is the taper in the fiber itself. At the taper, losses can occur because of the nonperfect matching of higher-order modes to the taper structure. The details of the change in the beam parameters in the taper is given in

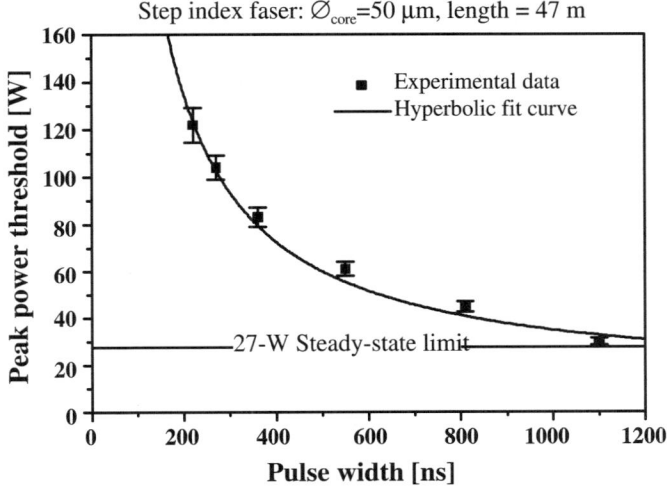

Figure 2.23. Peak power threshold of the applied step index fiber with a length of 47 m and a core diameter of 50 μm as a function of the pulse duration of the used pulses. The steady-state limit of this fiber will be 27-W peak power for the applied laser system.

Fig. 2.25. Fortunately, the losses in the taper are not important because the main reflection of the incident pump light takes place in the amplifier part of the fiber. So far, the losses in the taper are finally less than a few percent of the total light.

For reducing the coherence length demands of the incident pump light, the length of the SBS amplifier and SBS generator part was chosen to be 40 and 50 cm, respectively. The whole set of parameters of the used fiber is given in Table 2.4. As described in Ref. 86, the results with this tapered fiber structure are very promising. With the given parameters a threshold value of 15 μJ could be obtained for the input pulse with a wavelength of 1064 nm and a pulse duration of 30 ns. The maximum reflectivity was 92%, whereas the endfaces of the fiber were not antireflection-coated. If these losses are taken out, the resulting reflectivity would have been in the range of 99%. As a result of the SBS-generator–SBS-amplifier structure, this fiber showed a very large dynamic range. This occurs due to the generating of the first Stokes light in the generator part, meaning the small-diameter fiber part, and then the

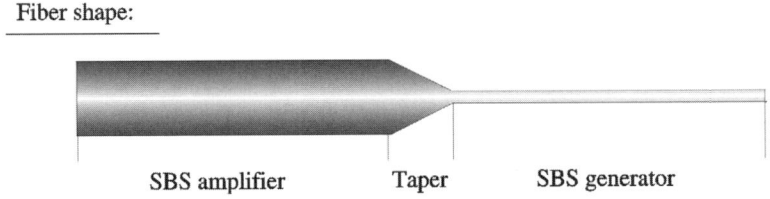

Figure 2.24. Structure of a tapered fiber consisting of an SBS generator and an SBS amplifier connected with a taper structure.

Beam propagation inside the taper:

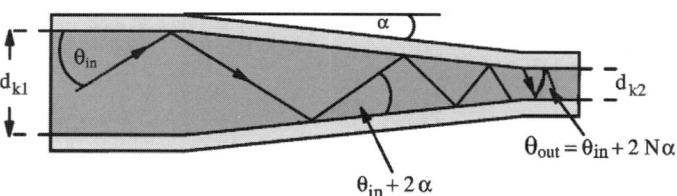

Figure 2.25. Beam propagation of incident beam in the taper region of a tapered fiber.

power reflectivity in the amplifier part, meaning the large-diameter part of the fiber. So far, the damage problem of the fiber will be released by a factor of 25. This type of fiber showed a dynamic range of 1:260, and so far it can be used over a range of 1:200 with very high reflectivity above 90% as shown in Fig. 2.26. The measured fidelity was above 90% over the whole dynamic range. Because of the short length of this tapered fiber below 1 m, the polarization of the incident light was conserved for the reflected light. Therefore, this type of optical phase conjugator can be used in double-pass amplifier schemes using a polarizing element to take out the phase conjugated SBS signal after the second pass through the amplifier system.

2.3.4 Liquid waveguides (capillaries)

If really low thresholds are required for the SBS phase conjugators in combination with low coherence demands for the incident light, the best geometry (which is the waveguide geometry) has to be combined with the best SBS material (which is the liquid CS_2, to our knowledge) [90–93]. So far, capillaries have to be constructed for this purpose. The scheme of the capillary is shown in Fig. 2.27. To stabilize the capillary, it was mounted inside a tube that was completely filled with the used SBS material CS_2.

These capillaries show very good performance with respect to threshold because it cannot be reached with any short length SBS-mirror, to our knowledge. But these capillaries are not easy to handle. The first problem is to get these capillaries

TABLE 2.4. Parameters of the Tapered Fiber

Total length	100 cm
Generator length	50 cm
Amplifier length	45 cm
Taper length	5 cm
NA	0.1
Generator core diameter	20 μm
Amplifier core diameter	100 μm

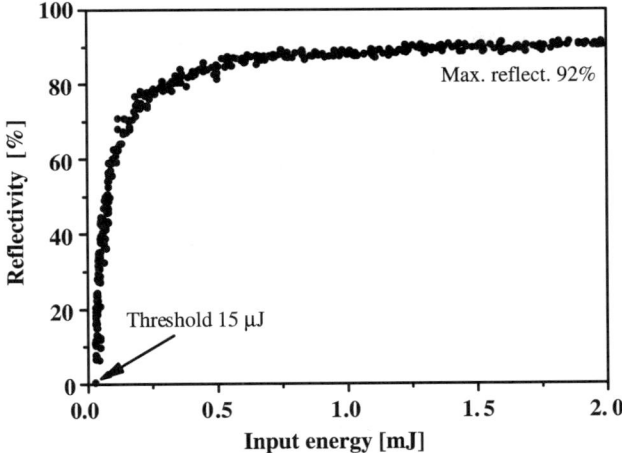

Figure 2.26. Experimentally obtained reflectivity of a tapered fiber structure with the above given parameters for an Nd:YAG laser pulse with 30-ns pulse duration.

homogeneously filled over the whole length. Secondly, if the pulse power is only slightly too high, little bubbles occur in the capillary which cannot be taken out very fast. So far, capillaries need a lot of attention for a save operation.

Nevertheless, the performance can be very good, as shown in Fig. 2.28. The SBS reflectivity in this capillary was obtained with a Q-switched Nd:YAG laser operated at a repetition rate of 20 Hz with a pulse duration of 25 ns. The obtained threshold energy was 1.64 µJ, which corresponds to a peak power of these pulses of 64 W. This low threshold value was obtained with a comparably short waveguide structure of only 50-cm length. So far, the coherence demands for the laser were not very high. The maximum reflectivity achieved with this capillary was above 80%.

But if the same capillary is used with a quasi-cw laser with 5-kHz repetition rate, a very low average power threshold of 80 mW can be obtained as shown in Fig. 2.29. But in this case the heating and generation of bubbles inside the CS_2 was quite disturbing. The dynamic range of this capillary was very small of 1:10. Therefore, these capillaries seem to be not very useful for high average power lasers.

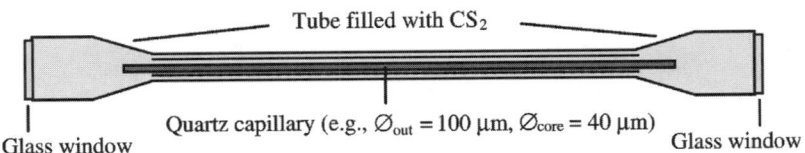

Figure 2.27. Scheme of the applied capillary with inner diameters of 40–100 µm and a length of below 1 m.

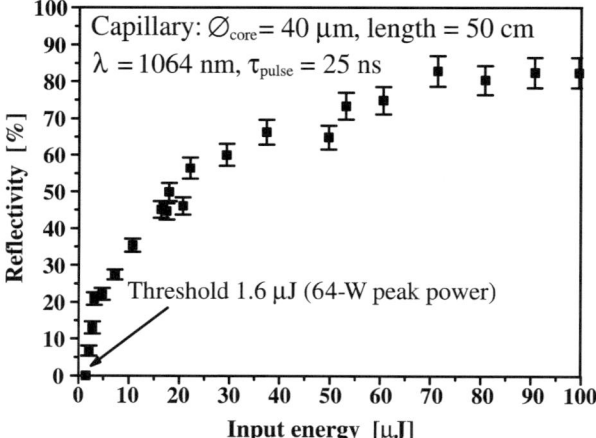

Figure 2.28. SBS reflectivity as obtained with a CS_2 filled capillary with a diameter of 40 μm and a length of 50 cm.

2.4 SUMMARY

Phase conjugating SBS mirrors are very easy to realize and are reliable nonlinear optical devices that can be applied in oscillator or master-oscillator–power-amplifier setups. Because of the power or pulse energy demands of the nonlinear optical process, these devices can be applied to pulsed laser systems. In this case the

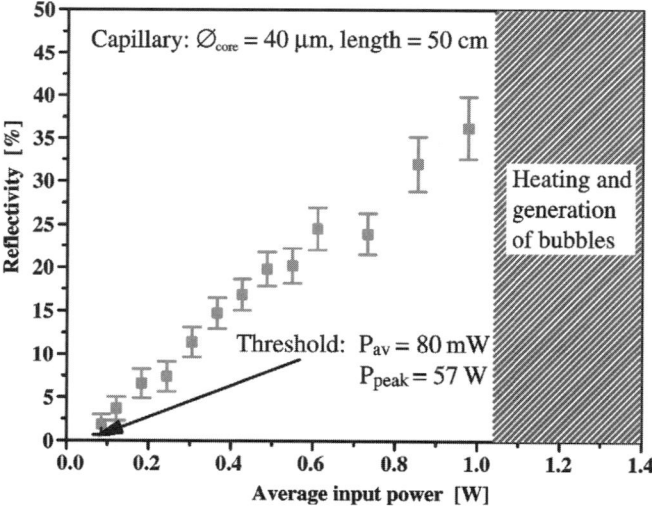

Figure 2.29. Reflectivity as a function of the average input power using a quasi-CW Yb:YAG laser with 5-kHz repetition rate.

geometrical conditions of the focusing and the material have to be chosen with respect to the demands and properties of the laser system. In particular, coherence, light power or pulse energy, and wavelength are the important features of the incident light. The design can be based on theoretical calculations that are possible in different steps of complexity. In the simplest case the stationary solutions can be used in rough approximations even analytically to give a first idea about the requirements for the SBS phase conjugation. More detailed analysis can be done in numerical calculations based on a two-dimensional or a three-dimensional model. The two-dimensional model gives some ideas about the focusing and the expected fidelity properties of the SBS phase conjugating mirror as well as the coherence length demands of the used pump light. The full three-dimensional model is a new solution that demands high computer power. But it is able to calculate the full set of the transversal and temporal profile of the reflected light as a function of all parameters with a good accuracy.

If low thresholds are required, waveguide structures can be applied. The threshold values of today's waveguide structures are in the range of a few tens of watts peak power. In the simplest case, commercial available waveguides can be applied for this purpose. In this case, usually a long coherence length is required for the applied pump light source. If low coherence is necessary, then the tapered waveguide configuration can be applied. With this configuration, reflectivity and fidelity can be above 90% over a dynamic range of larger than 1:200. If very low thresholds are required, capillaries filled with CS_2 are a suitable solution. In this case, reflectivity and dynamic range, as well as any impurity that will disturb the process, are crucial. Therefore, capillaries are more or less useful for scientific applications.

For many available commercial laser systems, SBS phase conjugating mirrors are available now. The design concepts are developed especially for pulsed systems, and these nonlinear devices can be applied without much effort. For quasi-CW and CW lasers the thresholds of the today's available phase conjugating mirrors based on SBS may be still a little bit too high. It can be expected that in the next few years this gap will be closed and then phase conjugating SBS mirrors should be available for all interesting laser systems with high average output powers. Thus, it can be expected that phase conjugating SBS mirrors may in the future be one of the important nonlinear devices applied in commercial laser systems. High brightness and efficiency will be the result.

REFERENCES

1. D. Rockwell, A review of phase-conjugate solid-state lasers, *IEEE J. Quantum Opt.* **24**, 1124 (1988).
2. R. Menzel and H. J. Eichler, Temporal and spatial reflectivity of focused beams in stimulated Brillouin scattering for phase conjugation, *Phys. Rev. A* **46**, 7139 (1992).

3. H. J. Eichler, A. Haase, and R. Menzel, High beam quality of a single rod neodym amplifier by SBS-phase conjugation up to 140 W average output, *Opt. Quantum Electron.* **28**, 261 (1996).

4. C. B. Dane, L. E. Zapata, W. A. Neuman, M. A. Norton, and L. A. Hackel, Design and operation of a 150 W near diffraction-limited Laser amplifier with SBS wavefront correction, *IEEE J. Quantum Electron.* **31**, 148 (1995).

5. H. L. Offerhaus, H. P. Godfried, and W. J. Witteman, All solid-state diode pumped Nd-YAG MOPA with stimulated Brillouin phase conjugate mirror, *Opt. Commun.* **128**, 61 (1996).

6. M. Ostermeyer, and R. Menzel, 50 Watt average output power with 1.2*DL beam quality from a single rod Nd:YALO laser with phase-conjugating SBS mirror, *Opt. Commun.* **171**, 85 (1999).

7. L. Brillouin, *Ann. Phys.* **17**, 88 (1922).

8. R. Y. Chiao, C. H. Townes, and B. P. Stoicheff, *Phy. Rev. Lett.* **12**, 592 (1964).

9. E. Garmire and C. H. Townes, *Appl. Phys. Lett.* **5**, 84 (1964).

10. N. M. Kroll, *J. Appl. Phys.* **36**, 34 (1965).

11. C. L. Tang, Saturation and spectral characteristics of the Stokes emission in the stimulated Brillouin process, *J. Appl. Phys.* **37**, 2945 (1966).

12. W. Kaiser and M. Maier, Stimulated Rayleigh, Brillouin and Raman Spectroscopy, in *Laser Handbook*, Vol. 2, F. T. Arecchi and E. O. Schulz-Dubois, eds., North-Holland, Amsterdam (1972).

13. A. Yariv, Phase conjugate optics and real-time holography, *IEEE J. Quantum Electron.* **QE-14**, 650 (1978).

14. B. Ya. Zel'dovich, V. I. Popovichev, V. V. Ragul'skii, and F. S. Faizullov, Connection between the wave fronts of the reflected and exciting light in stimulated Mandel'shtam–Brillouin scattering, *Sov. Phys. JETP* **15**, 109 (1972).

15. On the history of the discovery of the phase conjugation effect, *Optics and Spectrosc.* **85**, 916 (1998).

16. D. M. Pepper and A. Yariv, Compensation for phase distortions in nonlinear media by phase conjugation, *Opt. Lett.* **5**, 59 (1980).

17. C. R. Guilliano, Applications of optical phase conjugation, *Physics Today* **34**, 27 (1981).

18. T. R. Moore, R. W. Boyd, Three-dimensional simulations of stimulated Brillouin scattering with focused Gaussian beams, *J. Nonlinear Opt. Phys. Mater.* **5**, 387 (1996).

19. R. W. Hellwarth, Theory of phase conjugation by stimulated scattering in a waveguide, *J. Opt. Soc. Am.* **68**, 1050 (1978).

20. R. H. Lehmberg, Numerical study of phase conjugation in stimulated Brillouin scattering an optical waveguide, *J. Opt. Soc. Am.* **73**, 558 (1983).

21. B. Y. Zel'dovich, N. F. Pilipetsky, and V. V. Shkunov, *Principles of Phase Conjugation*, Springer-Verlag, Berlin (1985).

22. P. Suni and J. Falk, Theory of phase conjugation by stimulated Brillouin scattering, *J. Opt. Soc. Am. B* **3**, 1681 (1986).

23. G. G. Kochemasov and V. D. Nikolaev, Reproduction of the spatial amplitude and phase distributions of a pump beam in stimulated Brillouin scattering, *Sov. J. Quantum. Electron.* **7**, 60 (1977).

24. N. B. Baranova and B. Y. Zel'dovich, Wavefront reversal of focused beams *Sov. J. Quantum Electron.* **10**, 555 (1980).
25. A. Kummrow, Hermite–Gaussian theory of focused beam SBS cells, *Opt. Commun.* **96**, 185 (1993).
26. V. E. Yashin and V. I. Kryzhanovskii, Apodisation and spatial filtering of light beams in stimulated Brillouin scattering, *Opt. Spectrosc.* **55**, 101 (1993).
27. G. G. Kochemasov and F. A. Starikov, Novel features of phase conjugation at SBS of beams passed through an ordered phase plate, *Opt. Commun.* **170**, 161 (1999).
28. P. Suni and J. Falk, Measurements of stimulated Brillouin scattering phase-conjugate fidelity, *Opt. Lett.* **12**, 838 (1987).
29. L. P. Schelonka and C. M. Clayton, Effect of focal intensity on stimulated Brillouin scattering reflectivity and fidelity, *Opt. Lett.* **13**, 42 (1988).
30. J. J. Ottusch and D. A. Rockwell, Stimulated Brillouin scattering phase-conjugation fidelity fluctuations, *Opt. Lett.* **16**, 369 (1991).
31. N. G. Basov, V. F. Efimkov, I. G. Zubarev, A. V. Kotov, S. I. Mikhailov, and M. G. Smirnov, Inversion of wavefront in SMBS of a depolarized pump, *JETP Lett.* **28**, 197 (1978).
32. V. N. Blaschuk, V. N. Krasheninnikov, N. A. Melnikov, N. F. Pilipetsky, V. V. Ragulsky, V. V. Shkunov, and B. Ya. Zeldovich, SBS wave front reversal for depolarized light: Theory and experiment, *Opt. Commun.* **27**, 137 (1978).
33. S. C. Matthews and D. A. Rockwell, Correction of phase and depolarization distortions in a multimode fiber at 1.064 μm with stimulated Brillouin scattering phase conjugation, *Opt. Lett.* **19**, 1729 (1994).
34. A. Kummrow and H. Meng, Pressure dependence of stimulated Brillouin backscattering in gases, *Opt. Commun.* **83**, 342 (1991).
35. R. W. Boyd, K. Rzazewski, and P. Narum, Noise initiation of stimulated Brillouin scattering, *Phys. Rev. A* **42**, 5514 (1990).
36. R. Menzel, *Photonics*, Springer, Berlin (2001).
37. G. P. Agraval, *Nonlinear Fiber Optics*, 2nd ed., Academic Press, San Diego (1995).
38. Y. E. D'yakov, Excitation of stimulated light scattering by broad-spectrum pumping, *JETP Lett.* **19**, 243 (1970).
39. P. Narum, M. D. Skeldon, and R. W. Boyd, Effect of laser mode structure on stimulated Brillouin scattering, *IEEE J. Quantum. Electron.* **22**, 2161 (1986).
40. Y. S. Kuo, K. Choi, and J. K. McIver, The effect of pump bandwidth, lens focal length and lens focal point location on stimulated Brillouin scattering threshold and reflectivity, *Opt. Commun.* **80**, 233 (1990).
41. K. D. Ridley, Stimulated Brillouin scattering with an infrared optical parametric oscillator, *J. Modern Optics* **45**, 1137 (1998).
42. G. C. Valley, A review of stimulated Brillouin scattering excited with broad-band pump laser, *IEEE J. Quantum. Electron.* **22**, 704 (1986).
43. R. A. Mullen, Multiple short pulse stimulated Brillouin scattering for trains of 200 ps pulses at 1.06 μm, *IEEE J. Quantum. Electron.* **26**, 1299 (1990).
44. G. K. N. Wong and M. J. Damzen, Enhancement of the phase-conjugate stimulated Brillouin process using optical feedback, *J. Mod. Opt.* **35**, 483 (1988).

45. G. K. N. Wong and M. J. Damzen, Investigations of optical feedback used to enhance stimulated scattering, *IEEE J. Quantum. Electron.* **26**, 139 (1990).
46. M. T. Duignan, B. J. Feldman, and W. T. Whitney, Threshold reduction for stimulated Brillouin scattering using a multipass Herriott cell, *J. Opt. Soc. Am. B* **9**, 548 (1992).
47. H. S. Kim, D. K. Ko, G. Lim, B. H. Cha, and J. Lee, The influence of laser gain on stimulated Brillouin scattering in an active medium, *Opt. Commun.* **167**, 165 (1999).
48. C. N. Pannell, P. St. J. Russell, and T. P. Newson, Stimulated Brillouin scattering in optical fibers: The effects of optical amplification, *J. Opt. Soc. Am. B* **10**, 684 (1993).
49. S. Afshaarvahid and J. Munch, A transient, three-dimensional model of stimulated Brillouin scattering, *J. Nonlinear Opt. Phys. Mater.* **10**, 1 (2001).
50. S. Afshaarvahid, A. Heuer, R. Menzel, and J. Munch, Temporal structure of stimulated-Brillouin-scattering reflectivity considering transversal-mode development, *Phys. Rev. A* **64**, 043803 (2001).
51. R. V. Johnson and J. H. Marburger, Relaxation oscillations in stimulated Raman and Brillouin scattering, *Phys. Rev. A* **4**, 1175 (1971).
52. S. Afshaarvahid, V. Devrelis, and J. Munch, Nature of intensity and phase modulations in stimulated Brillouin scattering, *Phys. Rev. A* **57**, 3961 (1998).
53. V. I. Kovalev, V. I. Popovichev, V. V. Ragulskii, and F. S. Faizullov, Gain and linewidth for stimulated Brillouin scattering in gases, *Sov. J. Quantum Electron.* **2**, 69 (1972).
54. V. I. Kovalev, M. A. Musaev, F. S. Faizullov, and A. K. Shmelev, Stimulated Brillouin scattering gains and decay times of hypersonic waves in optical crystals at the 10.6 μm wavelength, *Sov. J. Quantum Electron.* **14**, 110 (1984).
55. M. J. Damzen, H. Hutchinson, and W. A. Schroeder, Direct measurement of the acoustic decay times of hypersonic waves generated by SBS, *IEEE J. Quant. Electron.* **QE-23**, 328 (1987).
56. F. E. Hovis and J. D. Kelley, Phase conjugation by stimulated Brillouin scattering in CClF$_3$ near the gas–liquid critical temperature, *J. Opt. Soc. Am. B* **6**, 840 (1988).
57. S. T. Amimoto, R. W. F. Gross, L. Garman-DuVall, T. W. Good, and J. D. Piranian, Stimulated-Brillouin-scattering properties of SnCl$_4$, *Opt. Lett.* **16**, 1362 (1991).
58. N. F. Andreev, E. Khazanov, and G. A. Pasmanik, Applications of Brillouin cells to high repetition rate solid state lasers, *IEEE J. Quant. Electron.* **28**, 330 (1992).
59. M. R. Osborne and M. A. O'Key, Temporal response of stimulated Brillouin scattering phase conjugation, *Opt. Commun.* **94**, 346 (1992).
60. G. W. Faris, L. E. Jusinski, and A. P Hickman, High resolution stimulated Brillouin gain spectroscopy in glasses and crystals, *J. Opt. Soc. Am. B* **10**, 587 (1993).
61. H. J. Eichler, R. Menzel, B. Sander, M. Schulzke, and J. Schwartz, SBS at different wavelengths between 308 and 725 nm, *Opt Comm.* **121**, 49 (1995).
62. D. C. Jones, Characterization of liquid Brillouin media at 532 nm, *J. Nonlinear Opt. Phys. Mater.* **6**, 69 (1997).
63. H. Yoshida, V. Kmetik, H. Fujita, M. Nakatsuka, T. Yamanaka, and K. Yoshida, Heavy fluorocarbon liquids for a phase-conjugated stimulated Brillouin mirror, *Appl. Opt.* **36**, 3739 (1997).
64. M. S. Jo and C. H. Nam, Transient stimulated Brillouin scattering reflectivity in CS$_2$ and SF$_6$ under multipulse employment, *Appl. Opt.* **36**, 1149 (1997).

65. D.C. Jones, G. Cook, K. D. Ridley, and A. M. Scott, High reflectivity phase conjugation in the visible spectrum using stimulated Brillouin scattering in alkanes, *J. Nonlinear Opt. Phys. Mater.* **7**, 331 (1998).
66. H. Yoshida, M. Nakatsuka, H. Fujita, T. Sasaki, and K. Yoshida, High-energy operation of a stimulated Brillouin scattering mirror in an L-arginine phosphate monohydrate crystal, *Appl. Opt.* **36**, 7783 (1997).
67. H. Yoshida, H. Fujita, M. Nakatsuka, and K. Yoshida, High resistant phase-conjugated stimulated Brillouin scattering mirror using fused silica Glass for Nd:YAG laser system, *Jpn. J. Appl. Phys.* **38**, 521 (1999).
68. Y. F. Kiryanov, G. G. Kochemasov, N. V. Maslov, and I. V. Shestakova, Influence of thermal defocusing on the quality of phase conjugation of Gaussian beams by stimulated Brillouin scattering, *Quantum. Electron.* **28**, 58 (1998).
69. J. Munch, R. F. Wuerker, and M. J. LeFebvre, Interaction length for optical phase conjugation by stimulated Brillouin scattering: An experimental investigation, *Appl. Opt.* **28**, 3099 (1989).
70. D. T. Hon, Pulse compression by stimulated Brillouin scattering, *Opt. Lett.* **5**, 516 (1980).
71. M. J. Damzen and H. Hutchinson, Laser pulse copression by stimulated Brillouin scattering in tapered waveguides, *IEEE J. Quantum. Electron.* **QE-19**, 7 (1983).
72. S. Schiemann, W. Ubachs, and W. Hogervost, Efficient temporal compression of coherent nanosecond pulses in a compact generator-amplifier setup, *IEEE J. Quantum. Electron.* **33**, 358 (1997).
73. S. Schiemann, W. Hogervost, and W. Ubachs, Fourier-transform-limited laser pulses tunable in wavelength and in duration (400–200 ps), *IEEE J. Quantum. Electron.* **QE-34**, 407 (1998).
74. V. Kmetik, H. Fiedorowicz, A. A. Andreev, K. J. Witte, H. Daido, H. Fujita, M. Nakatsuka, and T. Yamanaka, Reliable stimulated Brillouin scattering compression of Nd:YAG laser pulses with liquid fluorocarbon for long-time operation at 10 Hz, *Appl. Opt.* **37**, 7085 (1998).
75. D. Neshev, I. Velchev, W. A. Majevski, W. Hogervorst, and W. Ubachs, SBS pulse compression to 200 ps in a compact single cell setup, *Appl. Phys. B* **68**, 671 (1999).
76. R. G. Harrison, V. I. Kovalev, W. Lu, and D. Yu, SBS self-phase conjugation of cw Nd:YAG laser radiation in an optical fibre, *Opt. Commun.* **163**, 208 (1999).
77. V. I. Kovalev, and R. G. Harrison, Diffraction limited output from a cw Nd:YAG master oscillator/power amplifier with fibre phase conjugate SBS mirror, *Opt. Commun.* **166**, 89 (199).
78. V. I. Kovalev, R. G. Harrison, and A. M. Scott, The build-up of stimulated Brillouin scattering excited by pulsed pump radiation in a long optical fibre, *Opt. Commun.* **185**, 185 (2000).
79. E. A. Kuzin, M. P. Petrov, and A. A. Fotiadi, Phase conjugation by SMBS in optical fibers, in M. Gower (ed.), *Optical Phase Conjugation*, Springer Berlin, Heidelberg, London, New York (1994).
80. E. A. Kuzin, M. P. Petrov, and B. E. Davydenko, Phase conjugation in an optical fibre, *Opt. Quantum. Electron.* **17**, 393 (1985).
81. H. J. Eichler, J. Kunde, and B. Liu, Quartz fibre phase conjugators with high fidelity and reflectivity, *Opt. Commun.* **139**, 327 (1997).

82. E. M. Dianov, A. Y. Karasik, A. V. Lutchinikov, and A. N. Pilipetskii, Saturation effects at backward-stimulated scattering in the single-mode regime of interaction, *Opt. Quantum. Electron.* **21**, 381 (1989).
83. V. A. Krivoshchekov, N. F. Pilipetskii, and V. V. Shkunov, Dependence of the quality of wavefront reversal by stimulated scattering in a fiber waveguide on the coupling-in conditions, *Sov. J. Quantum. Electron.* **16**, 827 (1986).
84. Y. Aoki and K. Tajima, Stimulated Brillouin scattering in a long single-mode fiber excited with a multimode pump laser, *J. Opt. Soc. Am. B* **5**, 358 (1988).
85. Y. Imai and M. Yoshida, Polarization characteristics of fiber-optic SBS phase conjugation, *Opt. Fiber Technol.* **6**, 42 (1999).
86. A. Heuer and R. Menzel, Phase conjugating SBS-mirror for low powers and reflectivities above 90% in an internally tapered optical fiber, *Opt. Lett.* **23**, 834 (1998).
87. G. J. Crofts and M. J. Damzen, Steady state analysis and design criteria of two-cell stimulated Brillouin scattering systems, *Opt. Commun.* **81**, 237 (1991).
88. G. J. Crofts, M. J. Damzen, and R. A. Lamb, Experimental and theoretical investigation of two-cell stimulated Brillouin scattering systems, *J. Opt. Soc. Am. B* **8**, 2282 (1991).
89. M. S. Mangir and D. A. Rockwell, 4.5 J Brillouin phase-conjugate mirror producing excellent near and far-field fidelity, *J. Opt. Soc. Am. B* **10**, 1396 (1993).
90. V. R. Belan, A. G. Lazarenko, V. M. Nikitin, and A. V. Polyakov, Stimulated Brillouin scattering mirrors made of capillary waveguides, *Sov. J. Quantum. Electron.* **17**, 122 (1987).
91. P. Shalev, S. Jackel, R. Lallouz, and A. Borenstein, Low-threshold phase conjugate mirrors based on position-insensitive tapered waveguides, *Opt. Eng.* **33**, 278 (1994).
92. D. C. Jones, M. S. Mangir, and D. A. Rockwell, A stimulated Brillouin scattering phase conjugate mirror having a peak-power threshold <100 W, *Opt. Commun.* **123**, 175 (1995).
93. R. Mays and R. J. Lysiak, Observations of wavefront reproduction by stimulated Brillouin and Raman scattering as a function of pump power and waveguide dimensions, *Opt. Commun.* **32**, 334 (1980).

CHAPTER 3

Laser Resonators with Brillouin Mirrors

MARTIN OSTERMEYER and RALF MENZEL

University of Potsdam, Institute of Physics, Chair of Photonics, 14469 Potsdam, Germany

3.1 INTRODUCTION

As has been shown in the preceding chapter, nonlinear mirrors based on stimulated Brillouin scattering (SBS mirrors) can be used to produce the phase conjugate of a wavefront [1, 2] with a fidelity close to one. There are two major different ways of realizing high-power lasers with excellent beam quality (high-brightness lasers). One is the laser oscillator, and the second is the master oscillator power amplifier concept (MOPA). Depending on the specific application, either the laser oscillator or the MOPA concept might be more appropriate. Phase conjugating SBS mirrors can be applied in both designs to compensate for phase distortions in the laser-active medium. Especially in solid-state lasers where the thermal load due to the pump is a serious problem, these concepts to improve the beam quality can be useful. The phase conjugating SBS mirror is easy to use since it is self-pumped by the incident beam. However, if SBS-phase conjugation is applied inside a laser resonator, a more elaborate Brillouin-enhanced four-wave mixing (BEFWM) scheme is the appropriate description of the physical process. As a consequence, the threshold of bulk SBS mirrors is decreased if they are placed in a resonator.

The motivation to use a phase conjugating mirror (PCM) within a laser resonator or simply to replace one of the resonator mirrors by the PCM is at least twofold. Besides the compensation of phase distortions to achieve an excellent beam quality, also the stability of phase conjugate resonators (PCR) is guaranteed and independent from the specific resonator design (see Section 3.3). Thus, fluctuating or drifting optical parameters in the laser resonator such as a fluctuating thermal lens in a solid-state laser can be dynamically corrected in a properly designed PCR. As a result, the beam parameters of the solid-state laser will be independent of the pump power. A further consequence is that the misalignment sensitivity of the resonator is reduced.

If the PCM is based on SBS apart from these two complexes known as aberration compensation and resonator stability, there are other motives arising to use such an

---

Phase Conjugate Laser Optics, edited by Arnaud Brignon and Jean-Pierre Huignard
ISBN 0-471-43957-6   Copyright © 2004 John Wiley & Sons, Inc.

SBS mirror in addition. For instance, the frequency shifting effect of the SBS mirror can be used to generate a large bandwidth and short pulses in injected resonators [3, 4] (see Section 3.6.1) or it can be used to avoid gain competition for counterpropagating waves in the laser-active material [5, 6] (see Section 3.2). The nonlinear properties of the SBS mirror inside the resonator (see previous chapter) give the opportunity for two major applications. One is Q-switching the laser resonator (see Section 3.4). Secondly, amplified spontaneous emission (ASE) can be avoided, for example, in regenerative amplifiers [7–13], but regenerative amplifiers are not dealt with here. To decrease the threshold of bulk SBS mirrors, they were placed in resonators just by themselves. Again for these techniques the reader is referred to the original literature [14–18].

Different laser resonator schemes containing SBS mirrors (SBS-laser oscillators) have been demonstrated and presented in the literature. An overview of the different types will be given in Section 3.2. The most prominent of the SBS-laser oscillators is probably also the most straightforward realization of a linear SBS-laser oscillator (sf-SBS laser). The description of this kind will take most of the space of this chapter. Section 3.3 deals with the transverse eigensolutions in SBS-PCRs.

The passive Q-switch of the SBS-laser oscillator is described in detail, and the variation of the pulse energy is explained and calculated from the properties of the start resonator in Section 3.4. It turns out that gaseous SBS mirrors are very efficient Q-switch devices for laser oscillators with a high gain factor and high average output power. Q-switched lasers with an SBS cell can reach even slightly higher efficiencies than comparable free running lasers as will be shown in Section 3.7.

Inside a laser resonator the SBS threshold can be reduced by a resonant generation of the sound wave [19, 20] from the longitudinal laser modes. It will be shown in Section 3.5 that the resonance condition for the so-called start resonator is a key precondition for reliable operation of a laser oscillator with phase conjugating SBS mirror.

The reflection at the moving sound wave in the SBS mirror leads to a Doppler shift (Brillouin shift) of the backscattered light. This causes a transient longitudinal mode spectrum of the SBS laser as will be shown in Section 3.6.

There are other possibilities than SBS to realize a phase conjugating cavity mirror. In the 1970s and 1980s, four-wave mixing and BEFWM were more in the focus to realize phase conjugating laser mirrors (see, e.g., Refs. 21–23). These techniques lead to PCMs without a threshold behavior. But high-quality pump beams are required to realize such a PCM. Photorefractive cavity mirrors are discussed in Chapter 9 of this book, gain gratings in Chapter 10 and other techniques in Chapters 10 and 11.

3.2 SURVEY OF DIFFERENT RESONATOR CONCEPTS WITH BRILLOUIN MIRRORS

The first laser resonator with an SBS mirror was realized in 1967 by Pohl [24]. This was even before experimental and theoretical evidence was found that SBS can be used for optical phase conjugation. Consequently, Pohl talks about a new Q-switching technique "only" and does not claim anything about a phase conjugate resonator or mirror.

This section gives a survey of different SBS-laser resonator concepts. Crucial points for the design of these resonators with emphasis on high average power operation specifically in the sf-SBS laser are addressed later in more detail in Section 3.7.

Due to the SBS threshold behavior, all the laser oscillators presented here are pulsed lasers, and almost all reported SBS-laser resonator designs initiate the laser process in a start resonator built by two conventional mirrors. The benefits or necessity of this will become clearer in Sections 3.4 to 3.6.3.

To our knowledge, there are only three SBS lasers reported which operate without a starter cavity. Jingguo and Hongwei [25] replaced the high reflectivity resonator mirror with a stimulated Rayleigh scattering mirror. The laser starts to "oscillate" as a super-radiator aided by the output coupling mirror only. With high pump rates they obtained emitted light to initiate the stimulated scattering process. In this case the gain of the stimulated Rayleigh scattering was greatly enhanced by absorption of light in the medium. Thus, this method is not suitable to realize lasers with high average output powers. The second laser of this kind was one of the lasers Pohl reported in Ref. 24. In his setup the SBS mirror acted as an output coupler. Also, Eichler et al. [26] reported on an excimer laser where a generator amplifier arrangement of two SBS cells acted as the high-reflecting mirror of the laser resonator.

SBS lasers with start resonator can be realized in many different ways. Figure 3.1 contains a scheme of the straightforward realization of a linear SBS-laser oscillator (sf-SBS laser). This concept exhibits a conventional start resonator between the mirrors M1 and M_{start}. M_{start} has a merely small reflectivity R_{start}. The telescope consisting of the lenses L1 and L2 reduces the coherence demands on the laser light for the SBS process (see Section 3.7).

The laser operates in the following way: At the beginning of optical pumping, there is no light power in the resonator, and the SBS mirror is transparent due to its threshold behavior. The laser starts to oscillate in the start resonator. While the intracavity power is increasing, the SBS-threshold power is reached at some time

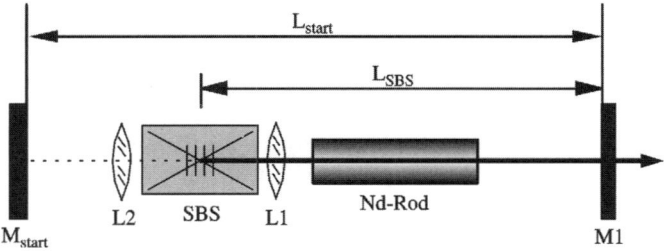

Figure 3.1. Scheme of the linear straightforward SBS-laser oscillator (sf-SBS laser).

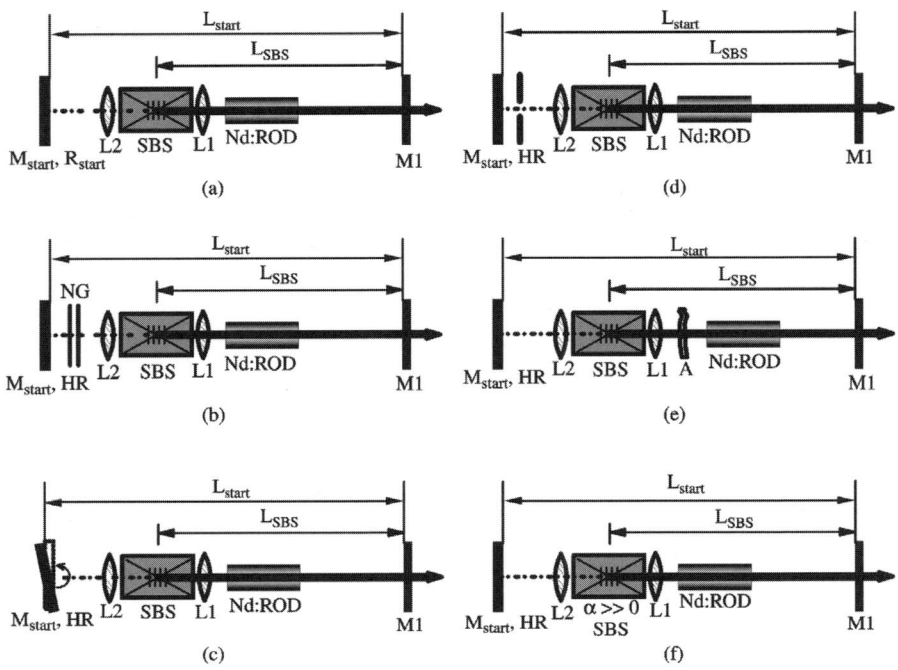

Figure 3.2. Concepts to realize a Q-difference between the start and the SBS resonator.

and then the reflectivity of the SBS mirror increases rapidly. Finally, the start resonator mirror M_{start} is nearly "switched off" and a giant pulse is emitted out of the SBS resonator between the output coupling mirror M1 and the SBS mirror. A smaller reflectivity of the mirror M_{start}, R_{start}, will lead to a higher pulse energy, and a larger R_{start} will lead to a smaller pulse energy. More inversion can be built up when the start resonator gets above threshold later. If the laser is further pumped after the pulse has been emitted, further Q-switched pulses are generated in the same manner. This way a whole burst of Q-switched pulses can be generated.

In order to switch the laser oscillation from the start to the SBS resonator, a difference in the quality Q between the two resonators is necessary. There are different possibilities to realize this Q difference in the sf-SBS laser (see Fig. 3.2). One easy and frequently used option is to choose the reflectivity of M_{start} smaller than the average reflectivity of the SBS mirror above threshold. The optimum value of M_{start} can be determined by inserting neutral density filters in front of the mirror (Fig. 3.2b) [19, 20, 27–29]. Moreover, the start resonator can be set up unstable (Fig. 3.2a) [18, 30, 31], elements of the start resonator can be slightly misaligned (Fig. 3.2c) [32], or a mode-selecting aperture between SBS mirror and M_{start} can be used (Fig. 3.2d) [30]. An aberrator also will lead to a higher Q for the phase conjugating SBS resonator than for the start resonator (Fig. 3.2e) [33]. The case shown in Fig. 3.2e is realized in every oscillator with an aberrated refractive index profile of the active laser material. Furthermore, absorbing SBS materials like CS_2 or acetone will lead to higher losses for the full transmission of the SBS cell in start

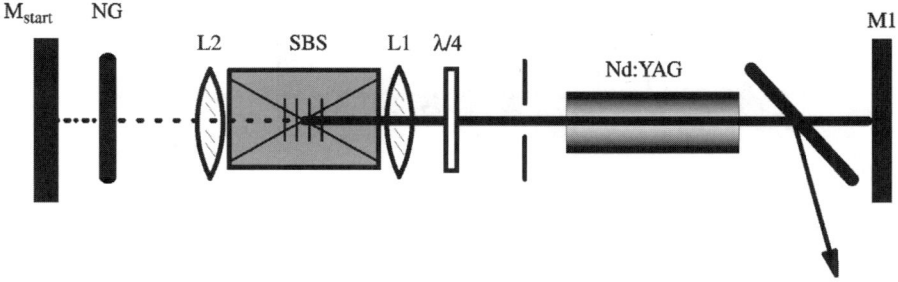

Figure 3.3. Polarization-dependent outcoupling.

resonator operation compared to the shorter path in the SBS resonator only (Fig. 3.2f) [34].

In a similar, slightly different concept the light is coupled out using a quarter-wave plate and a polarizer [35] (see Fig. 3.3). In the sf-SBS laser the output pulse energy can be varied by the reflectance of M_{start} or a filter in front of M_{start}. Here it can be done by rotating the quarter-wave plate to change the rotation angle of the intracavity field. In the first case solely the Q of the start resonator is varied. In the latter case the Qs of both the start resonator and the SBS resonator are changed.

A fundamentally different SBS-laser concept was invented by Chandra et al. [36] (see Fig. 3.4) and was later used by others also [35, 37]. Here the SBS mirror is placed outside the resonator. The concept uses the polarization status of the light to discriminate between a start and an SBS resonator. The Q ratio of the two resonators can be adjusted by rotating the quarter-wave plate. Placing the SBS-mirror extracavity makes the alignment of the focusing lens in front of the SBS mirror uncritical compared to the intracavity telescope (see above). Also, absorbing media intracavity leading to wavefront distortions or filamentation can be used with high average output powers in this scheme. On the other hand, one gives up the threshold

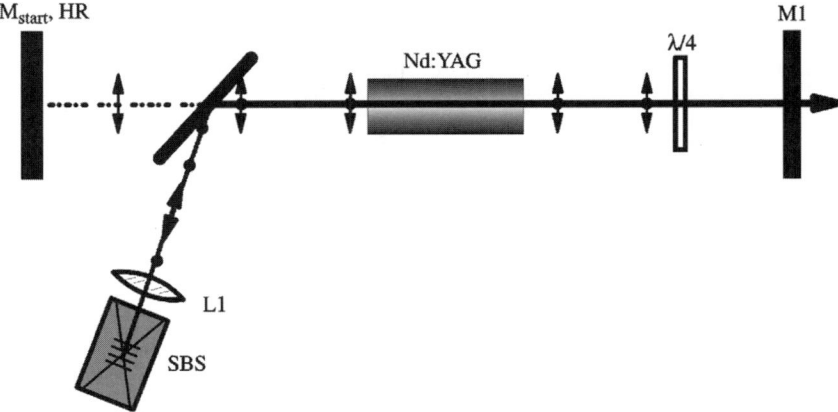

Figure 3.4. Phase conjugated resonator with sidearm SBS mirror.

Figure 3.5. Linear, seeded SBS laser.

reduction obtained for the intracavity SBS-mirror scheme, and a larger fraction of the conventional start resonator mode is superposed to the SBS-laser mode.

Another design with the SBS mirror outside the start resonator is shown in Fig. 3.5 [38, 39]. Pashinin and Shklovsky [38] realized a start resonator confined by two uncoated wedged glass substrates. This resonator was Q-switched with a saturable absorber. Apparently, this was necessary to overcome the threshold of the extracavity SBS mirror. The authors claim that the mode out of the conventional start resonator had a significantly smaller mode volume than the mode with an SBS mirror in operation. Both were diffraction-limited up to repetition rates of 3 Hz.

In other schemes a second SBS mirror was used for different purposes. Typical pulse lengths of around 30 ns are too short for many applications in material processing due to plasma generation at high peak powers. In Ref. 40, Seidel and Phillipps reported on an SBS laser with two intracavity SBS cells for stretching the Q-switch pulses (see Fig. 3.6) . The SBS laser is based on a polarization-dependent output coupling scheme, in which the losses can be adjusted by a quarter-wave plate ($\lambda/4_{out}$) as pointed out above. The effective reflectivity of the second SBS cell can be adjusted using the two additional quarter-wave plates ($\lambda/4_{stretch1}$ and $\lambda/4_{stretch2}$ in Fig. 3.6) to out-couple radiation via the polarizer $P_{stretch}$. With this additional output coupling, the pulse duration was stretched by about a factor of 5.

In the SBS laser presented by Anikeev and Munch [13] a second extracavity SBS cell is used to extract more power out of the SBS laser and simultaneously compress the Q-switch pulse (see Fig. 3.7). This way the original values of 30-mJ pulse energy and 65-ns pulse duration were transformed to 65 mJ and 7 ns.

SBS mirrors in ring resonators are mostly used to feedback one of the circulating directions [10, 11, 41–43]. Therefore they are positioned outside the actual

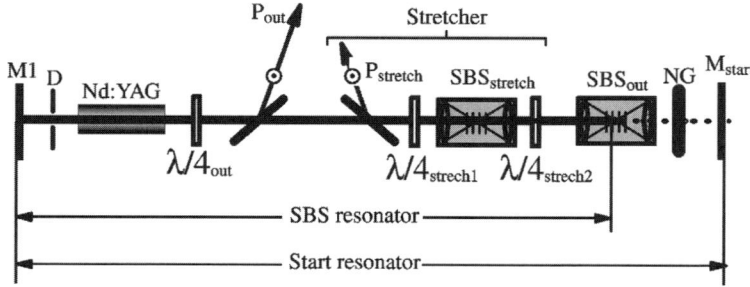

Figure 3.6. Two intracavity SBS cells to enlarge Q-switch pulse duration.

Figure 3.7. SBS laser with an additional extracavity SBS mirror to enhance extraction efficiency and compress pulses.

resonator. The ring resonator scheme presented by Lamb [6] is depicted in Fig. 3.8. It is similar to the other cited ones. It essentially represents an oscillator double-pass amplifier arrangement with one rod only. It exhibits good energy extraction due to the partial feedback of the output by the SBS mirror. Q-switching is initiated by the intracavity saturable absorber. The SBS cell contributes to the Q-switching process, and thus a shortening of the pulse duration from 22 ns to 7 ns and increasing the peak power by 32 times was observed. Since the feedback light with power P_c is frequency-shifted by the Brillouin frequency, spatial hole burning of the counterpropagating waves is prevented and stable single-frequency operation can be achieved.

Barrientos et al. [5] present a numerical investigation of an externally injected SBS ring laser. They mainly discuss the influence of the acoustic decay time of the SBS material. In Section 3.6.4 we present similar correlations for a standing-wave sf-SBS laser according to investigations of the same authors [44]. Another different ring resonator incorporating a loop like Brillouin enhanced for wave mixing mirror

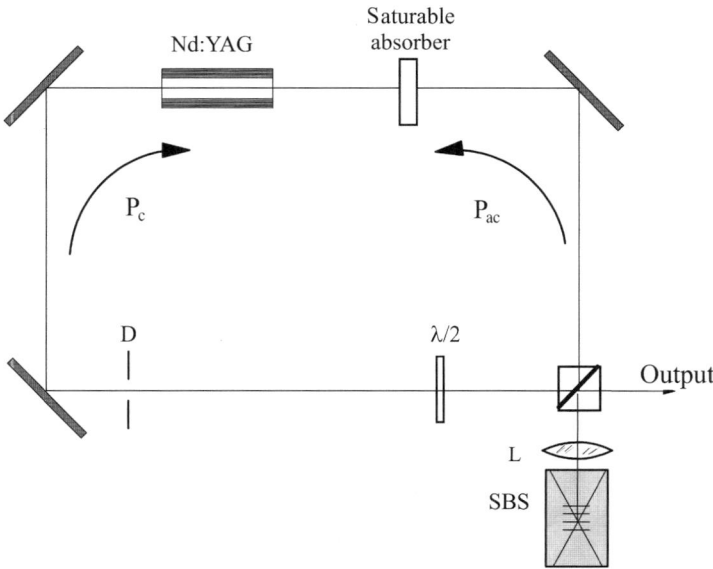

Figure 3.8. Ring laser with extracavity SBS mirror.

is reported in Ref. 45 and theoretically investigated in Ref. 46. With this scheme the subsequent frequency downshifting as it is obtained with simple SBS mirrors (see Section 3.6.1) in resonators can be avoided. Instead, in this scheme the light frequency is alternately shifted up and down.

A large number of resonator schemes with an SBS mirror to encounter certain problems of conventional resonators is known. Depending on the specific application, one of the presented schemes may be favorable. For efficient high-power Q-switch operation, the straightforward linear SBS-laser oscillator (sf-SBS laser) is probably the most simple and elegant design. It combines the advantage of a small number of resonator elements with the SBS self-Q-switching and the mechanism of threshold reduction due to the intracavity setup of the SBS mirror. When, for example, polarization and single-frequency operation become an issue, other schemes (e.g., the ring schemes) become more relevant.

3.3 STABILITY AND TRANSVERSE MODES OF PHASE CONJUGATING LASER RESONATORS WITH BRILLOUIN MIRRORS (SBS-PCRs)

A few general aspects on phase conjugating resonators (PCRs) will be discussed briefly here before we point out more precisely the theoretically predicted mode behavior in an SBS-PCR. These considerations in this section do not include an impact of any start resonator on the PCR. The phase conjugating resonator is thought to be confined by a conventional mirror (CM) as output coupler and the phase conjugating mirror (PCM). Only the term of perturbation stability, introduced in this section, allows for some consideration of the start resonator influence, however.

3.3.1 General aspects of an ideal PCR [48]
For simplicity, first, we suppose an ideal PCM with the fidelity of 1, with no threshold behavior and no frequency shift. This PCM operates in a configuration as depicted in Fig. 3.9. We consider only one transverse coordinate x without restricting the general case. This propagation of the field through the phase perturbation $\rho(x)$, along with the optical

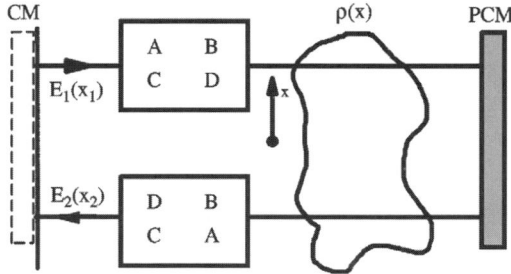

Figure 3.9. Out- and back-propagation of an arbitrary field using a phase conjugating mirror.

ABCD system to the PCM and back, can be formulated using the Collins integral:

$$E_2(x_2) = \rho(x_2) \cdot \frac{k}{2\pi|B|}$$

$$\times \int_{-w_{PCM}}^{w_{PCM}} \left\{ dx_{PCM} \left[\int_{-a}^{a} dx_1 \rho(x_1) E_1(x_1) \cdot \exp\left(i\frac{k}{2B}(Ax_1^2 - 2x_{PCM}x_1 + Dx_{PCM}^2)\right) \right]^* \right.$$

$$\left. \times \exp\left(-i\frac{k}{2B}(Dx_{PCM}^2 - 2x_2 x_{PCM} + Ax_2^2)\right) \right\} \tag{1}$$

$2w_{PCM}$ is the width of the PCM mirror; $2a$ is the width of any aperture in the reference plane.

We can absorb the width of the conventional mirror into the phase perturbation function $\rho(x)$ and extend the integration to $\pm$ infinity. If we further assume that we have real *ABCD* elements only (no transverse losses or gain variations), the integration kernel becomes more simple, and we can perform the integration at the PCM plane and write [47]

$$E_2(x_2) = \rho(x_2) \cdot \int_{-\infty}^{\infty} \rho(x_1)^* E_1^*(x_1) \cdot \exp\left(-i\frac{kA}{2B}(x_2^2 - x_1^2)\right)$$

$$\times \frac{\sin(kw_{PCM}/B)(x_2 - x_1)}{\pi(x_2 - x_1)} dx_1 \tag{2}$$

In case of an unbound phase conjugate mirror, w_{PCM} becomes infinite, the $\sin(x)/x$ function acts as a Dirac-like function, and Eq. (2) simplifies to

$$E_2(x) = |\rho(x)|^2 \cdot E_1^*(x) \tag{3}$$

Since $\rho(x)$ is a phase perturbation only, its square is equal to 1. Then, Eq. (3) expresses the fact that after the back-and-forth journey via a phase conjugate reflection, the phase conjugate of the initial field is yielded again. Now we can complement the whole arrangement by a conventional mirror (CM) in the reference plane to make it a PCR. Any phase shift (e.g., from a curved mirror) can be absorbed in the phase perturbation term $\rho(x)$, so that a plane conventional mirror remains. The resonator eigenvalue equation then reads

$$E_2(x) = E_1^*(x) = \gamma E_1(x) \tag{4}$$

This is fulfilled by any field distribution with a plane wavefront $E_1(x) = |E_1(x)| * \exp(i\phi)$, where ϕ is independent of x. Thus γ becomes 1 and we can conclude that any field with a wavefront matching the conventional mirror curvature is an eigensolution of a PCR. It should be noticed that this means that arbitrary phase

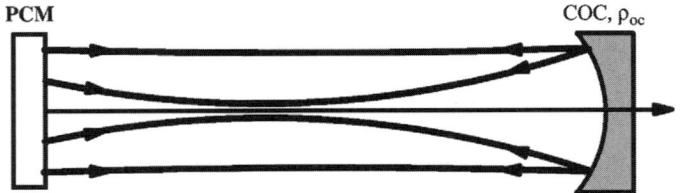

Figure 3.10. Example of double round-trip eigensolution in ideal PCR.

perturbations are perfectly compensated in an ideal unbounded PCR as it is suspected, and one of the major motivations to use PCRs is hereby justified. Furthermore, this means that the ideal PCR is always stable-independent from its specific design.

A double round trip can be treated by applying Eq. (3) two times in a row, yielding a field $E_4(x) = |\rho(x)|^4 * E_1(x)$. Again $|\rho(x)|^4$ equals 1, and therefore after two round trips any field distribution without any further condition is reproduced in such an ideal PCR (see Fig. 3.10).

3.3.2 Gaussian mode analysis of PCRs

Now, we neglect any phase distortion in the resonator for a while to learn more about the lowest-order mode of an SBS-PCR and its perturbation stability. In this case, we can work with a simple *ABCD* matrix formalism and the SBS-PCRs can be treated within the complex paraxial matrix formalism.

In searching for the eigensolutions of PCRs, we have to obey special calculation rules for optical systems containing phase conjugate optics. The *ABCD* law for an optical system has to be rewritten, so that the condition for the self-consistent solution reads

$$\frac{1}{q_E} = \frac{C + (1/q_E)^* D}{A + (1/q_E)^* B} \tag{5}$$

The roundtrip in the resonator may start and end at the conventional mirror (see Fig. 3.11). Assuming a flat conventional mirror is not a restriction of the general

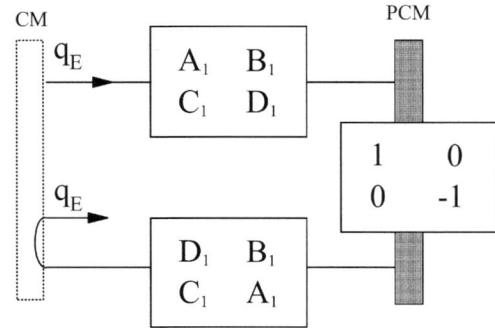

Figure 3.11. Round-trip model to calculate the eigensolutions of a general complex PCR.

case, since any curved mirror can be absorbed in two halves into the transfer matrices for the back-and-forth journey.

Taking care of the coefficient exchange of the A and D element for the back-and-forth journey the $ABCD$ matrix for the round trip reads

$$M_{\text{rt}} \equiv \begin{pmatrix} A & B \\ C & D \end{pmatrix} = M_{\text{back}} M_{\text{PCM}} M_{\text{out}} M_{\text{CM}}$$

$$= \begin{pmatrix} D_1 & B_1 \\ C_1 & A_1 \end{pmatrix} \begin{pmatrix} 1 & 0 \\ 0 & -1 \end{pmatrix} \begin{pmatrix} A_1^* & B_1^* \\ C_1^* & D_1^* \end{pmatrix}$$

$$= \begin{pmatrix} A_1^* D_1 - C_1^* B_1 & B_1^* D_1 - D_1^* B_1 \\ A_1^* C_1 - C_1^* C_1 & B_1^* C_1 - D_1^* A_1 \end{pmatrix} \quad (6)$$

Note that the matrix for the path toward the PCM is conjugated to take the phase conjugate reflection for the complex matrix elements into account (see Chapter 1). The C and B elements are purely imaginary. In case of real matrix elements, meaning no transverse gain or aperturing profiles in the resonator, the round-trip matrix becomes the unity matrix. In this case, Eq. (5) reads $1/q_E = (1/q_E)^*$. This is consistent with the result from the last subsection. There we obtained that the single round-trip solution in an unbound PCR has no restrictions for the beam radius except to match the conventional mirror curvature that was assumed to be flat.

Introducing the purely real elements A_m, B_m, C_m, D_m,

$$\begin{aligned} A &= A_m \cdot e^{-i\theta_m} \\ B &= -iB_m \\ C &= iC_m \\ D &= -A_m^* \end{aligned} \quad (7)$$

Siegman et al. [48] showed that in the general case of *complex* matrix elements the eigensolution is discrete and the beam radius w_{CM} and wavefront curvature R_{CM} read as follows for the reference plane at the conventional mirror:

$$\frac{1}{R_{\text{CM}}} = \pm \sqrt{\frac{C_m}{B_m}} \sin \theta_m$$

$$\frac{\lambda}{\pi w_{\text{CM}}^2} = \pm \sqrt{\frac{C_m}{B_m}} \cos \theta_m \quad (8)$$

Thus, for example, an introduction of an arbitrary weak aperture will lead to a discrete solution in the PCR. The eigensolution for the general complex PCR is discussed further in Refs. 48 and 49. We will now proceed and consider the perturbation stability. The eigensolution for the specific example of a PCR that is a reasonable model for a real SBS-PCR is discussed later in more detail.

Apart from being confined, to be useful the transverse mode also has to be stable against perturbations. The stability of the PCR against perturbations can be tested considering a growth rate per round trip of the deviation Δq_2 from the self-consistent complex beam parameter q_E. We consider an injected mode with a complex beam parameter deviating by Δq_1 from the beam parameter q_E of the nonperturbed eigenmode of the PCR [50]. After one round trip the deviation has become Δq_2. And the growth rate results in

$$\frac{\Delta q_2}{\Delta q_1^*} = \frac{1}{(A + B \cdot (1/q_E^*))^2} \tag{9}$$

This leads to a perturbation stable mode if $|(\Delta q_2)/(\Delta q_1^*)| < 1$. Whether this inequality holds or not in the case of an SBS-PCM will be discussed below.

3.3.3 Example: SBS-PCR as PCR with soft aperture

The phase conjugating SBS mirror exhibits a transverse graded reflectivity if operated below saturation of the Brillouin gain, since it is self-pumped by a transversely varying intensity. This is equivalent to a combination of an ideal PCM with a soft aperture. The aperture shall be shaped according to the incident field. It is reasonable for near-diffraction-limited beams to assume a Gaussian reflectivity distribution of the SBS mirror. The *ABCD* matrix for the SBS-PCM then becomes the product of an ideal PCM with an apodized Gaussian aperture of transmission $T(r) = T_0 \cdot \exp(-r^2/w_a^2)$, where w_a is the beam radius of the incoming Gaussian beam.

$$M_{\text{SBS-PCM}} = \begin{pmatrix} 1 & 0 \\ -\frac{i\lambda}{\pi w_a^2} & -1 \end{pmatrix} \tag{10}$$

There are two possibilities to describe such an SBS-PCR. Giuliani et al. [51] developed a model based on a back-and-forth propagation through the resonator starting at the SBS-PCM and ending there. This interrupted round trip is closed with separate equations describing the reflectivity at the SBS mirror. Secondly, the complete uninterrupted round-trip matrix used in the self-consistency condition for a full round trip leads to the same solution. In the latter case the round-trip matrix contains the matrix for the phase conjugating mirror with radial graded reflectivity. No extra equations are needed to close the round trip (see Fig. 3.12). To make the calculation easier, the reference plane is now located at the PCM rather than at the

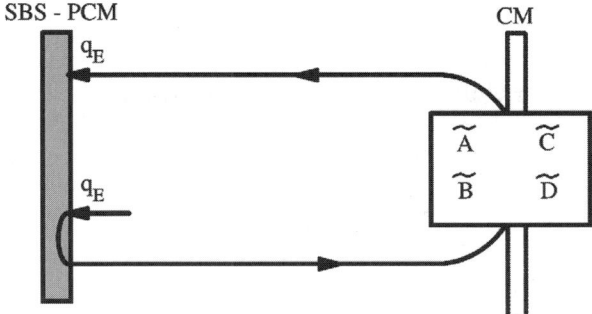

Figure 3.12. Round-trip model to calculate the eigensolutions of an SBS-PCR. The $\tilde{A}\tilde{B}\tilde{C}\tilde{D}$ matrix describes the entire resonator round trip including the SBS-PCM.

conventional mirror as before. The complete round-trip matrix reads

$$M_{rt} \equiv \begin{pmatrix} A & B \\ C & D \end{pmatrix} = M_{back} M_{CM} \cdot M_{out} \cdot M_{SBS-PCM}$$

$$= \begin{pmatrix} \tilde{A} & \tilde{B} \\ \tilde{C} & \tilde{D} \end{pmatrix} \cdot \begin{pmatrix} 1 & 0 \\ -\dfrac{i\lambda}{\pi w_a^2} & -1 \end{pmatrix} \quad (11)$$

The radius of the Gaussian aperture a translates to a factor β ranging between 0 and 1. β indicates the ratio by which the incident beam radius w_i is reduced by the SBS-PCM:

$$w_r = \beta \cdot w_i, \qquad w_a^2 = \frac{\beta^2}{1-\beta^2} \cdot w_i^2 \quad (12)$$

w_r indicates the beam radius reflected off the PCM. The wavefront radius of the reflected beam is inverted $R_r = -R_i$. These equations for the beam radius and wavefront radius are derived from the general formulas in [52] for the special case of sharp focusing [52].

The round-trip self-consistency condition in the laser resonator has to be based on the *ABCD* law for optical systems containing phase conjugate optics, as described in Eq. (5). The eigensolution of this SBS-PCR consists of the following discrete eigenmode parameters in the plane of the SBS-PCM:

$$w_i = \sqrt{\frac{\lambda \tilde{B}}{\beta \pi}}, \qquad R_i = \frac{\tilde{B}}{\tilde{A}} \quad (13)$$

Let us consider this result for w_i for a β of 1 and compare it to the eigensolution for a conventional resonator where the PCM is exchanged with a flat conventional mirror

(CM). If we assume a typical stability parameter of $m = (A + D)/2 = 0.5$, a beam radius of $w_{\text{conventional}} = \sqrt{(\sqrt{4/3}\lambda B)/\pi}$ is yielded. It can be seen that this value is very similar to the beam radius at the PCM in the SBS-PCR. A closer look [53] confirms that the mode volume in a real PCR is comparable to that of corresponding stable resonators. Thus, different from common considerations (see, e.g., Ref. 30) a good extraction efficiency of the stored inversion by a transverse fundamental mode with a big cross section in the laser active material is not self-evident for PCRs. On the other hand, the wavefront curvature radius at one of the resonator mirrors resulting for the Gaussian beam with the lowest losses in a complex paraxial resonator is independent from the specific resonator design always provided by a PCM [53]. Thus, when large-diameter transverse fundamental mode eigensolutions are realized, they will suffer from lower losses in a PCR as compared to a conventional resonator. The experimental results of Skeldon and Boyd [22] and Giuliani [51] point out that this simple theory for the transverse mode works quite well at least when the injected start resonator mode is already close to the SBS-resonator theoretical eigensolution.

Figure 3.13 shows the eigensolution for an SBS laser with a variable lens (e.g., the laser rod in a solid-state laser) in two different configurations. It can be seen that for maintaining constant beam parameters the position of the PCM should be as close as possible to the variable lens. Perfect matching cannot be obtained, since the position of variable lens and PCM are not identical and the variable lens is a lens duct with a finite length itself. Both points lead to a change of the $\tilde{B}$ element in the round-trip matrix and therefore a change of w_r. Nevertheless, practically constant beam parameters are obtained for such an SBS-PCR—for example, for the full pump power range of solid-state lasers with a variable thermal lens.

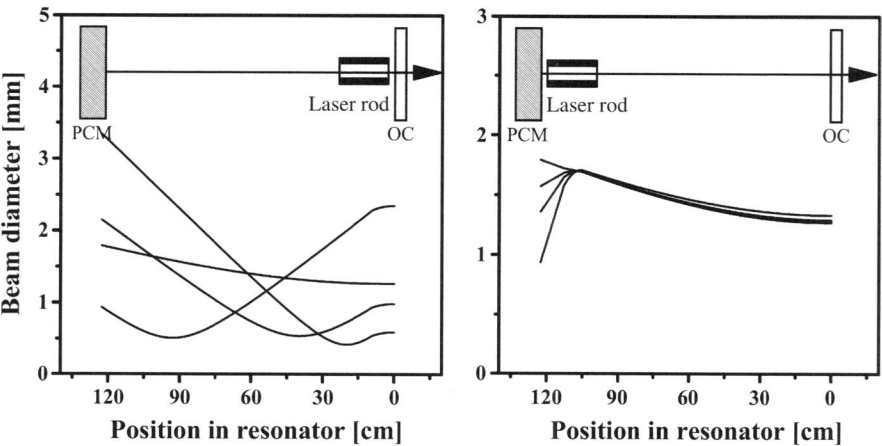

Figure 3.13. Eigensolution in an SBS-PCR with a $\beta = 1$ for four different dioptric powers of a pumped laser rod treated as a lens duct calculated with Eq. (13) for two extreme positions of the rod.

The perturbation stability of such an SBS-PCR can be characterized by using the results of Eq. (11) and Eq. (13) in Eq. (9). The growth rate results in

$$\left|\frac{\Delta q_E}{\Delta q^*_{\text{start}}}\right| = \beta^2 \qquad (14)$$

Thus, perturbation stability is yielded for the SBS-PCM with a $\beta < 1$. The confined solution in the case of an infinitely wide aperture with $\beta = 1$ will be marginally perturbation stable only. From Eq. (14) one can deduce that a deviation Δw_1 from the beam radius w_i and ΔR_1 for the wave front radius R_i will yield $w_i + \Delta w_2$ and $R_i + \Delta R_2$, respectively, after one round trip.

$$\Delta w_2 = -\Delta w_1$$
$$\Delta R_2 = -\Delta R_1 \beta^2 \qquad (15)$$

The wavefront radius will converge toward the eigenmode radius R_r, but the beam radius once perturbed will oscillate between the two solutions $w_i + \Delta w_2$ and $w_i - \Delta w_2$ around the eigensolution w_r. If this can be observed, it is dependent on the influence of a possibly existing superposed start resonator mode. Giuliani et al. [51] found at least some indication for this peculiar double round-trip mode. This should not be mixed up with the type of double round-trip eigenmodes which is obtainable in ideal PCRs.

Another special feature of SBS-PCRs is that the sensitivity of the mode radius against fluctuations is independent of its actual size. The eigenmode beam radius w_i at the PCM for a $\beta = 1$ is expressed as follows for a resonator of the length L; $w_i^2 = (2\lambda L g_2)/\pi$ using the equivalent g parameters of the resonator. The mode size sensitivity can then be expressed as

$$\frac{\partial w_2^2}{w_i^2} = \frac{\partial g_2}{g_2} \qquad (16)$$

Stable conventional resonators where big mode sizes are typically achieved at the rim of stability regions suffer from strong fluctuations if geometrical resonator parameter changes occur. However, the mode size sensitivity of an SBS-PCR is directly proportional only to the mode size variation, and it does not depend on the geometrical resonator parameters.

There is not much literature on the experimental observation of higher order modes in PCRs (see, e.g., Refs. 48 and 54 for theoretical investigations). In Ref. 54 it is shown that PCRs with complex elements—for example, a Gaussian aperture in front of an PCM as the SBS-PCR—have real confined Gaussian–Hermite higher-order eigenmodes.

Of course most resonators will exhibit hard apertures as for limited mirror sizes or boundaries of the active material. In these cases only numerical investigations yield satisfactory results for the eigenmodes. Numerical calculations were presented, for example, in Refs. 47 and 55. We will not discuss this issue here, but it should be mentioned that PCRs with hard apertures show lower round-trip losses and better beam quality compared to conventional resonators with hard apertures.

A PCR can compensate for intracavity phase distortions. Therefore it is a good concept especially for high-brightness solid-state lasers where the strong thermal load can be a big obstacle when realizing excellent beam qualities. The degeneracy of modes of an ideal PCR does not apply in real PCRs with apertures. An SBS-PCR exhibits a soft aperture in the SBS mirror generated by the transversely varying incident field. The beam radius of the discrete eigensolution of an SBS-PCR has a dimension similar to that of modes in stable resonators of the same interior.

A PCR always is the resonator with minimum loss as compared to conventional resonators with the same interior but arbitrary curvature of the replacing conventional mirror. With apertures the desired mode in the PCR can be determined. The sensitivity of the realized eigenmode then is independent from the actual size of the mode itself in contrast to the situation in stable resonators.

3.4 Q-SWITCH VIA STIMULATED BRILLOUIN SCATTERING

The following three sections concentrate on the principle of function of the sf-SBS laser introduced in Section 3.2 and give design rules for its setup. The performance data given in these sections were mostly measured with the laser setup depicted in Fig. 3.14. The laser heads were flash lamp pumped solid-state laser heads inhibiting either a Nd:YALO or a Nd:YAG rod.

The length of the SBS resonator is given by the distance of the focus between the two telescope lenses and the output coupler M1. The length $L_B = c/2\nu_B$ is called Brillouin length, where ν_B denotes the Brillouin frequency of the hypersound wave. This definition will become more evident in Section 3.5. Most of the discussed experiments were performed with SF_6 as SBS medium. It showed a Brillouin frequency of $\nu_B = 240\,\text{MHz}$ [56] and thus a corresponding Brillouin length of $L_B = 62.5\,\text{cm}$.

Solid-state lasers typically start to oscillate in the spiking regime, because solid-state active materials typically have considerably longer fluorescence lifetimes τ than the lifetime of the photons in the resonator τ_R. Therefore in solid-state SBS lasers a spike will grow up in the start resonator after pumping for time t_s. At this time the active material contains the threshold inversion density $\Delta n_{t,\text{start}}$ related to the start resonator:

$$\Delta n_{t,\text{start}} = \Delta n(t_s) = \frac{-\ln(\sqrt{R_{\text{start}}R_{\text{oc}}} \cdot V)}{\sigma l} \qquad (17)$$

The spike power continues to grow in the start resonator until the SBS threshold is reached. Then the SBS reflectivity increases rapidly. This increase happens in a

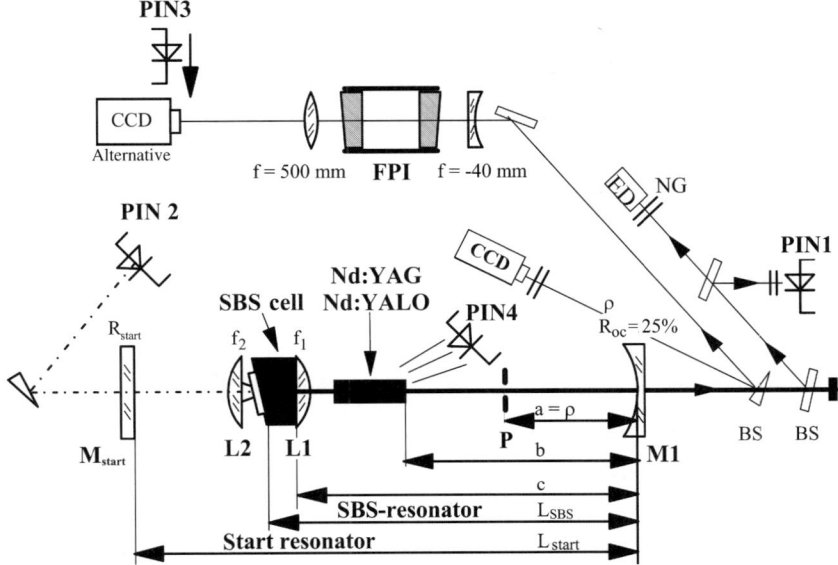

Figure 3.14. Scheme of the experimental setup. PIN denotes photodiodes, CCD a CCD camera, FPI a Fabry–Perot interferometer.

much shorter time than the Q-switch-pulse duration. Thus, a switch of the laser action occurs from the start resonator to the SBS resonator, and a Q-switch pulse is emitted out of the SBS resonator. If the SBS threshold is not reached by the spike power, the SBS mirror is not activated and thus the laser operates in the spiking regime only. Since the duration of the spike is short compared to the pump duration and assuming only negligible inversion depletion at the leading edge of the spike, the initial inversion for the Q-switch pulse is approximately $\Delta n_{t,\text{start}}$. Thus the Q-switch pulse energy E can be derived as given in Ref. 57:

$$E(R_{\text{start}}) = \frac{h\nu \cdot \pi w^2}{2\sigma} \cdot \ln\left(\frac{1}{\sqrt{R_{\text{SBS}}R_{\text{oc}}}}\right) \cdot \ln\left(\frac{-\ln(\sqrt{R_{\text{start}}R_{\text{oc}}} \cdot V)}{\Delta n_f \cdot \sigma \cdot l}\right) \quad (18)$$

$$\Delta n_i - \Delta n_f = \Delta n_{t,\text{SBS}} \ln\left(\frac{\Delta n_i}{\Delta n_f}\right) \quad (19)$$

$$\Delta n_{t,\text{SBS}} = \frac{-\ln(\sqrt{R_{\text{SBS}}R_{\text{oc}}} \cdot V)}{\sigma l}$$

σ denotes the emission cross section of the laser active material, ν the laser frequency, R_{SBS} the reflectivity of the SBS mirror, R_{oc} the reflectivity of the output coupler, V the single-pass loss of the start resonator, l the geometrical length of the laser crystal, w the radius of the laser mode in the laser rod, and Δn_f the residual inversion density after the decay of the Q-switch pulse. Δn_f is connected with the initial inversion and the threshold inversion density by a transcendental equation where $\Delta n_{t,SBS}$ denotes the threshold inversion density of the SBS resonator with a reflectivity R_{SBS} of the SBS mirror. Equation (18) shows the correlation of the Q-switch pulse energy with the reflectivity R_{start} of the mirror M_{start} and the spike onset time. Smaller reflectivities R_{start} lead to longer onset times of the first spike, because a higher inversion density is needed to come above threshold. Consequently, this bigger amount of stored inversion leads to higher pulse energies.

The Q-switch pulse energy can be calculated from Eq. (18) together with the solution of the transcendental Eq. (19) (calculation method I in Fig. 3.15). Another way to compute the pulse energies is the numerical solution of the rate equations for the inversion density in Nd:YAG together with the rate equation for the light intensity (calculation method II in Fig. 3.15) (see Ref. 58 for details). A constant reflectivity of the SBS mirror $R_{SBS} = 80\%$ was assumed, following former measurements with beam splitters inside the resonator.

The deviation of the calculated pulse energy from the measured one for bigger values of R_{start} above 40% probably stems from the higher effective reflectivity of the combination of SBS mirror together with mirror M_{start}.

A theoretical description of the Q switch in a slightly different SBS laser was given in Ref. 35, too.

The Q-switch process can be observed in detail by measuring the synchronized signals at PIN1 and PIN2 (see. Fig. 3.16). The time when the SBS mirror starts to

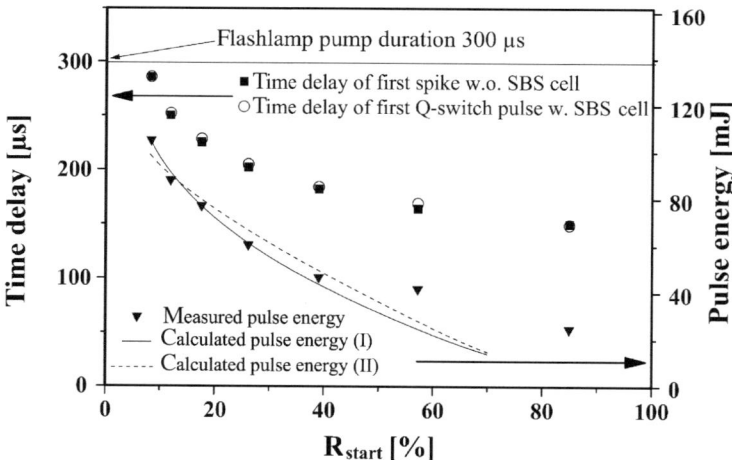

Figure 3.15. Measured onset times of the first spike of the start resonator without SBS cell and of the Q-switch pulse with SBS cell filled with SF_6 at 20 bar as a function of the reflectivity R_{start} of M_{start}. Additionally, the measured and calculated pulse energies are shown. The position of the resonator elements were: $a = 200$ cm, $b = 451$ cm, $c = 477$ cm.

reflect remarkably can be determined from the periodicity of the modulation of the two signals of the start and SBS resonator. In the diagram at the bottom of Fig. 3.16, first one period can be addressed to four modulation peaks in the signal measured at PIN2. After the switch to the SBS resonator, the fourth intensity maximum appears together with the first in the signal at PIN2. The four-fold modulation within one period is changed to a three-fold modulation within one period because of the change in the effective resonator length from $4L_B$ to $3L_B$ and the corresponding round-trip times, respectively.

As can be seen from Fig. 3.17, the Q-switch-pulse energy seems to be independent of the threshold of the used SBS material. The SBS thresholds of xenon, CO_2, and SF_6 were measured with a single longitudinal Q-switched Nd:YAG laser

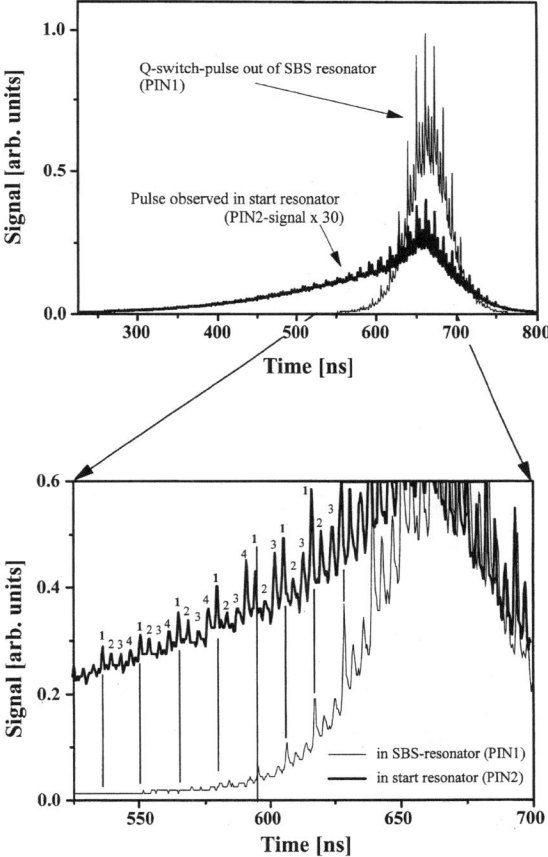

Figure 3.16. Signals detected at PIN1 and PIN2. The Q-switch is clarified by the periodicity in the round-trip times of the signals. This measurement was made with xenon as SBS material. The lengths of the start and SBS resonator were tuned to the Brillouin length of xenon, $L_B = 53.6$ cm: $L_{start} = 4L_B$ and $L_{SBS} = 3L_B$. The reflectivity of the start resonator mirror was 50%.

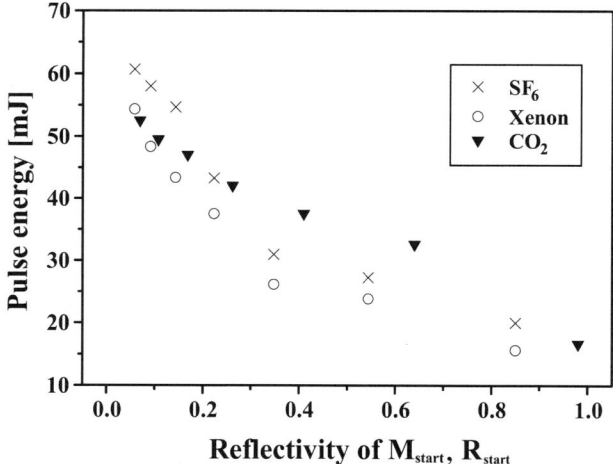

Figure 3.17. Measured pulse energies for three SBS materials as a function of the reflectivity R_{start} of M_{start}.

with a pulse duration of 30 ns and a bandwidth of 100 MHz to be 3.9 mJ, 6 mJ, and 10 mJ (Ref. 56 and previous chapter). On the other hand, from Eq. (18) we can see that the pulse energy is a function of R_{SBS}, which is a function of the threshold of the SBS mirror. But because the SBS threshold inside a resonator is reduced as explained in the next section by Brillouin enhanced four-wave mixing (BEFWM), these SBS mirrors are operated far above threshold. In this case, R_{SBS} becomes more or less the same for the three materials.

3.5 RESONANCE EFFECTS BY INTERACTION OF START RESONATOR MODES WITH THE SBS SOUND WAVE

The SBS threshold can be reduced by operating the SBS cell inside a resonator. This reduction is even necessary to operate most of the gaseous SBS materials in a laser oscillator since the SBS Q switch starts from spiking with low powers. The reduction of the threshold can be explained by a resonant generation of the sound wave via Brillouin enhanced four-wave mixing (BEFWM) of at least two longitudinal start resonator modes of the frequencies v and $v + \Delta v$ with $\Delta v = v_B$. They act as two pump fields, one signal field, and the conjugate field (Fig. 3.18). This BEFWM takes place during the beginning of the spike oscillation in the start resonator.

The driving force of the sound wave is proportional to $\nabla^2(E^2)$, with E being the total electrical field in the SBS medium. For an effective energy transfer from the electrical field to the sound wave, a beat signal with a frequency within the bandwidth of the sound wave and with the velocity of the sound wave has to be generated. Thus, only $\mathcal{E}_2^* \cdot \mathcal{E}_3$ and $\mathcal{E}_1 \cdot \mathcal{E}_4^*$ out of the six beat terms of E^2 transfer energy to the sound wave (see Fig. 3.18, see Ref. 23 for details). The velocity of the

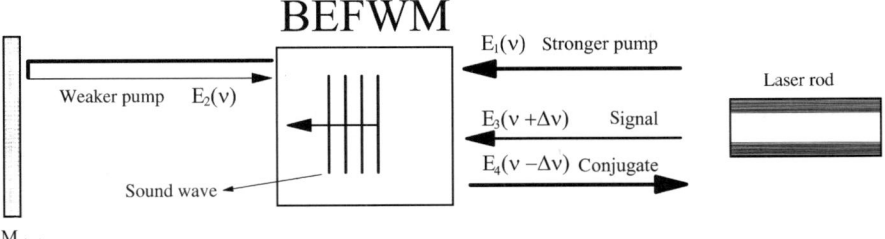

Figure 3.18. Scheme of the Brillouin enhanced four-wave mixing (BEFWM) in the start resonator.

beating ν_{beat} of E_2 and E_3 reads

$$\nu_{beat} = \frac{\Delta\nu_{start}}{2\nu_L}\frac{c}{n} = \frac{\lambda_{L,n}}{2}\Delta\nu_{start} \tag{20}$$

ν_L denotes the light frequency, $\lambda_{L,n}$ denotes the light wavelength in the SBS medium with the refractive index n, and $\Delta\nu_{start}$ denotes the spectral distance of the longitudinal modes in the start resonator. This leads to a wavelength of the beat signal λ_{beat}:

$$\lambda_{beat} = \frac{\nu_{beat}}{\Delta\nu_{start}} = \frac{\lambda_{L,n}}{2} \tag{21}$$

The corresponding quantities for the sound wave read [59]

$$\nu_{sound} = \frac{\nu_B}{2\nu_L}\frac{c}{n} \tag{22}$$

$$\lambda_{sound} = \frac{\nu_{sound}}{\nu_B} = \frac{\lambda_{L,n}}{2} \tag{23}$$

The beating of E_2 and E_3 and the back traveling sound wave have the same direction as well as the same velocity and wavelength if the spectral distance of the start resonator modes $\Delta\nu_{start}$ equals the Brillouin frequency $\Delta\nu_{start} = \nu_B$. In this case the beating of the longitudinal modes of the start resonator can couple to the sound wave most efficiently. Hence the resonant start resonator length should fulfill the condition

$$L_{start} = k\frac{c}{2\nu_B} \quad \Rightarrow \quad \Delta\nu_{start} = \frac{c}{2L_{start}} = \frac{\nu_B}{k} \tag{24}$$

Then the spectral distance $\Delta\nu$ given in Fig. 3.18 can become $\Delta\nu = k^*\Delta\nu_{start} = \nu_B$. The case of $k = 1$ represents the elementary Brillouin length $L_B = c/2\nu_B$. Higher numbers of k require beating between every kth longitudinal mode to drive the sound wave resonantly.

The sound wave generated by the fields E_2 and E_3 scatters the pump beam 1 into the conjugate beam 4. The beating of the fields 1 and 4 amplifies the sound wave and leads to an increasing intensity of the conjugate beam. This is equivalent to an increasing reflectivity of the SBS mirror. A distinct attenuation of E_2 takes place. This enables the switching from BEFWM to SBS in the description of the PCM. Then, only the fields E_1 and E_4 are of significant importance in the resonator. The scattering of the weaker pump E_2 into the signal beam E_3 is also possible, but this process has only a low efficiency as pointed out in Refs. 23 and 60.

Figure 3.19 shows how the ignition of the SBS mirror and consequently the Q-switch operation of the SBS-laser oscillator can be achieved best by choosing a resonant-start resonator length that is a multiple of the particular Brillouin length. If the start resonator length is too far from this resonant length, the laser operates in spiking only. In those cases the beat frequencies of the start resonator modes are spectrally too far away from v_B to stimulate the sound wave in the SBS mirror. For the measurements of Fig. 3.19 the length of the start resonator in an Nd:YALO SBS-laser oscillator was varied. The pulse energy, its standard deviation, the longitudinal mode frequencies, and their beat frequencies were measured. While varying the start resonator length, it was assured that the start resonator remained stable. Also, an almost constant fundamental mode volume in the laser rod was used by setting the second telescope length appropriately (see Fig. 3.20). The suspension of the Q switch at the resonator lengths $L_{start} \approx 4L_B$ (SBS cell in the middle of the start resonator) can be explained by two counterpropagating sound waves in the SBS cell [8, 61] which are the result of a BEFWM without an effective reflectivity [62].

Working with liquid SBS materials, this resonance behavior of the start resonator length does not have to be that distinct and can even vanish. For the same start resonator length the ratio of the Brillouin line width and spectral width of the longitudinal modes in the start resonator will be bigger. An efficient generation of the sound wave becomes possibly independent of the start resonator length L_{start} if the Brillouin gain line overlaps with the beat frequency of the longitudinal modes which is closest to the Brillouin frequency: $k^* \Delta v_{start}(L_{start}) \approx v_B$.

3.6 LONGITUDINAL MODES OF THE LINEAR SBS LASER

The longitudinal mode spectrum of the sf-SBS laser is based on two major mechanisms. One is the Doppler shift at the moving sound wave grating in the SBS mirror, and the other is ordinary constructive interference of modes with frequencies $n*c/2L$. Section 3.6.1 concentrates on the first mechanism, whereas Sections 3.6.2 and 3.6.3 discuss interactions between the two mechanisms. Section 3.6.4 presents the influence of SBS materials with different acoustic decay times.

3.6.1 Transient longitudinal mode spectrum

The main structure of the longitudinal mode spectrum can be seen from Fig. 3.21 and Fig. 3.22. In Fig. 3.21, two typical longitudinal mode patterns detected with a

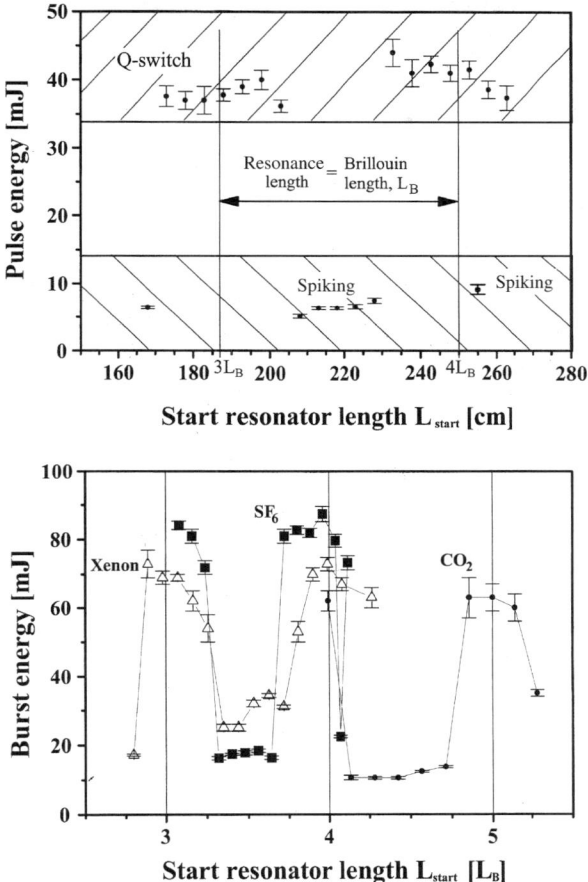

Figure 3.19. *Top:* Measured pulse energy of an sf-SBS laser as a function of the optical start resonator length for SF_6 as SBS material. Every data point is an averaged value over 500 measured single pulse energies. The error bars give the standard deviation. The length of the SBS resonator was $2L_B$. The two vertical lines indicate the third and fourth resonant start resonator length equal to $3L_B$ and $4L_B$. *Bottom:* Measured burst energy for the same laser as a function of the optical start resonator length normalized to the Brillouin length L_B of the used SBS materials xenon SF_6 and CO_2. Each data point is an averaged value over 500 measured burst energies. The error bars give the standard deviation. The lengths of the SBS resonator were $2L_B$ for SF_6 and xenon and $3L_B$ for CO_2.

Fabry–Perot interferometer (FPI) for SF_6 as SBS material is depicted. Figure 3.22 illustrates the results of such an FPI measurement for xenon ($L_{start} = 3L_B$, $L_{SBS} = 2 L_B$) and CO_2 ($L_{start} = 4L_B$, $L_{SBS} = 3L_B$) as SBS materials. The FP interferograms show one free spectral range only. For all the three materials the dominant spectral distance between the longitudinal modes is the Brillouin frequency ν_B of the particular SBS material. In a few cases apart from the Brillouin frequency, other

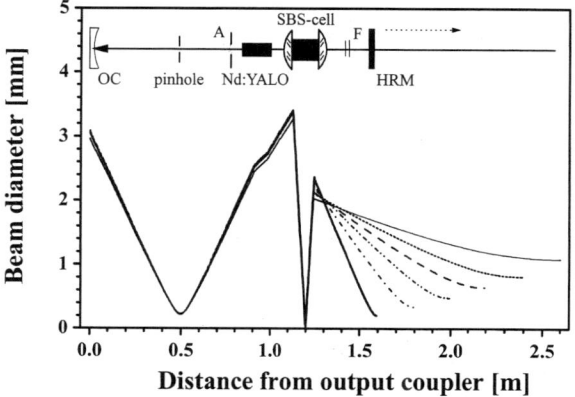

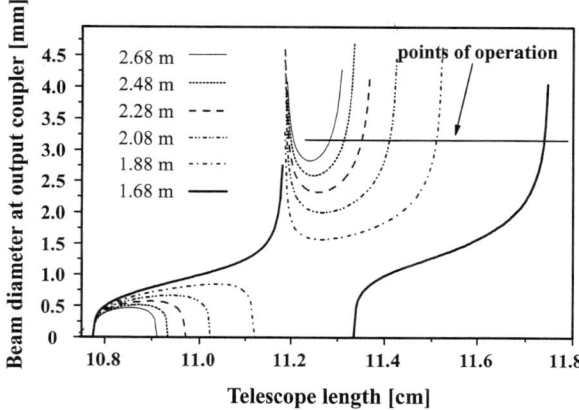

Figure 3.20. Calculated beam paths (*top*) and stability ranges (*bottom*) for different start resonator lengths of the SBS-laser oscillator using SF_6. The different realized points of operation for each resonator length are given by the horizontal line. The given resonator lengths are optical lengths. The beam path is calculated for the lowest-order eigensolution, a Gaussian beam, inside the stable start resonator. With the same formalism the beam diameter at the output coupler was calculated as a function of the telescope length. For details see Ref. 50.

frequency distances corresponding to the start resonator mode spacing $c/2L_{start}$ were determined. A measurement of such a case is shown in Fig. 3.21 on the right-hand side. The appearance of such pulses is statistical.

Figure 3.23 shows that the detected frequency distances in the interferograms are caused by the Brillouin shift only and are independent of the length of the SBS-resonator L_{SBS}. The also presented beat frequencies will be discussed in the next subsection. However, a more detailed observation reveals that the Brillouin shift of the SBS mirror is a function of the frequency of the driving force which is the beat signal of the longitudinal modes of the start resonator. If the start resonator is not

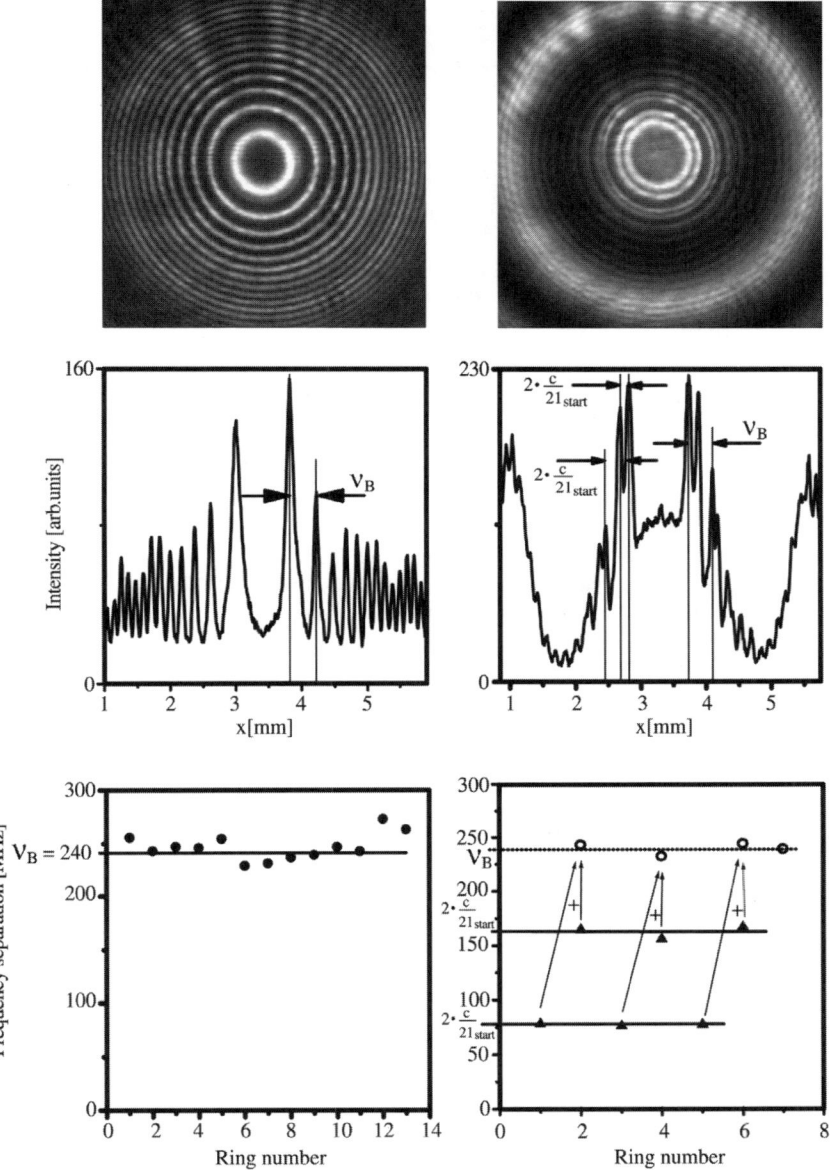

Figure 3.21. *Top* and *middle:* Typical Fabry–Perot interferograms (FPI) and their intensity cross sections taken for two different *Q*-switch pulses. The digrams at the bottom show the frequency distances of neighboring interference fringes (solid points). The SBS material was SF_6. Start and SBS resonator were tuned to the Brillouin frequency ($L_{start} = 3L_B$, $L_{SBS} = 2L_B$) *Left:* Only spectral distances equal to the Brillouin frequency ν_B can be detected. The shown range in the interferogram is totally within one free spectral range (3 GHz) of the FPI. *Right:* Additionally to the Brillouin frequency (every second ring), also frequency distances equal to the fundamental start resonator mode spacing $c/2L_{start}$ can be detected.

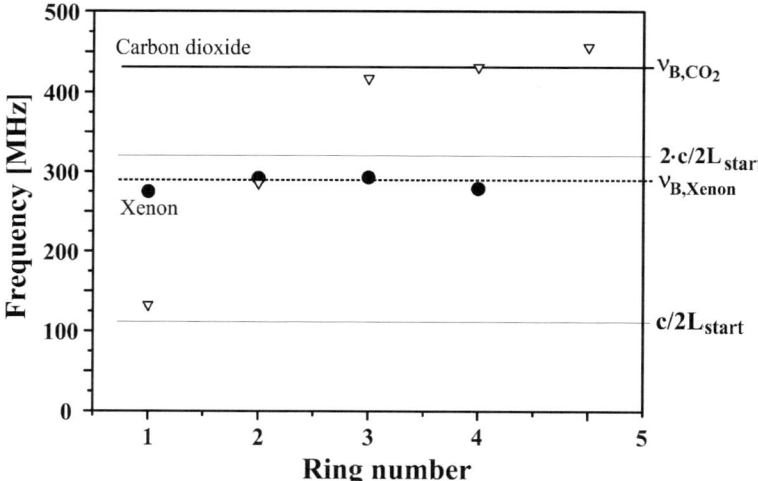

Figure 3.22. Frequency distances of neighboring interference fringes during one Q-switch pulse measured with the FPI similar to that in Fig. 3.21. The SBS materials were xenon and CO_2. Their Brillouin frequencies are marked as horizontal lines (CO_2 ———; xenon ----). For CO_2 also the spectral distance of the longitudinal modes of the start resonator are marked as thinner horizontal lines.

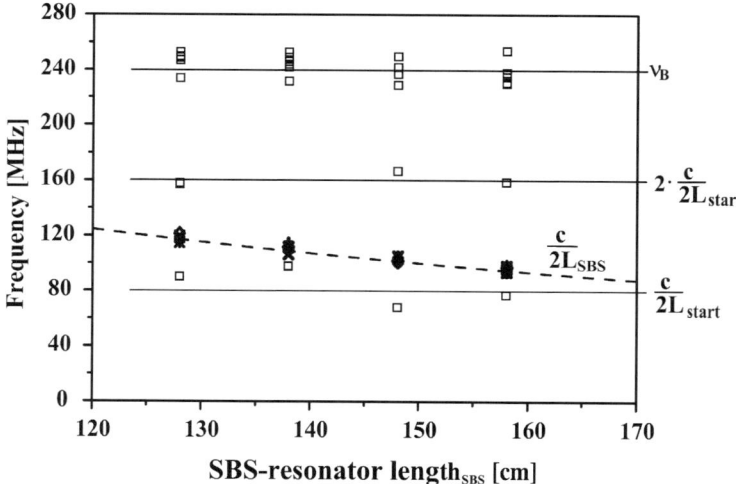

Figure 3.23. Frequency distances of neighboring interference fringes ($\square$) during one Q-switch pulse as a function of different optical SBS-resonator lengths, measured with FPI. The horizontal lines denote the Brillouin frequency ν_B for the used SBS material SF_6 and the frequency distances of the start resonator modes $c/2L_{start}$ and $2 * c/2L_{start}$. The measured frequency distances ($*$, Δ, $\times$) of the beat frequencies from a Fourier transformation of the time distribution of the Q-switch pulses are also printed. The broken line corresponds to the calculated mode spacing of the SBS resonator $c/2L_{SBS}$.

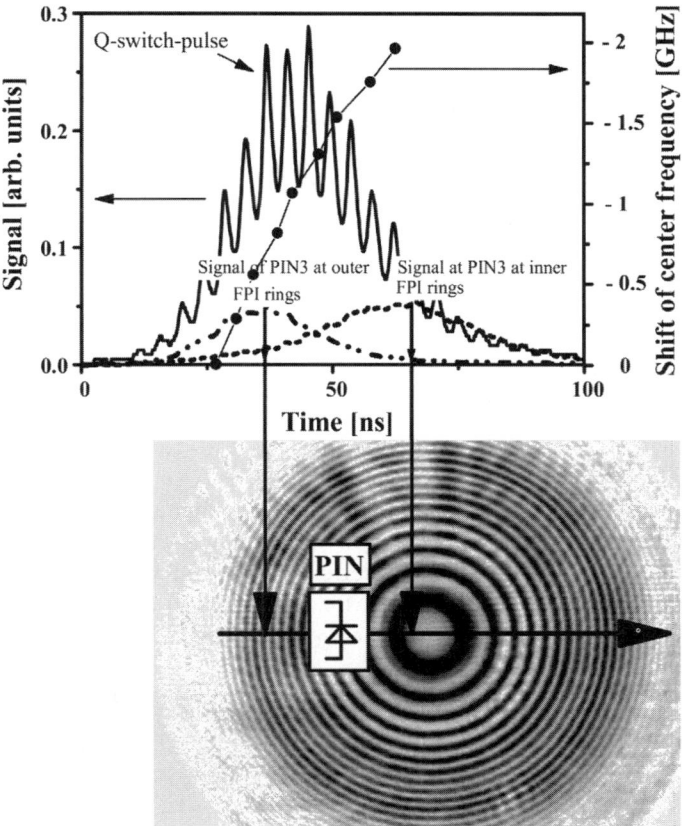

Figure 3.24. Measurement of the light frequency as a function of time during one Q-switch pulse. *Top:* The modulated time distribution belongs to the Q-switch pulse measured at the photodiode PIN1 (see Fig. 1). The curves in broken lines are the measured signals at the photodiode PIN3 at outer and inner FPI rings illustrated by the interferogram at the bottom. For nine positions of PIN3 the maximum of 100 averaged signals was determined and yielded a solid point in the diagram. The positions of PIN3 determined the frequency shift, which is depicted at the vertical axis.

exactly tuned to the Brillouin frequency, the spacing of the laser modes becomes a function of time. It is starting with the longitudinal mode beat frequency of the start resonator and is relaxing toward the Brillouin frequency of the used SBS material.

The longitudinal modes spaced by the Brillouin frequency are generated round trip by round trip due to the Doppler shift from the moving sound wave grating in the SBS mirror. One would expect that the spectrum becomes either wider and wider or just becomes shifted round trip by round trip. The latter description is in this case the appropriate one in contrast to the injected oscillators investigated by Damzen and co-workers [3, 4]. This will be clarified in the next subsection. Figure 3.24 shows the effect of the shifting mechanism. The mid-frequency of the longitudinal mode spectrum was measured during the Q-switch pulse (see Ref. 58 for details of the measurement). In the top diagram of Fig. 3.24, each solid circle marks the temporal position of the maximum signal of a measurement with PIN3 of the setup of Fig. 3.14 at a certain position in the ring system corresponding to a certain frequency. It can be seen that the mid-frequency of the mode spectrum is shifted almost linearly during the Q-switch pulse.

The reasons for the temporal width of the signal detected with PIN3 are the averaging over 100 pulses and the size of the photodiode detector on one hand. But also several longitudinal modes exist simultaneously, leading to a "temporal overlap" of the modes, as will be shown in the next section.

3.6.2 Mode locking

Next to the effect of the SBS-threshold reduction by tuning the start resonator length, a resonant length $n \cdot L_B$ of the start resonator also leads to a common phase of the longitudinal modes. During the leading edge of the spike but before the SBS mirror is switched to high reflectivities, a generation of new longitudinal modes takes place. These new modes are generated via BEFWM with a spectral distance of ν_B. This mechanism can be observed in an interferogram taken with a highly reflecting ($R_{\text{start}} = 100\%$) start resonator mirror M_{start} as shown in Fig. 3.25. Although the laser was operated in the spiking regime without Q-switching, the influence of the

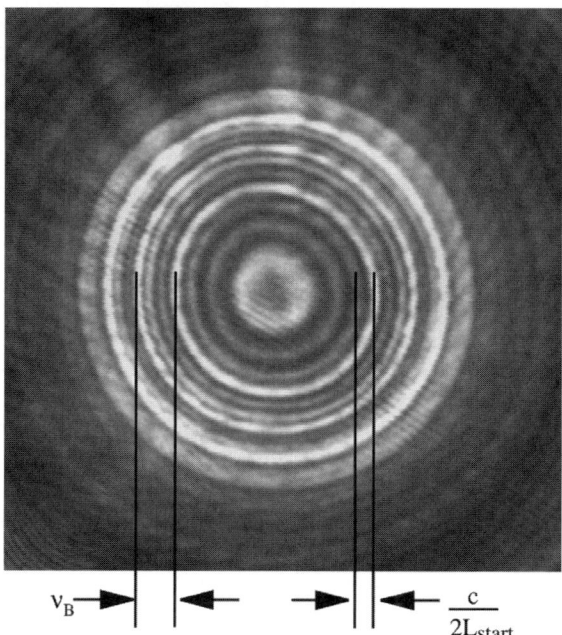

Figure 3.25. Interferogram of the longitudinal modes in the start resonator when using a highly reflecting start resonator mirror M_{start}. The lengths of the start and the SBS resonator were $L_{\text{start}} = 3L_B$, $L_{\text{SBS}} = 2L_B$. The measurement was performed with SF_6 as SBS material.

SBS cell can be clearly deduced from the higher intensities of those rings in Fig. 3.25 that have a spectral distance equal to the Brillouin frequency v_B. Due to the chosen length of the start resonator $L_{\text{start}} = 3L_B$, v_B equals the spectral distance of every third start resonator mode in this case.

As these strong modes with frequencies spaced by v_B have arisen from each other, their phases are locked. After the Q switch, these locked phases can be preserved if the length of the SBS resonator is tuned to the Brillouin frequency, too. In this case the round-trip time in the SBS resonator is an integer multiple of the period of the sound wave. After one round trip of the light in the SBS resonator the maxima of the sound wave will be at the same position again as they have moved by the same multiple of the sound wave wavelengths. This phase locking leads to a regular modulation of the pulses in the tuned SBS resonators. The effect of seeding the SBS-resonator modes with a spectral spacing of the Brillouin frequency can also be seen from the dominant Brillouin beat frequency in the Fourier transform of the averaged pulses (see Fig. 3.26).

Due to an averaging over 50 pulses and a little jitter in the trigger time of the oscilloscope, the modulation is not as deep as for a single pulse. For a single pulse the modulation depth can be up to 90%. The longer the SBS resonator, the stronger the influence of the elementary round-trip time in the start resonator on the amplitudes of the beat frequencies in the pulse. In contrast to the pulses in the length-tuned SBS resonators of Fig. 3.26 in Fig. 3.27, the temporal profile of 50 averaged pulses of a nontuned SBS resonator is shown. The strongest beat frequency in this case is the inverse SBS resonator round-trip time of 92 MHz = 1/11 ns.

The interferograms showed that the Brillouin frequency v_B is the dominant frequency spacing in the mode ensemble during one Q-switch pulse independent of the length of the SBS resonator (see Fig. 3.23). In some cases even the elementary mode spacing of the start resonator was obtained. The mode spacing of the SBS resonator, $c/2L_{\text{SBS}}$, is never been observed in the interferograms of any Q-switch pulse. The reason for this will be explained in the following section. However, the modulation periods in the pulses correspond well to the chosen SBS-resonator length as can be seen from the Fourier transform of the pulses in Fig. 3.23 (measured points along the broken theoretical curve).

The Brillouin frequency was not observed in the Fourier transform of the pulse intensities in Fig. 3.23. The Fourier transform contains the frequency distances of the absolute frequencies, and these beat frequencies can only be generated from simultaneously existing absolute frequencies. But outside the resonator the Brillouin shifted signals do not exist simultaneously in time and space. They are separated by the SBS-resonator round-trip time and the double SBS-resonator length, respectively. This is distinctively different from the injected oscillators realized by Damzen and co-workers [3, 4]. They injected the original frequency over the entire pulse duration into the oscillator so that a real increase in bandwidth is yielded. Due to the mode locking mechanism they observed short ps-pulses. But in the self-starting SBS laser case the start frequency is not continuously supplied so that the whole spectrum is shifted during the pulse duration but not broadened.

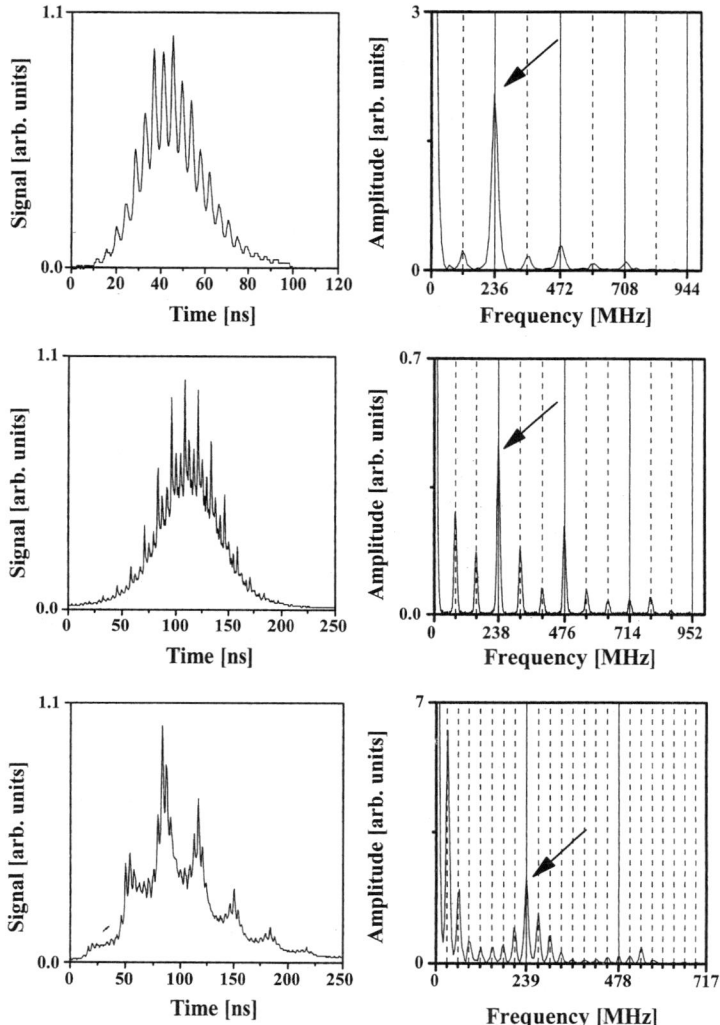

Figure 3.26. Measurement of averaged Q-switch pulses of SBS-laser oscillators with SF_6 as SBS material but different resonant resonator lengths. The ratios of the lengths of start and SBS resonator in units of $L_B = 62.5$ cm were: *Top:* 3:2 (Nd:YAG), *Middle:* 4:3 ($a = 50$ cm, $b = 132$ cm, $c = 162$ cm $= d$). *Bottom:* 9:8 ($a = 200$ cm, $b = 451$ cm, $c = d = 477$ cm). On the right-hand side the Fourier transform of the time distributions are shown. The broken lines denote multiples of the frequency spacing of the SBS resonator $\Delta\nu_{SBS} = c/2L_{SBS}$.

Besides the impact of the SBS-resonator length on the longitudinal mode structure, the longitudinal mode structure itself has an impact on the efficiency of the SBS reflectivity [61, 63]. We will not discuss this issue here in detail. But Fig. 3.28

LONGITUDINAL MODES OF THE LINEAR SBS LASER

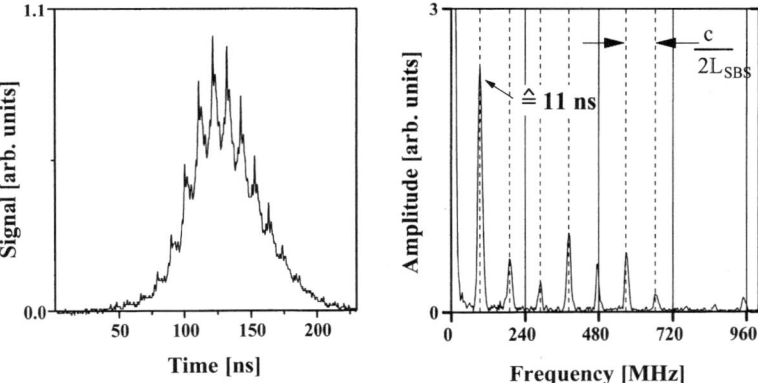

Figure 3.27. Time distribution and its Fourier transform of 50 averaged Q-switch pulses measured with an SBS resonator length $L_{SBS} = 2.5 L_B$. The SBS material was SF_6.

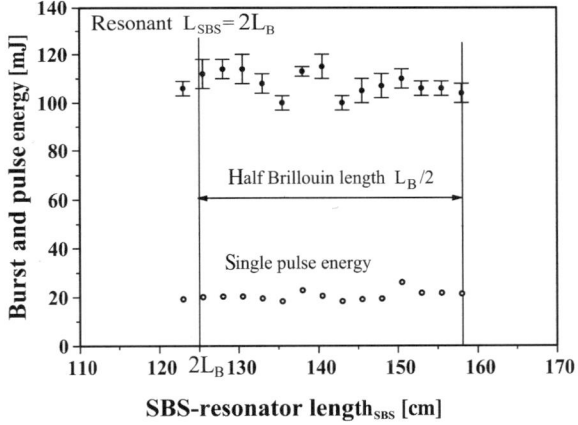

Figure 3.28. Measured pulse energies for different optical SBS-resonator lengths. Each point is an energy average over 100 pulses. The error bar denotes the measured standard deviation.

shows some experimental evidence that there is no significant impact on the laser efficiency found while tuning the start resonator length (see Fig. 3.28).

3.6.3 Analytical pulse shape description

The complex electrical field $\mathcal{E}(t)$ of the Q-switch pulse can be described by the sum of plane wave fields with equidistant frequencies $\omega_m = \omega_0 + m \cdot \Delta\omega$:

$$\mathcal{E}(t) = \sum_{m=1}^{n} a_m \exp\left(i(\omega_m t + \theta_m)\right), \qquad E := \frac{1}{2}(\mathcal{E} + \mathcal{E}^*) \qquad (25)$$

θ_k denotes the phase of the field component with the frequency ω_k and the real amplitude a_k. Time averaging of the electric field amplitude over the period $T_0 = 1/\omega_0$ leads to the following intensity of the Q-switch pulse with a Gaussian pulse shape of the duration t_w (FWHM):

$$I(t) = \epsilon_0 c_0 \overline{E^2} = \epsilon_0 c_0 \mathcal{E}(t) \cdot \mathcal{E}^*(t) \cdot \exp\left(-\frac{t^2}{t_w^2}\ln(2)\right) \tag{26}$$

where c_0 denotes the vacuum light velocity whereas ϵ_0 is the vacuum permittivity. For a spectrum of n frequencies the intensity is given as

$$I(t) = \epsilon_0 c_0 \left\{ \sum_{m=1}^{n} a_m^2 + 2 \sum_{m=1}^{n-1} \left[\left(\sum_{j=1}^{n-m} a_j a_{j+m} \cos(\theta_j - \theta_{j+m}) \right) \cos(m \cdot \Delta \omega t) \right] \right\}$$

$$\times \exp\left(-\frac{t^2}{t_w^2}\ln(2)\right) \tag{27}$$

It can be seen that only the difference frequencies (beat frequencies) and no absolute frequencies ω_k are contained in this intensity anymore. Consider a mode series consisting of every kth mode of the longitudinal modes of a resonator. The frequencies of this mode series $\omega_{k\times p}$ have a spectral distance of $\omega_{k\times p} - \omega_{k\times(p+1)} = k \cdot \Delta\omega$ (k and p being integers). If there exists a locking mechanism for their phases, it will apply $\theta_{k\times p} - \theta_{k\times(p+1)} =$ const. for every integer p and every pulse. In this case in Eq. (27), $\cos(\theta_{k\times p} - \theta_{k\times(p+1)})$ has the same value for every Q-switch pulse (e.g., 1). Other terms like $\cos(\theta_{k\times p+1} - \theta_{k\times(p+1)+1})$ in front of $\cos(k\Delta\omega)$ have different values since they are not supported by the locking mechanism.

If there is no locking mechanism for a mode series containing every kth mode, the phase angles are determined by the spontaneous laser emission. Then, for every mode pair and each pulse the difference $\theta_{k\times p} - \theta_{k\times(p+1)}$ has different values. The sum in front of $\cos(k \cdot \Delta\omega t)$ in Eq. (27) has different values for every pulse, too. Therefore the average for many pulse intensities of all the sums in front of $\cos(k\Delta\omega)$ approaches zero as a result. Consequently, in the Fourier transform the amplitude for $k \cdot \Delta\omega$ is approaching zero, too. On the other hand, in the case of locked modes the amplitude for $k \cdot \Delta\omega$ is strong since there is no cancellation. This is the explanation for the dominant beat frequency equal to the Brillouin frequency $k \cdot \Delta\omega = \nu_B$ and corroborates the explanation of the mode locking of the modes spaced by ν_B given in the last section.

Assuming that after every round trip with the duration t_{rt} a constant frequency shift ω_B is applied for all components of the field $\mathcal{E}(t)$, a time-dependent frequency $\omega_m(t)$ for the mth mode can be introduced. This frequency alters its value after each round trip in the resonator with the time duration t_{rt}.

$$\omega_m(t) = [\omega_0 + m \cdot \Delta\omega] - [\omega_B \cdot (N(t))] \tag{28}$$

where $N(t)$ is the number of the round trip at the time t counted from the beginning of the Q-switch pulse. But since the whole spectrum is shifted by the same amount ω_B, the frequency differences of the longitudinal modes remain the same. Therefore it can be shown by a straightforward calculation that under the assumption of such a shifting mechanism the intensity still has the form of Eq. (27). Consequently, the calculated pulse shapes of Eq. (26) are very similar to the measured pulse shapes if the correct amplitudes and phases are applied (see Figs. 3.29 and 3.30).

As has been explained, there is an identical phase angle for the longitudinal modes spaced by the Brillouin frequency if start and SBS resonator are tuned to the Brillouin frequency. If there are beat frequencies with a different spectral distance, their corresponding fields should have any arbitrary phase angle since there is no locking mechanism for them. Therefore, in the calculation the phases of the fields with an even index (it is even because $L_{SBS} = 2L_B$) and a spectral distance of ν_B were chosen identical and the phase angles of the remaining fields with uneven index were chosen randomly.

It turns out that every observed pulse shape can be calculated in an optimal way following one of the two assumptions for two different cases. Either there exists only one strong amplitude at one frequency ω_k in a non-length-tuned SBS resonator or in

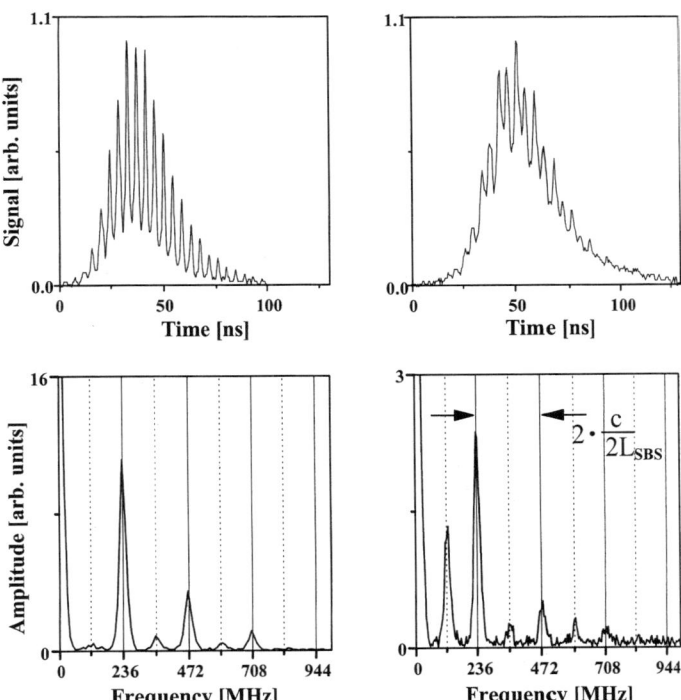

Figure 3.29. Two measured time distributions of Q-switch pulses from Fig. 3.21 and its Fourier transforms.

96 LASER RESONATORS WITH BRILLOUIN MIRRORS

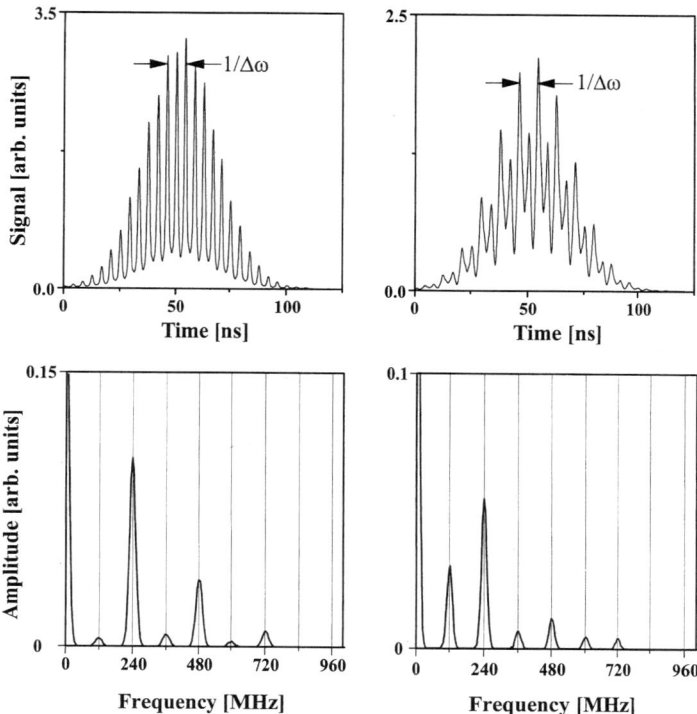

Figure 3.30. Two calculated time distributions of pulses and their Fourier transforms. To adapt the shapes of the pulses in Fig. 3.29, the amplitudes of the electrical fields are as follows: *Left:* $a_1 = 1, a_2 = 0.011, a_3 = 0.5, a_4 = 0.03, a_5 = 0.2, a_6 = 0.015, a_7 = 0.05$. *Right:* $a_1 = 1, a_2 = 0.15, a_3 = 0.27, a_4 = 0.03, a_5 = 0.03, a_6 = 0.015, a_7 = 0.02$. The phases of fields with an even number k are chosen randomly. For the uneven fields, $\Theta_k = 0$ applies.

a length-tuned SBS resonator there are one or two additional strong amplitudes with a spectral distance of the Brillouin frequency ν_B.

If there are longitudinal modes with frequency distances other than ν_B, they show such small amplitudes that they cannot be measured with the FPI. Because of the spectral resolution of 59 MHz of the FPI and the squaring of the fields in the measured intensity, these small amplitudes are part of the background of the measured signal (notice that the modulation of the FPI signals does not reach the 0 level).

3.6.4 Impact of acoustic decay time on longitudinal modes

So far we have not taken into account the specific threshold of the SBS material and the acoustic decay time of the sound wave grating. Low thresholds are especially interesting for CW applications. To the best of our knowledge, there has not been

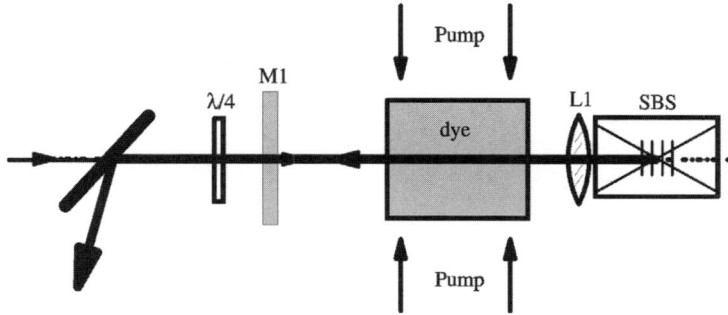

Figure 3.31. Layout of injected SBS-dye laser for investigations of impact of acoustic decay time on lasing process.

any CW-SBS-laser oscillator realized yet, and we are not going to discuss concepts for CW SBS lasers here. Up to now, typical materials for cw operation are photorefractive crystals (see Chapter 8). Low-threshold SBS mirrors are discussed in the previous chapter.

However, the acoustic decay time of the SBS material is also an issue for Q-switched SBS laser. Especially the impact of this decay time on the lasing process was investigated theoretically by Barrientos et al. [44] for an injected dye laser (see Fig. 3.31). They used a set of SBS equations assuming plane waves, narrow bandwidth for the SBS, and an SVE approximation for the electrical fields (see Section 2.2.1 in Chapter 2). The laser process is described by rate equations assuming absence of excited-state absorption, triplet-state formation, and amplified spontaneous emission as well as no scatter or diffraction losses.

The main characteristics of this injected oscillator in terms of impact of the decay time are also valid for the self-starting SBS laser. Four different decay times τ_B corresponding to four different SBS materials were investigated: 0.5 ns (n-hexane), 1.5 ns ($TiCl_4$), 3.5 ns (C_2F_6), and 8.5 ns ($CClF_3$). The round-trip time of the cavity was 7.3 ns. The duration of the injected pulses was 12 ns and the pulse energy was 300 mJ for the first three materials. For $CClF_3$ with the longest acoustic decay time the pulse energy was risen to 500 µJ.

The modulation period of the out-coupled intensity corresponds to the round-trip time t_{rt} of the SBS resonator. The shorter the decay time τ_B, the more distinct became the modulation (see Fig. 3.32). For decay times shorter than about one-fourth of the round trip, time termination of SBS and hence no laser output was found. For long decay times comparable to the round-trip period, steady-state oscillation was approached. Also, the shorter the decay time, the faster the reflectivity of the SBS mirror reacts. When the decay time becomes too long (in this investigation 4.4 ns), SBS becomes too transient. The SBS-threshold intensity is not reached, and no lasting laser oscillation takes place. If the injected pulse energy was elevated, the threshold would have been reached again.

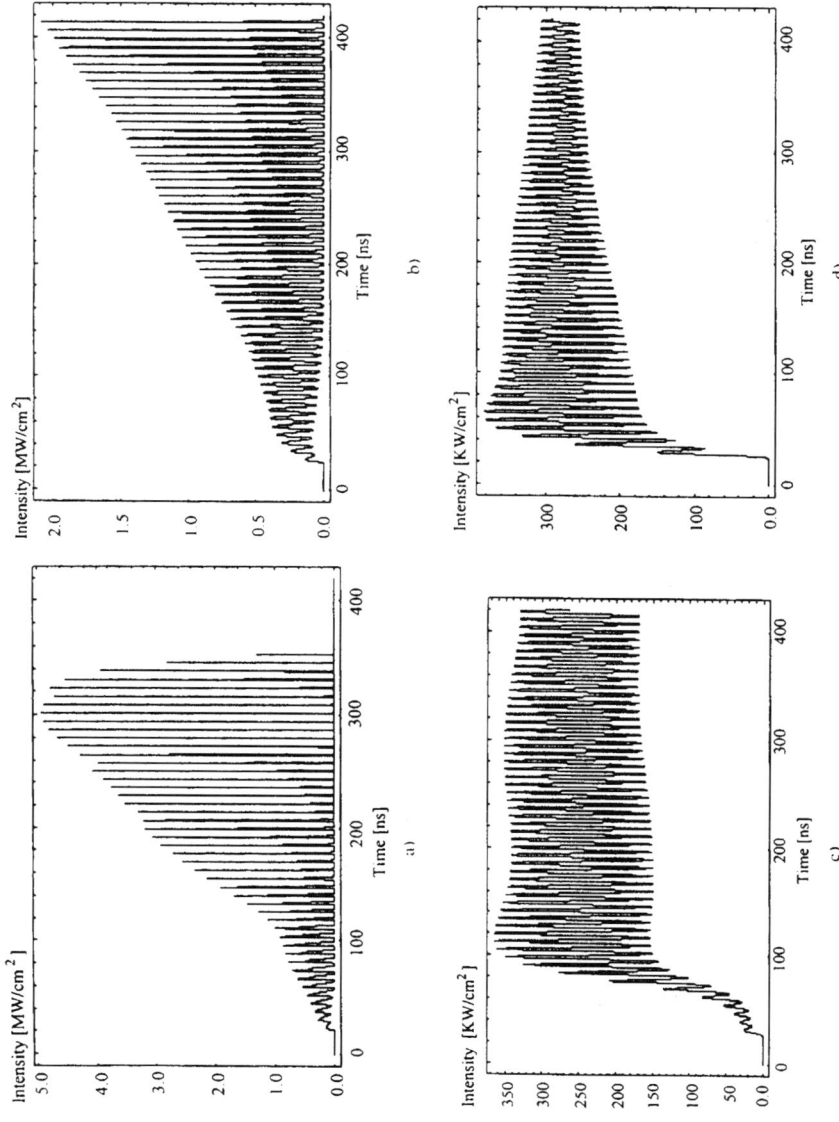

Figure 3.32. Calculated out coupled intensity as a result of a numerical simulation according to Ref. 44 for four different decay times: 0.5 ns (n-hexane), 1.5 ns (TiCl$_4$), 3.5 ns (C$_2$F$_6$), and 8.5 ns (CClF$_3$).

3.6.5 Summary

In Sections 3.4–3.6 the SBS mirror as part of the laser resonator has been explained with regard to its function as a simple device for Q-switching of laser oscillators. The Q-switch pulse parameters can be easily varied by a loss variation in the start resonator since the laser oscillation starts with the leading edge of a spike out of the start resonator. A length tuning of the start resonator is necessary to achieve a resonant generation of the sound wave grating in the SBS mirror. This way a threshold reduction for the SBS mirror via Brillouin enhanced for wave mixing (BEFWM) of the start resonator modes takes place. Without this length tuning, an SBS-laser oscillator with a gaseous SBS material will not operate in Q-switch regime but in spiking. The length of the SBS resonator does not seem to be important for the efficiency and stability of the Q switch, but has influence on the phases of the longitudinal modes. Due to the seeding process in the start resonator via BEFWM, phase locking can be achieved during the Q-switch pulse if a length tuning of start and SBS resonator has been performed. Due to the moving sound wave grating in the SBS mirror, the SBS-laser oscillator embodies a time-dependent mode spectrum with a decreasing center frequency of the longitudinal mode spectrum during the Q-switch pulse. Within a simple model the shape of the Q-switch pulses can be calculated using a few longitudinal modes only. Their frequencies are related to the length of the SBS resonator, the correct phase angles must be considered, but the Doppler shift of the SBS mirror has no influence on the pulse shape. Shorter acoustic decay times lead to stronger modulations in the out coupled intensity. Laser oscillation may break down if the acoustic decay time is shorter than one-fourth of the round-trip time of the SBS resonator.

3.7 HIGH BRIGHTNESS OPERATION OF THE LINEAR-SBS LASER

The two major reasons to use SBS mirrors in high-power laser oscillators are their phase conjugating properties on one hand and their nonlinear reflectivity to realize a very efficient Q switch for higher pulse energies on the other hand.

The thermal load in solid-state lasers limits their brightness. Due to the temperature dependence of the heat conductivity and the specific distribution of the heat generation inside the laser crystal, the deviation from the ideal parabolic refraction index profile will increase with increasing pump powers [64]. This leads to an increasing deterioration of the beam quality. Furthermore, in stable solid-state laser resonators with varying thermal lenses, the highest output power realizable at a given beam quality is limited by the strength of the thermal lensing [65]. The stability of the resonator becomes more and more problematic for higher pump powers. Phase conjugating resonators (PCRs) can compensate for these problems [49]. In this section we discuss the ability of the sf-SBS laser to realize high average output powers with a near diffraction-limited beam quality at a specific example.

This specific PCR with SBS mirror for high brightness output is a flashlamp-pumped Nd:YALO laser oscillator [27]. The schematic of the laser resonator and the

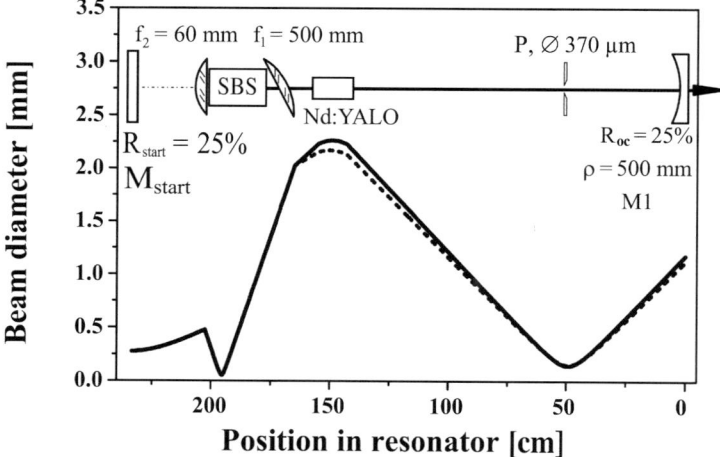

Figure 3.33. Scheme of the laser resonator. The distances of the resonator elements from the output coupler are as follows: pinhole P, 50 cm; laser rod's endface, 143 cm; lens L1, 165 cm; lens L2 depending on pump power; mirror M_{start}; 234 cm. The SBS cell is 30 cm long and 1 cm in diameter and is positioned directly in front of lens L2. The laser rod dimension is 7 mm × 120 mm with a doping level of 0.8 at%. The overall pump energy of the two flashlamps was 100 J at a pump pulse duration of 1 ms. The solid and broken curves, respectively, show the eigensolution for the stronger and weaker dioptric power of the laser rod thermal lens.

calculated eigenmodes regarding to the two dioptric powers of the thermal lens of Nd:YALO are shown in Fig. 3.33.

The design of this resonator is to be found in the peculiarities of PCRs: The degeneracy of modes in PCRs can lead to eigensolutions that build focuses on surfaces of the optical elements in the resonator which would lead to damage of these elements. This resonator design with a pinhole at the position of the beam waist and the rods aperture as second mode-selecting aperture does not allow for other than the transverse fundamental mode to come above threshold [66]. It particularly does not allow for any mode with a focus on one of the SBS-resonator elements. The mode diameter on the start resonator mirror is less crucial since the intensity in this resonator part is low. Furthermore, this set of apertures lifts the mode degeneracy in a way that a large transverse fundamental mode volume is realized, since this is the eigensolution with the lowest round-trip loss.

The SBS resonator is not an ideal PCR. To some extent, it is an SBS-PCR as discussed in Section 6.3 with a mode injected from the start resonator. This mode is more or less perturbed compared to the eigenmode of the SBS resonator. But guaranteed by apertures the eigenmode of the SBS resonator is close to the one of the start resonator. The pinhole in the resonator acts at least partially as a spatial filter. A fraction of the lens duct will have a focal length equal to the distance between laser rod and pinhole. Therefore, in addition to the condition of matching the curvatures at the output coupler, it realizes a second loss mechanism for nonparabolic wavefront distortions of the transverse fundamental mode and supports the convergence of the eigensolution in the SBS resonator toward the high beam quality transverse fundamental mode.

3.7.1 Crucial components of the SBS laser Anisotropic materials allow for operation with linear polarized laser light without any additional polarizers, since the thermally induced birefringence becomes negligible compared to their naturally birefringence. Therefore in this example Nd:YALO was chosen as laser-active material.

The telescope enclosing the SBS cell benefits in four different ways. First, it reduces the coherence demands on the laser light for the SBS. The sharper the beam is focused, the stronger the reflectivity at each plane of the sound wave grating. Thus, less grating planes are necessary to reach a certain reflectivity, and hence a shorter interaction length results when using a sharper focusing [67]. Second, the telescope makes it possible to set up the start resonator with a large transverse fundamental mode beam diameter in the laser rod. For reaching a high efficiency and achieving a good discrimination of higher-order modes, a beam diameter of 4.4 mm in the laser rod was calculated and realized in this laser. The mode selection is achieved by the rod aperture of 7-mm diameter in combination with the pinhole P. The start resonator was set up for dynamically stable operation (see Fig. 3.34). Third, despite the large beam diameter inside the laser rod, a wide pump power range can be used by varying the distance of the two telescope lenses [28]. Fourth, using an anisotropic material such as Nd:YALO, the astigmatism of the thermally induced dioptric power of the laser rod (0.70 dpt/kW and 0.83 dpt/kW in this case) can be compensated by tilting the first telescope lens as a function of the pump power.

Although we have a phase conjugating resonator, the compensation of the astigmatism is necessary because the astigmatic thermal lens leads to a separation of the stability range of the start resonator in two ranges, one for each dioptric power. If these two stability ranges have no overlap, the start resonator becomes unstable. As a consequence, if the diffraction losses become too big, the start resonator will not reach threshold at all. This is avoided by the astigmatism compensation.

The focal lengths of the telescope lenses should be chosen for spot sizes not too small to avoid optical breakdown in the SBS material. Optical breakdown for gases is known to occur at intensities of $10^{11}-10^{12}$ W/cm^2 [68] for nanosecond pulses.

To avoid an optical breakdown at the pinhole plane, this resonator type can be evacuated around the pinhole. The pinhole was made of quartz glass. To obtain the maximum intracavity SBS-threshold reduction as pointed out in Section 3.5, the start resonator length should be chosen regarding to the Brillouin frequency. Here we have chosen the optical length of the start resonator to be $L_{\text{start}} = 4 * c/2v_B = 250$ cm ($v_B = 240$ MHz for the SBS material used, SF$_6$). The reason for choosing four times the mode distance is related to the design of the start resonator and the transverse mode size: The beam diameter at the pinhole should not be too small to accommodate high intensities. Consequentially, for this kind of a start resonator design, the large

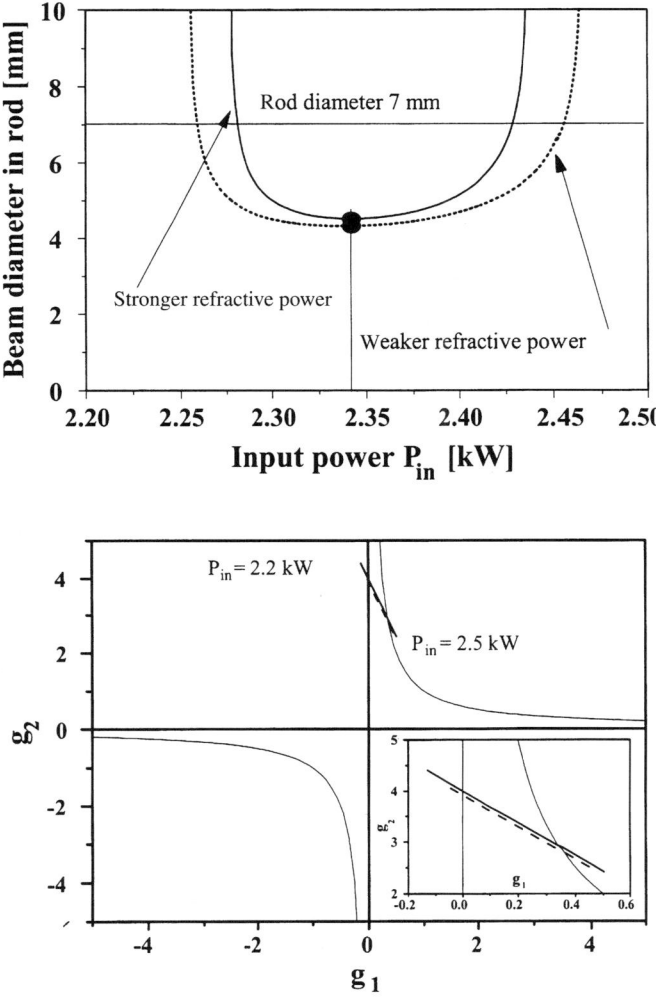

Figure 3.34. *Top:* Calculated stability ranges for the two dioptric powers of the Nd:YALO crystal in the start resonator. The resonator is designed for dynamically stable operation. Due to the compensation of the astigmatic thermal lens of the Nd:YALO, the stability ranges are centered to each other. *Bottom:* g-Diagram for the two dioptric powers. Both top and bottom are calculated for an optimal telescope length adjustment at $P_{in} = 2.34$ kW.

transverse fundamental mode diameter in the laser rod required for efficient operation leads to a longer resonator length.

3.7.2 Output data of the flashlamp-pumped SBS laser

This laser was pumped by two flashlamps with a total electrical pulse energy of 100 J and a pump pulse duration of 1 ms. A maximum pump power of 3.2 kW was reached at the

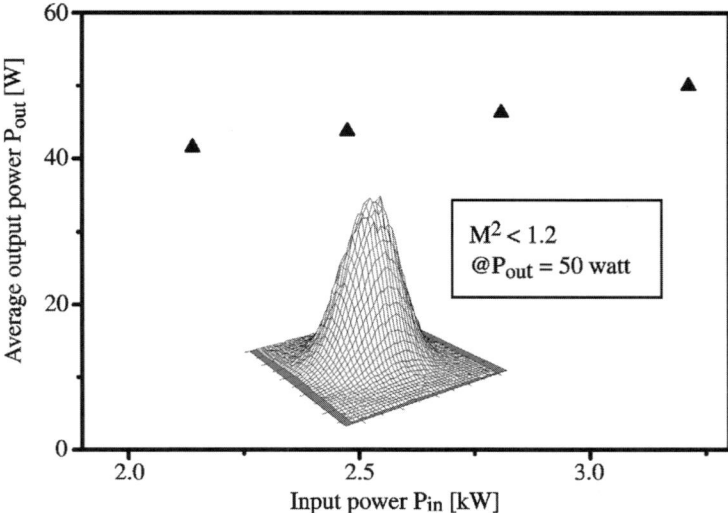

Figure 3.35. Average output power of SBS-laser oscillator as a function of pump power. Four measurements are shown with telescope lengths changed in steps of 3 cm. The depicted beam shape is measured at an average output power of 50 W of the SBS laser.

maximum repetition rate of 33 Hz. Using this laser head in a short multimode cavity, the total efficiency $\eta_{tot} = P_{out}/P_{in}$, the ratio of output to pump power, was $\eta_{tot} = 4.3\%$.

A maximum average output power of 50 W was achieved operating this SBS laser (see Fig. 3.35). It was reached with a telescope length of 32 cm at a pump power of 3.2 kW, which is equivalent to a total efficiency of 1.6%. At a pump power of 2.1 kW, an efficiency of 1.9% and an output power of 41 W was obtained. Unlike other Q-switch techniques, the SBS mirror is even able to enhance the efficiency of the free running laser because of its phase conjugating properties. The total efficiency of the same laser without the SBS mirror but with a 100% reflecting mirror M_{start} was measured to be 1.8% at a pump power of $P_{in} = 2.1$ kW. The maximum output power of this conventional free running laser without an SBS mirror was 43 W at a pump power of 2.7 kW. The beam quality was measured to be better than $M^2 < 1.2$. Within the measurement accuracy the beam qualities were the same for both the conventional and the SBS laser. As a result of the astigmatism of the Nd:YALO, the beam is slightly elliptical with a ratio of the two main axes of 1.1 at the highest output power.

During the pump pulse of duration of 1 ms, a burst of 22–16 Q-switch pulses related to the pump power range of 2–3.2 kW was emitted (see Fig. 3.36). In this pump power range the burst energy decreases from 2 J to 1.6 J. The Q-switch pulses within in the burst were separated by 40–53 μs. The single pulse energy was about 95 mJ. The Q-switch pulse duration was 60 ns (full width at half-maximum).

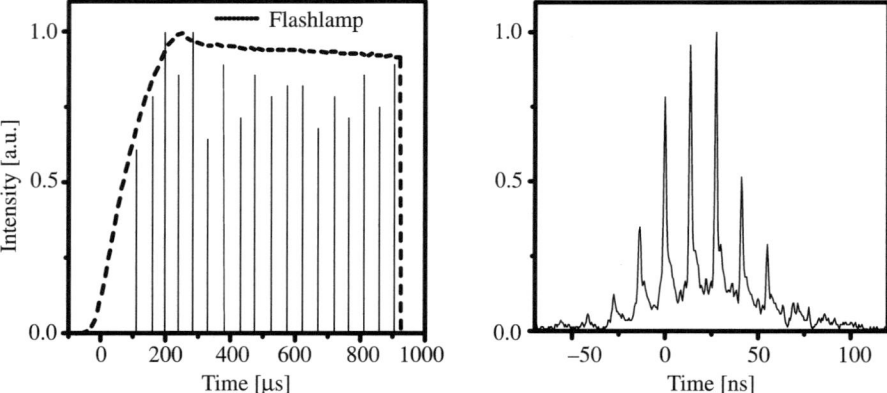

Figure 3.36. Temporal structure of the SBS-laser pulses. *Left:* Burst of Q-switch pulses of the SBS laser. *Right:* Single Q-switch pulse of the burst.

The strong modulation in the Q-switch pulse shown in Fig. 3.36 originates from the coupling of the longitudinal modes by the nonlinear SBS mirror (see Section 3.6.3). The basic period of the modulation is 13.7 ns and corresponds to an optical resonator length of 2.05 m. This is the length of the SBS resonator defined by the distance of mirror M1 and the focus in the SBS cell. For the high-power experiments presented in this section, no attempts were made to reach a certain pulse shape neither a strongly modulated nor a smooth pulse shape. The laser was optimized for a high average output power only. At the highest output power the stability of the burst energy fluctuated with a standard deviation of 10%. At an output power of 46 W the standard deviation is less than 5%.

In summary the efficiency of the SBS laser is even slightly higher than the efficiency of the corresponding laser with identical resonator but removed SBS mirror (to our knowledge there is no other fair comparison reported in the literature between an SBS- and equivalent conventional laser). Thus, the phase conjugating SBS mirror was confirmed to be a very efficient Q-switch device for higher pulse energies. Because the pinhole is designed for a high transmission of an undistorted TEM_{00} only, to a first approximation, aberrations of the TEM_{00} will not lead to a worse beam quality but will decrease the efficiency. The aberrations will be filtered out by the pinhole. But up to a pump power level of 2.7 kW the efficiency of the laser without the SBS mirror is only 0.1% smaller than the SBS laser.

Instead of the aberrations in the active material, it was found that the stability of the conventional start resonator is the limiting factor for the power scaling of the sf-SBS laser. At higher pump power levels the stability of the start resonator became more and more problematic. If a further increase in the average output power is desired, the fluctuating start resonator stability is the first problem that has to be solved by developing further advanced start resonator concepts.

3.7.3 Conclusion The realization of high-power SBS-laser oscillators leads to high-brightness lasers with a very efficient passive Q switch. So far the stability of the conventional start resonator limits the realized SBS-laser oscillators in their average output power. Phase conjugating MOPA systems do not have any stability problem because of their separation of low-power oscillator and high-power amplification. Either an improvement of the start resonator or a new concept for the entire SBS-laser oscillator is necessary to take full advantage of the promises of the theoretical analysis of SBS-PCRs. However, if single-frequency radiation is not required or may be even broad bandwidth desired, the SBS-laser oscillator is a simpler approach to realize a high-brightness laser compared to MOPA concepts for diffraction-limited beams up to a certain power level.

REFERENCES

1. B. Y. Zeldovich, V. I. Popovichev, V. V. Ragulskii, and F. S. Failzullov, Connection between the wave fronts of the reflected and exciting light in stimulated Mandelshtam–Brillouin scattering, *ZhETF Pis. Red.* **15**, 109–112 (1972).
2. G. C. Valley, A review of stimulated Brillouin scattering excited with a broad-band pump laser, *IEEE J. Quantum Electron.* **22**, 704–711 (1986).
3. R. A. Lamb and M. J. Damzen, Phase locking of multiple stimulated Brillouin scattering by a phase-conjugate laser resonator, *J. Opt. Soc. Am. B* **13**, 1468–1472 (1996).
4. M. J. Damzen, R. A. Lamb, and G. K. N. Wong, Ultrashort pulse generation by phase locking of multiple stimulated Brillouin scattering, *Opt. Commun.* **82**, 337–341 (1991).
5. B. Barrientos, V. Aboites, and M. Damzen, Temporal dynamics of a ring dye laser with a stimulated Brillouin scattering mirror, *Appl. Opt.* **35**, 5386–5391 (1996).
6. R. A. Lamb, Single longitudinal-mode, phase conjugating ring master oscillator power amplifier using external stimulated-Brillouin-scattering Q-switching, *J. Opt. Soc. Am. B* **13**, 1758 (1996).
7. S. Seidel, Improvement of extraction efficiency by regenerative amplification in an Nd-YAG-MOPA with a phase-conjugating SBS cell, *Opt. Quantum Electron.* **27**, 625–632 (1995).
8. A. D. Case, P. J. Soan, M. J. Damzen, and M. H. R. Hutchinson, *J. Opt. Soc. Am. B* **9**, 374 (1992).
9. A. D. Case, M. R. Osborne, M. J. Damzen, and M. H. R. Hutchinson, *Opt. Commun.* **69**, 311 (1989).
10. M. J. Damzen and H. R. Hutchinson, *Opt. Lett.* **5**, 282 (1984).
11. M. R. Osborne, W. A. Schroeder, M. J. Damzen, and H. R. Hutchinson, *Appl. Phys. B* **48**, 351 (1989).
12. S. A. Lesnik, M. S. Soskin, and A. I. Khizhnyak, Laser with a stimulated-Brillouin-scattering complex-conjugate mirror, *Sov. Phys. Tech. Phys.* **24**, 1249–1250 (1979).
13. I. Y. Anikeev, and J. Munch, Improved output power performance of a phase conjugated laser oscillator, *Opt. Quantum Electron* **31**, 545–553 (1999).
14. V. I. Odintsov and L. F. Rogacheva, *Pis'ma Zh. Eksp. Theor. Fiz.* **36**, 281 (1982).

15. G. K. N. Wong and M. J. Damzen, Enhancement of the phase conjugate stimulated Brillouin scattering process using optical feedback, *J. Mod. Opt.* **35**, 483 (1988).
16. G. K. N. Wong and M. J. Damzen, Investigations of optical feedback used to enhance stimulated scattering, *IEEE J. Quantum Electron.* **26**, 139 (1990).
17. K. D. Ridley, and A. M. Scott, Stimulated Brillouin scattering in a transverse resonator, *J. Opt. Soc. Am. B* **11**, 1361–1366 (1994).
18. A. M. Scott, and W. T. Whitney, Characteristics of a Brillouin ring resonator used for phase conjugation at 2.1 μm, *J. Opt. Soc. Am. B* **12**, 1634–1641 (1995).
19. H. Meng and H. J. Eichler, Nd:YAG laser with a phase-conjugating mirror based on stimulated Brillouin scattering in SF_6 gas, *Opt. Lett.* **16**, 569–571 (1991).
20. A. Kummrow, R. Menzel, D. Schumann, and H. J. Eichler, Length tuning effect in SBS-lasers, *Int. J. Nonlinear Opt. Phys.* **2**, 261–266 (1993).
21. J. Auyeng, D. Fekete, D. M. Pepper, and A. Yariv, A theoretical and experimental investigation of the modes of optical resonators with phase-conjugate mirrors, *IEEE J. Quantum Electron.* **15**, 1180–1188 (1979).
22. M. D. Skeldon and R. W. Boyd, Transverse-mode structure of a phase-conjugate oscillator based on Brillouin-enhanced four-wave mixing, *IEEE Quantum Electronics* **25**, 588–594, (1989)
23. A. M. Scott and K. D. Ridley, A review of Brillouin-enhanced-four-wave-mixing, *IEEE J. Quantum Electron.* **25**, 438–459 (1989).
24. D. Pohl, A new laser Q-switch-technique using stimulated Brillouin scattering **24A**, 239 (1967).
25. Y. Jingguo and J. Hongwei, Self-Q-switching Nd:YAG laser operation using stimulated thermal Rayleigh scattering, *Opt. Quantum Electron.* **26**, 929–932 (1994).
26. H. J. Eichler, S. Heinrich, and J. Schwartz, Self-starting short-pulse XeCl laser with a stimulated Brillouin scattering mirror, *Opt Lett.* **21**, 1909–1911 (1996).
27. M. Ostermeyer and R. Menzel, 50 Watt average output power with 1.2*DL beam quality from a single rod Nd:YALO laser with phase-conjugating SBS mirror, *Opt. Commun.* **171**, 85–91 (1999).
28. M. Ostermeyer, A. Heuer, and R. Menzel, 27 Watt average output power with 1.2*DL beam quality from a single rod Nd:YAG-laser with phase conjugating SBS-mirror, *IEEE J. Quantum Electron.* **34**, 372–377 (1998).
29. H. J. Eichler, R. Menzel, and D. Schumann, 10-W single rod Nd:YAG laser with stimulated Brillouin scattering Q-switching mirror, *Appl. Opt.* **31**, 5038–5043 (1992).
30. N. N. Il'ichev, A. A. Malyutin, and P. P. Pashinin, Laser with diffraction-limited divergence and Q switching by stimulated Brillouin scattering, *Sov. J. Quantum Electron.* **12**, 1161–1164 (1982).
31. I. Y. Anikeev and J. Munch, Variation in the coherence length of a phase conjugating oscillator, *Opt. Commun.* **178**, 449–456 (2000).
32. A. Drobnik and L. Wolf, Influence of self-focusing on the operation of a neodymium glass laser, *Sov. J. Quantum Electron.* **8**, 274–275 (1978).
33. V. I. Bezrodnyi, F. I. Ibragimov, V. I. Kislenko, R. A. Petrenko, V. L. Strrizhevskii, and E. A. Tikhonov, Mechanism of laser Q switching by intracavity stimulated scattering, *Sov. J. Quantum Electron.* **10**, 382–383 (1980).
34. A. Z. Grasyuk, V. V. Ragul'skii, and F. S. Faizullov, *JETP Lett.* **9**, 6 (1969).

35. A. Agnesi and G. C. Reali, Passive and self-Q-switching of phase-conjugation Nd:YAG laser oscillators, *Opt. Commun.* **89**, 41–46 (1992).
36. S. Chandra, R. C. Fukuda, and R. Utano, Sidearm stimulated scattering phase-conjugated laser resonator, *Opt. Lett.* **10**, 356–358 (1985).
37. M. R. Perrone and Y. B. Yao, Phase conjugated XECL laser resonator, *Opt. Lett.* **19**, 1052–1054 (1994).
38. P. P. Pashinin and E. J. Shklovsky, Laser with a stimulated Brillouin scattering mirror switched on by its own priming radiation, *Sov. J. Quantum Electron.* **18**, 1190 (1988).
39. B. I. Denker, I. Kertes, P. P. Pashinin, V. S. Sidorin, and E. J. Shklovsky, Compact laser with a stimulated Brillouin scattering mirror operated at pulse repetition frequency up to 150 Hz, *Sov. J. Quantum Electron.* **20**, 770 (1990).
40. S. Seidel and G. Phillipps, Pulse lengthening by intracavity stimulated Brillouin-scattering in a Q-switched, phase-conjugate Nd-YAG laser-oscillator, *Appl. Optics* **32**, 7408–7417 (1993).
41. A. B. Vasilev, O. M. Vokhnik, L. S. Korntsenko, V. A. Mikhailov, V. A. Spazkin, and I. A. Sherbakov, Lasing in apolarization-closed cavity with stimulated Brillouin scattering mirror, *Opt. Spektrosk.* **75**, 877–880 (1993).
42. O. M. Vokhnik, V. A. Mikhailov, V. A. Spahakin, I. V. Terent'eva, and I. A. Sherbakov, Solid state laser with a loop SBS-mirror, *Opt. Spektrosk.* **78**, 303 (1995).
43. P. P. Pashinin, E. J. Shklovsky, C. Y. Tang, and V. V. Tumorin, Passivly Q-switched single frequency Nd:YAG ring laser with feedback and phase conjugation, *Laser Phys.* **99**, 340 (1999).
44. B. Barrientos, V. Aboites, and M. J. Damzen, Temporal dynamics of an external-injection dye laser with a stimulated Brillouin scattering reflector, *J. Opt. (Paris)* **26**, 97–104 (1995).
45. M. J. Damzen, M. H. R. Hutchinson, and W. A. Schroeder, Single-frequency phase conjugate laser resonator using stimulated Brillouin scattering, *Opt. Lett.* **12**, 45 (1987).
46. W. A. Schroeder, M. J. Damzen, and M. H. R. Hutchinson, Studies of a single-frequency stimulated-Brillouin-scattering phase conjugate Nd:YAG laser oscillator, *J. Opt. Soc. Am. B* **6**, 171 (1989).
47. J. F. Lam and W. P. Brown, Optical resonators with phase conjugate mirrors, *Opt. Lett.* **5**, 61 (1980).
48. A. E. Siegman, P. A. Belanger, and A. Hardy, in *Optical Phase Conjugation*, R. A. Fischer (ed.), Academic Press, New York, (1983), Chapter 13, Optical Resonators Using Phase-Conjugate Mirrors.
49. P. A. Bélanger, A. Hardy, and A. E. Siegman, *Appl. Opt.* **19**, 602 (1980).
50. A. E. Siegmann, *Lasers*, Chapter 21, University Science Books, Mill Valley, California (1986).
51. G. Giuliani, M. Denariez-Roberge, and P. A. Belanger, Transverse modes of a stimulated scattering phase-conjugate resonator, *Appl. Opt.* **21**, 3719–3724, (1982).
52. G. G. Kochemasov and V. D. Nikolaev, Reproduction of the spatial amplitude and phase distributions of a pump beam in stimulated Brillouin scattering, *Sov. J. Quantum Electron.* **7**, 60–63 (1977).
53. P. A. Belanger and C. Paré, in *Optical Phase Conjugation*, M. Gower and D. Proch (eds.), Springer-Verlag, Berlin, (1994), Chapter 10, Phase Conjugate Resonators.

54. P. A. Belanger, A. Hardy, and A. E. Siegman, Resonant modes of optical cavities with phase-conjugate mirrors: Higher order modes, *Appl. Opt.* **19**, 479 (1980).
55. A. Hardy, Sensitivity of phase conjugate resonators to intracavity phase perturbations, *IEEE J. Quantum Electron.* **17**, 1581 (1981).
56. J. Schultheiss, Diploma thesis, University of Potsdam, 1996.
57. W. Koechner, *Solid State Laser Engineering*, 3rd ed., Springer-Verlag, (1992), Chapter 8.1.
58. M. Ostermeyer, K. Mittler, and R. Menzel, Q switch and longitudinal modes of a laser oscillator with a stimulated-Brillouin-scattering mirror, *Phys. Rev. A* **59**, 3975–3985 (1999).
59. M. J. Damzen, M. H. R. Hutchinson, and W. A. Schroeder, Direct measurements of the acoustic decay times of hypersonic waves generated by SBS, *IEEE J. Quantum Electron.* **23**, 328–334 (1987).
60. A. M. Scott, Brillouin induced four wave mixing, *Laser Wavefront Control, Proceedings of the SPIE*, Vol. 1000 (1988).
61. G. K. N. Wong and M. J. Damzen, Multiple Frequency Interaction in Stimulated Brillouin Scattering, Internal Report, The Blackett Laboratory, Imperial College, 1990.
62. A. Kummrow, R. Menzel, and D. Schumann, Resonant emission of a solid state laser with stimulated Brillouin scattering mirror, *J. Opt. Soc Am. B*, submitted.
63. P. Narum, M. D. Skeldon, and R. W. Boyd, Effect of laser mode structure on stimulated Brillouin scattering, *IEEE J. Quantum Electron.* **22**, 2161–2167 (1986).
64. N. Hodgson and H. Weber, *IEEE J. Quantum Electron.* **29**, 2497, (1993).
65. V. Magni, *J. Opt. Soc. Am. A* **4**, 1962 (1987).
66. R. Menzel, and M. Ostermeyer, Fundamental mode determination for guaranteeing diffraction limited beam quality of lasers with high output powers, *Opt Commun.* **149**, 321–325 (1998).
67. R. Menzel and H. J. Eichler, *Phys. Rev. A* **46**, 7139 (1992).
68. R. A. Mullen, *IEEE J. Quantum Electron.* **26**, 1299 (1990).

CHAPTER 4

Multi-kilohertz Pulsed Laser Systems with High Beam Quality by Phase Conjugation in Liquids and Fibers

THOMAS RIESBECK, ENRICO RISSE, OLIVER MEHL, and HANS J. EICHLER

Technische Universität Berlin, Optisches Institut, 10623 Berlin, Germany

4.1 INTRODUCTION

High-brightness pulsed solid-state laser sources are of great interest for application in industry and science such as high-precision materials processing, visible beam generation by nonlinear frequency conversion, plasma production, and X-ray production. The brightness B depends linearly on the average output power but quadratically on the beam quality $K = 1/M^2$. Thus an improvement of beam quality increases the brightness considerably:

$$B = \frac{P}{\lambda^2 (M^2)^2} \qquad (1)$$

where B is brightness, P is power, λ is wavelength, and M^2 is times-diffraction limit factor.

In the case of materials processing, higher beam quality (lower M^2 value) leads to smaller beam waists of focused laser beams and higher intensities in the focal region. In conjunction with high average output power the required processing time decreases. The increased Rayleigh length of the laser beam results in deep and narrow cuts with high aspect ratio and low material losses.

Due to phase distortions of laser crystals, like thermal lensing and thermally induced stress depolarization, the beam quality of conventional solid-state lasers decreases rapidly with average output power [1]. To reduce thermal loading, diode pumping can be applied, resulting in improved beam quality compared to flashlamp pumped laser systems [2]. In addition, advanced crystal geometry (e.g., slab or disk laser) can be applied to reduce the induced phase distortions [3, 4]. However, for

---

Phase Conjugate Laser Optics, edited by Arnaud Brignon and Jean-Pierre Huignard
ISBN 0-471-43957-6 Copyright © 2004 John Wiley & Sons, Inc.

average output powers above 100 W, the residual phase distortions lead to beam qualities far away from the diffraction limit. In any case the remaining phase distortions have to be compensated with adaptive mirrors to facilitate near diffraction-limited beam quality at average output powers of several hundred watts. Such mirrors are realized by self-pumped phase conjugation based on stimulated Brillouin scattering (SBS) which are compact and possess a fast response time without additional electronic equipment (e.g., mechanically deformed mirrors).

Such SBS mirrors can be used in oscillators [5–7] and oscillator amplifier arrangements (MOPA) [8–26]. In the case of oscillators a resonator design with high extraction efficiency and stability in the fundamental transversal mode becomes complicated at 1-μm wavelength. Thus only a few oscillators with SBS mirrors have been realized. In the case of amplifier systems the beam diameter can be adapted to the amplifier. Here the phase conjugating mirror compensates phase distortions of the amplifier crystal after double or multiple passes through the amplifiers. Therefore a beam with high quality but low average power can be scaled up to the kilowatt range.

In the following sections the development of flashlamp- and diode-pumped MOPA systems with phase conjugating mirrors will be reviewed. Different amplifier arrangements are discussed with respect to the average output power. Then a brief survey about the used active media, design rules for MOPA systems, and the measurement of beam quality is given. After the description of characterization methods of the used fiber PCM, the developed MOPA systems are presented in detail: (a) two pulsed flashlamp-pumped laser systems with fiber PCM based on Nd:YAG and Nd:YALO as active medium which leads to average output powers up to 315 W and a beam quality of $M^2 = 2.6$ in both directions in space and (b) a diode-pumped, active Q-switched system with fiber PCM for low power threshold and a system with SBS based on CS_2, with an average output power of 520 W.

4.2 AMPLIFIER SETUPS

All setups discussed below are based on the MOPA concept, where a nearly diffraction-limited master oscillator beam is increased in power within an amplifier setup (Fig. 4.1). After the first amplification pass the beam quality is reduced due to thermally induced phase distortions. The spatial distorted beam enters the SBS mirror and becomes phase conjugated. As a result the reflected beam propagates along the same trajectory through the amplifier and its phase distortions. The initial beam quality of the master oscillator can be nearly reproduced if the phase distortions remain constant during the beam propagation through the amplifier arrangement. The amplified beam can be extracted with help of an optical isolation, which in this case consists of a Faraday rotator and a polarizer.

With a single amplifier setup (Fig. 4.1a), an average output power of 140 W could be extracted with a beam propagation factor of $M^2 = 1.1$ [9]. This became possible due to the use of the polarization preserving active medium Nd:YALO (see

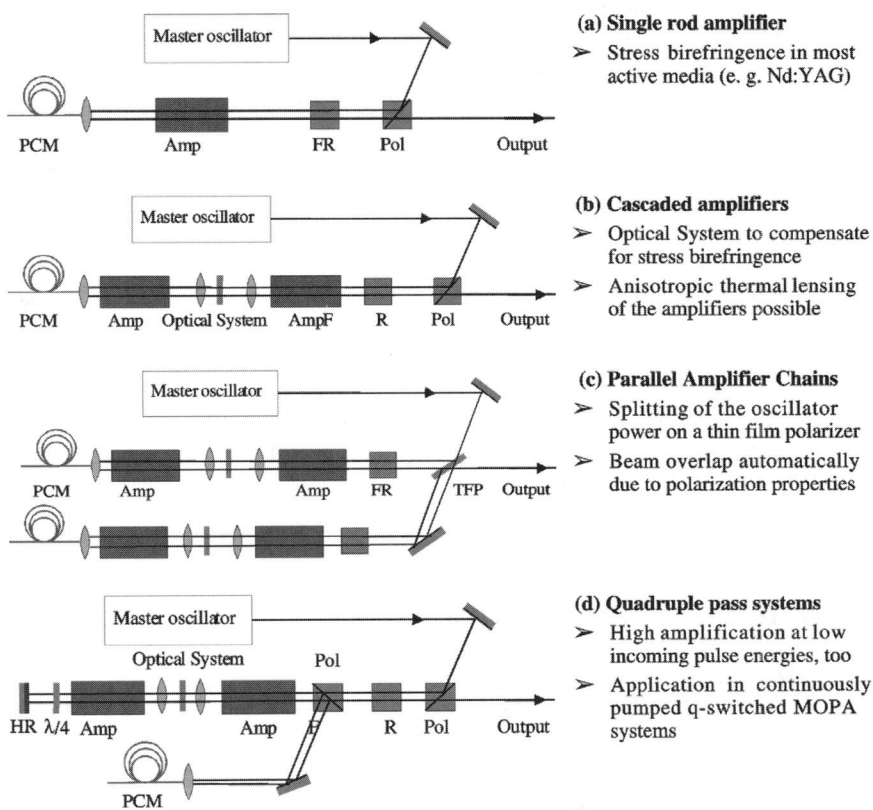

Figure 4.1. Different amplifier arrangements for MOPA systems with phase conjugation (Amp, amplifier; Pol, polarizer; FR, Faraday rotator; HR, high reflective mirror).

Section 4.3). To run the amplifier in the saturated regime, a high power oscillator with an advanced resonator design is required [27].

The power requirements of the oscillator can be reduced significantly with a serial arrangement of two amplifiers (Fig. 4.1b). With the depicted system but with a CS_2 cell as PCM, an average output power of 210 W with near-diffraction-limited beam quality has been realized [12]. The optical system between both amplifiers is important to reduce the influence of thermal lensing [28]. Thus the average output power can be tuned easily by changing the average pumping power. The optical system guarantees high extraction efficiency and protects the second amplifier against damage. The active medium Nd:YALO shows astigmatic thermal lensing, but this can be compensated by the rotation of one amplifier by 90°.

Due to a reduction of the required oscillator power, a parallel setup of two cascaded amplifier arrangements becomes feasible and the amplifiers do not suffer from the maximum average output power (Fig. 4.1c). In case of phase conjugation in two independent SBS cells, the phases of both channels are not coupled. Thus the number of parallel channels is restricted to two because superposition requires orthogonal polarized beams. The beams could be combined on a thin-film polarizer (TFP), and the resulting beam has statistical polarization due to the phase jitter between both channels.

For many applications [e.g., fast material processing continuous-wave (CW)], exciting lasers with several-kilohertz repetition rates and high brightness become of great interest. To reach high gain in amplifiers, multipass setups are necessary. Assuming a maximum average oscillator power of around 10 W and a repetition rate of 10 kHz, the pulse energy cannot exceed 1 mJ. Taking a saturation energy density of 500 mJ/cm^2 for Nd:YAG into account, the amplifiers would operate far below their saturation fluence, resulting in a small extraction efficiency. The extraction efficiency can be improved by increasing the number of amplifier passes. Therefore a four-pass amplifier arrangement was applied (Fig. 4.1d).

All four explained setups have been already realized and will be presented in detail in the following sections.

4.3 ACTIVE LASER MEDIA Nd:YAG AND Nd:YALO

4.3.1 Nd:YAG as active medium The active laser medium is of great importance for the beam quality and the average output power. Nd:YAG has the advantage of a lower price and the availability of high-quality crystals, but it shows thermally induced stress birefringence that leads to depolarization of the laser beam during the first amplification pass. The polarization properties of a scalar phase conjugating mirror are the same as those of a conventional mirror. Thus, depolarization cannot be compensated by scalar SBS. This results in a reduction of beam quality [29] and losses in the output power. Vector phase conjugating schemes were developed to compensate the birefringence, too. In a serial arrangement of two amplifiers the depolarization can nearly be compensated with help of two lenses and a 90° quartz rotator (Fig. 4.2). This scheme requires two identical active Nd:YAG media in identical pumping chambers under the same pumping conditions and a symmetric beam propagation through the amplifier [30].

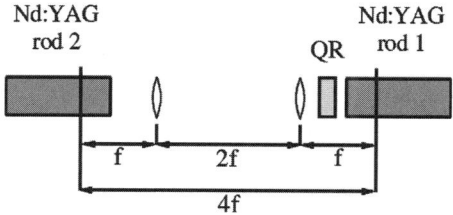

Figure 4.2. Scheme of birefringence compensation in Nd:YAG rods (QR: 90° quartz rotator).

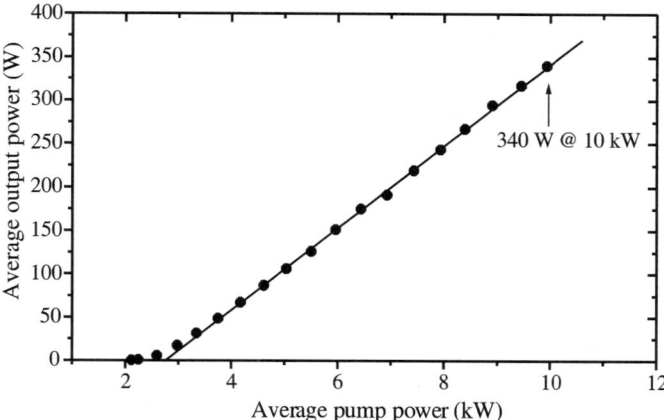

Figure 4.3. Average output power versus average pump power of continuously pumped Nd:YAG-oscillator.

Figure 4.3 shows the average output power for a free-running continuously pumped oscillator with an Nd:YAG rod with a diameter of 6.35 mm and a length of 178 mm pumped by two arc lamps in a specular cavity.

4.3.2 Nd:YALO as a polarization preserving active medium

To preserve the polarization of the beam, Nd:YALO (also called Nd:YAP) can be used as an active medium. It shows no stress birefringence, but it has an efficiency and durability comparable to Nd:YAG. Figure 4.4 shows the extracted average output power versus the average pump power for a free-running oscillator with an Nd:YALO rod (9 mm × 152 mm) in a diffuse pumping chamber with two

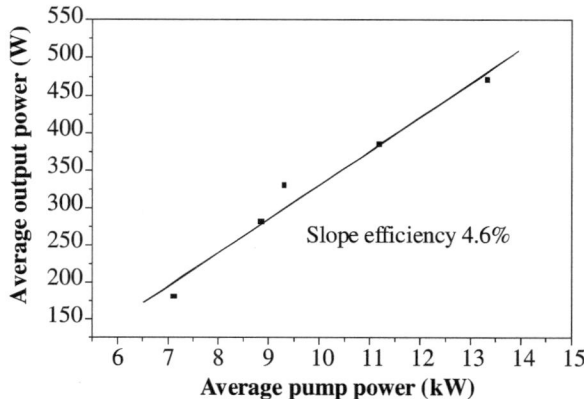

Figure 4.4. Average output power versus average pump power at a repetition rate of 100 Hz for a free-running Nd:YALO multimode oscillator.

flashlamps. The repetition rate is 100 Hz with a typical excitation time of 600 μs. No damage of the crystals was observed, indicating a robustness comparable with YAG. Due to a maximum average pump power of 13.5 kW from the power supply, the average output power was limited to 470 W (Fig. 4.4).

4.4 DESIGN RULES FOR MOPA SYSTEMS

To build up MOPA systems with high beam quality at high average output, the power oscillator and the amplifier have to fulfill several requirements. The major beam properties like pulse width or spatial and temporal distribution of the pulse intensity are determined by the master oscillator. In the amplifier the oscillator's output power should be boosted without losses of the beam quality.

4.4.1 Master oscillator To successfully design a master oscillator, four parameters have to be considered: beam quality, average output power, coherence length, and temporal structure. The beam quality should be diffraction-limited, and the average output power has to be in the range of several watts to guarantee high extraction efficiency in the amplifier stages. To achieve high reflectivity from the phase conjugator, the interaction length should be extended to several 10 cm inside the SBS mirror. The used master oscillators are described in detail in the respective chapters, together with the whole systems.

4.4.2 Cascaded amplifier setup In an amplifier the beam distorted by its thermal lensing. In the case of a cascaded setup (Fig. 4.1b) the influence of amplifier 1 onto amplifier 2 has to be considered. The phase conjugating mirror compensates for phase distortions of the entire single pass and not the individual thermal lensing of each amplifier. An optical system consisting of two lenses can be inserted in order to avoid damage of optical components even if the average pump power is varied strongly. The optical system should be optimized with respect to the maximum extraction efficiency in the amplifier rods independent of thermal lensing.

Figure 4.5 shows the amplifier parameters and the optical system: n is the refractive index, l the amplifier length, and Γ the propagation constant describing the

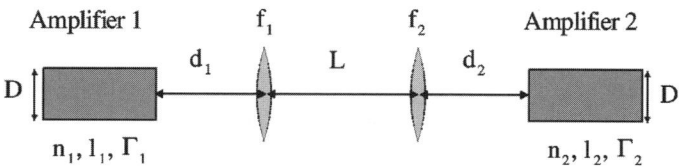

Figure 4.5. Parameters of the optical system for a two-amplifier system.

effect of thermal lensing. D is the diameter of the amplifier rods. d_1 and d_2 are the distances between the lenses and the amplifier crystals. f_1 and f_2 are the focal lengths.

For identical amplifier dimensions the formulas for an optimized optical system become relatively simple [12]. The d_1 is chosen to avoid small beam diameters on the lenses. For optimal extraction efficiency the optical system is a telescope that is adjusted for high-beam diameters in the second amplifier.

The optical system is optimized by calculating the beam propagation of the fundamental mode. The propagation constant Γ ranges from $2\,\text{m}^{-1}$ up to $5\,\text{m}^{-1}$, corresponding to an average pumping power from 1.5 kW up to 10 kW for one pumping chamber. The optical system contains two lenses with focal lengths of 100 mm. A further lens serves to attain the required intensity for the stimulated Brillouin scattering. The optical system fixes the position of the thermal lens focus between the amplifier rods and thus avoids damages on optical components. Because of the phase conjugation properties, the beam passes back on the same trajectory. High extraction efficiency of the amplifiers is achieved by matching the beam radii to the amplifier radii.

To achieve average output powers above 100 W, three design problems have to be solved: amplified spontaneous emission (ASE) as a limiting effect, astigmatism of thermal lensing, and parasitic reflexes that enter the phase conjugator. Amplified spontaneous emission from the amplifier rods can be suppressed using an aperture in the focal region of the optical system. The astigmatism of thermal lensing (Fig. 4.6) does not affect the amplifier's extraction efficiency, but leads to lower reflectivity from a phase conjugating SBS cell due to bi-focusing (Fig. 4.7). To compensate for astigmatism, the second amplifier rod can be rotated by 90°. Then an additional half-wave plate has to be inserted between the amplifiers to match the polarization of the beam to the second cavity.

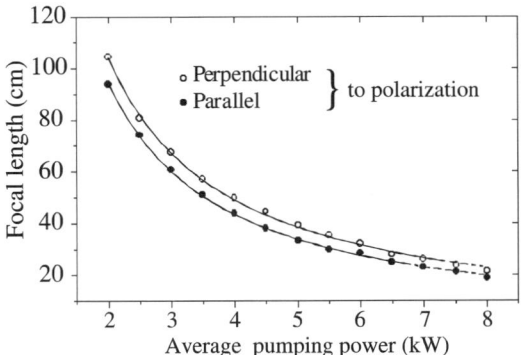

Figure 4.6. Astigmatic thermal lensing. Focal length of Nd:YALO amplifier rod for two directions in space.

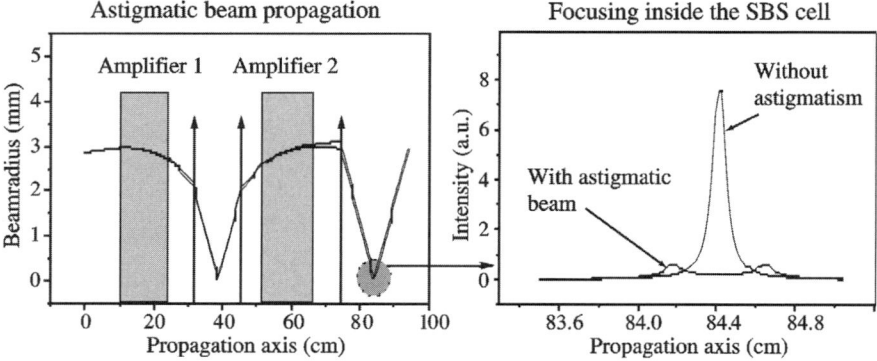

Figure 4.7. Calculated astigmatic beam propagation (*left*) and calculated intensity distribution inside an SBS cell with and without astigmatism (*right*).

4.5 BEAM QUALITY MEASUREMENT

An exact determination of beam quality requires careful beam diameter measurements. The beam quality measurements were carried out according to the international standard "Test methods for laser beam parameters," ISO/CD 11 146 [31], which define the beam diameter by the second moment of the intensity distribution. The setup shown in Fig. 4.8 was used for a simultaneous measurement of beam quality and average output power. Fresnel reflection from two uncoated wedge plates supplies an attenuated beam that is focused with a long focal length lens yielding an external beam waist. A digital CCD camera with a resolution of 10 bits was used to record the beam's intensity distribution in the caustic. The beam's diameter was determined with the help of the laser beam analyzer software "LBA-PC" from Spiricon.

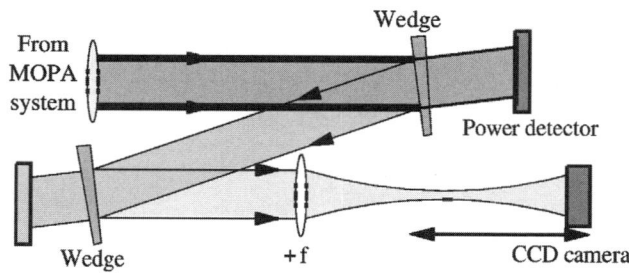

Figure 4.8. Setup for beam quality measurements.

The variation of the beam diameter d of a Gaussian beam is given by the equation:

$$d^2(z) = d_0^2 + (z - z_0)^2 (2\theta)^2 \qquad (2)$$

The beam propagation is approximated using a parabolic fit for the square of the determined beam diameter in dependence of the location z:

$$d^2 = A + B \cdot z + C \cdot z^2 \qquad (3)$$

From the fit parameters A, B, and C the far-field divergence θ (half-angle) and the beam waist radius w_0 can be calculated:

$$\theta = \frac{1}{2} \cdot \sqrt{C} \qquad (4)$$

$$w_0 = \frac{1}{2} \cdot \sqrt{A - \frac{B^2}{4C}} \qquad (5)$$

This results in a beam propagation factor K:

$$K = \frac{\lambda}{\pi \cdot \theta \cdot w_0} \qquad (6)$$

From the beam propagation factor the diffraction limit factor M^2 follows:

$$M^2 = \frac{1}{K} \qquad (7)$$

To demonstrate the ability of phase conjugating mirrors to compensate phase distortions, the beam quality was measured at different locations inside the laser system and for different operating parameters.

4.6 CHARACTERIZATION OF FIBER PHASE CONJUGATE MIRROR

For the design of SBS mirrors the properties of the used laser have to be considered. In many cases the mirror consists still of a cell filled with liquids or gases and a focusing lens to increase the intensity. Focal length and scattering material have to be chosen suitable to achieve a high SBS reflectivity and a good reproduction of the wavefront. Side effects in the material such as absorption, optical breakdown, or other scattering processes have to be avoided. In more recent works, solid-state materials like quartz are investigated as SBS media, too [32, 33].

To reduce the power threshold of SBS, a waveguide geometry can be used [34, 35]. The beam intensity inside the waveguide is high within a long interaction length, resulting in low power thresholds. To avoid toxic liquids and gases under high pressure, multimode quartz fibers can be used [36]. The lower Brillouin gain of quartz glass compared to suitable SBS gases and liquids can be overcome using fibers with lengths of several meters, resulting in SBS thresholds down to 200-W peak power [21].

Stimulated backscattering leads to efficient phase conjugation in multimode quartz fibers with core diameters between 25 μm and 200 μm. Such fiber phase conjugators with lengths of several meters show a reflectivity of more than 80%, a maximum fidelity over 90%, and a power threshold of less than 200 W at a wavelength of 1 μm. Depolarization of phase conjugated light can be reduced to less than 0.5% by proper adjustment of the fiber. High power densities of more than 500 MW/cm^2 do not damage the fiber phase conjugator. Furthermore, fiber phase conjugators are attractive due to their nontoxicity, in contrast to the often-used SBS liquids, and are easier to handle than gases under high pressure. Finally, they facilitate highly reliable all-solid-state laser systems.

Because of the high interaction length inside the fiber, the power threshold for SBS is reduced significantly. The power threshold can be calculated from Eq. (8), where A_{eff} is the effective mode field area inside the fiber core, L_{eff} is the effective interaction length, which depends on the coherence length of the incident beam, and g_B is the Brillouin gain coefficient; for quartz, g_B is in the range of 5 cm/GW [37].

$$P_{\text{th}} = \frac{21 A_{\text{eff}}}{L_{\text{eff}} g_B} \tag{8}$$

As Eq. (8) shows, the power threshold decreases with lower core diameters. Using a step index fiber with a core diameter of 25 μm, the power threshold was reduced down to 200 W [21]. However, the beam quality $1/M^2$, which can be coupled into the fiber, is limited by the core diameter D, Eq. (9). NA is the numerical aperture and λ is the wavelength.

$$M^2 \leq \frac{\pi D N A}{2\lambda} \tag{9}$$

Table 4.1 shows the calculated SBS threshold, the M^2 limit, and the power limit for fiber phase conjugators with different core diameters between 25 μm and 200 μm.

TABLE 4.1. SBS Threshold, M^2 Limit, and Power Limit for Different Fiber Core Diameters (Coherence Length 0.5 m)

Core diameter (μm)	SBS threshold (kW)	M^2 limit (NA = 0.22)	Power Limit (kW) ($I_{\text{max}} = 500$ MW/cm^2)
200	27	64	160
100	6.8	32	40
50	1.7	16	10
25	0.4	8	2.5

TABLE 4.2. Measured Reflectivity and SBS Threshold for Different Fiber Core Diameters

Core Diameter (μm)	Focal Length of Coupling Lens (mm)	Maximum SBS int. Reflectivity (%)	SBS Threshold (kW)
200	100	80	26
100	80	87	8.3
50	40	88	1.6
25	25	89	0.3

For the calculation of the M^2 limit, a standard numerical aperture of 0.22 was assumed. The upper power limit can be approximated regarding a damage threshold above 500 MW/cm^2 for nanosecond pulses.

An important feature of a fiber phase conjugator, in contrast to a Brillouin cell, is the SBS threshold behavior for different M^2 values of the incoming beam. In a fiber the SBS threshold is nearly independent of the incoming beam quality. This is caused by mode conversion inside the fiber, resulting in a homogeneous illumination and therefore in constant SBS reflectivity. In a Brillouin cell the reflectivity depends strongly on the far field of the incoming beam due to the focusing lens. Here, phase distortions result in strong amplitude fluctuations in the focal region. Thus the nonlinear response of SBS leads to a decrease of fidelity. A comparison between a diffraction-limited beam ($M^2 = 1.0$) and a highly distorted beam ($M^2 = 10$) showed an increase of the SBS threshold of 300% in the Brillouin cell. For the fiber phase conjugator, no remarkable changes of the power threshold were observed [15].

The reflectivity of the optical fibers is one of the most important properties for their usage as a PCM. The reflectivity of the fiber PCM has been experimentally optimized and compared to an analytical plane wave model. Figure 4.9 shows the system used to study the reflectivity of the fibers. It consists of a master oscillator, two amplifiers, an optical isolator, and a lens to couple the laser beam into the fiber. The Nd:YAG oscillator is flashlamp-pumped and Q-switched using a KDP–Pockels cell and a double polarizer. The output mirror is plane with 70% reflectivity, and the rear mirror has a 3-m radius of curvature and more than 99% reflectivity. The length of the resonator is 85 cm. An aperture of 1.5 mm is inserted into the cavity to achieve operation in the fundamental transversal TEM$_{00}$ mode. The spectral width of the output beam is reduced using two etalons consisting of a glass plate (20-mm thickness) with 70% reflectivity on both sides and one (2-mm thickness) with 50% reflectivity. The maximum output energy of the oscillator was 0.4 mJ in a nearly diffraction-limited beam with reduced spectral width and a duration of 30-ns FWHM. Stable single-longitudinal mode output was difficult to obtain because the two etalons were not temperature-controlled. The coherence length of the laser radiation achieved with two etalons in the resonator was determined (with a Fabry–Perot interferometer) to be 50 cm.

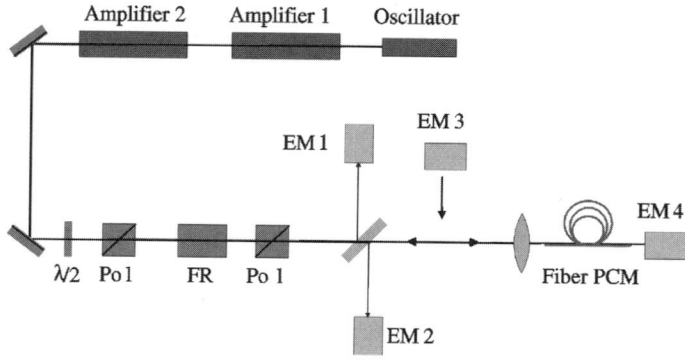

Figure 4.9. Experimental setup for fiber characterization (Pol, polarizer; FR, Faraday rotator; EM, energy meter).

In this system, two flashlamp-pumped amplifiers with Nd:YAG crystals were used. The optical isolator is composed of two Glan polarizers and a Faraday rotator. The second polarizer has been introduced to keep the depolarized Stokes beam from entering the oscillator. A half-wave plate is used to attenuate the energy from the oscillator. The reflectivity of the fiber is measured by placing a mirror into the beam to obtain a reference beam for the pumping energy and the backreflected Stokes energy. The energies of the beams are measured using pyroelectric detectors.

To measure the pumping and the backreflected energies with respect to the reference value, we have calibrated the system consisting of the energy meters EM1, EM3, and EM4 and the beam splitting mirror and deduced the calibration factors for the pumping and Stokes energies. In order to determine the SBS reflectivity, we took into account corrections due to the fiber coupling losses and the Fresnel loss at the end of the fiber (Fig. 4.10).

The relation between the pumping energy E_{pump}, the energy coupled into the fiber E_{in} and the transmitted energy E_{trans} is

$$E_{\text{in}} = C_{\text{eff}} E_{\text{pump}} = \frac{E_{\text{trans}}}{T} \tag{10}$$

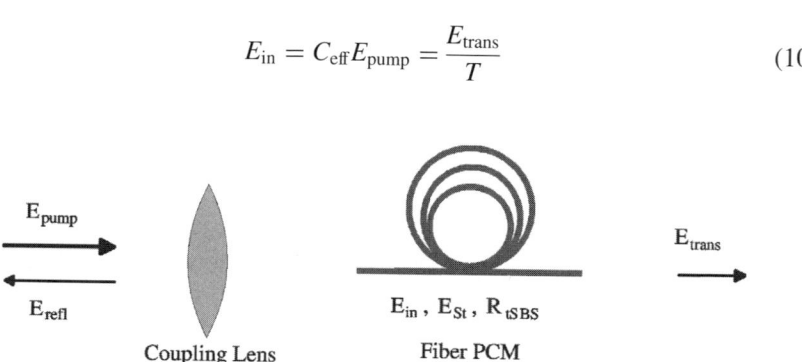

Figure 4.10. Schematic representation of the quantities measured in the experiment.

where C_{eff} is the coupling efficiency and $T \approx 0.96$ is the transmission of the fiber endface. The value of the coupling efficiency is related to the focal length of the lens used to focus the pumping energy into the fiber and the Fresnel losses at the end of the fiber.

We determined the coupling efficiency from Eq. (10) by measuring the transmitted energy E_{trans} through the fiber and the pumping energy E_{pump}. We maximized the transmitted energy by aligning the longitudinal and transversal position of the fiber endface and also its tilt angles with respect to the incoming beam. The measurements were performed at low energies in comparison with the Brillouin threshold energy. The SBS reflectivity in the fiber (the internal reflectivity) is the ratio between the Stokes energy E_{St} and the energy coupled into the fiber E_{in}. Taking into account the Fresnel losses, the formula for the reflectivity is

$$R_{SBS} = \frac{E_{St}}{E_{in}} = \frac{E_{refl}}{TC_{eff}E_{pump}} \qquad (11)$$

The external reflectivity, which is important for practical applications of fibers as phase conjugating mirrors, is simply given by the ratio E_{refl}/E_{pump} and is smaller than the internal reflectivity defined by Eq. (10).

A comparison of the dependence of the experimental fiber reflectivities on the pulse peak power is shown in Fig. 4.11.

To deduce the theoretical SBS reflectivity, the analytical plane wave model of the SBS process from Ref. 38 was modified considering Gaussian beam propagation in

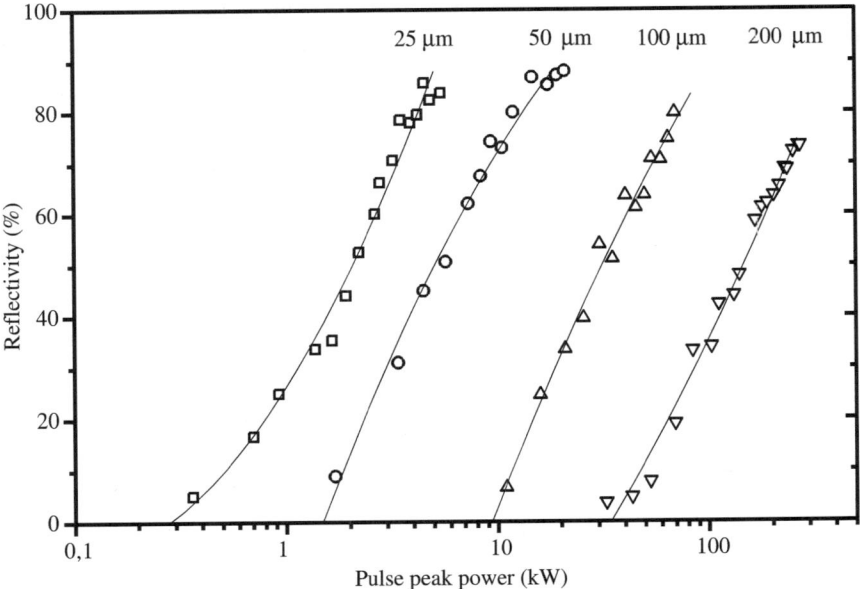

Figure 4.11. Reflectivity of the fibers investigated versus pulse peak power.

the fiber and a finite pulse duration. Assuming that the pump and the Stokes fields are uniform in the transversal cross section and that laser pulse compression is negligible, the SBS reflectivity is [31]

$$R_{\text{SBS}}(t) = \frac{I_S(t)}{I_{L0}(t)} = \frac{\exp[g_B \cdot I_{L0}(t) \cdot l(z_c) - \alpha z_c - G]}{1 + \exp[g_B^e I_{L0}(t) \cdot l(z_c) - G]} \tag{12}$$

where g_B is the Brillouin gain, $I_{L0} = E_{\text{pump}}/(\pi(d/2)^2 t_L)$ the pumping intensity at the fiber entrance, d the core diameter of the fiber, and t_L the laser pulse duration. $L(z_c)$ is the maximum interaction length. We have taken the length $l(z_c)$ for the SBS interaction in the fiber to be equal to the coherence length of the pump laser. α is the linear optical loss in the nonlinear medium. In the case of the fibers used, the linear optical loss is very small (the attenuation constant at 1.06 μm is about $\alpha = 2$ dB/km), therefore the factor αz_c is neglected. G is a parameter related to the threshold reflectivity. The value of $G = 25$ resulted from the expression [Eq. (12)] considering 2% SBS reflectivity at threshold.

The dependence of the internal SBS reflectivity for the 50-μm fiber on pumping energy is illustrated in Fig. 4.12. The behavior of the other fibers with core diameters of 25, 100, and 200 μm is similar.

The steep curve represents the reflectivity calculated from Eq. (12), while the boxes represent the experimental SBS reflectivity. The gentler curve results from the introduction of the intensity distribution in Eq. (12) and numerical integration over time and the beam cross section [39].

Equation (12) gives a very steep rise of the SBS reflectivity in comparison with the experimental results obtained for all of the multimode fibers investigated. Taking into account the temporal and spatial distribution of the input intensity results in the slower rise of the reflectivity depicted in Fig. 4.12.

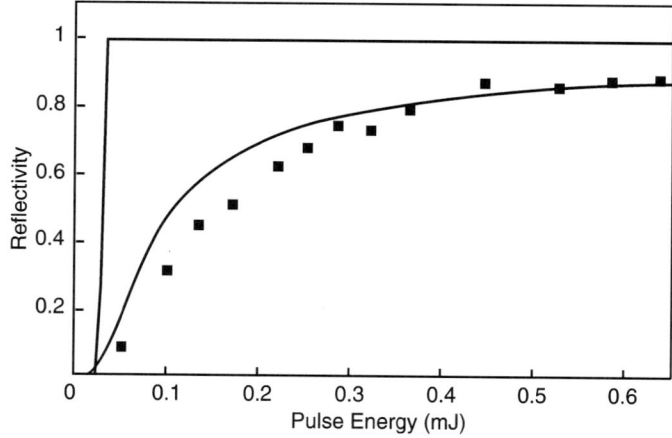

Figure 4.12. SBS reflectivity against pumping energy for the 50-μm fiber.

4.7 FLASHLAMP-PUMPED Nd:YALO MOPA SYSTEMS WITH FIBER PHASE CONJUGATOR

Taking into account the considerations made in Section 4.4, a flashlamp-pumped Nd:YALO MOPA system was built up in three stages. Starting from a single amplifier, the system was enhanced to a setup with two parallel chains consisting of four amplifiers overall.

4.7.1 Passively Q-switched flashlamp-pumped ring resonator

All conditions claimed in Section 4.4 for a suitable master oscillator can be satisfied using a flat mirror ring resonator as shown in Fig. 4.13. The system is flashlamp-pumped at a repetition rate of 100 Hz with a pumping energy of 17 J per shot. The pumping chamber contains a europium-doped glass filter to achieve high efficiency and to protect the YALO crystal (4 mm × 79 mm) against color center formation. The unidirectional operation is realized using an external feedback mirror.

The energy stability from shot to shot is about 1%. An intracavity tilted etalon with a thickness of 2 mm and a reflectivity of 73% reduces the spectral bandwidth, resulting in a coherence length of approximately 30 cm. The average output power is 4.7 W in a diffraction-limited beam.

The master oscillator is not optimized with respect to maximum average output power. However, the resonator geometry is chosen for high stability. Regarding possible applications, the temporal pulse structure is important. To run the PCM at high reflectivity, the pulse peak power has to exceed the power threshold several times. On the other hand, the upper power limit is given by the damage threshold of the optical fiber.

For variable pulse peak powers, chromium-doped YAG crystals are used as saturable absorbers to obtain passive Q-switch operation. Due to a saturation intensity of about 9 kW/cm^2, the oscillator emits usually a burst of pulses per flash. With an absorber length of 2 mm and a doping concentration of 0.15 at %, Cr^{4+}

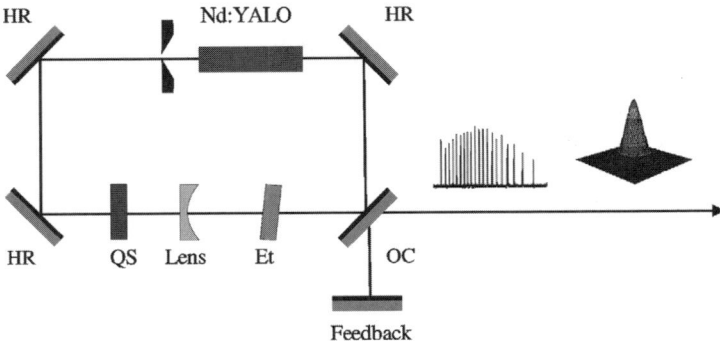

Figure 4.13. Nd:YALO master ring resonator with passive Q switch (HR, high reflective mirror; QS, passive Q-switch; Et, etalon; OC, outcoupling mirror).

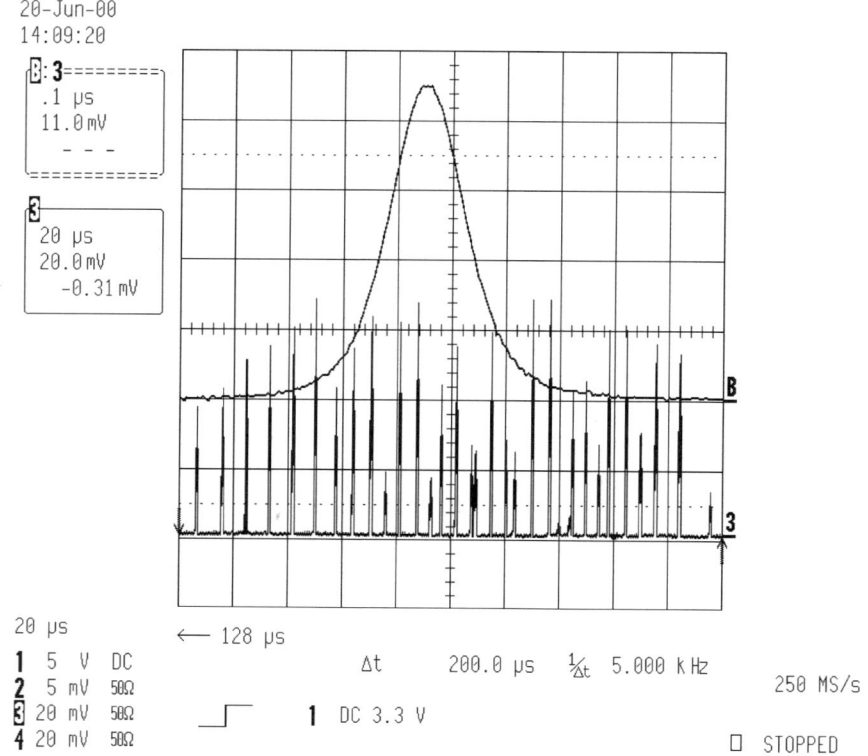

Figure 4.14. Pulse structure of the master oscillator: one burst and a typical Q-switch pulse.

about 28 pulses are generated per flashlamp shot (Fig. 4.14). The single pulses show half-widths between 150 ns and 170 ns (FWHM). Using different absorber lengths, the number of pulses and therefore the peak power can be adjusted. For an average output power of 4.7 W, the burst energy is 47 mJ at 100-Hz repetition rate. Assuming 28 pulses in one burst, the single pulse energy is 1.7 mJ and the peak power is about 10.6 kW.

The unidirectional operation of the ring resonator is performed with help of an external feedback mirror instead of an intracavity optical diode to reduce the complexity and system costs. As a side effect, a few percent of the total intracavity power propagates along the wrong direction (anticlockwise, Fig. 4.13). Therefore, residual spatial hole burning occurs in the active medium, which leads to a small coupling between different longitudinal oscillator modes. This is indicated by a modulation of Q-switched pulses, which can be observed for about 10% of the pulses.

Because of Q-switching and residual mode locking, the single pulse energy and half-width are subject to fluctuations. However, the overall burst energy is constant within a standard deviation of 1% from flash to flash.

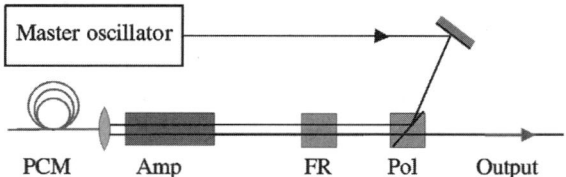

Figure 4.15. Single-amplifier MOPA system with fiber phase conjugator.

Figure 4.15 shows the schematic setup of a single-amplifier MOPA system, equipped with a quartz fiber phase conjugator. The oscillator beam is coupled into the amplifier by a polarizer. The dimensions of the amplifier crystal are 9.5 mm × 152 mm. It is pumped by two flashlamps in a specular cavity with a total average pump power of 7 kW at maximum. Because Nd:YALO is an anisotropic crystal, thermal lensing depends on the direction perpendicular to the propagation axis. Therefore, the amplified beam becomes astigmatic and beam coupling into the fiber core requires extensive adjustment.

The phase conjugator is a low-absorption (attenuation: 2 dB/km) quartz–quartz step-index multimode fiber with a core diameter of 200 μm, a numerical aperture of 0.22, and a length of 3.5 m. After phase conjugation inside the fiber and the second amplification, pass phase distortions are compensated and the beam with high quality can be coupled out by the polarizer. Due to the anisotropic gain of Nd:YALO, the beam that passes the amplifier must be linearly polarized in a specific direction to obtain the maximum gain. Thus a 45° Faraday rotator was used as a part of the optical isolation to maintain the beam linearly polarized in a fixed direction for the whole double pass. Due to the fiber coupling with a single lens, the amplifier has to be driven at constant thermal lensing to preserve a high coupling efficiency. However, the average output power can be tuned by variation of the oscillator power using a half-wave plate.

With such a single amplifier setup, an average output power up to 84 W can be extracted at an average pump power of 7 kW for the amplifier (Fig. 4.17). The average output power was limited by the increasing astigmatism which reduces the coupling efficiency in case of higher average pumping power.

To increase the average output power we substituted the amplifier head in the MOPA setup of Fig. 4.15 by a serial arrangement of two amplifier rods (Fig. 4.16).

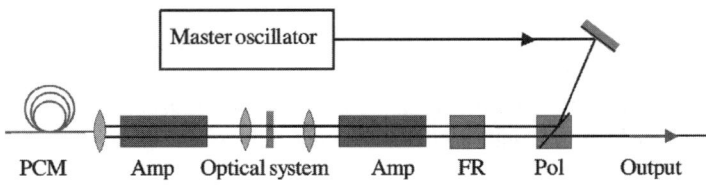

Figure 4.16. Setup for a cascaded amplifier setup.

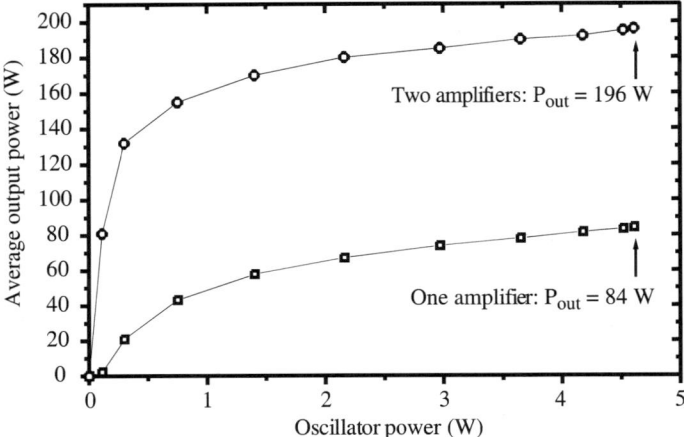

Figure 4.17. Average output power versus oscillator power for one amplifier and for two cascaded amplifiers.

To allow a variation of average pumping power, an optical system consisting of two lenses was used between the amplifier rods (see Section 4.6). Anisotropic thermal lensing of the first amplifier can be compensated by rotating the second amplifier rod by 90°. Due to the anisotropic gain of Nd:YALO, the polarization direction also has to be rotated by 90° using a half-wave plate between the amplifier heads. Therefore, the beam shows no astigmatism after the first pass and can be coupled into the fiber with high efficiency using a single lens. With this double amplifier setup, an average output power up to 196 W was obtained at an average pump power of 7 kW for each amplifier (Fig. 4.17). The average output power was limited due to parasitic oscillations that occur above 6 kW of average pumping power. In future the threshold for parasitic oscillations will be increased applying an antireflection coating on the fiber tip.

4.7.2 315-Watt average output power MOPA system

To increase the average output power, parallel amplifier arrangements are preferable (see Section 4.3). In case of serial arrangements, the output power is limited due to several reasons: On one hand, a single amplifier (the last one) suffers from the total output power. Consequently, large apertures are required to prevent damage of optical components. On the other hand, the beam quality decreases after the first pass with an increasing number of amplification stages. Therefore the coupling efficiency drops (in case of a fiber phase conjugator) or the SBS threshold strongly increases (when using an SBS cell with focusing geometry).

With regard to the advantages of a parallel amplifier setup, a solid-state laser system containing four amplifiers and a master oscillator was developed (Fig. 4.18). The system consists of a parallel arrangement of two double pass amplification

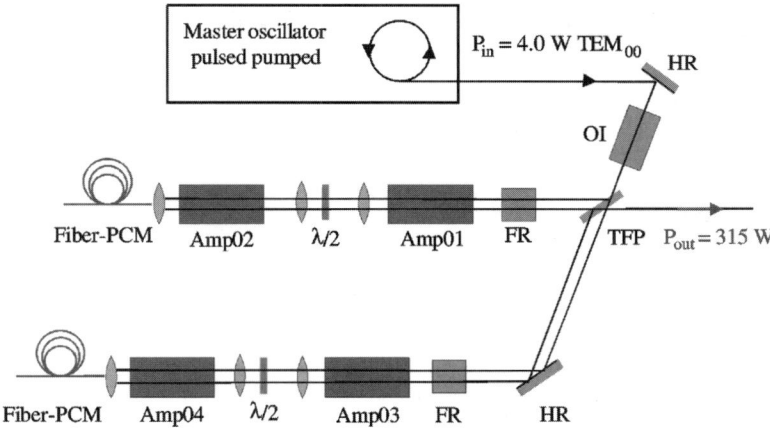

Figure 4.18. Setup of a six-amplifier MOPA system with phase conjugating mirrors.

stages with phase conjugating mirrors. After double pass amplification the two beams are combined again using a thin film polarizer. Optical systems between the amplifiers guarantee the variability of average pumping power without damaging optical components.

Four commercially available pumping chambers from an industrial laser system are used as amplifiers in two parallel chains. The Nd:YAP rods with a diameter of 8 mm and a length of 160 mm are pumped by two flashlamps in specular cavities, with an average pumping power of 6 kW each. To achieve a high extraction efficiency, the amplifiers are driven in a double-pass arrangement (see Fig. 4.3). The oscillator beam passes an optical isolation, which prevents perfectly polarized backtraveling parts of the beam from entering the oscillator. Due to the additional components in the beam path, the power input in front of the amplifiers is reduced to 4.0 W.

Then the beam is splitted at the TFP in two perpendicular polarized parts onto the two chains. By passing a Faraday rotator, the polarization plane of each beam is rotated by an angle of 45° and is then amplified by the two serial amplifier arrangements. To compensate for astigmatic thermal lenses, the amplifier rods are rotated by 90° to each other and the polarization is matched again by a half-wave plate. After the single pass, the beam is coupled into the fiber phase conjugator. The beam is phase conjugated by a multimode silica step index fiber with a core diameter of 200 μm, a numerical aperture of 0.22, and a length of 2 m. After an additional amplifier pass, the initial beam quality is almost reproduced and the beam is rotated again by an angle of 45° after passing the Faraday rotator a second time. After the two passes through the Faraday rotator, the polarization plane of each beam is rotated by 90° and therefore could be extracted at the TFP. Here the amplified beams of both chains are superposed automatically due to the optical properties of the phase conjugated signal.

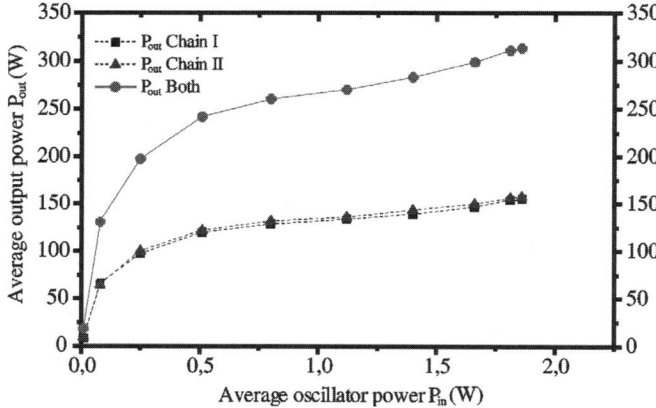

Figure 4.19. Measured average output power as a function of the oscillator power.

4.7.3 Experimental results Applying a parallel amplifier arrangement, an average output power up to 315 W has been achieved at 2-kHz average repetition rate. Figure 4.19 shows the measured output power as a function of the oscillator power.

The beam quality measurements are performed, as described in Section 4.2. The beam propagation factor M^2 was determined to be smaller than 2.6 for both directions in space (see Fig. 4.20).

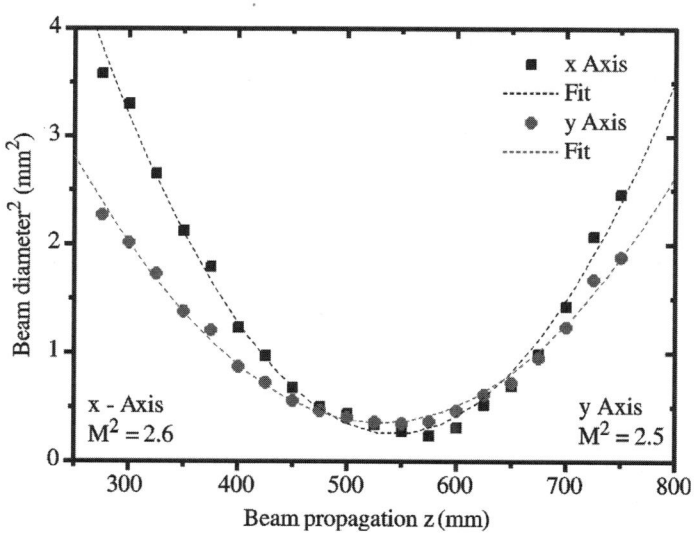

Figure 4.20. Determination of the beam propagation factor M^2.

4.8 ACTIVELY Q-SWITCHED FLASHLAMP-PUMPED Nd:YAG MOPA SYSTEMS WITH FIBER PHASE CONJUGATOR

Activily Q-switched twisted mode linear resonator In laser media that preserve a circular state of polarization, like the cubic crystal Nd:YAG, SLM operation could be fulfilled in linear resonators, too. Spatial hole burning in linear resonators leads to low coherence length because of several longitudinal modes occurring in the laser process. This problem can be solved elegantly by using a "twisted mode" design. In such a setup the active medium is positioned between two quarter-wave plates, and an additional polarizer is placed in the resonator (Fig. 4.21). This causes a circular state of polarization and thus an uniform intensity distribution inside the active medium. So spatial hole burning doesn't occur, and the resonator works in single longitudinal mode operation [40]. All other requirements stipulated in Section 4.4 can be satisfied in such a linear setup, too.

The system is flashlamp-pumped at a repetition rate of 100 Hz with a pump energy of 25 mJ per shot. The pumping chamber contains a europium-doped glass reflector to achieve high efficiency and to protect the YAG crystal (5 mm × 140 mm) against color center formation. The average output power is 6 W in a diffraction-limited beam.

The master oscillator is optimized with respect to high temporal stability [41]. For variable pulse peak powers, an acousto-optical modulator (AOM) is used; thus it is possible to generate pulses from one single pulse to 35 pulses during one pump pulse (Fig. 4.22). So the average repetition rate could be varied from 100 Hz to 4 kHz. The single pulses show half-widths between 27 ns (single pulse) and 120 ns (FWHM). Varying the number of pulses and thereby the pulse energy and pulse peak power make it possible to adjust the proper conditions for phase conjugation.

The output from this master oscillator is coupled into a cascaded amplifier setup based on two amplifiers designed like the system described in Section 4.6. Here the amplifiers consist of two Nd:YAG rods (9.5 mm × 152 mm) in the same cavities as in the system above. As mentioned above, it is necessary to compensate for the birefringence occurring in Nd:YAG during strong optical pumping—for example, by a telescope together with a 90° rotator (see Section 4.3). With such a setup, it was possible to keep the part of depolarized light below 4% over the whole pumping range.

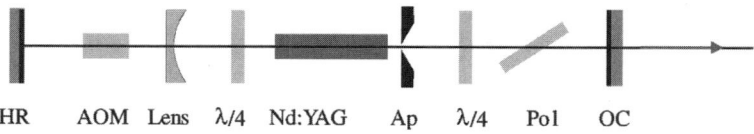

Figure 4.21. Linear Nd:YAG resonator in "twisted mode" operation (HR, high reflective mirror; AOM, acousto-optic modulator; λ/4, quarter-wave plate; AP, aperture; Pol, polarizer; OC, outcoupling mirror).

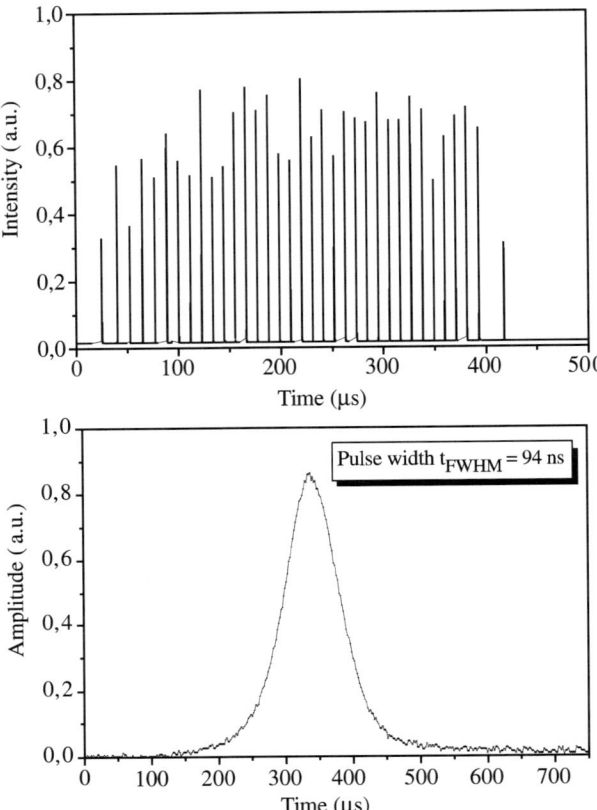

Figure 4.22. Pulses during one burst of the master oscillator (*left*) and one single pulse (*right*).

As a result of the better crystal qualities of YAG in comparison to YALO, it is possible to operate such a MOPA system in single pass with acceptable beam quality. With the system depicted in Fig. 4.23 (without fiber PCM), we achieved an average output power of 84 W (35 burst pulses lead to an average repetition rate of

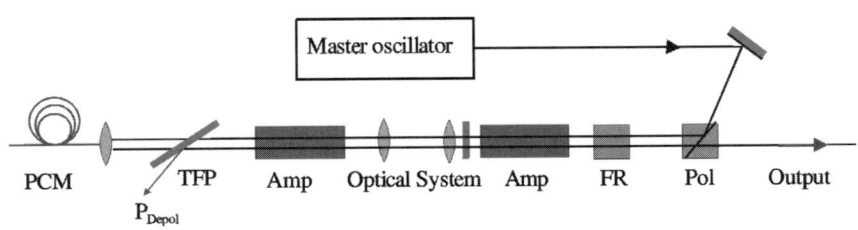

Figure 4.23. Setup for a cascaded amplifier.

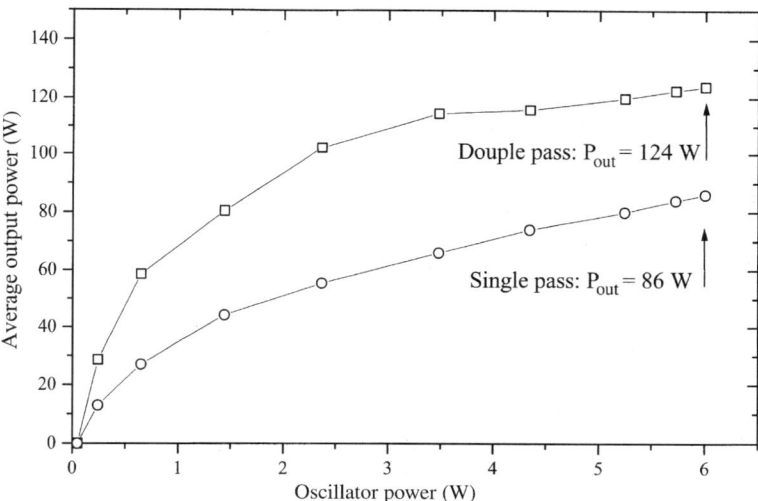

Figure 4.24. Average output power versus oscillator power for one amplifier and for two cascaded amplifiers.

3.5 kHz) with the great advantage that the flexibility given by the active Q switch can be used fully regardless of the phase conjugating conditions in the optical fiber. This is very important for material processing or high-power frequency conversion. The beam quality was $M^2 < 2.4$ in both spatial directions.

However, it is possible to increase both average output power and beam quality by applying a fiber PCM in the single-pass system (Fig. 4.23). The output power as a function of the oscillator power is shown in Fig. 4.24.

The beam quality measurements are performed, as described in Section 4.5. The beam propagation factor M^2 was determined to be smaller than 2.2 for both directions in space.

4.9 CONTINUOUSLY PUMPED Nd:YAG MOPA SYSTEMS WITH FIBER PHASE CONJUGATOR

4.9.1 Diode pumped Nd:YAG master oscillator
The developed master oscillator is depicted in Fig. 4.25. It consists of a diode pumping chamber, an acoustooptic modulator (AOM), an etalon, a thin-film polarizer (TFP), an output coupler, and high reflectivity mirrors. The ring resonator is applied to avoid spatial hole burning. The external feedback mirror facilitates clockwise unidirectional operation. As active medium a Nd:YAG rod is used with a length of 56 mm and a diameter of 4 mm.

An additional intracavity etalon with a thickness of 2 mm and a reflectivity of 50% reduces the spectral bandwidth to increase the possible interaction length inside

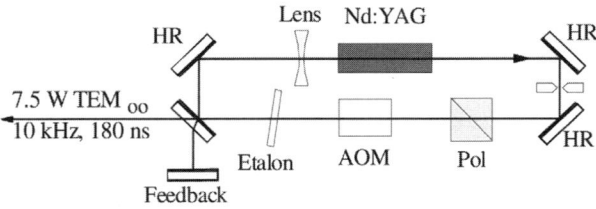

Figure 4.25. Scheme (*left*) and picture (*right*) of the developed CW diode pumped Nd:YAG master oscillator.

the fiber. Figure 4.26 shows the spectral bandwidth without and with the intracavity etalon of the ring resonator.

A lens with a focal length of $f = -800$ mm increases the fundamental mode diameter. Q-switching operation is obtained using an AOM with a repetition rate of 10 kHz and an RF driver power of 100 W. The oscillator provides an average output power of 7.5 W with a nearly diffraction-limited beam quality (see Fig. 4.27). Pulse widths of 180 ns lead to pulse peak power of 4 kW (see Fig. 4.28).

4.9.2 Nd:YAG amplifier arrangement

Two commercially available pumping chambers from an industrial laser system are used as amplifiers. The identical Nd:YAG rods with a diameter of 4 mm and a length of 56 mm are diode-pumped. Assuming a maximum average oscillator power of 7.5 W and a repetition rate of 10 kHz, the pulse energy cannot exceed 1 mJ. Taking a saturation energy density of 500 mJ/cm^2 for Nd:YAG into account, the amplifiers would operate far below their saturation fluence, resulting in a small extraction efficiency. The extraction efficiency can be improved by increasing the number of amplifier passes. Therefore a four-pass amplifier arrangement was applied (see Fig. 4.29).

The oscillator beam passes an optical isolation and is amplified in a serial amplifier arrangement. To compensate for thermally induced stress birefringence in the active media, the well-known relay-imaging setup containing two lenses and a

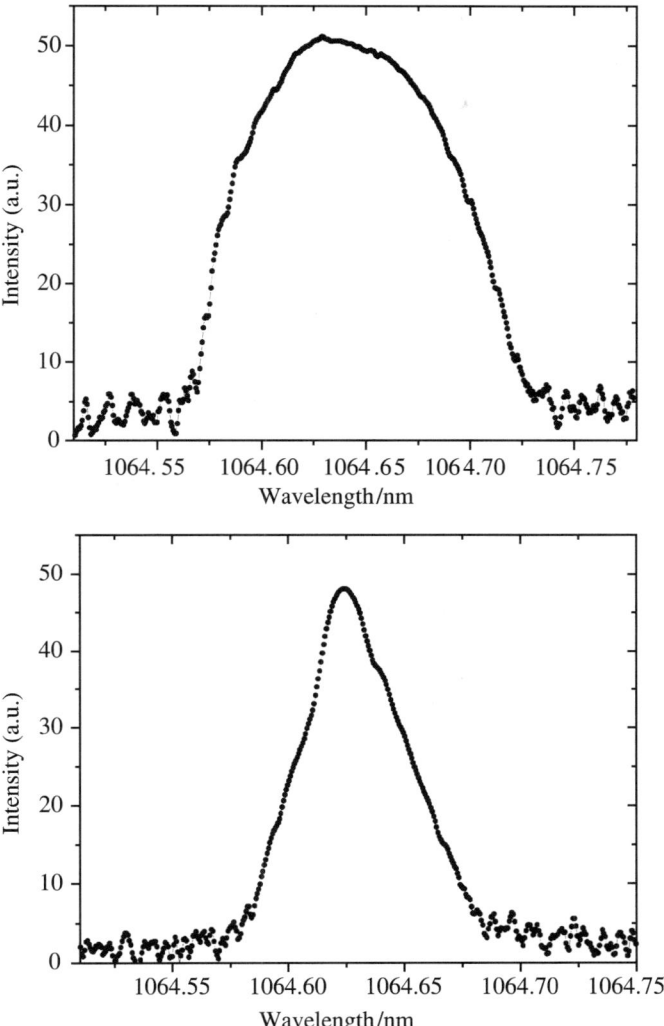

Figure 4.26. Spectral bandwidth without intracavity etalon (*left*) and reduced spectral bandwidth with intracavity etalon (*right*) of the ring resonator.

90° quartz rotator is used [30]. Birefringence compensation results in depolarization losses below 1.3% after passing both amplifiers in a single pass. After the first amplification pass, the beam is collimated and reflected using a conventional curved HR mirror.

The polarization direction is rotated by 90°, passing the quarter-wave plate twice. During the second pass, the beam is amplified again and coupled into the fiber phase conjugator. After the double pass are the depolarization losses below 2.3% (see Fig. 4.30). The beam is phase conjugated in a multimode silica step index fiber with

134 MULTI-KILOHERTZ PULSED LASER SYSTEMS WITH HIGH BEAM QUALITY

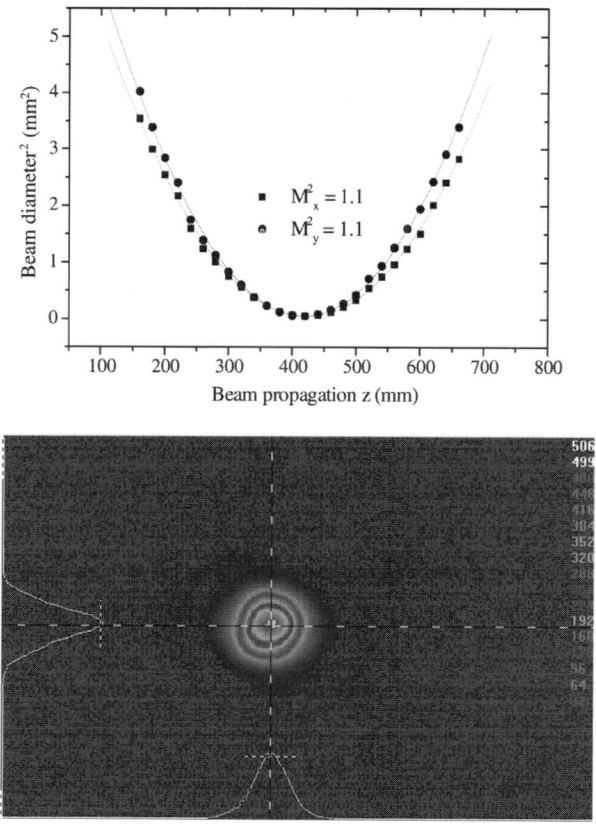

Figure 4.27. Determination of the times diffraction limit factor M^2 and measured beam profile of the ring resonator.

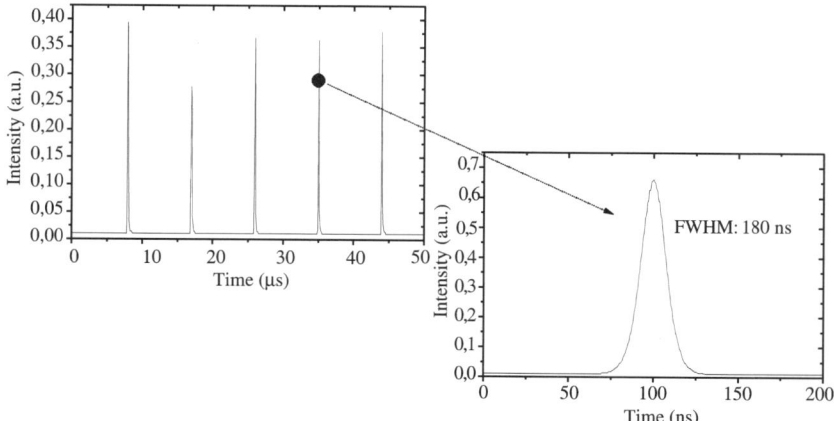

Figure 4.28. Pulse structure of the master oscillator.

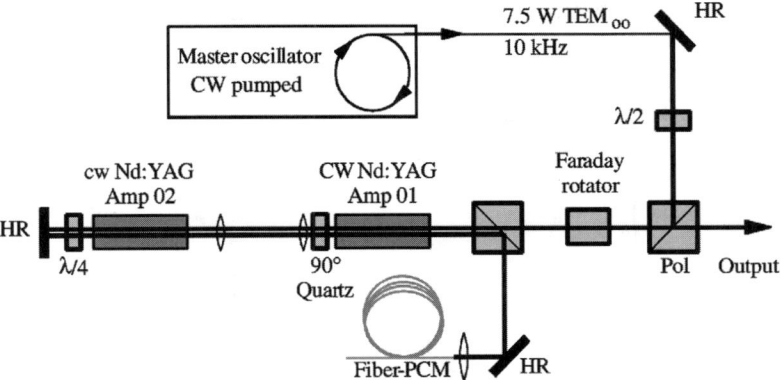

Figure 4.29. Continuously diode-pumped MOPA system with four-pass amplifier arrangement and fiber phase conjugator.

a core diameter of 50 μm, a numerical aperture of 0.22, and a length of 10 m. After two additional amplifier passes, the initial beam quality is reproduced (see Fig. 4.31) and the beam is extracted with help of the optical isolation. Applying the four-pass scheme to the serial amplifier arrangement, an average output power up to 31 W has been achieved at a 10-kHz repetition rate. Figure 4.30 shows the measured output power as a function of the oscillator power.

Currently, output power scaling about 100 W is under development. In addition, two further pumping chambers are to be used. This new setup is shown in Fig. 4.32. After the second amplification pass, the beam is reflected using a conventional HR

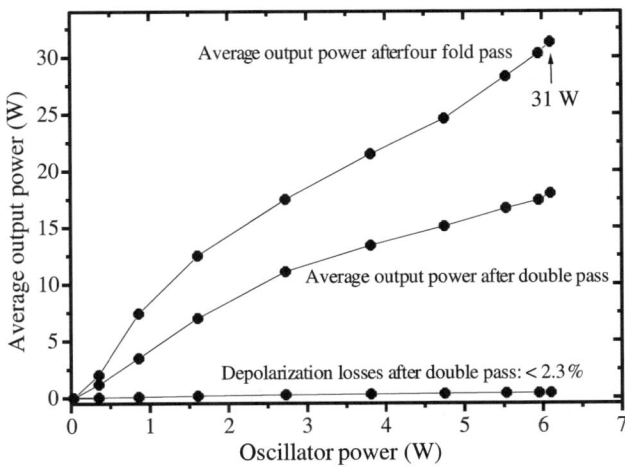

Figure 4.30. Measured average output power versus oscillator power of the four-pass MOPA system.

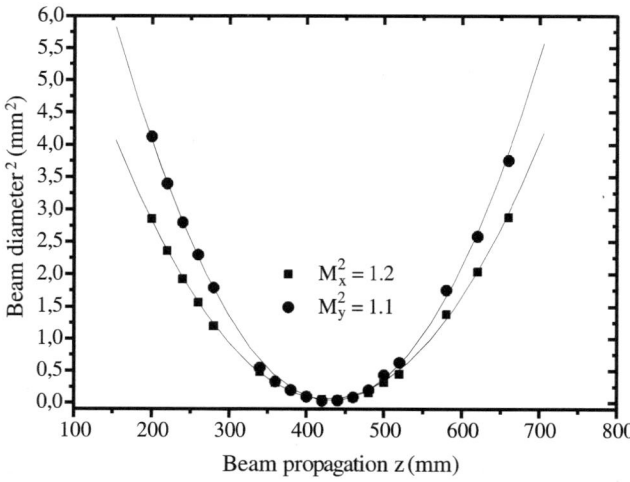

Figure 4.31. Determination of the times diffraction limit factor M^2.

mirror and is then coupled in a further serial amplifier arrangement with birefringence compensation. After this amplification pass, the beam is coupled into a fiber phase conjugator. The beam is phase conjugated in a multimode silica step index fiber with a core diameter of 100 μm, a numerical aperture of 0.22, and a length of 15 m. After three additional amplifier passes, the beam is extracted with the help of the optical isolation. Applying this scheme an average output power up to 71 W has been achieved at a 10-kHz repetition rate with a nearly diffraction-limited beam quality (see Fig. 4.33).

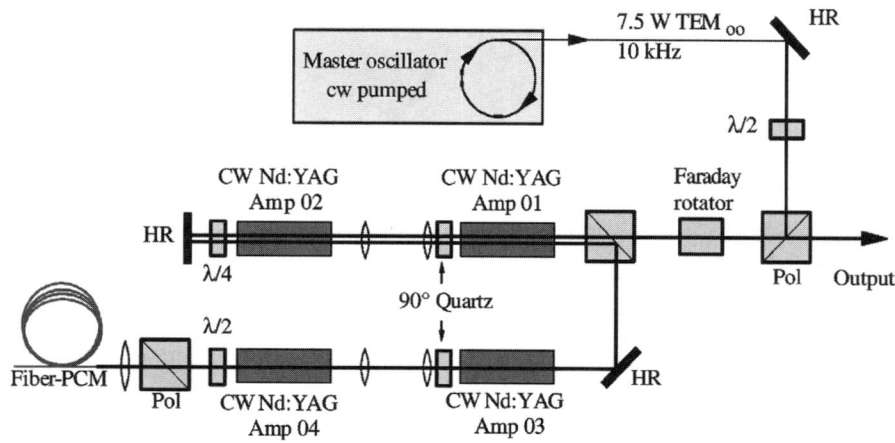

Figure 4.32. Continuously pumped MOPA system with four amplifiers in a serial amplifier arrangement.

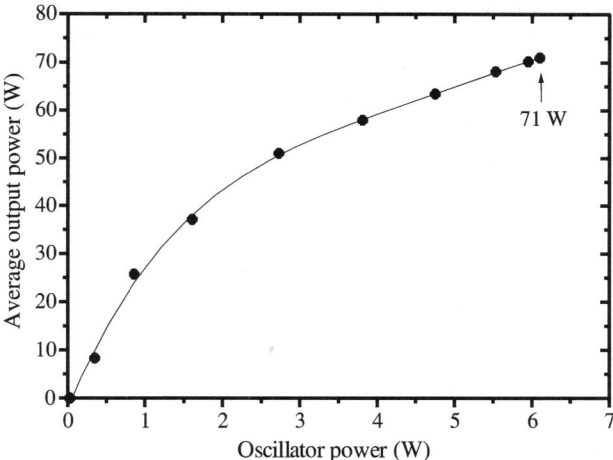

Figure 4.33. Measured average output power versus oscillator power of the continuously pumped MOPA system with four amplifiers in a serial amplifier arrangement.

4.10 500-WATT AVERAGE OUTPUT POWER MOPA SYSTEM WITH CS$_2$ AS SBS MEDIUM

A MOPA system with a serial arrangement of two amplifiers was realized with regard to the above-mentioned design rules (Fig. 4.34). The beam from the master oscillator is coupled into the amplifier chain by a polarizer. It passes through both amplifiers before it is focused in the phase conjugating mirror (SBS cell). The SBS cell is a glass tube filled with carbon disulfide (CS$_2$). This liquid shows an SBS threshold in the range of 20 kW. The beam is phase conjugated and passes back. Then the amplified signal is coupled out by the combination of a 45° Faraday rotator

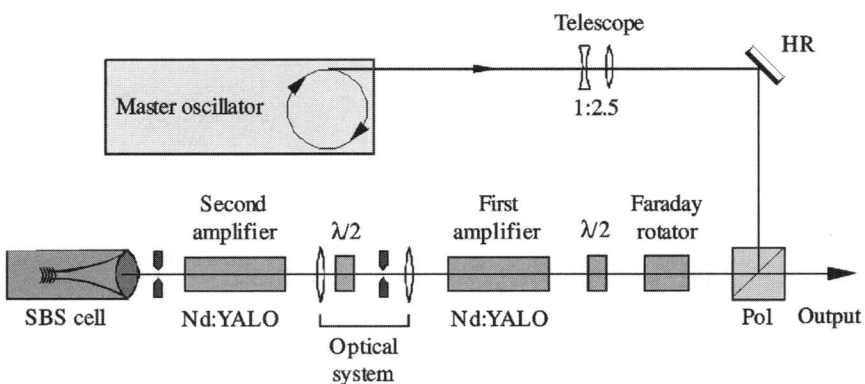

Figure 4.34. Setup of the two-amplifier system.

and the polarizer. Due to the anisotropic gain of YALO, the beam must be linearly polarized in a specific direction to obtain the maximum gain. Thus a Faraday rotator is used as a part of the optical isolation to preserve linear polarization for the whole double pass. Due to the absence of depolarization in the amplifier rods, no second optical isolation for the master oscillator is necessary.

The system is flashlamp-pumped with a repetition rate of 100 Hz and an excitation time of about 140 µs. The master oscillator is based on the system described in Section 4.4. However, the saturable absorber has a length of 3 mm; thus 13 pulses with a half-width of 90 ns are generated per flash within one burst. The amplifiers contain Nd:YALO rods with a dimension of 9.5 mm × 152 mm. The total average pump power for both amplifiers is 18 kW. Diffuse reflecting pump chambers were used in combination with cerium-doped flowtubes to protect the YALO crystals against color center formation.

The average output power is tunable from 1 W up to 211 W (Fig. 4.35). To increase the average output power, the master oscillator was driven in a TEM_{00}–TEM_{01} mixed mode. The beam diameter is constant within a maximum deviation of 10% for the whole tuning range. Figure 4.36 shows three beam profiles for different average output powers.

With regard to the advantages of a parallel amplifier setup, a solid-state laser system containing six amplifiers and a master oscillator was developed (Fig. 4.37). The system consists of a parallel arrangement of two double pass amplification stages with phase conjugating mirrors. To increase the average output power, both beams are amplified in additional single-pass amplifiers again and combined using a thin-film polarizer. Optical systems between the amplifiers guarantee the variability of average pumping power without damaging optical components.

All active media are flashlamp-pumped at a typical repetition rate of 100 Hz. The oscillator beam is divided in two parts, and each of them is coupled into a serial arrangement of two amplifier rods. Each rod is excited in a diffuse pump chamber by two flashlamps with an average pumping power of up to 7 kW. Additional cerium-

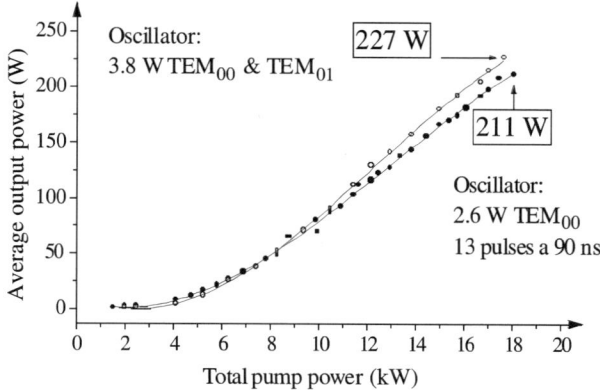

Figure 4.35. Average output power versus total pump power for both amplifiers.

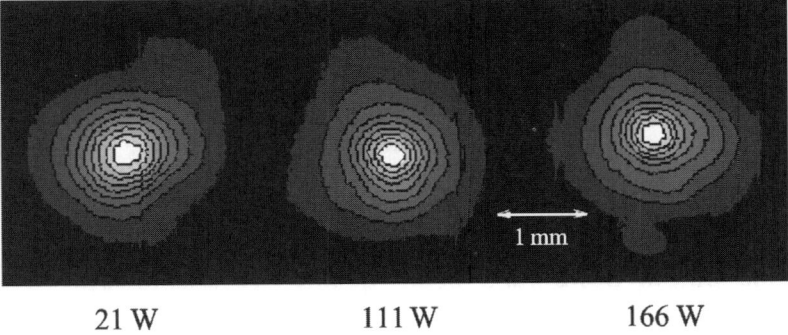

Figure 4.36. Measured beam profiles in a distance of 2.5 m from the polarizer for different average output powers.

or europium-doped flowtubes protect the YALO crystals against color center formation. The amplifier rod dimensions are (9 mm × 152 mm) or (8 mm × 160 mm). Using optical systems between the rods the average pump power can be varied from 1 kW up to 7 kW without changes in the extraction efficiency or risk to damage optical components.

After the first amplifier pass, the phase conjugating mirror (PCM) compensates the phase distortions of both amplifier rods. The SBS cell consists of a glass tube filled with carbon disulfide (CS_2). After the second amplification pass, the beam profile of the oscillator is reproduced at the polarizer (Pol). Due to the anisotropic gain of YALO, the optical isolation has to consist of a 45° Faraday rotator and the polarizer. Both beams are coupled out by the polarizer. To increase the average output power, the beams are amplified additionally in single-pass stages (amplifier 3 and 6). This results in an acceptable decrease of beam quality with regard to the obtainable average output

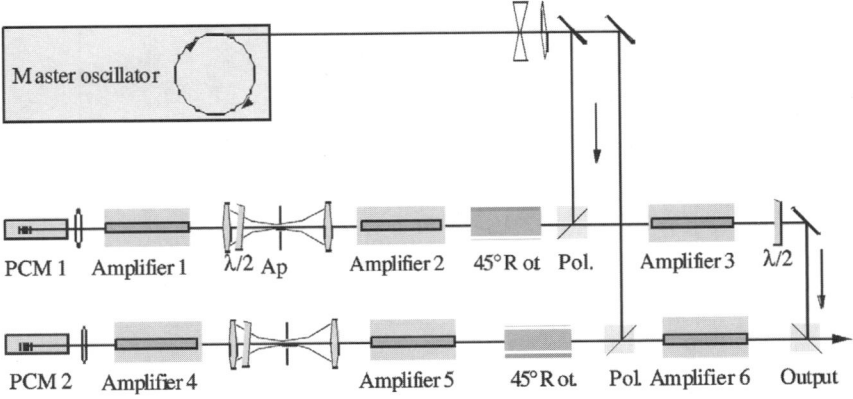

Figure 4.37. Setup of a six-amplifier MOPA system with phase conjugating mirrors.

power. Using a thin-film polarizer, we can combine both beams, resulting in one beam with statistical polarization direction.

A further improvement of beam quality could be achieved using additional SBS mirrors located behind the single pass of amplifier 3 and amplifier 6. Experiments with a system containing more than two amplifiers in a serial arrangement in conjunction with a single SBS cell have shown low SBS reflectivity due to the reduced beam quality after passing three or more amplifiers.

Average output power and beam quality To achieve high extraction efficiency in the double-pass amplification regime, the laser system is pulsed. A shot rate of 100 Hz and an excitation time of 200 μs lead to a high flashlamp lifetime. The required moderate peak power is obtained by burst mode operation of the master oscillator.

The average output power can be tuned over a wide range by variation of the amplifier pump power. Figure 4.38 shows the measured average output power versus total average pump power for all amplifier rods. The average output power was measured without attenuation using a water-cooled power detector.

For a total average output power up to 260 W, each double-pass chain supplies 25 W in front of the single-pass stages. In this case the average pumping power of the single-pass amplifiers (3 and 6) was varied. For average output powers above 260 W the single pass amplifiers were driven at constant average pump power with an average input power from 25 W up to 130 W supplied by the double-pass subsystems with SBS mirror (amplifier 1/2 and 3/4). Due to constant average pumping power for the single-pass amplifiers, the beam diameter and the beam divergence do not depend on the average output power above 260 W.

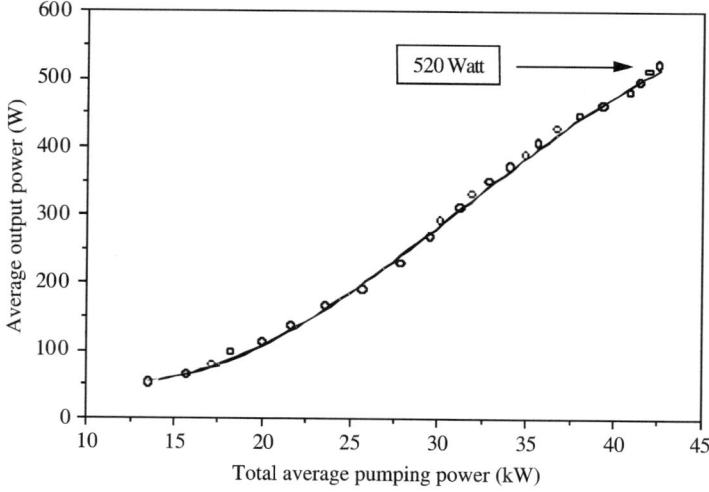

Figure 4.38. Average output power for different total average pumping power.

500-WATT AVERAGE OUTPUT POWER MOPA SYSTEM   141

Both double-pass systems with phase conjugating mirrors, consisting of amplifier 1/2 and amplifier 4/5, are each able to produce an average output power up to 210 W at an average pumping power of 18 kW. However, they are driven at an average pumping power of 13 kW to guarantee reliable long-time operation, to decrease the flashlamp consumption, and to reduce the master oscillator requirements.

The beam quality was measured as mentioned in Section 4.5, but with the moving edge method instead of using a CCD camera. A part of the beam is coupled out by a wedge plate and is focused with a lens to obtain a beam waist suitable for the measurement. The focal length was chosen to minimize the lens aberration and therefore the influence on beam quality. The measurements were carried out with the moving edge method, according to the beam quality ISO draft from 1996. A computer-controlled translation stage and a Gentec energy detector were used to determine the beam diameter at 12 positions on both sides of the beam waist. For the measurement of each beam diameter, 40 energy values were measured as a function of the edge position in x and y direction. Therefore the calculated beam quality is an

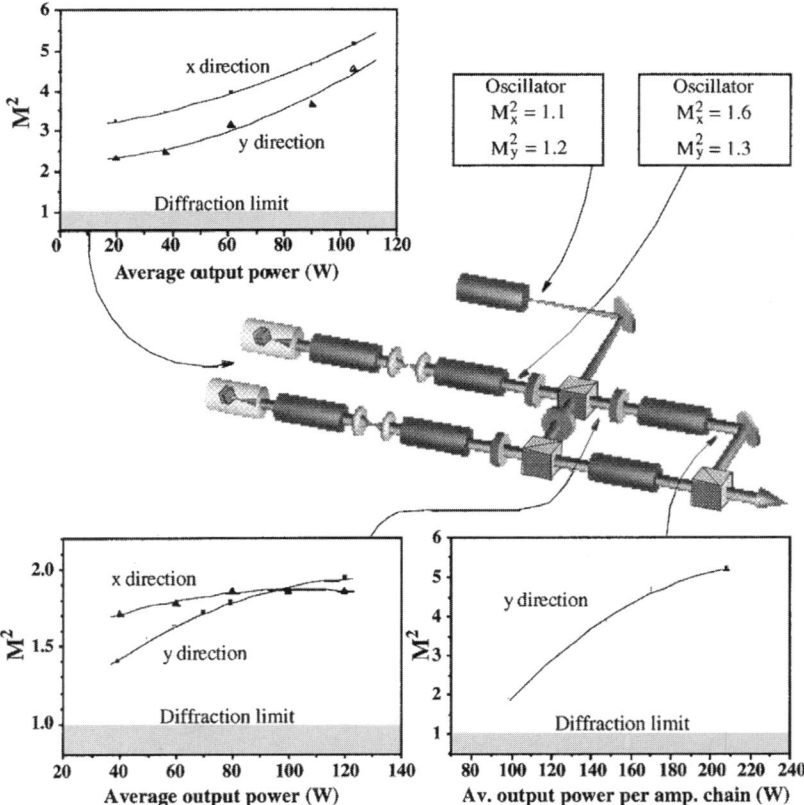

Figure 4.39. Measured beam qualities at different locations and average output powers.

averaged value and lower than the beam quality for one shot due to the finite pointing stability of the beam.

To demonstrate the ability of phase conjugating mirrors to compensate phase distortions, the beam quality was measured at different locations inside the laser system and for different operating parameters (Fig. 4.39).

The oscillator beam has a near-diffraction-limited beam quality ($M_x^2 = 1.1$ and $M_y^2 = 1.2$) that is already reduced in front of the first amplifier ($M_x^2 = 1.6$ and $M_y^2 = 1.3$). This is due to phase distortions introduced by the telescope and the nonperfect polished entrance window of the polarizer. After single-pass amplification the beam quality decreases rapidly due to phase distortions introduced by both pumped amplifier rods. At low average pumping power of 3.2 kW for each amplifier the diffraction-limited factors are $M_x^2 = 3.2$ and $M_y^2 = 2.3$; these values increase up to $M_x^2 = 5.2$ and $M_y^2 = 4.5$ at maximum average pumping power (6.5 kW for each amplifier).

After double-pass amplification the beam quality can be nearly reproduced at the polarizer ($M_x^2 = 1.7–1.8$ and $M_y^2 = 1.4–1.9$) due to the phase conjugating SBS mirror. Differences between the initial and final beam quality are caused by nonperfect compensation of phase distortions and amplitude distortions, like gain saturation and apertures.

After single-pass booster amplification by amplifier 3 and amplifier 6, the beam suffers again from phase distortions that are not yet compensated by a PCM ($M_y^2 = 1.8–5.2$). The beam quality was measured only for one direction until now. Assuming an optimized beam superposition of both channels, an average output power of 400 W can be supplied with a beam quality of $M_y^2 = 5$.

Further measurements are planned after introducing additional phase conjugating mirrors to compensate phase distortions of the last single amplification stage.

4.11 CONCLUSION AND OUTLOOK

We have reported about multi-kilohertz diode and flashlamp-pumped MOPA systems with high beam quality which supply average output powers from several tens up to 520 W. Due to phase conjugation by stimulated Brillouin scattering in optical fibers and liquids, the master oscillator beam quality can be largely reproduced after double-pass amplification. Taking several design rules into account, the cascaded setup of two amplifiers supplies an average output power up to 227 W. In contrast to high-pressure gas cells or liquid cells, novel fiber phase conjugators facilitate reliable all-solid-state setups. An all-solid-state system with an average output power of 315 W was demonstrated.

In parallel amplifier arrangements the average output power can be extended to several hundred watts. Two serial amplifier chains in a parallel arrangement supply average output powers up to 315 W with high beam quality, corresponding to an M^2 value better than 2.6 for both directions in space. Due to optical systems between the amplifiers, the average output power can be tuned easily by the variation of pumping power.

A further scientific and technical challenge is the extension of phase conjugation to continuous-wave systems. With an advanced setup it was possible to achieve 71-W average output power in a continuous pumped system.

REFERENCES

1. N. Hodgson and H. Weber, *Optical Resonators*, Springer, (1997).
2. W. Schöne, S. Knoke, A. Tünnermann, and H. Welling, Efficient diode-pumped cw solid-state lasers with output powers in the kW range, *CLEO '97, Technical Digest Series*, Vol. 11, talk CFE2.
3. H. Kiriyama et al., Laser diode-pumped eight pass Nd:YAG slab laser amplifier, *High-Power Lasers, San Jose (California), Proceedings of the SPIE*, Vol. 3264, (1998), pp. 30–36.
4. U. Brauch, K. Contag, A. Giesen, I. Johannsen, M. Karszewski, C. Stewen, and A. Voss, Thin disk laser design for high-power diode-pumped solid-state lasers, *CLEO '97, Technical Digest Series*, Vol. 11, talk CFE1.
5. H. J. Eichler, R. Menzel, and D. Schumann, 10-W single rod Nd:YAG laser with stimulated Brillouin scattering Q-switching mirror, *Appl. Opt.*, **31**(24), 5038–5043 (1992).
6. M. Ostermeyer, N. Hodgson, and R. Menzel, High-power, fundamental mode Nd:YALO laser using a phase-conjugate resonator based on SBS, *CLEO '98 Technical Digest Series*, Vol. 6, talk CThH4.
7. A. Mocofanescu and V. Babin, Nd:YAG laser resonators using external stimulated Brillouin scattering Q-switching mirror, *Optical Resonators—Science and Engineering*, R. Kossowsky et al. (eds.), (1998), pp. 453–462.
8. H .J. Eichler, A. Haase, and R. Menzel, 100-watt average output power 1.2 diffraction limited beam from pulsed neodymium single-rod amplifier with SBS phase conjugation, *IEEE J. Quantum Electron.*, **31**, 1265–1269 (1995).
9. H. J. Eichler, A. Haase, and R. Menzel, High beam quality by SBS phase conjugation of a single rod Nd-amplifier up to 140 watts average output power, *Opt. Quantum Electron.* **28**, 261–265 (1996).
10. Th. Riesbeck, E. Risse, O. Mehl, A. Mocofanescu, and H. J. Eichler, Continuously pumped all solid-state laser system with fiber phase conjugation, *CLEO 2000, OSA Technical Digest*, pp. 404–405 (2000).
11. H. J. Eichler, O. Mehl, Th. Riesbeck, and E. Risse, CLEO Europe 2000, Nice (France), September 10–15, High-brightness laser systems with fiber phase conjugation. *IEEE Conference Digest*, p. 177 (2000).
12. E. Risse, O. Mehl, Th. Riesbeck, A. Mocofanescu, and H. J. Eichler, Continuously pumped all-solid-state laser system with fiber phase conjugate mirror, GCLHPL Florenz 2000, *Proceedings of the SPIE*, Vol. 4184, pp. 179–182 (2001).
13. Th. Riesbeck, E. Risse, and H. J. Eichler, "Pulsed solid-state laser system with fiber phase conjugation and 315 W average output power, *Appl. Phys. B (Laser and Optics)* **73**, 847–849 (2001).

14. A. Dehn, H. J. Eichler, A. Haase, B. Liu, and O. Mehl, Phase Conjugation for High-Power Solid State Lasers with Repetition Rates in the kHz-Range, *Solid State Lasers VI, San Jose (California), Proceedings of the SPIE*, Vol. 2986, pp. 74–83 (1997).
15. H. J. Eichler, A. Haase, J. Kunde, B. Liu, and O. Mehl, Fiber phase-conjugator as reflecting mirror in a MOPA-arrangement, *Solid State Lasers VI, San Jose (California), Proceedings of the SPIE*, Vol. 2986, pp. 46–54 (1997).
16. H. J. Eichler, A. Haase, B. Liu, and O. Mehl, Phase Conjugation Techniques, *Optical Resonators—Science and Engineering* (NATO ASI Series 3. High Technology, Vol. 45, ISBN 0-7923-4962-8), R. Kossowsky, M. Jelinek, and J. Novak (eds.), (1997), pp. 103–117.
17. H. J. Eichler, A. Haase, B. Liu, and O. Mehl, High average power neodymium MOPA systems with high beam quality by fiber phase conjugator, *CLEO 1997, Technical Digest Series*, Vol. 11, pp. 283–284 (1997).
18. H. J. Eichler, B. Liu, A. Haase, O. Mehl, and A. Dehn, Fiber phase conjugators for high power lasers, *Nonlinear Optical Engineering, San Jose (California), Proceedings of the SPIE*, Vol. 3263, pp. 20–31, (1998).
19. H. J. Eichler, A. Haase, and O. Mehl, Solid state lasers with phase conjugation approaching the kW-range, *High-Power Lasers, San Jose (California), Proceedings of the SPIE*, Vol. 3264, pp. 9–17 (1998).
20. H. J. Eichler, A. Dehn, A. Haase, B. Liu, O. Mehl, and J. Schwartz, Phase conjugation for improving the beam quality of solid state and excimer lasers, *Laser Resonators, San Jose (California), Proceedings of the SPIE*, Vol. 3267, pp. 158–169 (1998).
21. H. J. Eichler, A. Dehn, A. Haase, B. Liu, O. Mehl, and S. Rücknagel, High repetition rate continuously pumped solid state lasers with phase conjugation, *Solid State Lasers VII, San Jose (California), SPIE*, Vol. 3265, pp. 200–210 (1998).
22. H. J. Eichler, A. Haase, and O. Mehl, 500-W average power MOPA system with high beam quality by phase conjugation, *CLEO 1998, Technical Digest Series*, Vol. 6, pp. 353–354 (1998).
23. R. Pierre, D. Mordaunt, H. Injeyan, et al., Diode array pumped kilowatt laser, *IEEE J. Select. Topics Quantum Electron.*, **3**, 53–58 (1997).
24. C. B. Dane, L. E. Zapata, W. A. Neuman, M. A. Norton, and L. A. Hackel, Design and operation of a 150 W near diffraction-limited laser amplifier with SBS wavefront correction, *IEEE J. Quantum Electron.* **31**, 148–163 (1995).
25. D. S. Sumida, D. C. Jones, and D. A. Rockwell, An 8.2 J phase-conjugate solid-state laser coherently combining eight parallel amplifiers, *IEEE J. Quantum Electron.* **30**, 2617–2627 (1994).
26. A. A. Betin, S. C. Matthews, and M. S. Mangir, 1-ms-long pulse Nd:YAG laser with loop PCM, *CLEO 1997, Technical Digest Series*, Vol. 11, p. 283 (1997).
27. H. J. Eichler, A. Haase, and R. Menzel, Cr^{4+}:YAG as passive Q-switch for a Nd:YALO oscillator with an average repetition rate of 2.7 kHz, TEM_{00} mode and 13 W output, *Appl. Phys. B* 58, pp. 409–411 (1994).
28. H. J. Eichler, A. Haase, and O. Mehl, Serielle Laserverstärker-Anordnung mit Optischem System zur Kompensation starker thermischer Linsenvariation, German patent pending, DE19609166A1 (1996).
29. Q. Lü and H. Weber, Beam quality degradation of polarized, circularly symmetric beams caused by a birefringent Nd:YAG rod, *Opt. Commun.* **118**, 457–461, (1995).

30. Q. Lü, N. Kugler, H. Weber, S. Dong, N. Müller, and U. Wittrock, A novel approach for compensation of birefringence in cylindrical Nd:YAG rods, *Opt. Quantum Electron.* **28**, 57–69 (1996).
31. Optics and optical instruments—Lasers and laser related equipment—Test methods for laser beam parameters; Beam widths, divergence angle and beam propagation factor (ISO/DIS 11146:1995).
32. W. Kaiser and M. Maier, Stimulated Rayleigh, Brillouin and Raman Spectroscopy, *Laser Handbook*, F. T. Arecchia, and E. O. Schulz-DuBois (eds.), North-Holland, Amsterdam (1972).
33. Yoshida et al., SBS phase conjugation in a bulk fused-silica glass at high energy operation, *CLEO 1997, Technical Digest Series*, Vol. 11, pp. 117–118 (1997).
34. P. Shalev, S. Jackel, R. Lallouz and A. Borenstein, Low-threshold phase conjugate mirrors based on position-insensitive tapered waveguides, *Optical Eng.* **33**, 278–284 (1994).
35. S. Jackel et al., Low threshold, high fidelity, phase conjugate mirrors based on CS_2 filled hollow waveguide structures, *Nonlinear Opt.* **11**, 89–97 (1995).
36. H. J. Eichler, J. Kunde, and B. Liu, Quartz fibre phase conjugators with high fidelity and reflectivity, *Opt. Commun.* **139**, 327–334 (1997).
37. G. P. Agrawal, *Nonlinear Fiber Optics*, Academic Press, New York, 1989.
38. V. Babin, A. Mocofanescu, V. Vlad, and M. Damzen, *J. Opt. Soc. Am. B* **16**, 155 (1999).
39. H. J. Eichler, A. Mocofanescu, Th. Riesbeck, E. Risse, and D. Bedau, Stimulated Brillouin scattering in multimode fibers for optical phase conjugation, *Opt. Commun.* **208**, 427–431 (2002).
40. S. Seidel, Multi-Pass Festkörper-Laser im Q-switch Betrieb mit phasenkonjugierenden SBS-Zellen und deren numerische Modellierung, Ph. D. thesis, TU Berlin, 1995.
41. V. Magni, Multi-element stable resonators containing a variable lens, *J. Opt. Soc. Am. A.* **4**, 1962–1968 (1987).

CHAPTER 5

High-Pulse-Energy Phase Conjugated Laser System

C. BRENT DANE and LLOYD A. HACKEL

Lawrence Livermore National Laboratory, Livermore, California 94550, USA

5.1 INTRODUCTION

There are a number of important applications that require both high pulse energy (>25 J/pulse) and high average power (>100 W) from a solid-state laser system. Among examples of these are the generation of X rays for photolithography [1] and the coherent illumination of distant objects for high-resolution imaging [2]. An important and growing application of these lasers is the strengthening of metal parts against stress corrosion cracking and fatigue failure by laser shock processing, often referred to as laser peening [3]. High pulse energies are also useful for Raman frequency conversion [4] and SBS pulse compression [5]. Another effective use is the high-throughput optical conditioning and damage testing of large optics required for large fusion driver lasers [6]. Outside of the devices with up to 100 J/pulse presented in this chapter, high-pulse-energy solid-state laser systems have very low pulse repetition frequencies (<0.1 Hz) or can only be operated in a single-shot mode. For example, a diode-pumped 100-Hz, 1-kW system has been demonstrated, but even in this case the pulse energies were limited to <10 J [7]. In this chapter we will describe laser systems with pulse energies in the range of 25 to 100 J and average powers spanning 100 to 1000 W. These systems use SBS phase conjugation in master oscillator power amplifier (MOPA) geometries and have pulse durations between 15 ns and 1 μs with near diffraction-limited divergence and transform-limited bandwidth.

High-pulse-energy, high-average-power operation of a solid-state laser system is a unique regime with very specific design requirements. The key enabling technologies for the successful development of these systems are the face-pumped

Phase Conjugate Laser Optics, edited by Arnaud Brignon and Jean-Pierre Huignard
ISBN 0-471-43957-6   Copyright © 2004 John Wiley & Sons, Inc.

zigzag slab amplifier architecture [8] and SBS phase conjugation. Accurate wavefront correction is not just a requirement dictated by the divergence needs of an intended application of the laser. High-beam-quality operation is first necessary for high-average-power operation with high pulse energies, providing optimal amplifier beam fill and optical extraction as well as preventing damage to optical components. In addition to the low output divergence enabled by nonlinear phase conjugation, other benefits include very high pulse-to-pulse and long-term beam pointing stability as well as the ability to phase-lock multiple amplifier apertures to generate even higher-energy beams. High beam quality also provides the option of very efficient harmonic wavelength conversion.

Although the Nd:glass gain medium exhibits a lower fracture strength and lower thermal conductivity than do crystalline materials such as Nd:YAG, its availability in large volumes and its small cross section for stimulated emission (3.5×10^{-20} cm^2) make it well-suited to the storage requirements of high energy per pulse operation. At the same time, the high saturation fluence for Nd-doped phosphate glass and the potentially large thermally induced distortions resulting from its low thermal conductivity offer a significant challenge for the operation of a high-average-power amplifier system. Good thermal management in the form of uniform pumping deposition, uniform cooling, and an optimized amplifier slab design is crucial to successful high-power operation. Even under ideal conditions, the high fluence required for efficient energy extraction requires carefully designed beam transport and the accurate correction of the remaining thermally induced wavefront aberrations in the amplifier medium in order to avoid damage to optical components. For this reason, the flashlamp-pumped Nd:glass amplifier system has been a valuable test bed for the development of high-average-power solid-state amplifier architectures. The result has been robust optical designs that are also readily applicable to high-average-power diode-pumped crystalline amplifier systems having lower optical fluences. We believe that through this development work, a number of valuable design principles have resulted which are directly applicable to SBS phase conjugated MOPA architectures in general, and these will be reviewed at the end of the chapter.

To illustrate the design principles for a high-pulse-energy, high-average-power, SBS phase conjugated MOPA, we have chosen to present summarized technical descriptions of three laser systems. Each of these shares a common flashlamp-pumped Nd:glass amplifier design operating at 1053 nm [9]. The first system is a 25-J/pulse, 15-ns laser using a liquid SBS phase conjugate mirror. The second extends this pulse duration to 500 ns and uses a high-pressure-gas SBS cell with a self-pumped four-wave-mixing loop geometry to lower the phase conjugation threshold and to provide narrow spectral bandwidth. Finally, the third system coherently combines four amplifier apertures using SBS phase-locking to increase the pulse energy to 100 J/pulse. Each builds on the technology of the previous system, and emphasis will be placed on the novel developments required for the laser under discussion.

5.2 HIGH-ENERGY SBS PHASE CONJUGATION

As we began our first efforts to develop a high-average-power solid-state laser system with an output energy of >25 J/pulse, we recognized that the practical design of a MOPA geometry would require the SBS phase conjugate mirror to reflect energies in the range of 1–3 J/pulse. With input pulse durations of around 15 ns, this represents up to 1000 times the SBS threshold for nonlinear media such as carbon tetrachloride (CCl_4) and high-pressure nitrogen (N_2). It had been proposed that SBS phase conjugators could be scaled to high peak powers using two-stage amplifier-generator designs [10–12]. However, when possible, we believed that it was desirable to employ a single-cell, simple-focused conjugator geometry operated many times above threshold, from the standpoint of achieving high reflectivity, accurate temporal pulse shape reproduction, and design simplicity. Even for the two-cell design, our experience indicated that it was advantageous to operate the focused generator cell high above threshold for optimal reversal of the high-frequency spatial content of the beam wavefront.

5.2.1 The question of fidelity versus input energy

The dependence of the wavefront fidelity of the Stokes return from an SBS phase conjugator on the input power had been extensively investigated both experimentally and theoretically. The lack of a complete understanding of this dependence was evidenced by the contradictory conclusions reached in many of these studies. Experimental measurements had indicated that the phase fidelity increased as the laser input power exceeded the SBS threshold [13–15], while other measurements showed a marked decrease in the correlation of input and output phase for high input powers [16, 17]. The results of numerical modeling also predicted that the phase fidelity should increase [18] as well as decrease [19] as the input power is increased to many times above threshold. In the latter report, it was concluded that the discrimination against other SBS noise modes by the phase conjugate mode should be largely suppressed for the case of strong pump depletion that results from operation of the conjugator with high optical reflectivity, at many times above threshold. Of particular note, soon after we began our work in phase-conjugated laser systems, the results of an experimental study were reported which supported this conclusion by demonstrating a large increase in the shot-to-shot SBS phase conjugation fidelity fluctuation for input energies that exceeded the SBS threshold by only five times [20]. Since a similar effect was observed in both nitrogen and titanium tetrachloride ($TiCl_4$), it was proposed that these fluctuations arose from a fundamental limitation of SBS phase conjugation fidelity at high power related to the loss of discrimination for the phase conjugate Stokes mode when the SBS threshold is greatly exceeded.

Recognizing the serious limiting implications toward our proposed high-pulse-energy amplifier designs, we undertook an experimental investigation aimed at further understanding the input energy limitations for SBS. In the course of this work, we were able to demonstrate good shot-to-shot SBS phase conjugation fidelity for input energies that were 100 times above threshold [21]. More importantly, it was also found that the strong fluctuations in the conjugated fidelity reported in other studies could be reproduced by using input pulses of the same temporal width but with steeper leading edges. It was well understood that poor wavefront reversal can result from input pulse widths that are short relative to the acoustic decay time, τ_B, in the SBS medium [17]. The results of our measurements, however, demonstrated that even for an input pulse width that is longer than τ_B, noise modes can be driven by a very fast leading edge. These can then compete for energy extraction by the normally favored phase conjugate (PC) mode for the duration of the pulse.

5.2.2 The experimental measurement of SBS wavefront fidelity

The laser used in the experiments was an Nd:YLF single-frequency TEM_{00} Q-switched laser system that delivered ~ 250 mJ at 1.053 μm in a 15-ns full width at half-maximum (FWHM) quasi-Gaussian pulse shape. Two SBS media were investigated: liquid carbon tetrachloride (CCl_4) and gaseous nitrogen (N_2) at 90 atm. A previously reported method of monitoring the SBS phase conjugation fidelity was used in which the transmission of the Stokes return through a pinhole located at the focus of a lens was measured for each laser shot [16, 20]. The size of this aperture was chosen as the $1/e^2$ diameter of the focused input beam, resulting in a transmission of $\sim 86\%$. By monitoring the pinhole transmission of the output of the SBS cell, an indication of the correlation of its wavefront to the input beam could be obtained. As shown in Fig. 5.1, the calculated reflectivity was the ratio of energy measured on energy meter 2 to that on energy meter 1 for each experimental point. The fidelity was defined as the ratio of the energy transmitted through the pinhole

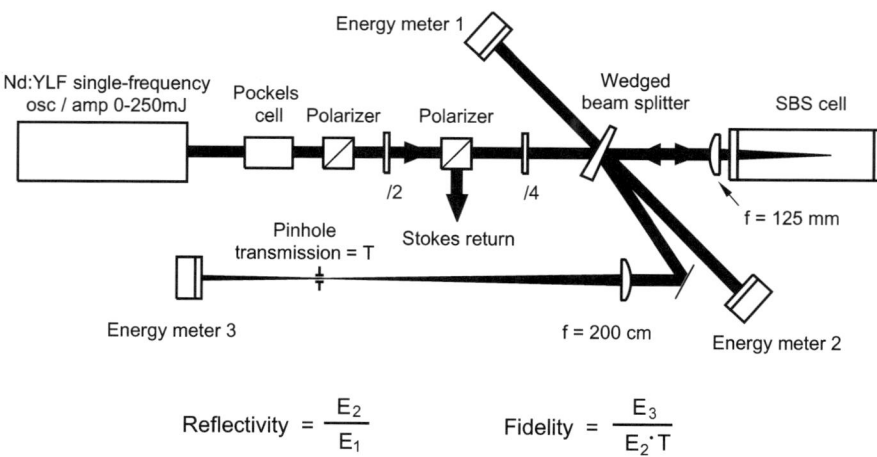

Figure 5.1. Experimental layout used to monitor the shot-to-shot fluctuations in the SBS phase conjugation fidelity.

(meter 3) to the total reflected energy (meter 2), normalized by the far-field transmission of the pump beam (86%), so that a fidelity of 1 corresponded to accurate replication of the input beam's angular divergence.

A Pockels cell with a 3-ns 20–80% rise time was used to truncate the leading edge of the laser pulse. The timing of the Pockels cell was adjusted so that the energy of the transmitted pulse was reduced by ~10% and the pulse width at FWHM remained approximately unchanged, as depicted in Fig. 5.2. The experimental pulse repetition frequency was varied between 2 and 10 Hz, and no dependence in the measured results was observed over this range. All of the data presented here were collected at a 5-Hz pulse repetition frequency.

Using this apparatus, no increase in the shot-to-shot fidelity fluctuations was observed at up to 100 times the measured 2.5-mJ threshold for CCl_4, resulting in a maximum energy reflectivity of 90%. These measurements are in contrast with those from Ref. 20 for a $TiCl_4$ cell in which the onset of large fluctuations in phase fidelity are recorded well below 10 times the SBS threshold. This difference in measured results prompted us to investigate the dependence of the conjugation fidelity on the input pulse shape. For an N_2 SBS cell at 90 atm, it was found that the large shot-to-shot fluctuations in phase fidelity could be reproduced if the leading edge of the input pulse was steepened as described previously. Figure 5.3 compares the measured results of the N_2 experiment with the full pulse shape having (a) a slow rising edge and (b) the truncated pulse shape with a 3-ns leading edge rise time. In the former case, the fidelity rose to a value of ~0.9 without a measurable increase in scatter for up to 17 times threshold. Large shot-to-shot fluctuations were observed,

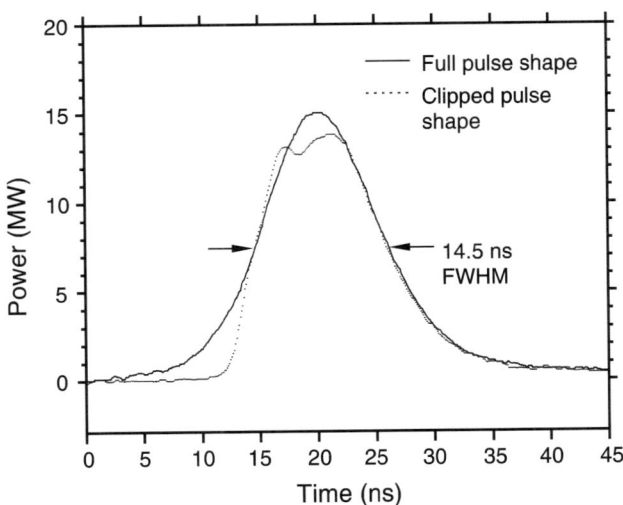

Figure 5.2. A comparison of the full quasi-Gaussian temporal profile with the truncated fast-rise-time temporal profile. The clipped pulse retained ~90% of the energy of the full pulse. The pulse shapes were measured using a 500-MHz oscilloscope so that the 3-ns observed rise time is an upper bound to the actual rise time.

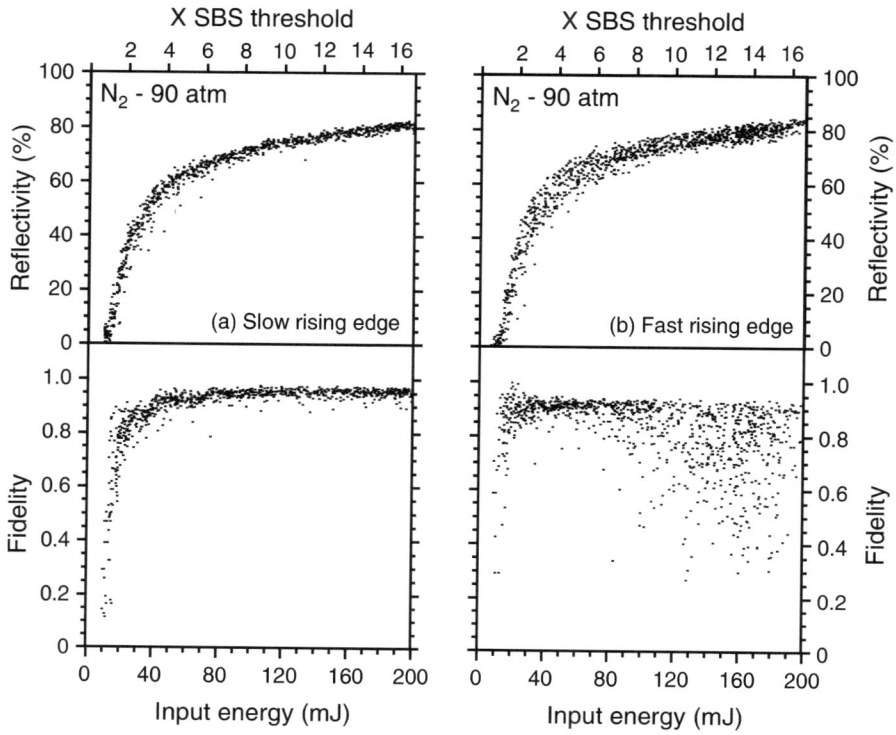

Figure 5.3. Experimentally measured reflectivity and fidelity versus input energy using a 90-atm N_2 SBS cell. The measurements were made for (a) the full quasi-Gaussian temporal profile and (b) the fast-rise-time truncated temporal profile.

however, with the fast rising edge input above only four times threshold. It should be noted that this dramatic difference in the Stokes return from the SBS cell was observed when only 10% of the total input energy was removed from the leading edge of the original quasi-Gaussian temporal profile. Figure 5.4 illustrates the qualitative nature of the observed fluctuations by presenting cross sections of the irradiance profiles at the focus of the 200-cm lens recorded with a CCD array in place of the pinhole. The shot-to-shot fluctuations in the far field were the result of a random combination of pointing error and variations in beam divergence. For the temporally clipped pulse, the far-field patterns often exhibited multilobed structures with two to three distinct peaks. However, when the wavefront reversal fidelity versus pulse shape experiment was repeated for CCl_4 with the steepened pulse, no increase in the scatter of the fidelity data at large input energies was observed.

5.2.3 The input pulse rise-time requirement

Although the poor wavefront reversal that results from a pulse whose width is significantly shorter than τ_B was recognized [16, 17], the results of this study suggested

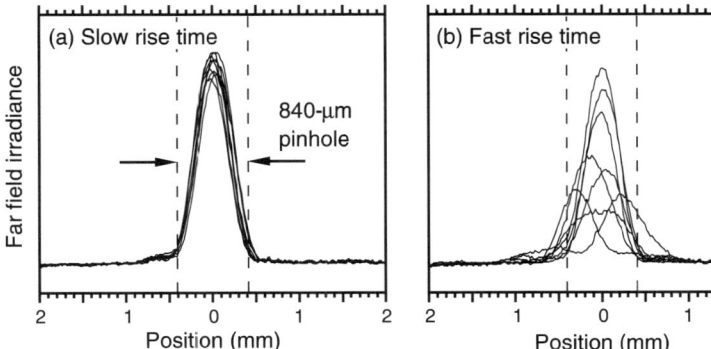

Figure 5.4. Line-outs of the spatial profiles at the focus of the 200-cm lens for eight consecutive return pulses from the N_2 SBS cell using (a) the full quasi-Gaussian temporal profile and (b) the fast-rise-time temporal profile. The input energy was ~ 200 mJ corresponding to $>16\times$ the SBS threshold.

that even wider pulses having the same characteristic leading-edge rise time associated with the short pulse can also produce a Stokes pulses with poor phase fidelity. Noise modes amplified early in the SBS extraction by the high pump beam irradiance reached at focus can dominate the SBS extraction from the incoming pump for the duration of the pulse. The competition by these modes would be expected to diminish after many τ_B in the case of a very long input pulse. For the case of slow rise times, the irradiance shielding of the focal volume that results from pump extraction by the PC mode initiated early in the pump pulse also decreases the possibility of fidelity degradation by other nonlinear mechanisms such as optical breakdown, Raman conversion, and self-focusing. Previous investigators have noted this pulse shape dependence of competition by other nonlinear processes [22]. It is of significance that the laser amplifier system used to generate the laser pulses reported in Ref. 20 incorporated an SBS mirror in the amplifier chain [23] since the threshold turn-on characteristics of such a mirror combined with gain saturation in the amplifiers would be expected to generate an input pulse with a steep leading edge. We believe that this explains the fidelity fluctuations observed in that report, even when using $TiCl_4$ with $\tau_B \sim 2$ ns.

It is important to note that these results do not completely resolve experimental and theoretical inconsistencies concerning the dependence of SBS phase conjugation fidelity on input energy. They demonstrate, however, that the laser pulse shape is also very significant and should be given careful consideration in designing SBS phase conjugate mirrors. The measurements suggest that for optimal wavefront reversal using an SBS focused oscillator cell and high input powers, the rise time of the incoming pulse should be comparable to or longer than the acoustic decay time in the medium. This can be accomplished by a combination of pulse-shape tailoring and the judicious selection of the SBS medium, and this criterion has been incorporated into our high-energy laser designs.

Measurements by other researchers have also confirmed good wavefront reversal fidelity using simple focus SBS high above threshold [24, 25]. Although theoretical questions remain about the precise requirements for optimal wavefront reversal in SBS, the fact that a phase conjugate mirror could be successfully operated at high input energies was of great practical significance to our laser designs. In the results reported in the rest of this chapter, SBS mirrors are shown to successfully operate with input energies that are hundreds of times the stimulated threshold, providing reliable, stable wavefront correction.

5.3 A 25-J, 15-NS AMPLIFIER USING A LIQUID SBS CELL

5.3.1 Design considerations for the 15-ns system

The initial motivation for the design of this laser system was for the production of 1-nm soft X rays for proximity-print photolithography [1]. The requirements were peak powers of >1 GW in a 10- to 20-ns pulse, good focusability, and a pulse repetition frequency of >5 Hz to provide acceptable processing throughput. Continuing development in this pulse duration regime, however, has more recently been driven by laser peening applications [3] which require 15- to 30-ns pulses. For effective laser peening, at least 20 J/pulse is required in order to minimize cold work and to minimize diffractive dissipation of the shock wave in the bulk of the metal part being treated.

5.3.2 Optical architecture of the 15-ns system

When the 25-J, 15-ns laser system was first designed, the intent was to minimize the thermally induced optical distortions introduced by the flashlamp-pumped amplifier and to then operate the amplifier beam train without wavefront correction. Provisions were made for the incorporation of a phase conjugate mirror in the event that a smaller focal spot were required for efficient X ray generation. However, during initial activation of the laser, it quickly became apparent that SBS phase conjugation provided significant increases in performance, apart from output divergence requirements. These included increased extraction efficiency, reduced optical component damage, increased interstage gain hold-off, and greatly improved alignment stability.

5.3.2.1 Nd:glass zigzag amplifier
The goal of the amplifier design is to maximize the efficiency and uniformity of the upper-laser-level activation in the gain medium while minimizing thermally induced distortion and depolarization of the transmitted optical wavefront [26, 27]. The major components of the amplifier design are the zigzag slab, the cooling system, and the flashlamp and reflector assemblies. The choice of a slab geometry requires analyses of the gain and energy storage required for the desired output energy, the ASE energy storage losses, and the thermally induced optical aberration and depolarization. A $1 \times 14 \times 40$ cm^3

Nd:glass zigzag slab is face-pumped by four xenon flashlamps using specially shaped diffuse reflector cavities. The extraction beam is 7–8 mm wide and 120 mm tall. A detailed description of the optical, mechanical, and electrical design, including pumping uniformity, ASE control, and cooling, can be found in other reports [9, 28, 29]. This amplifier is a proven and reliable design and, with minor variations, is the amplifier used in all of the high-energy laser systems reported in this chapter.

5.3.2.2 Relay imaging Optimal extraction efficiency from a zigzag slab solid-state gain medium requires the amplification of a near-top-hat spatial beam profile. However, the high-aspect-ratio rectangular slab aperture combined with this spatial profile results in a beam with poor near-field propagation characteristics, particularly in the presence of thermally induced phase aberrations in the solid-state amplifier. For this reason, the architecture presented here uses a 1:1 relay imaging telescope inside of the regenerative amplifier ring which re-images each plane inside the zigzag slab back onto itself, preserving uniform irradiance spatial profiles in the amplifier for multiple ring passes. A significant part of the design of this system is the use of a double-pass relay telescope as illustrated in Fig. 5.5. The telescope consists simply of two lenses mounted with a confocal spacing as pressure windows on a vacuum chamber. The counterpropagating beams pass through the lenses offset

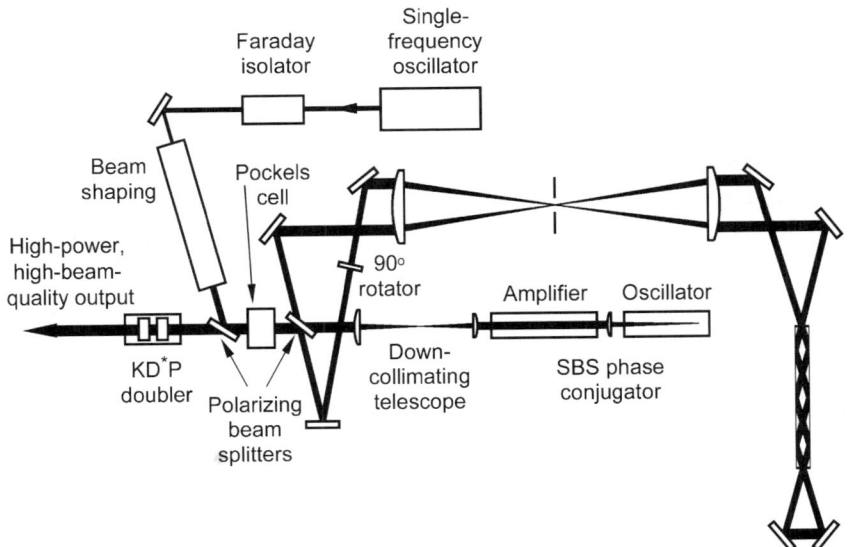

Figure 5.5. A schematic diagram of the regenerative amplifier geometry incorporating an SBS phase conjugate mirror. In the original optical layout, without SBS phase conjugation, the Pockels cell was placed inside the ring at the position of the 90° quartz rotator and the output emerged on the beam path that is shown to enter the SBS cells.

from the optical axis, thereby allowing two separate optical paths through the telescope.

5.3.2.3 Unidirectional non-phase-conjugated operation
When the laser system is operated as a conventional regenerative amplifier system without SBS phase conjugation, an input pulse from the oscillator enters the system by a reflection off of a polarizing beam splitter placed in the ring as shown in Fig. 5.5. Although the optical layout shown in Fig. 5.5 is that for the phase conjugated system, only minor differences exist between the layout for conventional unidirectional operation. The Pockels cell moves from the location shown to a position inside the amplifier optical ring, replacing the 90° quartz rotator. The deuterated potassium diphosphate (KD*P) gas-cooled Pockels cell uses 45° quartz rotators as input and output windows so that it also has a passive 90° polarization rotation with no voltage applied [30]. Therefore the s-polarized input beam is rotated to p-polarization after passing through the Pockels cell. During the first ring revolution, the half-wave ($\lambda/2$) voltage is applied to the Pockels cell crystals so that further polarization rotation is prevented and the input pulse is trapped in the regenerative amplifier ring until the voltage is removed. In this zigzag amplifier design, near-normal input and output faces are used on the glass slab with either sol–gel [31] or conventional dielectric antireflection (AR) coatings. This allows the slab to be double passed in a "diamond" pattern as shown in Fig. 5.5 and either s- or p-polarized beams can be amplified. When the Pockels cell voltage is switched back to zero, the injected pulse makes one more ring pass in s-polarization, corresponding to two more gain passes. The Pockels cell is therefore not required to transmit the fully amplified output pulse energy, reducing the possibility of optical damage to the KD*P. In this non-phase-conjugated geometry, the amplified output pulse exits the system along the beam line that is shown in Fig. 5.5 to lead to the SBS mirror.

5.3.2.4 Operation with an SBS phase conjugator
By replacing the Pockels cell in the regenerative amplifier ring with a 90° quartz rotator as depicted in Fig. 5.5, an injected pulse from the oscillator makes two revolutions through the ring, for a total of four gain passes, before it is reflected by the polarizing beam splitter. This reflected beam, formerly the output beam in the unidirectional configuration, is now directed into an SBS phase conjugation cell. The wavefront-reversed return from the conjugator retraces the two ring passes, accumulating four more gain passes, before it is again reflected out of the ring. The high-power pulse now retraces the input beam path and must be isolated from the oscillator. This is accomplished by placing the Pockels cell removed from the ring into the input beam as illustrated in Fig. 5.5. The $\lambda/2$ voltage is applied to the Pockels cell as the injected seed pulse is transmitted into the ring, canceling its passive 90° rotation. During the following four ring transit times, the Pockels cell voltage is switched back to zero so that the high-energy output pulse experiences a 90° rotation and is transmitted through the input polarizing beam splitter.

The most important advantage of operating the regenerative amplifier system with an SBS mirror lies, of course, in its wavefront reversal properties. As the input beam makes its first four passes through the amplifier, it accumulates wavefront phase errors due to thermal aberrations in the solid-state amplifier slab as well as to residual optical figure errors in the other optics in the ring. The SBS cell effectively conjugates the phase of the input beam, producing a wavefront-reversed output Stokes beam with a small frequency shift (<5 GHz for carbon tetrachloride, CCl_4). As this output beam retraces the path of the input beam through four additional gain passes, the phase errors cancel and the high-quality wavefront of the injected beam is recovered in the high energy output.

However, it is important to note that there are additional very distinct advantages to the operation of the amplifier system with the SBS phase conjugator. Eight gain passes through the zigzag slab amplifier can be achieved using passive polarization switching in the regenerative amplifier ring. The fact that the SBS cell provides interstage gain isolation makes this possible since, if it were replaced with a mirror, the small signal gain through eight consecutive gain passes would result in possible parasitic oscillation from the small reflective losses of AR-coated optical surfaces in the ring or in the output beam. When the Pockels cell is located inside the regenerative amplifier ring, the length of the laser pulse to be amplified is limited to the transit time inside the ring minus the high-voltage switching time for the Pockels cell. Now, with passive polarization switching, this time limit is increased to be the transit time for four ring revolutions. Longer pulses can be amplified, and the relative timing accuracy requirement between the injected pulse and the Pockels cell voltage pulse is greatly reduced. In fact, the replacement of the Pockels cell with a full-aperture Faraday isolator has been shown to allow the amplification of laser pulses of over 500 ns in duration [32], as will be described in a following section of this chapter.

Another advantage is that the SBS phase conjugator very effectively conjugates the first-order aberration of tilt. The combination of this tilt correction and relay imaging greatly reduces the sensitivity of the system performance to small changes of optical alignment in the ring. No change in output power or pointing direction during operation is observed for large mirror misalignments in the ring, limited only to those angular excursions that result in vignetting of the beams at the edges of the amplifier slab.

5.3.2.5 Single-frequency oscillator The 1053-nm transition in Nd:YLF closely matches the fluorescence curve peak in the Nd-doped phosphate glass used in this amplifier system. It is desirable to have a single-frequency oscillator input pulse both for generating reproducible smooth temporal input profiles without modulation from multilongitudinal modes and for achieving the best wavefront reversal fidelity in the SBS phase conjugator [33, 34]. Although single longitudinal mode output from an Nd:YLF Q-switched oscillator has been demonstrated by injection locking to a low-power single-frequency master oscillator, this technique requires careful mode matching between the master and slave oscillators and active cavity length stabilization of the Q-switched slave oscillator [35]. A simpler method of achieving single-frequency output was chosen for this application by using a variation of the electronic linewidth narrowing technique [36, 37]. In this method, very low power oscillation is allowed to build up in the Q-switched cavity over an extended period of time (1–2 μs). By placing a frequency-selective element such as an etalon in the cavity, this long buildup time and the corresponding many passes through the etalon results in single longitudinal mode oscillation. The intracavity power is monitored by the leakage through the high reflectivity (HR) mirror; when it peaks in a weak relaxation oscillation, the Q-switch is opened. The low-power single-frequency flux then serves to seed the buildup of a high-power Q-switched pulse.

The output of the single-frequency Nd:YLF Q-switched oscillator is amplified to ∼100 mJ using additional Nd:YLF preamplification. The beam is then anamorphically expanded to match the aspect ratio of the extracted aperture of the zigzag slab. A slightly oversized elliptical beam is clipped by a hard rectangular aperture and relayed to a point conjugate to an image plane in the slab as depicted in Fig. 5.5. In order to efficiently extract the stored energy in the slab amplifier in eight gain passes, an injected seed energy of ∼60 mJ is required.

5.3.3 SBS phase conjugation with a liquid cell

The basic concept of SBS phase conjugation is well understood and is described in more detail elsewhere in this book. However, we believe that there are a number of important considerations to be made when incorporating an SBS phase conjugate mirror into a high-power laser system in order to accommodate the often-required high input energies and to maintain good near- and far-field fidelity in the amplified output.

5.3.3.1 Wavefront reversal fidelity As we described in the previous section, the conclusion that good wavefront reversal fidelity in the SBS Stokes waves is obtainable for input powers high above threshold is very significant since there are strong motivations for operating the SBS phase conjugator with large input energies. In a somewhat oversimplified but still useful understanding of the SBS phase conjugation process, an input wave with a uniform irradiance spatial profile and flat phase profile is transmitted through an aberrating medium. If this beam is then focused into an SBS cell, the aberrated wavefront is transformed into a distorted irradiance spatial profile in the far field. The Stokes wave is seeded from random acoustic fluctuations in the medium and is selectively amplified by stimulated scattering in the regions of peak irradiance, hence peak gain. The diffraction of the resulting counterpropagating wave then produces a reversed, or phase conjugated, wavefront at the focusing lens. Based on this description, albeit a overly simple one, it can be reasoned that it is desirable to operate far above threshold with a highly aberrated input beam in order for the highest frequency modes, which appear in spatially extended distributions at focus, to reach threshold and be reversed.

5.3.3.2 SBS amplifier/oscillator configuration For input energies >1 J/pulse, we have found that a two-cell system configured in an oscillator/amplifier arrangement becomes useful [11, 12]. As depicted in the optical layout of Fig. 5.5, the collimated input beam is directed through the first SBS cell and then focused into a second cell. In this way the peak optical power in the focused oscillator cell is reduced by the depletion of the pump wave in the collimated amplifier cell. The beam size in the collimated cell is nominally adjusted such that the peak Brillouin gain, irradiance, and length product (gIL) is <30, the threshold for stimulated return from that cell alone. In practice, however, it is found that very efficient amplification in the first cell is achieved with a calculated peak gIL product of ~ 45. Even at this large value, no spontaneous SBS contribution is observed from the amplifier cell since depletion of the incoming laser pump pulse prevents the gain product from reaching the estimated value.

Another consideration in configuring the SBS oscillator/amplifier system is the round-trip optical propagation length in the SBS medium. This should not greatly exceed τ_B for the SBS material. In this case, when the SBS system is driven high above threshold, strong pulse reshaping in the Stokes return in the form of pulse compression can result. For input pulses of sufficient duration, periodic modulation with a period corresponding to the round-trip time through the SBS cells is observed. In this amplifier system, the input beam was configured with an area of $\sim 1 \text{ cm}^2$ ($8 \times 12 \text{ mm}^2$) through a 33-cm amplifier cell. It was then focused into the oscillator cell with a 14-cm focal length lens. Although saturation in the amplifier cell allows attenuation to be added in front of the oscillator cell without a great penalty in energy efficiency, it was not found to be necessary up to an input energy of 3 J in a pulse width of 15 ns.

5.3.3.3 Relay imaging to the SBS cell

A highly relayed optical system, as described in previous sections, is essential for the successful operation of a phase conjugated amplifier system. As wavefront aberration accumulates on the laser pulse circulating in the amplifier, the system optics must collect, relay, and deliver to the SBS cells all components of the aberrated beam. In addition to this, we have found that the irradiance spatial profile delivered to the SBS oscillator is also significant. It is important that aberrations on the wavefront not be allowed to freely propagate, producing a distorted irradiance spatial profile. The near-field reproduction in the SBS conjugator return is observed to be poor for a strongly modulated input irradiance profile. If the distorted phase information is lost to near-field distortion in the beam, then this poor near-field reproduction will degrade the phase conjugation fidelity of the SBS mirror. By image relaying the amplifier aperture to the focusing lens of the SBS oscillator, a smooth spatial profile is presented to the phase conjugator. Any errors in the near-field irradiance distribution in the SBS return in the form of small amplitude modulation are then reduced by gain saturation in the final amplifier passes.

5.3.3.4 Temporal fidelity

It is also important to operate the SBS phase conjugator high above threshold to have high reflectivity and the best reproduction of the temporal profile of the input pulse. The energy efficiency of the phase conjugate mirror is determined primarily by the leading-edge portion of the input pulse that is transmitted through the cells and lost before the SBS threshold is

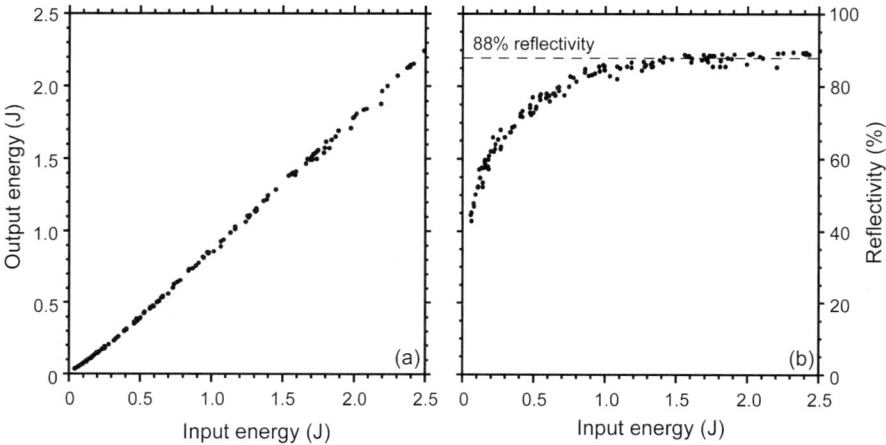

Figure 5.6. (a) The experimentally measured reflected energy versus input energy for the two-cell CCl_4 SBS phase conjugator arrangement with 15-ns FWHM input pulses. (b) The same data plotted as SBS reflectivity versus input energy.

reached. For this reason, maximizing the SBS efficiency also optimizes the temporal reproduction fidelity of the input pulse. Figure 5.6 shows the reflectivity versus input energy for the described two cell amplifier/oscillator configuration up to the nominal operating input level of 2.5 J. The reflectivity saturates at a maximum value of 88% beyond ~ 1 J. The insertion loss into the optical beam train is very small with this high value of the SBS conversion efficiency, and virtually no pulse shortening or steepening of the leading edge of the output temporal profile is observed.

5.3.3.5 Liquid SBS medium purity
The purity of the SBS medium is crucial with respect to minimizing both optical absorption and the possibility of optical breakdown at the focus of the SBS oscillator. The removal of dissolved impurities which can undesirably increase absorption can normally be accomplished by distillation techniques. In practice, however, we have determined that in the case of most liquid SBS media, it is sufficient to use the commercially available 99.9% spectroscopic-grade material. However, a more challenging problem was found to be the presence of suspended particulate impurities which, when intercepted by the focused input beam, lead to a significantly reduced optical breakdown threshold. The most effective way of removal of these particles has been found to be the use of a closed-loop filtration system installed on the SBS oscillator cell. A small centrifugal pump of stainless steel and teflon construction is used to circulate the CCl_4 through a 0.1-μm filter. Once the cell has been adequately filtered, it is not necessary to run the circulation system during operation of the laser amplifier. After

filtration, the optical breakdown threshold for the 10- to 15-ns input pulses is increased from ~40 mJ to over 1 J for the single-focused SBS cell.

5.3.4 Operation of the 25-J/pulse 15-ns laser system

5.3.4.1 *Energy storage and extraction efficiency*

A single-pass small signal amplification of 3.1 (1.1 Np) can be achieved in the amplifier with 2.7 kJ of lamp energy. Although this provides a potential total stored energy of almost 70 J in the laser slab, the maximum energy extracted from the amplifier in a 13- to 15-ns laser pulse is limited to 30 J due to the possibility of nonlinear self-focusing in the amplifier glass. Even for this pulse energy, the amplifier cannot be operated in a highly saturated extraction regime. The increased glass propagation distance at high irradiance required to achieve very large extraction efficiencies can result in beam filamentation and bulk glass damage. For this reason, an optimal operating regime for 25-J output results from storing 75 J in the amplifier slab with a single-pass amplification ratio of 3 and an optical extraction efficiency of ~33%. As illustrated in Fig. 5.7, 25 J can be extracted in a 15-ns pulse with an oscillator input energy of 60 mJ. This results in an input energy of ~2.5 J to the SBS cells. The B-integral, a scaled product of the nonlinear index and optical irradiance integrated over the propagation distance, reaches a value of ~0.5. This is a factor of 5–6 below the expected onset level of self-focusing for a beam with a smooth irradiance profile

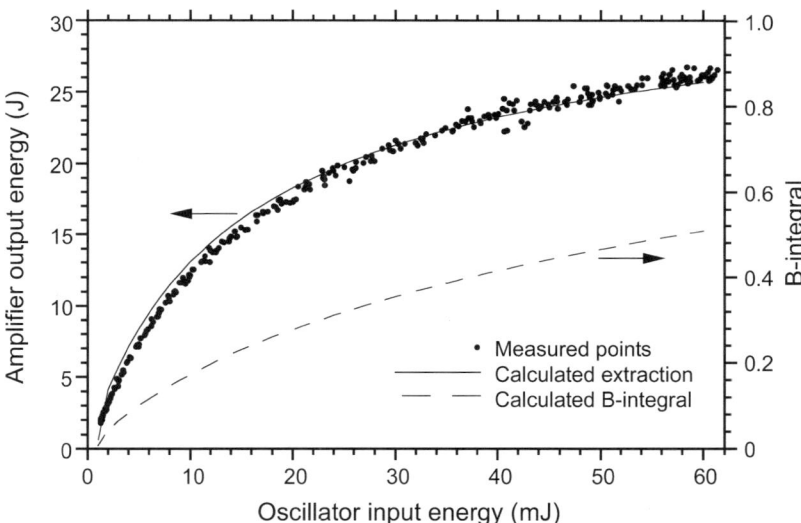

Figure 5.7. The experimentally measured amplified output energy versus input energy from the oscillator for 2500-J flashlamp energy and a single-pass small-signal amplification of ~3.5. The solid line is the numerically predicted extraction curve calculated using a Frantz–Nodvik model. The dashed line is the accumulated B-integral calculated over this output pulse energy range.

[38]. Also plotted in Fig. 5.7 is a theoretically calculated energy extraction curve using a Frantz–Nodvik numerical model [39]. When corrections for the zigzag overlap paths in the amplifier slab [40], the SBS phase conjugator reflectivity curve of Fig. 5.6, and an experimentally measured regenerative amplifier ring transmission of 90% are included in the model, good agreement is observed between the predicted and measured output energies.

5.3.4.2 Output beam quality

In order to extract high-beam-quality output from the phase conjugated amplifier system, a high-quality input pulse must be provided. In ideal operation, the wavefront reversal achieved in the SBS phase conjugator allows the spatial phase distribution of the input beam to be reproduced in the amplified output beam. Using the previously described single-frequency self-seeded Nd:YLF oscillator, excellent seed beam quality has been measured with less than 1.1 times the diffraction limited divergence. The 25-mJ output of this oscillator, however, must be further amplified and anamorphically expanded to provide the ~ 60-mJ, 8-$\times$ 120-mm input to the amplifier system. The divergence of this injected pulse is increased to 1.25 times the diffraction limit in this process, primarily due to optical aberrations in the Nd:YLF preamplifier.

As the pulse repetition frequency of the laser system is increased, the thermal aberrations in the glass amplifier significantly increase output divergence when the amplifier is operated in the unidirectional uncorrected mode without the SBS phase conjugate mirror. The measured output divergence versus repetition rate is shown in

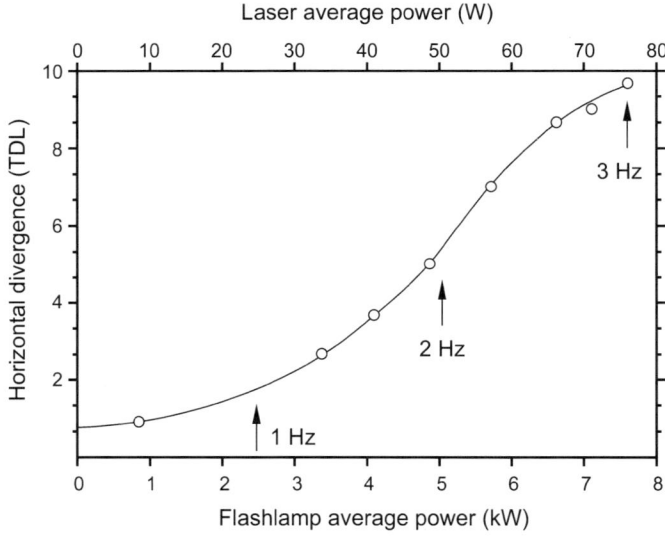

Figure 5.8. The measured output divergence in the narrow beam dimension versus repetition rate for 25-J output pulses. At 3 Hz, with an average flashlamp power of 7.5 kW, the divergence is increased to almost 10 times the diffraction limit.

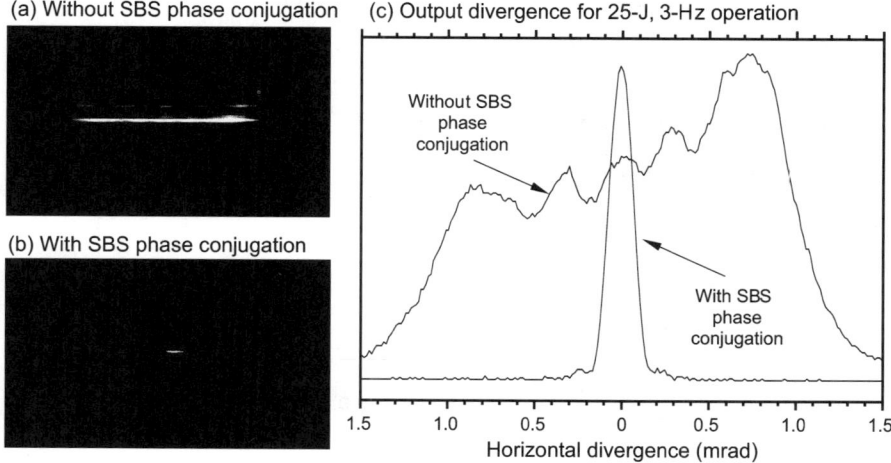

Figure 5.9. A comparison of the far-field profiles measured at the focus of a 120-cm focal length lens for 3-Hz, 75-W operation (a) without and (b) with SBS phase conjugation. Plot (c) overlays horizontal cross sections for each case.

Fig. 5.8 for 25-J output pulses. At 3 Hz with an average output power of 75 W the divergence in the narrow beam dimension is increased to almost 10 times the diffraction limit. However, when the amplifier system is configured as described previously with an SBS phase conjugator, the divergence does not measurably increase up to the maximum achievable pulse repetition frequency. Figure 5.9 compares the corrected and uncorrected far field profiles measured at the focus of a 120-cm corrected aspherical lens for 3-Hz, 75-W operation. With the phase conjugator, the width of the far-field pattern is dramatically reduced to a single central peak with small side lobes characteristic of the angular distribution from a near uniformly filled aperture.

The 8 × 120 mm output of the amplifier system is characterized by smooth reproducible beam profiles. Figure 5.10 shows (a) a CCD image of the near-field output imaged at the external relay plane conjugate to the amplifier aperture (Fig. 5.10a) and the far field recorded at the focus of a lens (Fig. 5.10d). In order to quantitatively evaluate the measured divergence of the amplifier output, the theoretically expected diffraction-limited divergence of a beam having the observed irradiance spatial distribution must be determined. A two-dimensional super-Gaussian distribution is first fitted to the near field profile. A comparison of this fit to line-outs of the measured profile is shown in Figs. 5.10b and 5.10c. The results of this numerical fit is then Fourier-transformed to generate an expected far-field distribution given a perfectly uniform phase profile. Figures 5.10e and 5.10f plot the predicted and experimentally measured far-field patterns for the horizontal and vertical dimensions. It is found that if the divergence scale for the measured data is reduced by a factor of 1.25, nearly exact overlap of the calculated and measured

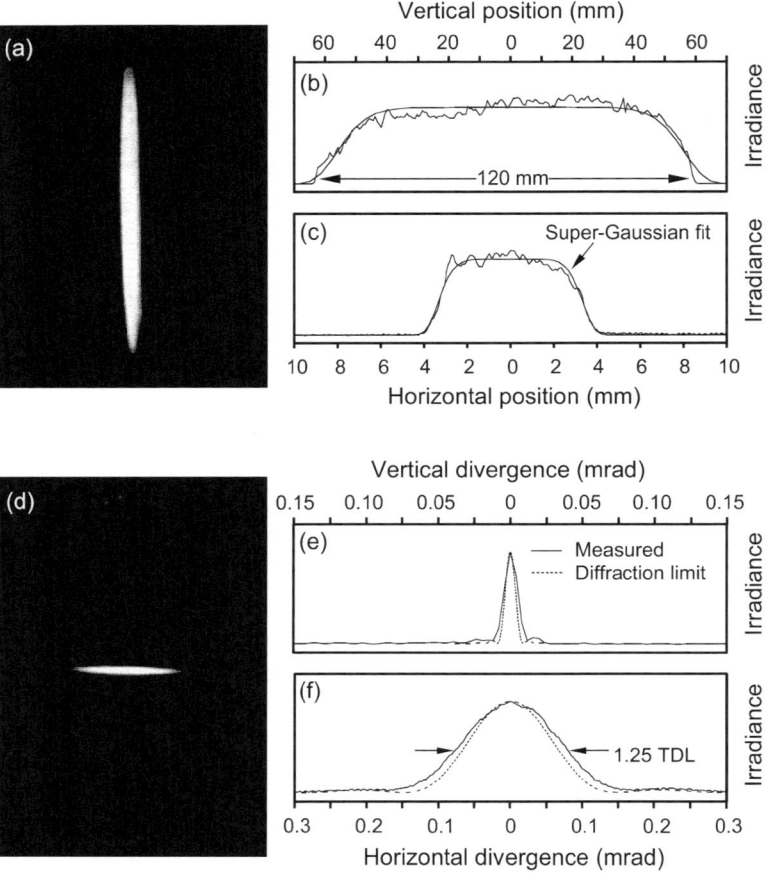

Figure 5.10. (a) Profile of the near-field imaged at the external relay plane conjugate to the amplifier aperture. Figures (b) and (c) show vertical and horizontal line-outs of this profile, respectively. The smooth curve plotted with each line-out is the result of a super Gaussian fit to this near-field profile. (d) Profile of the far field measured simultaneously with the near field profile. A comparison between (e) vertical and (f) horizontal line-outs of this profile and the Fourier transform of the super Gaussian fit of the near-field profile is shown. It is found that if the divergence scale for the measured data is reduced by a factor of 1.25, nearly exact overlap of the calculated and measured profiles results.

profiles results. It is significant to note that this is the same divergence measured for the input pulse to the amplifier as illustrated by the overlay of the far-field profiles of the input and amplified output beams in Fig. 5.11a. It can be therefore concluded that the beam quality of the high power output is presently limited by that of the low power input.

The results of the analysis of the far-field images indicate that there is a far-field output lobe whose width is 1.25 times that expected for a diffraction-limited beam.

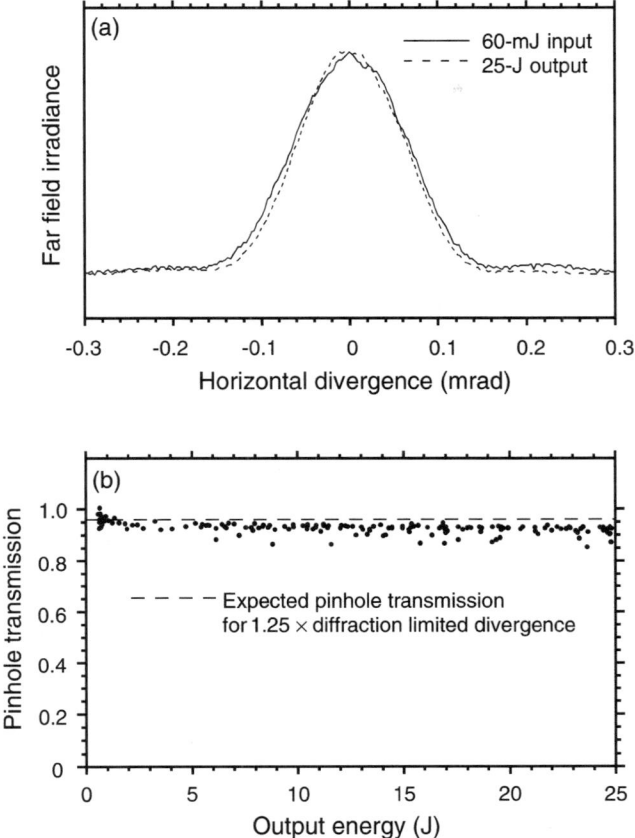

Figure 5.11. (a) A comparison of horizontal line-outs of the far-field profiles of the low-energy input pulse (solid line) and the amplified output pulse (dashed line). The divergence is unchanged to within the uncertainty of the experimental measurement. (b) Transmission of a 0.5-mm aperture placed at the focus of a 120-cm focal length lens. The expected transmission of this "bucket" for a 1.25× diffraction-limited beam was calculated to be 96%. A transmission of ∼92% was measured across the range of output energies up to 25 J.

However, it is also important to determine what fraction of the measured output energy is contained in this narrow peak. There is the need to rule out the possibility that high spatial frequency irradiance or phase modulation is scattering a significant but undetected portion of the radiation into larger divergence angles. The dynamic range of the CCD cameras is such that a very low irradiance distributed evenly across the image background could fall below the signal-to-noise detection limit of the camera but still account for a significant fraction of the total output power. A straightforward measurement that can be made in this regard is the transmission of an aperture placed at the focus of the analysis lens. In this case an aperture was chosen that was 0.5 mm in diameter, corresponding approximately to the distance between the first minima in the horizontal dimension of the measured far field profile. The expected transmission of this "bucket" for a 1.25× diffraction-limited beam is calculated to be 96%. As illustrated in Fig. 5.11b, a transmission of ~92% was measured across the range of output energies up to 25 J. This 5% reduction in the expected transmission is reasonable based on the fact that the fit between the super-Gaussian and measured near-field profiles shown in Figs. 5.10b and 5.10c is not perfect. The small-scale irradiance modulation on the beam profile is expected to contribute to a small reduction in Strehl, as was observed in the measurements.

5.3.4.3 Average power operation

In the present configuration of this flashlamp-pumped Nd:glass amplifier system, the lamps can be pulsed at repetition rates of up to 6 Hz with 2.5 kJ of deposited electrical energy per pulse without reaching the stress fracture limit of the amplifier glass. When the laser amplifier system is operated in the SBS phase conjugated configuration, the near-field irradiance distribution and far-field divergence described in the previous section is maintained up to this full pulse repetition frequency. In fact, in an experiment in which the amplifier slab was intentionally thermally loaded above 6 Hz, near-diffraction-limited output divergence was preserved up to the last laser pulses before the slab fractured. The amplifier system has been operated for periods exceeding 1 hr with average output power above 150 W. Figure 5.12 illustrates continuous records of laser pulse energies between 25 J and 27 J for 4-, 5-, and 6-Hz operation. The temporal profile of the amplified output pulses is characterized by a smooth, unmodulated envelope without a steep leading edge that would be

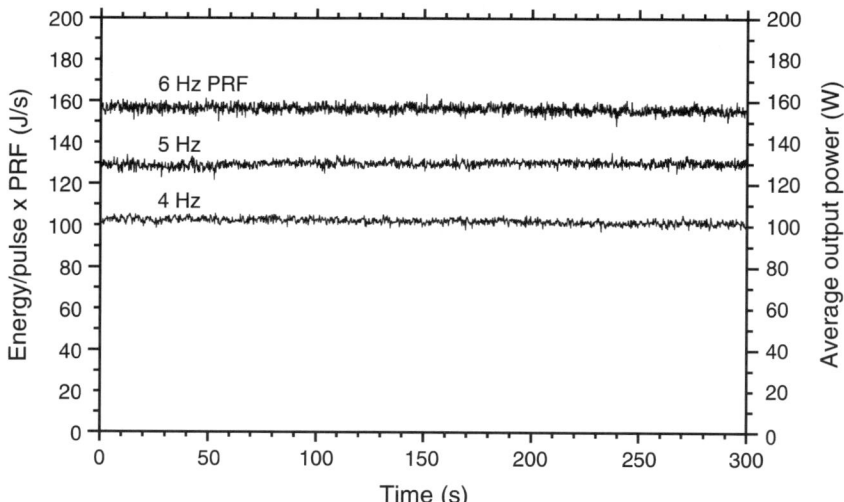

Figure 5.12. Measured energy for each laser pulse (25–27 J) for continuous operation at 4, 5, and 6 Hz. The amplifier system has been continuously operated for periods exceeding 1 hr with average output power above 150 W.

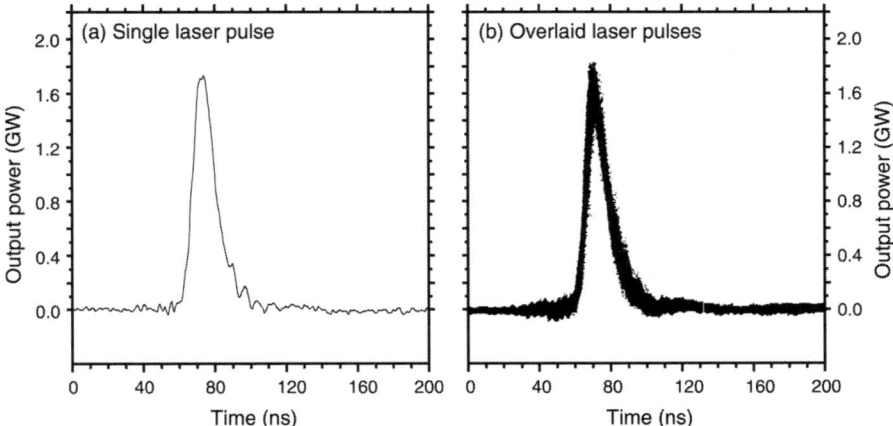

Figure 5.13. The plot of (a) a single temporal waveform as well as (b) that of ∼1800 superimposed waveforms collected at 4 Hz. The overlaid pulses were accumulated in the infinite persistence mode of a digital storage oscilloscope.

characteristic of significant temporal reshaping in the phase conjugator. The plot of a single temporal waveform as well as that of ∼1800 superimposed waveforms collected at 4 Hz are shown in Fig. 5.13.

5.3.4.4 Second harmonic conversion The high-beam-quality, high-energy output from the phase conjugated laser system is ideally suited for harmonic frequency conversion. The frequency doubler design philosophy used for this system is best summarized by "less is better." This extends to the choice of the doubler material itself, any required auxiliary optics, and any supporting mechanical hardware. As applied to the output of a slab laser, this means determining if efficient conversion is possible without any beam-shaping optics intervening between the amplifier output and the frequency converter. The nonlinear material parameters that were evaluated with this consideration included: the nonlinear coefficient, the optical damage threshold, the intrinsic absorption at the fundamental and the second harmonic wavelengths, and the sensitivity of the conversion efficiency to angle, thermal distortion, and temperature. To avoid a beam-shaping optical system, KD*P is the only material readily available in the required size and optical quality. Given this choice, a type II KD*P doubler was designed to achieve high conversion efficiency on both a per-pulse and average power basis using the principles outlined by Eimerl [41]. A detailed description of the optical design of the second harmonic converter and the optimization modeling can be found elsewhere [29].

A maximum external conversion efficiency of 82% was obtained at 21-J/pulse input, corresponding to a peak input irradiance of 185 MW/cm². The overall beam shape of the 1ω output is preserved in the 2ω beam with a reduction in the beam divergence of 2×. Operation of the doubler at an average input power of 75 W yielded an output at the second harmonic of 60 W with no reduction from the single shot conversion efficiency. This performance requires only a minor optimization of the angular orientation of the crystals for steady-state operation as predicted from the modeling. This high level of performance can only be achieved by maintaining high beam quality under full average power operation. The second harmonic conversion experiments were performed at an intermediate stage in the development of the full average power amplifier system. For this reason, operation with the doubler at the full 150-W average input power with the 15-ns laser system was not demonstrated but would be expected to pose no problems.

5.3.5 Summary of the 15-ns high-energy laser system

We have designed and constructed an Nd:glass amplifier system incorporating an SBS phase conjugate mirror that operates at 25–30 J/pulse with a pulse width of 15-ns FWHM and a pulse repetition frequency of 6 Hz. This results in a peak power of 2 GW and an average output power that significantly exceeds 150 W. The divergence of the system is measured to be 1.25 times the diffraction limit, and the output can be frequency doubled with an efficiency of 82%. The laboratory prototype has proven to be an extremely reliable system and has been in continuous operation for almost 10 years at the time of this writing. Even so, its average power and brightness performance is still unmatched for pulse energies over 10 J. It is presently in use as a laser source for a number of experimental investigations including high-throughput optical damage testing for the National Ignition Facility project [6] where the output beam frequency is being tripled, providing 10 J/pulse at 351 nm for testing.

5.4 A LONG PULSE 500-NS, 30-J LASER SYSTEM

5.4.1 Design considerations for the 500-ns system

In laser illumination applications such as long-range coherent radar or high-resolution speckle imaging [2], a narrow-bandwidth, high-pulse-energy output is needed, often with pulse widths in the range of 500 ns to 1 μs. These applications require the near-constant output power provided by a quasi-rectangular temporal pulse shape. The optimization of nonlinear frequency conversion processes such as harmonic generation or the pumping of an optical parametric oscillator (OPO) also benefits from this pulse shape. In addition to its use as a laser illuminator, the efficient operation of a high-pulse-energy, high-average-power, long-pulse laser system provides the potential for other applications for which the lower peak power improves radiation coupling and raises the optical damage threshold of materials. Examples include medical applications, the annealing of amorphous semiconductor coatings, and pumping other laser amplifiers such as Ti:sapphire or liquid dye.

The goal of >500-ns pulse duration is significantly longer than the 5–30 ns generally available from the Q-switched operation of solid-state lasers, although some work has been reported on the generation of long pulses. Long-pulse oscillators emitting near-rectangular pulses have been demonstrated by suppressing the relaxation oscillation behavior (temporal spiking) typical of solid-state lasers. This has been accomplished by actively varying the intracavity loss during the laser pulse either with closed-loop feedback on the output power [42] or with a preprogrammed temporal shape [43]. In both cases, pulse widths of up to 600 ns were achieved. However, the sensitivity and temporal bandwidth required for these control schemes are often difficult to achieve even in a carefully controlled laboratory environment. Long pulses have also been generated with passive control using an intracavity saturable absorber. In this manner, pulse widths of up to 500 ns were demonstrated but with a peaked temporal profile, large shot-to-shot pulse width variations (100-ns RMS), and a pulse energy of only 0.2 mJ [44].

The pulse energy requirements of a long-range illuminator (>10 J) make a master-oscillator power amplifier (MOPA) the most practical laser architecture. The design goal for the laser amplifier output that is presented here was 30 J/pulse at 1 μm. The large amount of amplification required to reach this energy from a low-energy injected oscillator pulse leads to severe temporal distortion of the input pulse shape due to amplifier gain saturation during efficient optical extraction. For this reason, the near-rectangular output of previously reported that long-pulse oscillators would not be suitable as an injection source. It has been recognized, however, that the smooth temporal profile of a relaxation oscillation pulse emitted by a free-running solid-state laser suffers much less pulse-width distortion when amplified [45]. Applying this concept, 1.1 J in a 110-ns pulse was demonstrated from a single amplifier aperture that was then frequency-doubled to 520 mJ at 532 nm [46]. The temporal profile remained, however, near Gaussian in shape.

For this laser system, we adopted a scheme that also relies on the amplification of the output of a free-running master oscillator [32]. However, only the leading edge of the pulse is used and it will be shown that the measured width of the injected pulse does not directly determine the width of the amplified output pulse. The amplification of the leading edge of the input pulse gives rise to a quasi-constant output power, the magnitude of which is dependent on the exponential time constant of this rising edge. A simple analytical theory allows the power, and hence pulse duration, to be tailored to a desired level by adjusting the optical buildup time in the oscillator.

5.4.2 Optical architecture of the 500-ns system

The long-pulse laser system is a direct adaptation of the optical architecture used for the 15-ns MOPA described in the previous section. The primary difference is the replacement of the Pockels with a permanent-magnet Faraday rotator. This allows pulses of arbitrary duration, constrained only by peak power requirements in the SBS phase conjugate mirror, to be amplified. While the 15-ns laser system is a laboratory development system built with standard components on an optical bench, an effort was made to package the 500-ns laser system in a compact, engineered

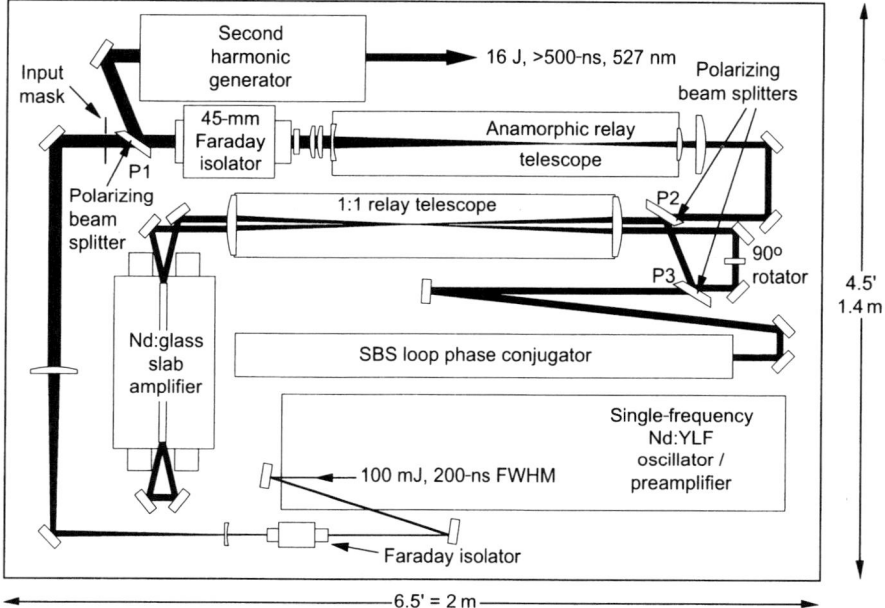

Figure 5.14. Optical layout of the 500-ns phase conjugated laser system. The Pockels cell from Fig. 5.5 has been replaced by a permanent-magnet Faraday rotator, allowing pulses of arbitrary duration to be amplified in the 8-pass zigzag slab amplifier.

assembly as illustrated in the optical layout of Fig. 5.14 and the photograph of the system in Fig. 5.15.

5.4.2.1 Long-pulse multipass polarization switching

The long-pulse laser is a MOPA design using an SBS phase conjugate mirror. The optical architecture and beam transport system are based on the design presented for the 15-ns laser system. The described considerations for relay imaging between gain passes and to the input aperture of the SBS conjugator are carefully incorporated into this design. In the previously described 15-ns pulse amplifier, a large-aperture Pockels cell is used as an active optical switch to isolate the phase conjugated and amplified output from the input beam train. However, in order to amplify pulses of up to 1 μs in duration, the Pockels cell switch was replaced with a 45-mm clear-aperture permanent-magnet Faraday rotator. This results in an eight-pass regenerative amplifier system with totally passive polarization switching. A schematic diagram of the amplifier optical design is depicted in Fig. 5.14. The oscillator output is expanded from a 1-mm beam to a 6-mm beam using a simple two-lens telescope. It is then amplified by a double-pass through an Nd:YLF preamplifier. After passing through a 10-mm clear-aperture Faraday isolator, the injection beam is expanded to a diameter of 30 mm. This beam is then shaped to a 25-mm square by a hard-edged

Figure 5.15. Photograph of the 500-ns laser system, with electro-optics technologist James Wintemute, during final activation and testing.

aperture before entering the amplifier system through polarizing beamsplitter P1. An anamorphic telescope takes the input from its 25×25-mm^2 size to the 7×120-mm^2 used in the glass amplifier aperture. The input beam enters the regenerative amplifier ring in p-polarization through polarizing beamsplitter P2 and undergoes two gain passes. The polarization is then rotated 90° by a quartz rotator and the beam reflects from the same beam splitter in s-polarization, undergoing two more gain passes. When the polarization is returned to the original p-state after the second pass through the rotator, the beam is coupled out through polarizer P3 and directed into the SBS phase conjugate mirror. The reflected beam from the phase conjugator retraces the path of the input beam, adding four more gain passes for a total of eight. The polarization rotation of the 45° Faraday rotator and the 45° quartz rotator cancel each other in the input direction, but now, in the output direction, they add, resulting in a full 90° rotation. The amplified beam is therefore reflected by polarizer P1 and enters the doubler. Since the output beam passes back through the anamorphic input telescope, the 25×25-mm^2 beam size is restored before reaching the doubler.

5.4.2.2 Long-pulse energy extraction For a given laser amplifier chain, it is possible in theory to generate any desired output temporal pulse shape by correctly tailoring the input pulse shape, taking into account the effects of energy extraction and total system gain. In order to determine this optimal input shape for a 500-ns rectangular pulse, a straightforward numerical treatment of the long-pulse extraction problem can be made. In the case of a solid-state energy storage medium such as Nd:glass, the fluorescence lifetime of ~ 300 μs and flashlamp pulse duration of

~250 μs are significantly longer than the total duration of the energy extraction process which, in the case of this laser system, is ≤1 μs. Based on this, it can be assumed that essentially no gain pumping occurs during the amplification of the injected pulse. It can then be shown [32] that the required input shape has an exponentially rising leading edge with a time constant τ defined by

$$\tau = \frac{F_{sat}}{nI_{out}} = \frac{AF_{sat}}{nP_{out}}$$

where F_{sat} is the saturation fluence (J/cm^2), I_{out} is the desired output irradiance (W/cm^2), A is the beam area in the amplifier (cm^2), P_{out} is the desired output power (W), and n is the number of amplifier passes. It is intuitively reasonable that to generate a constant output power from the amplifier system, an exponentially rising seed laser input power is required. As energy is extracted from the laser amplifier at a constant rate, the gain decreases exponentially. An exponentially rising input pulse profile can therefore be used to compensate this loss in gain, maintaining near-constant output power during the pulse. In the case of the laser system described here, our output energy goal was 40 J in a 500-ns pulse (80 MW) after the last of the eight gain passes. This would then ensure an energy of >30 J (60 MW) in the beam after it is reformatted in the anamorphic telescope and coupled out through the Faraday optical isolation stage to the harmonic converter. With an $F_{sat} = 5$ J/cm^2 for the Schott APG-1 laser glass used for the zigzag slab and a beam area of 0.7×12 cm^2, a simple application of the above formula yields an optimal exponential time constant $\tau = 65$ ns for the injected seed pulse. This very simple derivation does not take into account optical losses for the components of the beam train and linear absorption in the amplifier glass. However, it provides a very effective tool for configuring the pulse shape of the injected pulse. We experimentally verified that a time constant of 60ns provides the desired 60 MW over 500 ns (30 J/pulse) to the second harmonic converter.

5.4.3 Long-pulse SBS phase conjugation

A challenging aspect of operating an amplifier system with SBS phase conjugation for pulse durations of 500–1000 ns is the power threshold requirement of the phase conjugate mirror. It was found that an SBS medium of either high-pressure nitrogen (N$_2$, ~90 atm) or xenon (Xe, ~50 atm) produces good results for microsecond-duration input pulses. Ultrahigh-purity (99.999%) N$_2$ was chosen for its low cost, ready availability, and observed resistance to optical breakdown.

5.4.3.1 *The SBS loop geometry* The measured power threshold of >400 kW for a simple-focus N$_2$ SBS cell and the input pulse rise times needed for this laser system was found to be too high to generate the desired pulse durations in the amplifier output. The SBS cell receives an incident energy of >1 J under the operating conditions for 30-J/pulse extraction. This energy would easily exceed the SBS threshold if delivered in a rectangular 1-μs pulse. However, the temporal

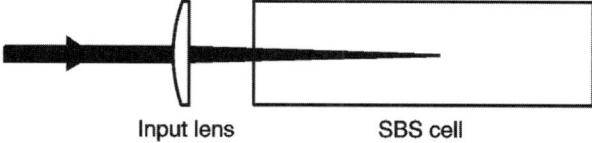

(a) Conventional SBS phase conjugate mirror

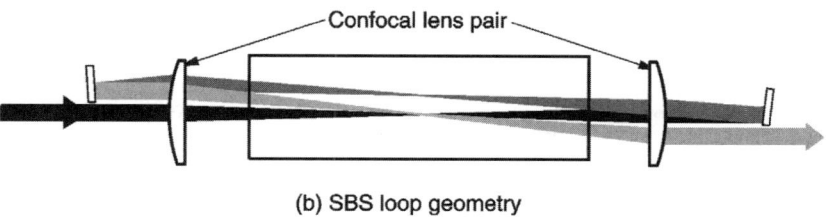

(b) SBS loop geometry

Figure 5.16. A comparison of (a) the simple single-focus SBS phase conjugate mirror with (b) the three-pass SBS loop geometry. The first and third foci in the SBS medium are overlapped, resulting in a self-pumped four-wave mixing interaction. The loop provides a lower nonlinear threshold and increased stability in the pulse shape and temporal phase.

profile at the phase conjugate mirror exhibits a large ratio in power from the end of the pulse to the beginning due to the required exponental buildup from the oscillator. This leaves a conventional, single-focus SBS mirror well below threshold for the first part of the pulse.

In order to lower the power threshold of the SBS phase conjugate mirror, a three-pass loop geometry was adopted [47]. This optical arrangement is compared schematically to the single focus SBS cell in Fig. 5.16. In the loop configuration, the light transmitted through the SBS cell is recollimated with a second lens of equal focal length. A mirror then directs the beam back through the SBS cell where it comes to a second focus. A second mirror sends the beam through the cell for a third and final pass. In this third pass the focal waist is overlapped with that of the first pass. These three consecutive optical foci extend the SBS nonlinear gain path for the long pulse input, reducing the power threshold by almost a factor of three. Figure 5.17 is a plot of the output Stokes power versus incident laser power for the single focus and the loop configurations. These were measured using input pulses with FWHM durations of 200 ns. As can be observed by the hysteresis in this data, there is a significant difference between the power threshold seen by the leading edge of the pulse and the steady state threshold observed at the end of the pulse. This is due to the transient build-up of the SBS acoustic grating and will be discussed in more detail in the following section.

There is significant benefit of the SBS loop architecture in addition to the reduction of the SBS threshold. As described in Ref. 47, once threshold is reached,

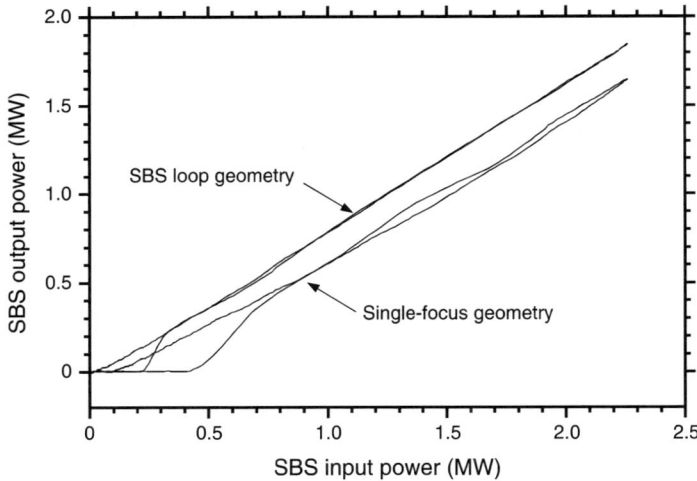

Figure 5.17. SBS output power versus input power measured for the SBS single-focus and loop geometries. 400 mJ pulses with a pulse duration of 200 ns FWHM were used. The single-focus curve was collected by simply blocking the transmitted beam after the SBS cell, interrupting optical feedback to the loop. Both the decreased threshold and increased temporal stability for the loop geometry are readily apparent.

the overlapping first and third foci in the SBS cell generate a four-wave mixing interaction that serves to strongly stabilize the Stokes return from the phase conjugate mirror. The effect of this interaction is to feed a portion of the generated Stokes signal back through the loop and into the phase conjugated wave. The stimulated scattering system then does not rely on input light scattered from acoustic noise at focus to sustain the Stokes return after the SBS threshold is reached. This increased stability is readily apparent upon comparing the input/output power plots for the single focus cell and the loop in Fig. 5.17. The power fluctuations observed in the single-focus SBS mirror were not reproducible from shot to shot but were always of comparable magnitude to the curve shown. This data were collected by simply placing a block in the beam after the first pass through the SBS cell, reducing the phase conjugator to a simple single focus. Besides the significant decrease in the power threshold, the Stokes power fluctuations are dramatically reduced when the loop is unblocked. The four-wave mixing interaction also plays a crucial role in maintaining temporal phase stability in the laser amplifier output, as will be described in the upcoming discussion on coherence length.

5.4.3.2 Bi-exponential oscillator buildup

As can be seen in the hysteresis of the data curves in Fig. 5.17, even when using the SBS loop geometry, the lowest possible threshold is not achieved with the 60-ns input pulse rise times required for 500- to 1000-ns pulse amplification. One way of lowering the transient SBS threshold and reducing the size of the observed hysteresis loop is to reduce the temporal slope of the input pulse. To accomplish this, a Pockels cell was inserted in the ring oscillator and was used to introduce a ~10% loss during the early part of the oscillator buildup. This increased the time constant for the exponential growth to 150 ns, which is approximately 10 times the 15-ns Brillouin time constant for the 90 atm N_2. The voltage from the Pockels cell is removed at an adjustable point during the oscillator buildup as monitored by a photodiode. This point is adjusted to occur just after the SBS phase conjugate mirror reaches threshold. When the loss from the Pockels cell is removed from the oscillator resonator, the buildup time then decreases from 150 ns to the 60 ns required for the previously described optimal extraction from the laser amplifier. The slower buildup also helps to ensure single-frequency output from the ring oscillator. Figure 5.18a shows a measured temporal

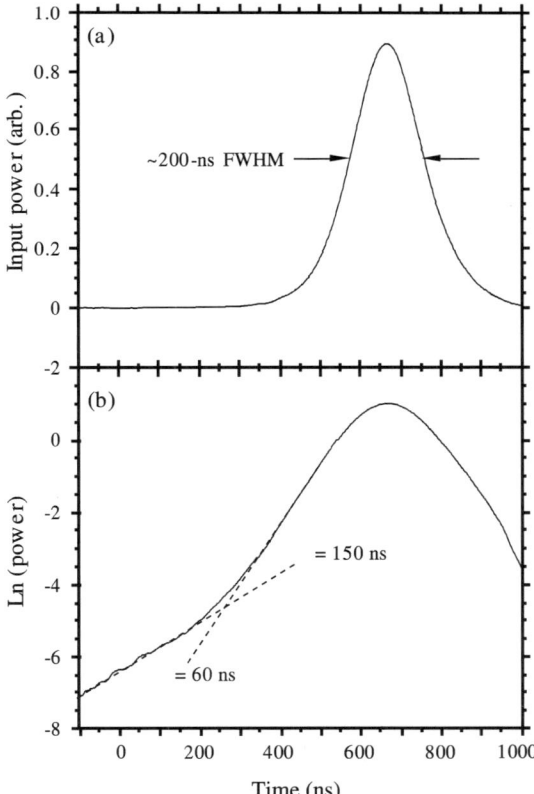

Figure 5.18. (a) Measured injected oscillator pulse used to generate 500-ns output pulses. The logarithmic plot shown in (b) illustrates the bi-exponential buildup used to lower the transient SBS threshold. The necessary dynamic range for the measured profile shown in this plot was achieved by dividing the curve into four temporal windows and simultaneously collecting each on separate channels of a digital oscilloscope with 10× steps in gain between each channel.

profile of a typical oscillator input pulse. The bi-exponential output growth is clearly seen in the logarithmic plot in Fig. 5.18b.

5.4.4 Output characteristics of the 500-ns laser system

At the pump levels required for generating 30 J in 500 ns, the laser amplifier system can be safely operated at a pulse repetition frequency of up to 3 Hz without risk of thermal fracture of the Nd:glass amplifier slab. We have demonstrated an average power of 96 W (32 J at 3 Hz) at 1 ω (1.053 μm). This is reduced from the 150 W reported earlier for the same glass amplifier with 15-ns pulses [29] due to the inherently lower extraction efficiency of the long pulse amplifier. The SBS phase conjugated mirror accurately reverses both (a) the static wavefront distortions arising from errors introduced by optical components and (b) those introduced by the thermally loaded glass amplifier. As a result, just as with the short pulse amplifier, the output beam characteristics of pulse energy, pulse shape, beam quality, pointing stability, and coherence length are essentially unchanged from single shot to full average power operation.

5.4.4.1 Long-pulse output energy and pulse shape To amplify pulses with durations from 500–1000 ns, the laser amplifier slab is operated with a single-pass gain of 1.8 Np corresponding to a small signal amplification of 6.2. At this laser gain, up to 32-J pulses are extracted with a injected seed energy of $\sim$ 100 mJ. The plot in Fig. 5.19a shows the temporal profile of an amplified 500-ns output pulse. An oscillator pulse is shown temporally aligned to the output pulse as it would be recorded after a full optical transit through the amplifier system without gain. Note that the bulk of the output pulse results from the amplification of the exponentially rising leading edge of the input pulse. This provides the somewhat counterintuitive result of the amplification of a 200-ns FWHM injected pulse to the full-energy, 500-ns output pulse. Just as the log of the input power, shown in Fig. 5.18b, begins to roll over from linear, the amplified output abruptly falls off. Also note that although 100 mJ is injected from the oscillator/preamplifier, less than half of the energy from the first part of the input pulse substantially contributes to the output pulse.

Figure 5.19b shows a progression of output pulse temporal profiles that readily illustrate the simple extraction theory presented earlier. Using a constant input seed pulse energy, as the gain in the amplifier is increased, the peak power of the amplified output pulses increases until a maximum value of 60 MW is reached. A further increase in amplifier gain beyond this point does not increase the peak power but only increases the pulse duration. In this manner, controlled extraction of long pulses is achieved from the amplifier resulting in near-rectangular, constant power output pulses.

5.4.4.2 Long-pulse beam quality As described earlier, the output beam from the amplifier system is approximately 25×25 mm^2. Figure 5.20 illustrates a near-field profile imaged at the input to the doubler assembly. Figure 5.21 shows the corresponding measured far-field profile for this beam. Although nearly square in

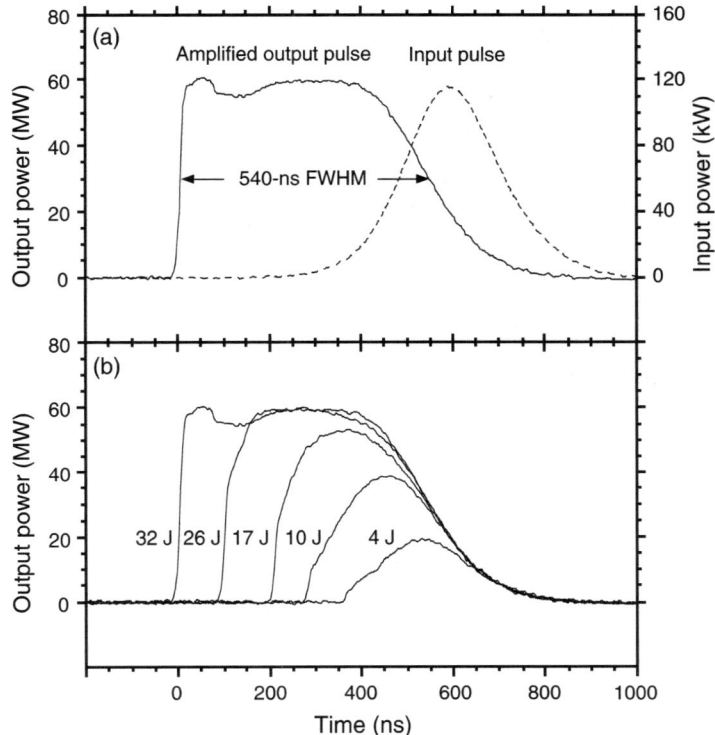

Figure 5.19. (a) A measured 540-ns output pulse shown temporally aligned with the 200-ns FWHM seed pulse. As can be seen, most of the long pulse is generated by the amplification of the exponential buildup on the leading edge of the input pulse. As the gain is increased in the Nd:glass amplifier with a constant seed energy, as shown in (b), the output power reaches a maximum value after which, further increases in gain only serve to extend the pulse duration.

shape, the near-field profile has somewhat rounded corners. In the plots of the lineouts of the far-field profiles shown in Figs. 5.21b and 5.21c, dashed lines indicate the expected zero irradiance nodes for both a square- and a round-top-hat near-field profile having the same transverse dimensions of the near field in Fig. 5.20. Comparing the predicted Airy disk widths to the data show that the divergence of the laser output is very near diffraction-limited. Beam transmission through a fixed $2\lambda/D$ aperture in the far field yields an estimated beam divergence of 1.1 times the diffraction limit. The far-field irradiance profiles for the long-pulse laser were measured for pulse repetition frequencies between 1 and 3 Hz with no measurable change in divergence.

5.4.4.3 Long-pulse second harmonic conversion

We adopted a design for the frequency doubler that is very similar to that used for the 15-ns system. It consists of a type II KD*P doubler using two crystals in an alternating-Z geometry

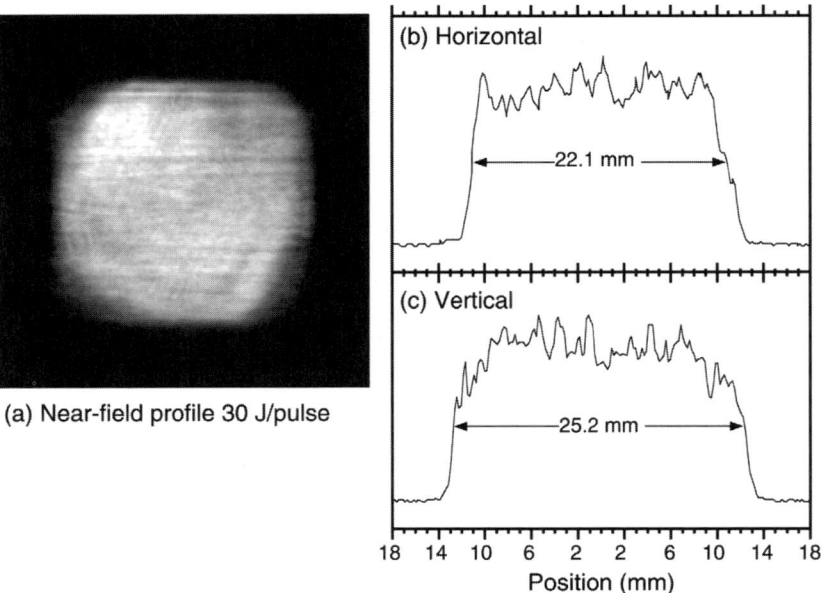

Figure 5.20. (a) A measured near field beam profile for the 30-J, 500-ns laser system showing both (b) horizontal and (c) vertical irradiance line-outs.

[41]. In the case of the 500-ns, 30-J output, however, it was necessary to add a set of beam-shaping optics between the output of the laser and the doubler in order to increase the irradiance incident on the crystals. Since the frequency converter is outside the SBS wavefront-corrected path, it is desirable to reshape the beam without concern for introducing wavefront error due to the imperfect alignment of a telescope. For this reason, a set of two 45° prisms are used to reduce the 25-mm output beam in one dimension by a factor of 2.5, resulting in a beam size of 10×25 mm^2 in the KD*P crystals. A second set of prisms restores the 527-nm beam to a 25-mm square after the crystals. A dichroic mirror outside of the laser enclosure is used to separate the unconverted 1053-nm light from the 527-nm output. A typical external conversion efficiency of 60% is obtained at up to 30 J/pulse corresponding to a peak input irradiance of 24 MW/cm^2. Under these conditions, routine operation of >45 W in 500-ns pulses at 3 Hz is achieved. Figure 5.22 shows a 10-minute portion of a run with an average input energy of 27 J and average output energy of 16 J/pulse at 527 nm.

Due to the increased length of the doubler crystals required for the long-pulse, low-irradiance laser output (60 mm total), the second harmonic converter introduces greater wavefront distortion than that used for the 15-ns laser system. Figure 5.23 shows the measured far-field profile at 16 J/pulse and 3 Hz. Because of weak absorption ($\sim 0.15\%$/cm) at 1053 nm, the converted green beam exhibits $\sim 0.2\lambda$ spherical wavefront error at full average power which can be compensated by a small adjustment of the far-field diagnostic system. As can be seen from the line-

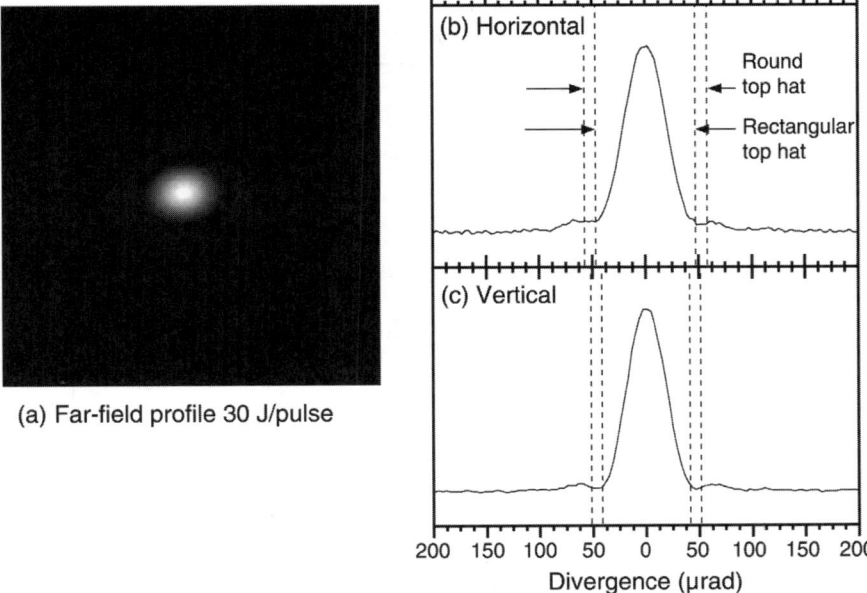

Figure 5.21. Far-field beam profile for the near field shown in Fig 5.20. The distribution represents a divergence of 1.1× the diffraction limit. The dashed vertical lines show the theoretically expected first minima in the line-outs for both a round and a square near-field distribution.

outs in Figs. 5.23b and 5.23c, higher-order distortions are also observed which lower the measured Strehl by 2× from the single shot value. In the final user application of the system, an external telescope was used to further expand the laser output before entering the beam director. The very small thermal focus introduced by the doubler was easily corrected in the focusing adjustment of this telescope.

5.4.4.4 Pointing stability As demonstrated by the measured beam quality of the laser output, SBS phase conjugation effectively corrects both the passive optical figure errors in the transport optics and the thermally induced aberrations in the glass amplifier. A very valuable aspect of this correction is also the cancellation of the first-order aberration of tilt in the amplifier system resulting from vibration, air disturbances, and long-term alignment drift. For precision, long-range illumination applications, it is critical that the laser system maintain reproducible shot-to-shot pointing. We have demonstrated that the incorporation of the SBS phase conjugate mirror is key to maintaining very tight beam pointing control. Rigorous relay-imaging through the amplifier system allows small drifts in pointing of the many optical components in the beam train without introducing aperture clipping in the amplifier. SBS phase conjugation then cancels these errors so that the output

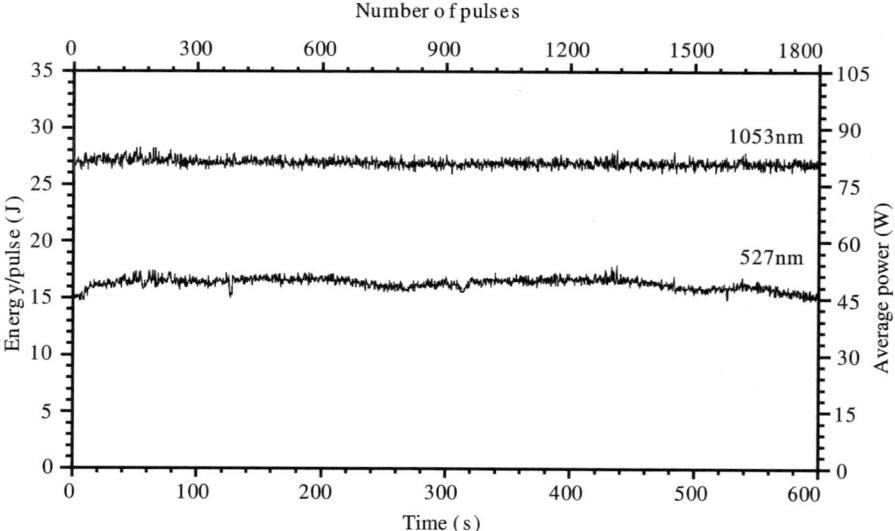

Figure 5.22. Measured 1053-nm and 527-nm pulse energies for the 500-ns laser system at 3 Hz. The slow variation in frequency conversion efficiency over the course of this 10-minute run was caused by a ±1°C fluctuation in the cooling water used to control the temperature of the second harmonic generator.

pointing stability of the high-power laser system is determined only by that of the low-power injection laser.

Figure 5.24 illustrates a measurement of the shot-to-shot pointing stability over the course of 100 full-energy laser pulses at 3 Hz. The RMS pointing error in each dimension is ~8% of the $2\lambda/D$ far-field spot size. While this is an excellent result for a laser of this pulse energy and average power and meets the <10% requirement of the intended application, measurements on the injection laser lead us to expect an RMS value of <5%. Further investigation showed that the lack of sufficient vibration isolation between the optical bench and the high-energy amplifier allows the flashlamp percussion and vibration from the high-speed water flow to couple to the oscillator, causing the increased instability in its output. This situation can be readily corrected, as will be shown in the phase-locked laser system design described in the next section.

5.4.4.5 Coherence length

An important motivation behind the generation of long-duration, high-energy pulses from this laser system was provided by coherent imaging applications that required not only the long pulse, but also the narrow spectral bandwidth that it could provide [48]. A paper published soon after we had begun work on the the 500-ns laser system called into question the feasibility of maintaining narrowband output and long coherence length from an SBS phase conjugate mirror [49]. In this work, the researchers showed that the temporal phase from a single-focus SBS mirror is characterized by both gradual drifts and abrupt $\lambda/$

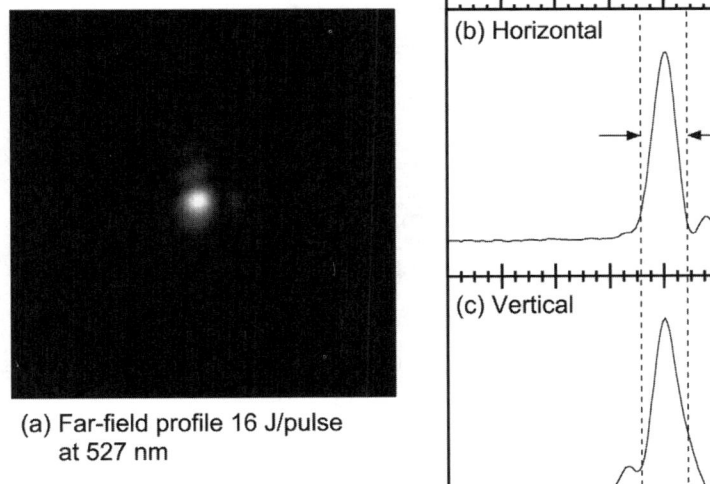

Figure 5.23. Far-field profile for the 527-nm output from the 500-ns laser system at 16 J/pulse and 3 Hz. The divergence is slightly degraded from the near-diffraction-limited 1053-nm profile shown in Fig. 5.21 due to refractive index inhomogeneities in the KD*P doubler crystals.

2 steps that are particularly pronounced for laser pulse durations longer than 10 times the Brillouin acoustic decay constant. The source of these instabilities was attributed to the noise that initiates and sustains the stimulated scattering process. At face value, this result would prevent the generation of transform-limited temporal coherence from an SBS phase conjugated amplifier system. However, our understanding of the four-wave mixing process in the SBS loop and its independence of noise at focus to sustain it [47] lead us to believe that we should not observe these phase drifts with the loop. Measurements of the high-energy pulses from the laser system confirmed this prediction. Similar conclusions have been reached by other researchers [50].

In order to measure the coherence length of the 500-ns output pulses, a stabilized single-frequency diode-pumped Nd:YLF laser operating with CW output was used as a reference source. Full-aperture attenuated samples of 25-J pulses at 1053 nm from the phase-conjugated amplifier system were heterodyned with the reference laser on a 1-GHz PIN photodiode and recorded using a digital oscilloscope. The reference laser was tuned to an offset of 100–200 MHz from the frequency of the high-energy laser, and the resulting heterodyne beat was measured. As seen in Fig. 5.25a, the waveforms exhibited good temporal contrast and a smooth temporal envelope, consistent with the laser pulse shape. Analysis of the beat frequency,

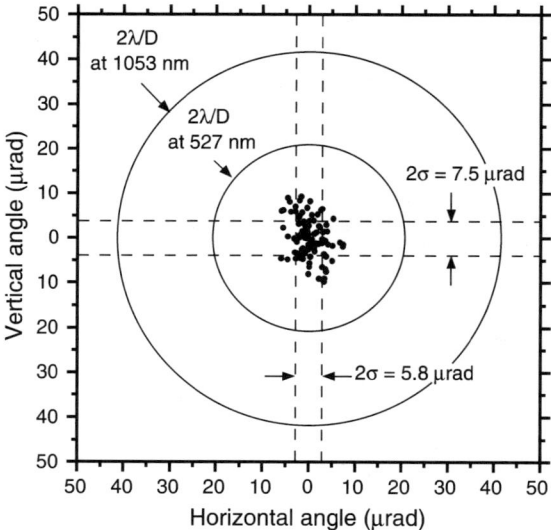

Figure 5.24. A pointing scatter plot measured for 100 pulses at 3 Hz for the 527-nm, 500-ns laser output showing a 2σ variation of 5.8 and 7.5 μrad in the horizontal and vertical dimensions, respectively.

shown in Fig. 5.25b, shows that variation in the output frequency of the laser generates a cumulative phase error of 0.1λ peak-to-valley over the entire 700-ns window, demonstrating excellent frequency stability. The effectiveness of the SBS loop in providing this phase stability can be readily illustrated by blocking the beam transmitted through the SBS cell, thus converting the phase conjugate mirror back to the conventional single-focus geometry. As shown in Fig. 5.25c, the heterodyne beat becomes unstable and exhibits phase jumps very similar to those reported in Ref. 49. Although the loop geometry was first adopted to lower the SBS threshold and increase the temporal amplitude stability, we found that it was also critical to maintaining near-transform-limited bandwidth in the high power output.

5.4.5 Summary of the 500-ns high-energy laser system

The incorporation of low-threshold SBS phase conjugation in this laser amplifier system provides near-diffraction-limited divergence, and the output pulses can be frequency-doubled to 527 nm with >50% total external conversion efficiency. This has resulted in a laser system with 500-ns full width at half-maximum (FWHM) output pulse duration in a near-rectangular temporal profile, 16 J/pulse at 527 nm, and a maximum pulse repetition frequency (PRF) of 3 Hz. These demonstrated performance specifications exceed previously reported long pulse green solid-state laser output [46] by 5× in pulse duration, 30× in pulse energy, and 18× in average power from a single amplifier aperture. This is also the first demonstration of narrow-

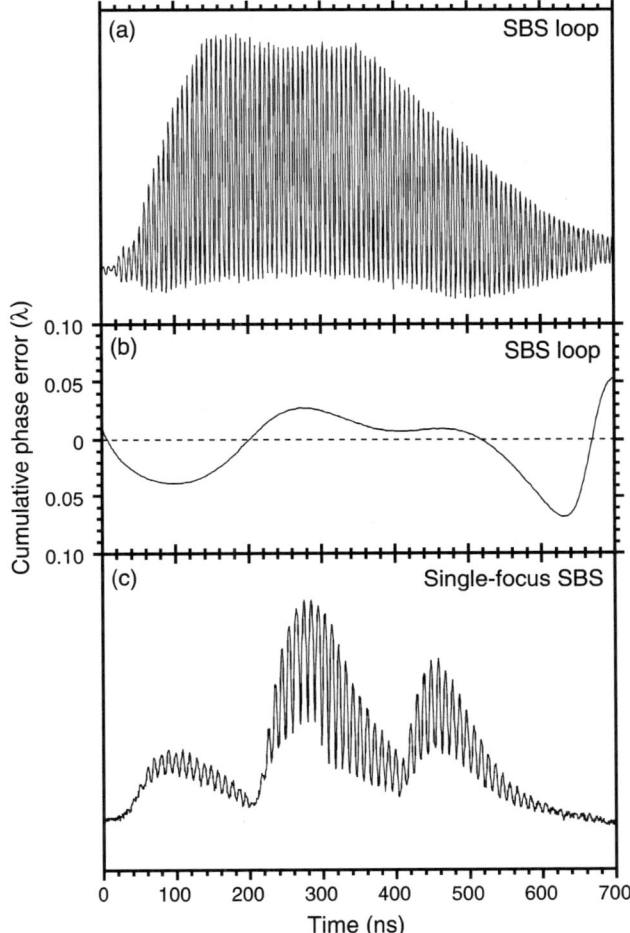

Figure 5.25. (a) The heterodyne beat between a sample of a 500-ns high-energy output pulse and a single-frequency 1053-nm diode-pumped CW laser. (b) The analysis of the beat profile demonstrates a peak-to-valley variation of the cumulative phase error of $<0.1\lambda$. When the SBS loop is blocked, returning the SBS mirror to the single-focus geometry, the heterodyne beat becomes very unstable and exhibits the large temporal phase steps that have been observed by other researchers [49].

bandwidth SBS phase conjugation in a long-pulse, long-coherence-length laser amplifier system. The output pulses have a measured coherence length that is transform-limited, readily exceeding our initial requirement of >60 m. Using the same laser system, pulses of almost 1 μs in duration can also be amplified with equal output energy at 1053 nm. These cannot be efficiently doubled, however, since the design point of the second harmonic converter is for 500-ns pulses.

5.5 A 100-J LASER SYSTEM USING FOUR PHASE-LOCKED AMPLIFIERS

An effective way of scaling the average power from a laser system is to coherently combine the outputs from a number of separate amplifiers that are operated in parallel. This allows the pulse energy to be increased without the need to increase the size of individual gain elements. The technique is particularly attractive in the case of solid-state laser systems since increasing the size of the amplifiers can greatly reduce the ability to efficiently pump and cool the gain medium. For this approach to be effective, the individual beams must be phase conjugated in such a way as to maintain the correct relative temporal phase relationships in the high-power output beam. This cannot be accomplished by using independent SBS phase conjugate mirrors for each beam since the output beams would then have arbitrary temporal phase alignment that would randomly vary from pulse to pulse. Each beam could individually have a well-corrected output wavefront with the effective correction of thermally induced distortions in the amplifiers. However, they could not be coherently combined into a single beam composed of a phase-locked array of beamlets that would propagate as if it had been amplified in a single optical aperture. For this reason, a method of individually referencing the absolute phases of each of the individually amplified beams must be achieved.

A number of approaches have been taken to the phase-locking of individual beams using SBS. Basov first dealt with the problem when addressing a technique to correct depolarization by dividing the two orthogonal polarization components into separate beams and then referencing them together in a single SBS cell [51]. In order to maintain sufficient overlap and mixing between the two beams, random phase plates were used in front of the cell. Unfortunately, the strong aberrations from the phase plates could not be completed conjugated and removed from the output beam. Other researchers have tried to overlap multiple beams at the focus of a simple SBS cell, without using phase plates. This approach, however, is very sensitive to the degree of beam overlap achieved, and phase-locking with adequate accuracy and stability cannot be attained. An example of the latter approach is work in which sections of a multi-segment flashlamp-pumped Cr:Nd:GSGG slab amplifier were phase-locked [52]. Although the divergence from four segments of the amplifier was shown to be significantly reduced from that expected from a single segment alone, the overall beam quality of the phase-locked array was limited to 2.5 times the diffraction limit.

The major concern in previous attempts to use SBS phase conjugation to successfully phase-lock multiple amplifier aperture was to improve the overlap, and hence nonlinear coupling, between the individual beams in the SBS cell. However, based on our previously discussed experience with extending the coherence length of the SBS Stokes output, we believed that the shortcoming of these attempts were primarily caused by temporal phase fluctuations in the SBS output which served to disrupt the phase-locking process. Based on this, the incorporation of the SBS loop architecture into the multiple beam phase conjugate mirror should significantly enhance the phase-locking peformance. This was also recognized by other researchers who have demonstrated the very accurate and stable low-power phase-locking of two laser beams [50]. The approach in that work, however, included the use of a third phase conjugated reference beam. To avoid the additional complexity and expense of a separate reference laser system, we have devised an approach that uses only the beams to be phase-locked in a modified SBS loop arrangement [53].

5.5.1 Design considerations for the phase-locked system

The requirements for the phase-locked laser system are 100 J/pulse in the long 500-ns FWHM pulse format. Just as with the single-beam, 500-ns laser system, the pulses are required to have near-transform-limited temporal bandwidth and need to be frequency-converted to the green (527 nm) with a goal of 50 J/pulse. Our approach to this laser design was to scale the single-beam, long-pulse laser system by adding an additional three parallel laser amplifier channels, thus increasing the 1053-nm output from 25 J/pulse to 100 J/pulse.

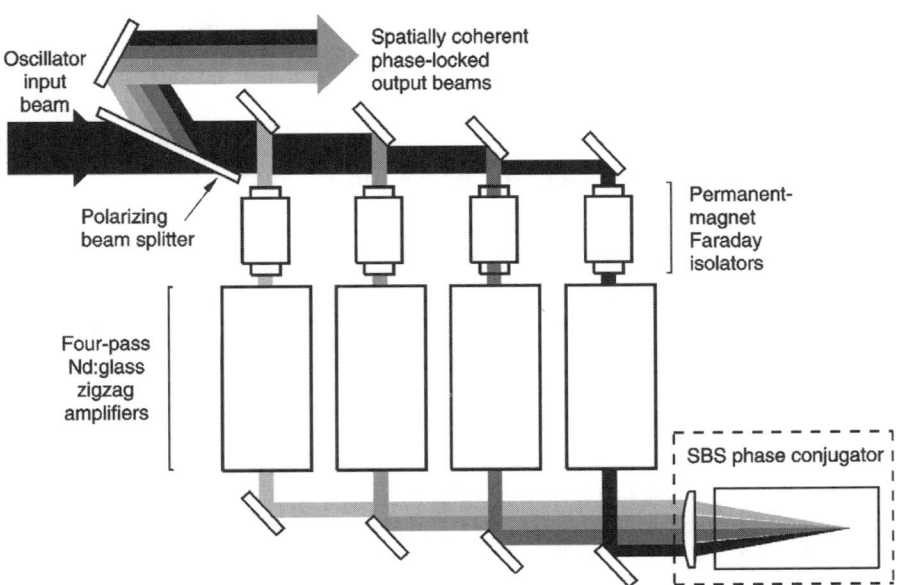

Figure 5.26. Conceptual design of the four-beam SBS phase-locked amplifier system. A single oscillator beam is spatially divided into four beamlets that are amplified in individual Nd:glass amplifiers before being rejoined in the SBS phase conjugate mirror.

186 HIGH-PULSE-ENERGY PHASE CONJUGATED LASER SYSTEM

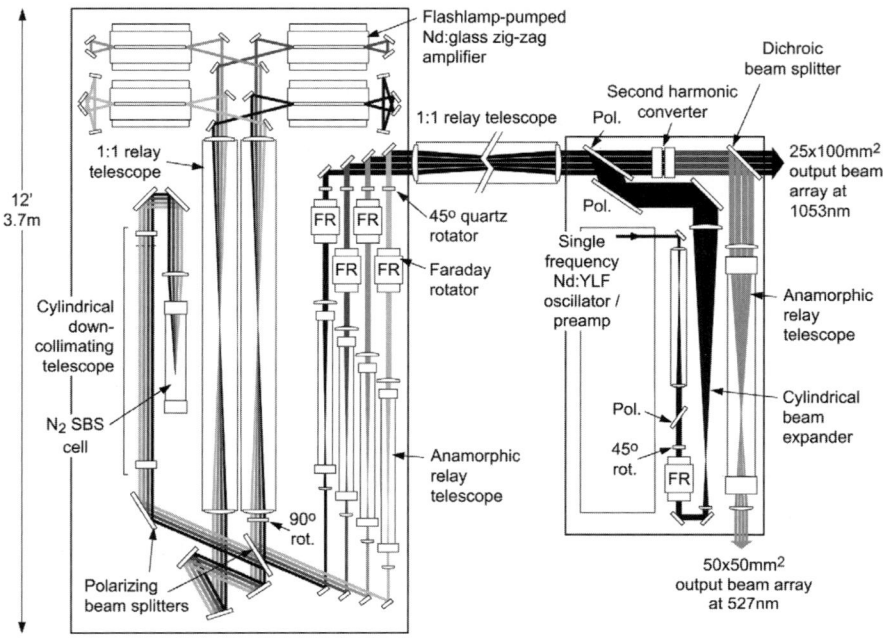

Figure 5.27. Detailed design of the phase-locked amplifier system. The four flashlamp-pumped Nd:glass amplifiers are placed on an optical bench separate from that for the low-power injection laser and the second harmonic converter. This provides mechanical isolation from flashlamp percussion and vibration from the amplifier cooling water.

5.5.2 Optical architecture of the phase-locked system

The optical design of the phase-locked amplifier system is very closely related to the 500-ns system described in the previous section. Each amplifier arm performs essentially just as for the single-channel long pulse laser. Since the pulse format goal of 500 ns remains, the same oscillator design with the bi-exponential rise and 200-ns FWHM is used. A second preamplifier stage consisting of a 9-mm Nd:YLF flashlamp-pumped rod was added to boost the injected signal to 400 mJ (100 mJ for each amplifier). Figure 5.26 illustrates the design concept in a very straightforward and schematic way. A single oscillator beam is spatially divided into four separate beams using simple "scraper" type mirrors at 45°. Each of these beams, as shown in the figure, is directed through a four-pass Nd:glass zigzag amplifier. Each amplifier box in the drawing represents an amplifier and all of the supporting multi-pass optical components associated with it. After four passes through the amplifiers, the four beams are brought back together, side by side, and directed into the SBS phase conjugate mirror. Figures 5.26 and 5.27 depict this mirror as a simple single-lens focused system for the purpose of clarity. The details of the three-pass SBS loop phase-locking system will be described in a section to follow. After reflection from the SBS cell, the four beams again split and follow separate paths through their respective amplifiers and are rejoined in the output beam of the laser. Just as with the single-channel long-pulse laser system, a permanent-magnet Faraday rotator located in each amplifier arm isolates the high-energy output pulses from the injected input beam.

Figure 5.27 takes on the slightly more daunting task of showing the actual beam paths and component placements used in the phase-locked laser system. Although an initial impression might be one of significant complexity, a careful comparison of this optical layout to that of the single-channel system shown in Fig. 5.14 easily reveals the direct correlation between the two designs. However, there are several unique features to note in the four-beam system.

5.5.2.1 Injection laser mechanical isolation
The long-range illumination application for which this laser system is intended demands very high beam point stability. As previously discussed for the single-channel system, phase conjugation effectively removes effects of vibration, thermal distortion, and component alignment drift in the amplifier section. However, any disturbances transmitted to the injection laser translate directly into the high-power output beam. For this reason, the optical bench supporting the injection laser system and that for the high-energy flashlamp-pumped amplifiers are physically separated. The injection laser is housed in a separate room and is mounted to a large isolated concrete slab along with the high-precision laser beam director. The larger amplifier optical bench is located in an adjacent room. Beam transport to and from the amplifiers through the adjoining wall is via a 2-m-long 1:1 relay-imaging telescope, providing a high degree of mechanical isolation from the flashlamp percussion and any vibration introduced by the high-flow cooling water through the amplifiers. No active alignment is needed between the two benches. This would not be possible except for the SBS wavefront correction and rigorous relay-imaging built into the laser system which fully compensate for relative optical bench motion caused by vibration or long-term mechanical drift between the isolated floor slabs. The net result is a high-power, 100-J/pulse laser system directly attached to the experimental apparatus that exhibits the spatial footprint and thermal and vibrational signature of the low-power injection laser system (see Fig. 5.28).

5.5.2.2 The four-beam optical transport system
The goal of the design of the optical beam train for the phase-locked amplifier system was to use shared optical components, as possible, to reduce the parts count and overall complexity of the system. The 1:1 image-relay telescopes inside the multipass regenerative amplifier train as well as the polarizing beam splitters that couple the beams in and out of the ring are shared by all four extraction beams. As can be seen in Fig. 5.27, it is only at the amplifier end of the telescopes that individual beams are picked off and directed through separate zigzag amplifiers. The output beam from the oscillator/preamplifier injection system is expanded to a 30-mm round beam before it passes through a 45-mm clear-aperture permanent-magnet Faraday rotator. The beam is then anamorphically expanded by a factor of four in the horizontal dimension using

Figure 5.28. Photograph of the 100-J/pulse phase-locked amplifier system with electro-optics technologists James Wintemute (*left*) and Balbir Bhachu (*right*). The low-power injection laser enclosure can be seen in the top right portion of the photo along with the vacuum relay telescope between the two optical benches.

a cylindrical telescope. A set of large-aperture polarizers is used to inject this 30×120-mm input beam into the amplifier beam train in s-polarization. A 25×100-mm mask and a set of high reflecting mirrors on the amplifier bench are used to divide the beam into four 25×25-mm^2 beamlets that are then directed into individual 45-mm Faraday rotators. An anamorphic image relay telescope of the same design used in the single-beam amplifier system (Fig. 5.14) then reshapes each of these square beamlets to the 7×120-mm^2 extraction beams for the Nd:glass zigzag amplifiers. After four amplifier passes, the beams are then coupled out of the ring on shared optics and directed to the SBS phase conjugate mirror. A cylindrical down collimating telescope reduces the beam sizes to 7×12 mm^2 for introduction into the SBS system. Just as with our previous designs, the four Stokes beams generated by the SBS mirror retrace their paths, accumulating an additional four gain passes. The high-energy output beams are coupled out in p-polarization through the injection polarizer and propagate to the second harmonic converter.

5.5.2.3 *The four beam SBS loop phase conjugate mirror*

As described in the introduction, we believe that previous attempts to reliably phase-lock multiple amplifier apertures using SBS were not entirely successful because of the disruption caused by temporal phase drift and random phase jumps that have

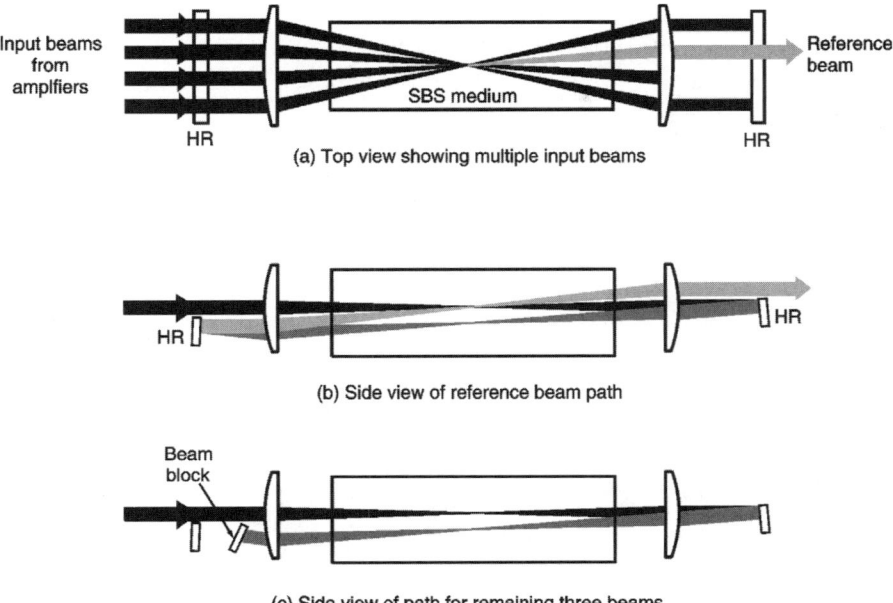

Figure 5.29. (a) Top view of our first version of the four-beam SBS loop geometry. As shown in (b), the beams are reflected out of the plane of the four beams to form the loop. In this configuration (c), it is generally beneficial to block all but one beamlet before the third SBS pass for maximum phase-locking stability.

been observed in the Stokes output. For this reason, we adopted a variation of the SBS loop architecture in which all four beams were focused into the same volume of the nonlinear medium. In previous work, we had shown that the SBS loop design stabilized the temporal phase for the single-beam laser amplifier system. Based on the successful operation of the single beam 500-ns laser system, we chose to use nitrogen at 90 atm as the nonlinear medium.

We took two different approaches to the details of the SBS loop phase-locking architecture as shown in Figs. 5.29 and 5.30. In both cases, the four beams were brought into the system, side by side, in a 1 × 4 linear array. For the first approach, shown in Fig. 5.29, the four beams were reflected out of the plane of the 1 × 4 array to form the SBS loop. In other words, the plane of the array was oriented at 90° to the plane of the SBS loop. Initial tests with this optical geometry did not exhibit the stability of the temporal phase-locking that was needed. We found that the best results were often achieved when three of the beamlets were blocked after the second pass through the SBS cell and only the fourth was allowed to reflect back through the cell and form the SBS loop, as shown in Fig. 5.29c. In this case the fourth beamlet formed the reference wave to which the remaining three beams were locked, in a manner similar to the architecture proposed by Ridley and Scott [50].

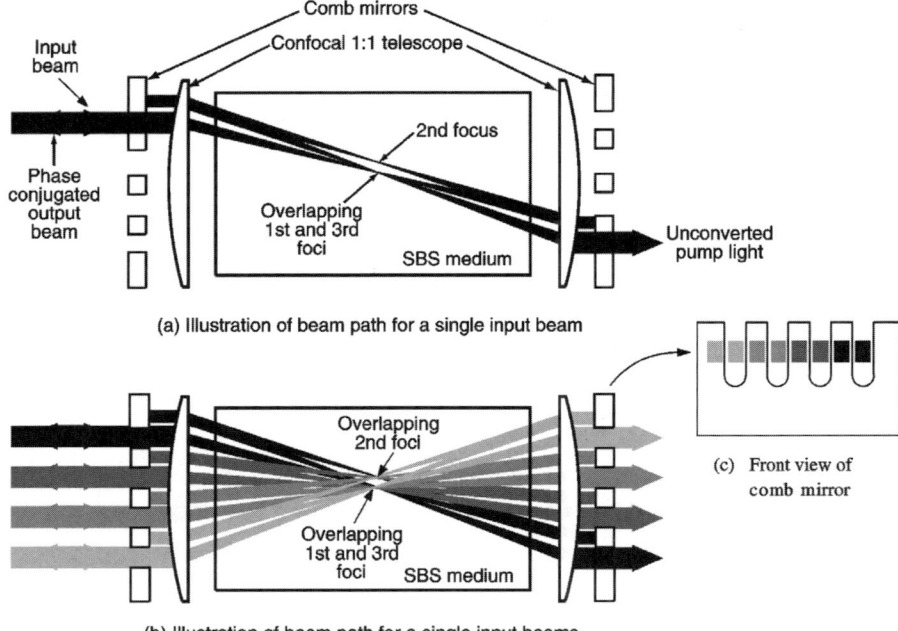

Figure 5.30. The greatly improved phase-locking optical configuration in which the four input beams remain in the same plane as they trace their paths through the SBS loop. Figure (a) shows the path for a single beamlet, and (b) shows the overlapping paths for all four beamlets. This geometry provides very high locking stability and is much less sensitive to relative misalignment between the beamlets. (c) Front view of comb mirror.

This approach was ultimately not satisfactory, however, since we observed that only the reference beam received the full benefit of the threshold reduction provided by the SBS loop four-wave interaction. Furthermore, optimal phase-locking performance often required that a different beam of the four be used as the reference from one operating period to another. It became a somewhat cumbersome "shell game" to experimentally identify the preferred reference beam for each run. Although our observations led us to believe that this was related to subtle differences in the co-alignment of the beams through the cells and small changes in integrated energy between the four channels, we did not determine with certainty the source of this fluctuating behavior.

Our second approach to the SBS loop beam combiner, shown in Fig. 5.30, completely eliminates the random behavior observed with the previous setup. Rather than orient the SBS loop path and the plane containing the four input beams orthogonally, they are made to overlap. In other words, all four input beams trace their way through the loop architecture in a single plane. This requires a specially shaped mirror that we have termed a "comb" mirror, shown in Fig. 5.30c. Vertical slots are ultrasonically milled in previously polished and coated HR mirrors. This alignment scheme provides not only beam overlap near focus, but also substantial overlap between adjacent beam paths away from focus. Although the optical architecture appears initially complex, its alignment is very straightforward. After one beam path is established through the SBS loop, the additional three beams need only to be pointed through the system using a far-field diagnostic looking at the transmitted beams to ensure collinearity. In practice, this system is very insensitive to relative beam alignment and provides consistent, long-term phase-locking performance between the four beams. Optimal phase-locking performance requires that the input beams only fall within 4–5 spot diameters between each other at focus in the SBS cell.

5.5.3 Output characteristics of the 100-J laser system

Each of the four beams of the 100-J/pulse phase-locked laser system is energetically equivalent to the single beam from the 500-ns system described in the previous

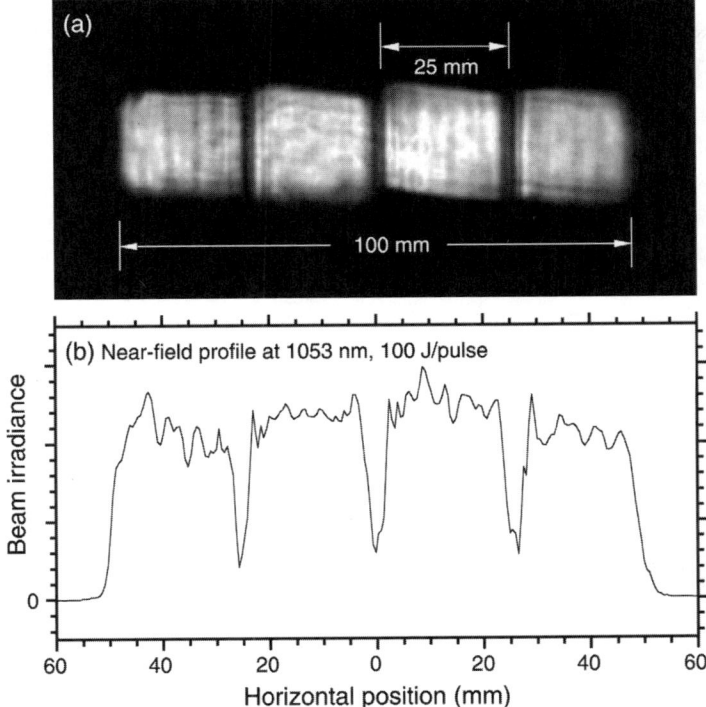

Figure 5.31. Near-field profile for a 100-J array of four beamlets from the phase-locked laser system. The outside beamlets are slightly less energetic than the central pair due to the 25% roll-off in injection energy from the center of the array to the edge.

section. Because of the increased beam formatting required to generate the four independent input beams with acceptable energy balance between each, the overall seed laser injection pulses are reduced from the 100 mJ per amplifier used in the single-beam system to ~60 mJ. This reduces the typical output energy per amplifier to 25 J/pulse, resulting in an overall energy of 100 J/pulse in the total output. This could be recovered by increased preamplification for the injection laser signal but was not pursued for this system.

5.5.3.1 Near-field beam pattern

Figure 5.31 shows the full-energy near-field beam patterns recorded from the phase-locked laser system at 1053 nm soon after the laser was first activated. Each beamlet is approximately a 25 mm square. Dark lines between the beamlets are the result of imperfect optical coating to the edges of the mirrors used to split the input beam into four beamlets as well as softening of the beam edges by the SBS phase conjugation process. Shortcomings in the alignment through the Nd:glass zigzag amplifier system are evidenced by beamlets that are somewhat shorter (vertical dimension) than the 25-mm goal and by the irregular clipping of the top and bottom edges from one beamlet to the next. Subsequent improvements in alignment techniques through the zigzag amplifier have largely eliminated these effects. At the time of these improvements, however, the output beam diagnostics were available only for the high-power frequency-doubled beam as shown in the more uniform near-field 527-nm profiles in Fig. 5.35.

5.5.3.2 Phase-locking performance

A straightforward way of visualizing the phase locking process for the four separate amplifier beamlets incident on the SBS mirror is to consider them as one beam having abrupt phase transitions in the wavefront profile located at the border between each beamlet. The SBS mirror conjugates the overall wavefront that includes the thermally induced distortion profile within each beamlet as well as the piston offset in phase between each beamlet. The great power of this technique lies in the correction of the piston phase offsets between each beam without the need to interferometrically control the pathlengths in the amplifier arms. If there were no Brillouin–Stokes shift in the output from the SBS phase conjugator, the phase-locking performance would be theoretically independent of the four propagation pathlengths through the amplifier arms, except for extreme cases in which temporal overlap of the pulses in the SBS cell would be lost. However, because of the Stokes frequency shift, the four pathlengths must be matched to some degree. The constant piston phase offset θ in the output between two given beamlets is given by [50]

$$\theta = \frac{2\pi \Delta L \nu_B}{c}$$

where ΔL is the pathlength difference between the beamlets, ν_B is the Brillouin frequency shift, and c is the speed of light. In the case of the 90 atm of N_2 used here, the Brillouin shift is approximately 750 MHz. This means that to maintain less than a one-tenth wave piston error between the beamlets, the pathlength in each amplifier

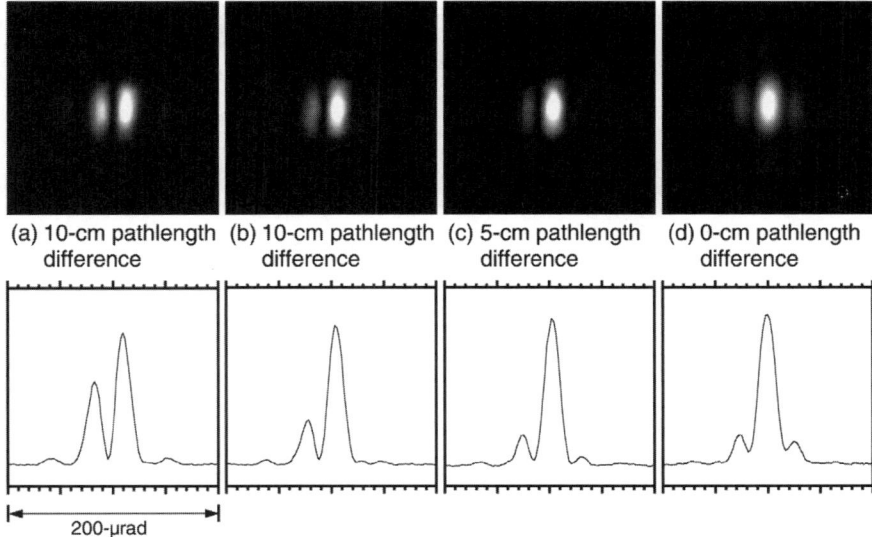

Figure 5.32. An illustration of the effect of pathlength differences in the amplifier beam trains using two adjacent beamlets. As shown, it is necessary to match the propagation distances to <2 cm when using high pressure nitrogen as the SBS medium. The 15 cm pathlength difference shown in Figure (a) corresponds to an 0.8π fixed phase offset between the beamlets.

beam train must be matched to within 40 mm. The only independent adjustment of pathlength, as shown in the optical layout of Fig. 5.27, is provided by the set of mirrors that take the beam that emerges from each zigzag amplifier head and directs it back into the slab for the counterpropagating path. Since each beam makes two passes through the regenerative amplifier loop on the way to the SBS mirror (for a total of four amplifier passes), the differences in propagation lengths for each beam line needs to be less that 20 mm, an easily achievable goal.

Figure 5.32 shows the far-field patterns of two adjacent beamlets for different relative pathlengths. The high-power output beams in the far field have near "textbook" profiles as would be expected from two temporally phase-locked beams. At a 15-cm pathlength difference, the two beamlets are out of phase by nearly one-half wave, resulting in an irradiance null on axis. As the pathlength difference approaches zero, a clean central lobe is achieved. In practice, this is the technique used to optmize the relative beam paths. The center, left, and right pairs of beamlets are sequentially examined and optical pathlength adjustments are made to achieve a symmetrical optical distribution in the far field for each pair.

When all four amplifiers are operated simultaneously, the phase-locked beamlets coherently sum in the far field to produce a single central lobe with one-fourth the divergence of each individual beamlet. A progression of one, two, three, and four 25-J beamlets at 1053 nm is shown in Fig. 5.33. Although the phase-locking performance is good, the previously described technique of optimizing the individual propagation

194 HIGH-PULSE-ENERGY PHASE CONJUGATED LASER SYSTEM

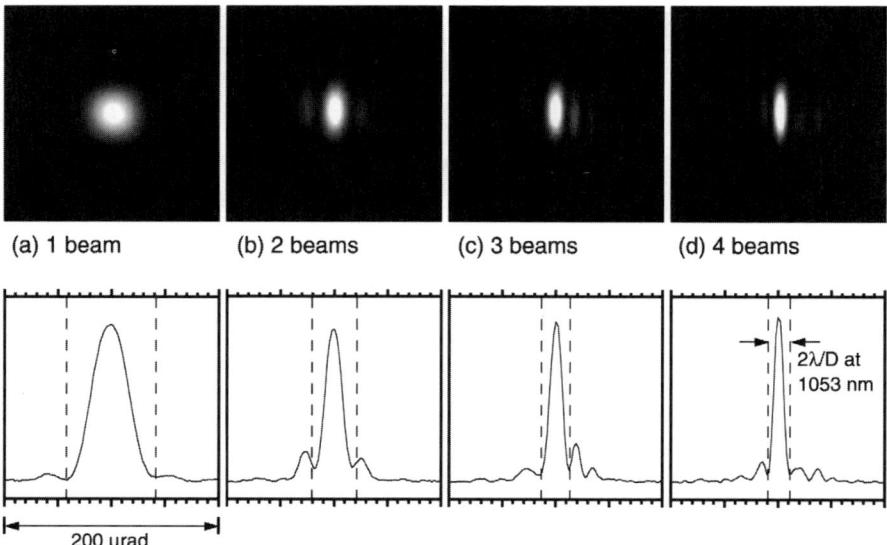

Figure 5.33. Measured far-field distributions for one, two, three, and four beamlets. The coherent combination (phase-locking) of the four beamlets in the far field generates a divergence profile 4× narrower than that for an individual beamlet.

lengths had not been implemented when these data were collected, so some residual asymmetry can be observed in the profiles. Later measurements at 527 nm show a further reduction in the relative piston phase offset errors (Fig. 5.36).

5.5.3.3 Second harmonic conversion of the phase-locked array

The frequency doubler for the 100-J/pulse phase-locked laser system is a scaled version of that used for the 30-J/pulse 500-ns laser. The width of the 45° beam shaping prisms and the KD*P doubler crystals was increased to accommodate the four phase-locked beams, side by side. A dichroic mirror outside of the laser enclosure is used to separate the unconverted 1053-nm light from the 527-nm beam. As shown in the optical layout of Fig. 5.27, the unconverted light is transmitted through the dichroic and emerges as a 25 × 100-mm² array of four beamlets. The 527-nm output is reformatted to a square 50 × 50-mm² array using a final anamorphic beam-reshaping telescope. This provides a circularly symmetric far field central lobe for the array of four phased-locked 12 × 50-mm² beamlets.

A typical external conversion efficiency of > 60% was obtained at up to 25 J/pulse with 500-ns pulse durations. Figure 5.34 shows an external frequency conversion efficiency curve measured for a single beamlet using one doubler crystal and two crystals in the alternating-Z geometry [41]. The measured efficiency includes the reflective losses from the prism and KD*P surfaces. Figure 5.35 shows a measured near-field profile at 527 nm, 50 J/pulse, and 3 Hz after reshaping by the final anamorphic telescope. Slightly less energy is typically present in the outside

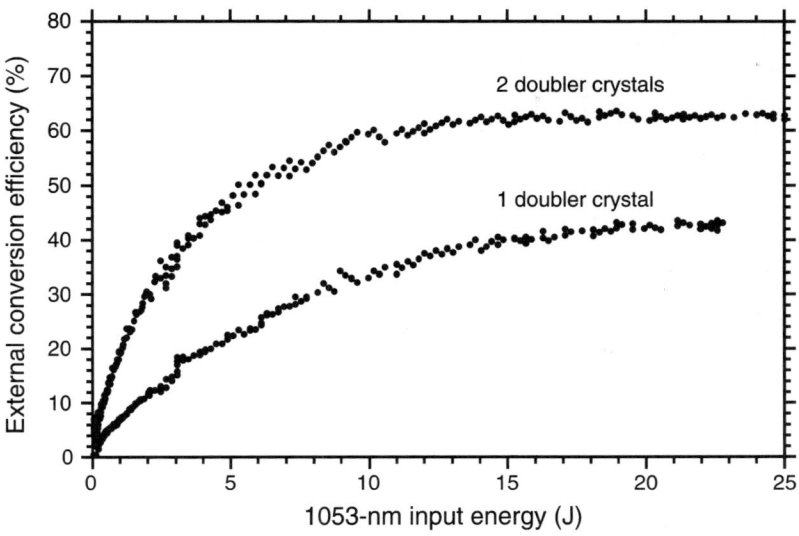

Figure 5.34. Measured second harmonic frequency conversion for one and two KD*P doubler crystals using a single beamlet. A conversion efficiency of >60% is readily achievable for the 500-ns, 25-J pulses.

two beams due to the roll-off of injected energy to the outside edges of the spatially truncated Gaussian oscillator beam profile.

Just as for the previously described laser designs, the second harmonic converter is not located in a phase-conjugated path so that any wavefront errors introduced by the doubler crystals or by the 527-nm anamorphic telescope are not corrected. Figure 5.36 shows the measured far-field profile for the 50-J pulse shown in Fig. 5.35. Small errors introduced by the doubler and telescope have increased the beam divergence by $\sim 1.5\times$ from the diffraction limit. Detailed measurements have confirmed a Strehl of ~ 0.5 for the 527-nm four-beam phase-locked output measured in single-shot mode. When the laser is taken from single shot up to 3-Hz steady-state operation, there is no measured increase in divergence for the 1053-nm beam. However, we discovered an unanticipated effect in the second harmonic converter. Weak linear absorption at 1053 nm (1.5–2% total) is known to introduce small wavefront distortions in the KD*P at full average power. As described earlier, this caused a $\sim 0.2\lambda$ focus term on the output of the long-pulse single-beam laser system that was easily corrected in a subsequent beam expansion telescope. However, in the case of the four-beam phase locked laser, the dark stripes that can be seen between each beamlet in the near field (Fig. 5.35) cause a $\sim 0.2\lambda$ wavefront ripple across the beam array which serves to break the far-field pattern into multiple lobes as shown in Fig. 5.37b, reducing the on axis Strehl by an additional $2\times$. We initially believed that the effect might be due to nonuniform heating in the crystals, causing piston phase errors from beamlet to beamlet. However, numerical simulation of the thermal effects in the KD*P have successfully explained the source of the wavefront error. It

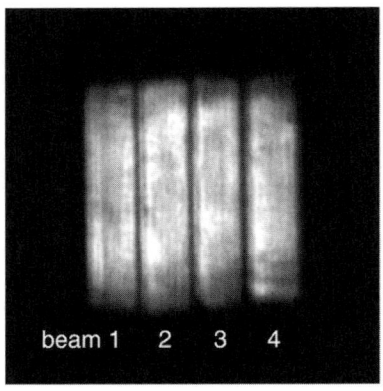
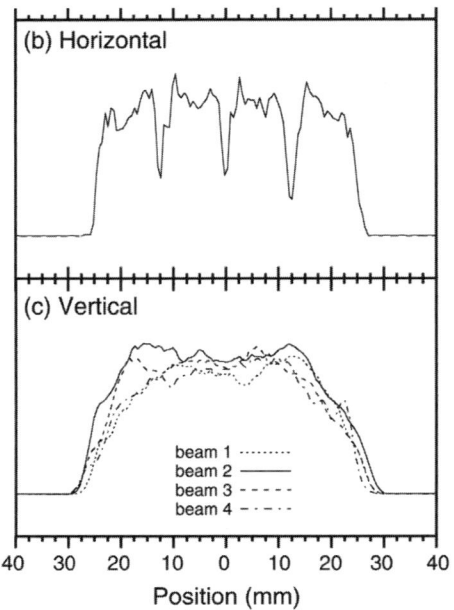

Figure 5.35. The near field distribution of the frequency converted, 527 nm phase-locked beamlet array. A final anamorphic relay telescope on the output of the laser system transforms the 25×100-mm^2 array shown in Fig. 5.31 to this 50×50-mm^2 square array in order to generate a round central lobe in the far-field.

can be effectively simulated, as shown in Fig. 5.37c by placing a 0.2λ focus term on each individual beamlet. Although the 527-nm brightness for the phase-locked array was sufficient to support the laser's intended mission, the ultimate solution for this problem may be a laser design that places the frequency doubler in a phase-conjugated beam path [7].

5.5.3.4 Phase-locked pointing stability The mechanical isolation of the injection laser system from the amplifier optical bench was expected to provide significantly increased pointing stability in the 100-J/pulse phase locked laser system as compared to that measured for the 30-J/pulse long-pulse laser (Fig. 5.24). In the vertical dimension, orthogonal to the phase-locking plane, this improvement was successfully realized, as shown in Fig. 5.38. The pointing jitter, normalized against beam dimensions, was decreased by a factor of two over the single-beam laser system, with an RMS variation of only 2% of $2\lambda/D$ at 1053 nm. However, in the horizontal dimension, a new source of beam pointing jitter was introduced in the phase-locked laser. Very small shot-to-shot variation in the locking accuracy (beamlet to beamlet piston phase variations) was found to introduce additional pointing fluctuation. While this variation does not significantly degrade the Strehl from pulse to pulse, the RMS pointing stability is reduced by $2\times$ in the phase-locking dimension as compared to the

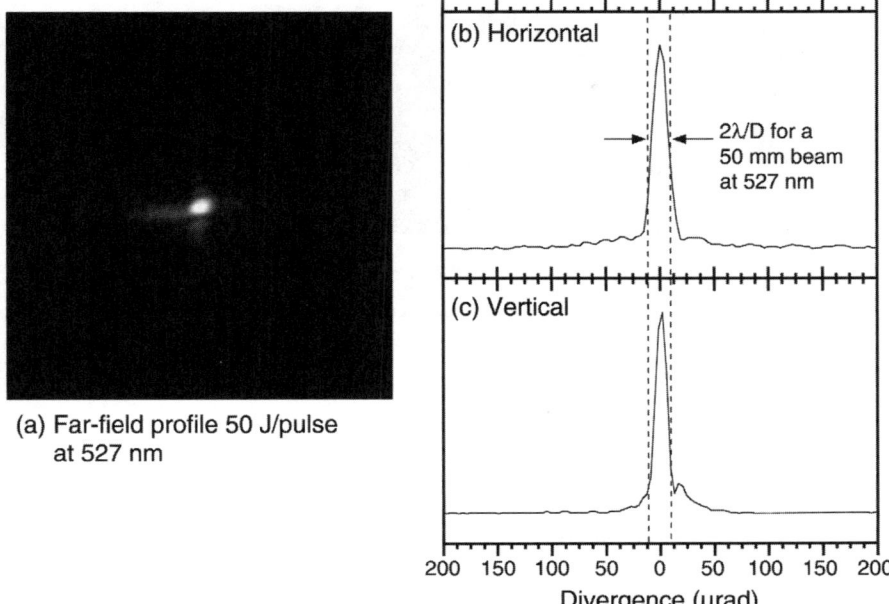

Figure 5.36. The far-field divergence profile for the near field distribution shown in Fig. 5.35 measured at 1 Hz. The divergence is increased to 1.25 times the diffraction limit by uncorrected aberrations introduced in the second harmonic converter and final anamorphic telescope.

vertical dimension. The measured pointing stability is nevertheless notable for a high-energy, high-average-power pulsed laser. The 2σ fluctuation of 4.4 μrad in the horizontal, phase-locking dimension for the phase-locked array is only 2.6% of the $2\lambda/D$ diameter of a single beamlet at 1053 nm.

5.5.4 Summary of the 100-J phase-locked laser system

The 100-J/pulse laser system described here represents the highest-average-power, highest-pulse-energy demonstration of SBS phase-locking of separate amplifier apertures. Four high-power beamlets are coherently combined in the near field to generate a near-diffraction-limited far-field pattern expected for the full beam array. The resulting high-quality beam can be efficiently frequency-doubled to 527 nm. This has resulted in a laser system with 500-ns full width at half-maximum (FWHM) output pulse duration with a near-rectangular temporal profile, >50 J/pulse at 527 nm, and a continuous pulse repetition frequency (PRF) of 3 Hz. The laser system can also be operated at 10 Hz for a 5-s burst, resulting in an average power of 1000 W and 500 W at 1053 and 527 nm, respectively. At this time of this writing, the laser system is in routine operation at the Starfire Optical Range in Albuquerque, New Mexico, to support the Active Imaging Testbed Program for the Air Force Research Laboratory.

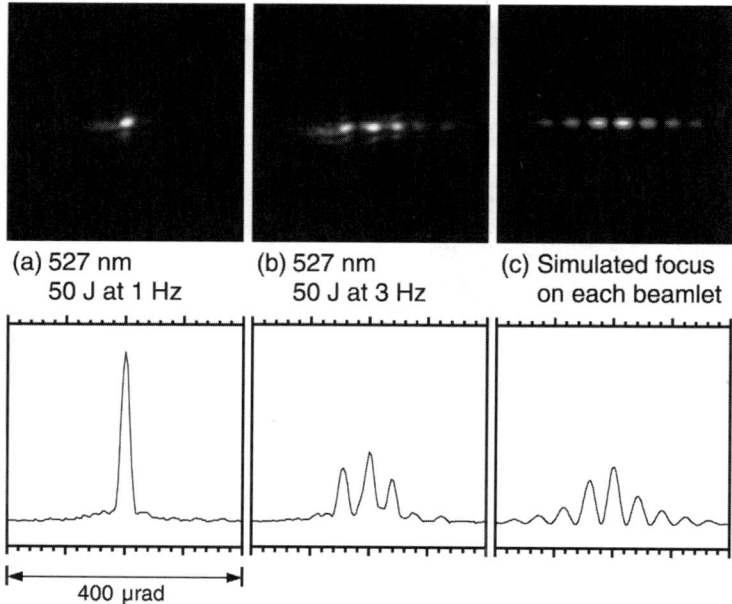

Figure 5.37. As the pulse repetition frequency is increased from (a) 1 Hz to (b) 3 Hz, the dark stripes between the individual beamlets in the array cause a spatially modulated aberration in the KD*P doubler crystals, reducing the peak irradiance by 2×. Figure (c) shows the results of a numerical simulation in which a 0.2λ focus error is introduced on each individual beamlet.

5.6 SUMMARY AND CONCLUSIONS

In this chapter, a number of design issues have been described that are important to the successful operation of a high-average-power amplifier system incorporating an SBS phase conjugate mirror. Although the operation of a solid-state laser system with high pulse energies offers unique challenges, we believe that there are valuable design principles that have arisen from this work that are very relevant to high-average-power SBS phase-conjugated lasers in general. Of these, the key considerations can be summarized as follows:

1. The amplifier design should minimize the thermally induced distortions by maintaining very uniform pumping and cooling. We believe the optimal amplifier architecture for solid-state amplifiers which minimizes optical distortions and depolarization continues to be the zigzag slab.
2. The multi-pass extraction geometry should be completely image relayed to maintain a uniform irradiance spatial profile in the amplifier, to avoid loss of aberrated components of the beam wavefront before reaching the phase

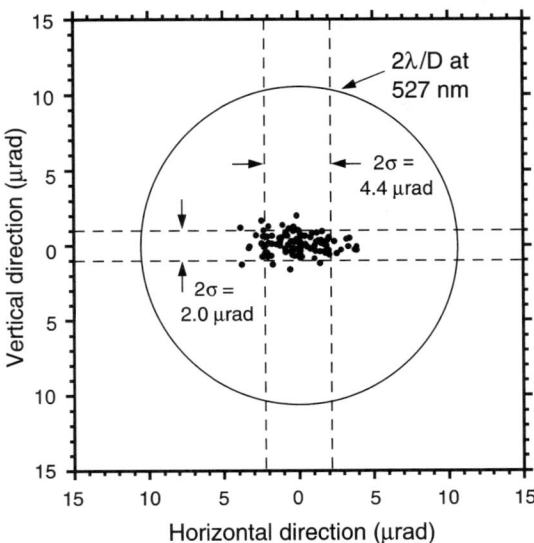

Figure 5.38. A pointing scatter plot measured for 100 pulses at 1 Hz for the 527-nm output beam demonstrating effective and consistent phase-locking between the four beamlets. The 2σ fluctuation of 4.4 μrad in the horizontal, phase-locking dimension is only 2.6% of the $2\lambda/D$ diameter of a single beamlet at 1053 nm.

conjugator, and to present a smooth spatial profile to the input of the phase conjugator.

3. The free propagation of the extraction beam should be minimized to reduce the irradiance distortions that result from the propagation of diffraction and thermal aberrations. The use of a double-passed refractive relay telescope places all optical surfaces near the relay image plane in the amplifier slab, minimizing potential optical damage problems that can accompany folded telescope designs using reflective optics.

4. The SBS phase conjugator should be operated high above threshold in order to ensure the best reproduction of the high-frequency components of irradiance and phase. This also results in high-energy reflectivity that minimizes the conjugator's insertion loss into the optical system and results in the accurate reproduction of the input temporal profile.

5. The rise time of the leading edge of the optical extraction pulse should be longer than the acoustic relaxation time in the Brillouin medium in order to maximize the wavefront reversal fidelity when the conjugator is operated high above the stimulated scattering threshold.

6. The SBS loop geometry not only provides a decreased SBS threshold, but also eliminates random temporal phase fluctuations that contribute to increased instability in the temporal profile and increased spectral bandwidth. These

same random phase fluctuations also disrupt the phase-locking of multiple beams in a single SBS cell; furthermore, an SBS beam combiner design, incorporating the loop geometry, is required for accurate and stable phase-locked laser output.

The successful operation of the high-energy laser systems presented in this chapter demonstrates the fact that SBS phase conjugation has, over the past decade, advanced beyond the level of an optical curiosity to become a reliable component in high-average-power laser systems. The most notable improvement in laser performance resulting from the wavefront correction provided by the conjugator is the near-diffraction-limited output divergence. However, the phase conjugator also contributes to the reliability of the amplifier system by greatly reducing the optical alignment sensitivity and by providing interstage gain isolation between amplifier passes. Finally, we feel that the detailed considerations required for this specific amplifier design presented here are applicable, in general, to the design of high-average-power, high-beam-quality solid-state laser systems using SBS phase conjugation.

ACKNOWLEDGEMENTS

The authors would like to recognize the outstanding effort of the technical team at Lawrence Livermore National Laboratory that was involved in the design, construction, and testing of the laser systems presented here. James Wintemute was responsible for the design, assembly, and alignment of the optical systems, William Manning and Steve Telford designed the high-energy pulsers and electronic controls, and Balbir Bhachu designed and oversaw the mechanical fabrication of the laser housing, cooling, and vacuum systems. We would like to thank Mary Norton for the optical design of the high-average-power second harmonic converters, Luis Zapata for contributions to the design of the Nd:glass zigzag amplifier, and William Neuman and Mark Hermann for computational support of optical modeling and SBS numerical simulations. We are also indebted to John Honig for the thermal analysis of the phase-locked doubler and to John Halpin for his support of the operation of these laser systems outside the laboratory.

This work is dedicated to the memory of Dr. Howard Powell for his inspiration, his encouragement, and his unfailing support of high-average-power solid-state laser development at LLNL.

This work was performed under the auspices of the U.S. Department of Energy by Lawrence Livermore National Laboratory under Contract W-7405-Eng-48.

REFERENCES

1. P. Celliers, L. B. DaSilva, C. B. Dane, S. Mrowka, M. Norton, J. Harder, L. Hackel, D. L. Matthews, H. Fiedorowicz, A. Bartnik, J. R. Maldonado, and J. A. Abate, Optimization of x-ray sources for proximity lithography produced by a high average power Nd:glass laser, *J. Appl. Phys.* **79**, 8258–8268 (1996).

2. D. G. Voelz, John D. Gonglewski, P. S. Idell, and D. C. Dayton, Coherent image synthesis using a Shack-Hartmann wavefront sensor, in *Digital Image Synthesis and Inverse Optics*, A. F. Gmitro, P. S. Idell, and I. J. Lahaie (eds.), *Proc. SPIE* **1351**, 780–786 (1990).

3. C. B. Dane, L. A. Hackel, J. Daly, and J. Harrison, Shot peening with lasers, *Advanced Materials and Processes* **153**, 37–38 (1998).

4. M. R. Herman, M. A. Norton, L. A. Hackel, and D. Twede, Efficient high power Stokes and anti-Stokes Raman frequency generation via polarization tuning enhancement, in *Conference on Lasers and Electro-Optics, 1993*, Vol. 11, OSA Technical Digest Series (Optical Society of America, Washington, DC, 1993), p. 392.

5. C. B. Dane, W. A. Neuman, and L. A. Hackel, High energy SBS pulse compression, *IEEE J. Quantum Electron.* **30**, 1907–1915 (1994).

6. S. G. Demos, A. Burnham, P. Wegner, M. Norton, L. Zeller, M. Runkel, M. R. Kozlowski, M. Staggs, and H. B. Radousky, Surface defect generation in optical materials under high fluence laser irradiation in vacuum, *Electron. Lett.* **36**, 566–567 (2000).

7. R. J. St. Pierre, D. W. Mordaunt, H. Injeyan, J. G. Berg, R. C. Hilyard, M. E. Weber, M. G. Wickham, G. M. Harpole, and R. Senn, Diode array pumped kilowatt laser, *IEEE J. Sel. Top. Quantum Electron.* **3**, 53–58 (1997).

8. W. S. Martin and J. P. Chernoch, Multiple internal reflection face pumped laser, U.S. Patent No. 3,633,126, 1972.

9. C. B. Dane, L. E. Zapata, W. A. Neuman, M. A. Norton, and L. A. Hackel, Design and operation of a 150 W near diffraction-limited laser amplifier with SBS wavefront correction, *IEEE J. Quantum Electron.* **31**, 148–163 (1995).

10. A. F. Vasil'ev and V. E. Yashin, Stimulated Brillouin scattering at high values of the excess of the pump energy above the threshold, *Sov. J. Quantum Electron.* **1**, 644–647 (1987).

11. G. J. Crofts and M. J. Damzen, Steady-state analysis and design criteria of 2-cell stimulated Brillouin scattering systems, *Opt. Commun.* **81**, 237–241 (1991).

12. G. J. Crofts, M. J. Damzen, and R. A. Lamb, Experimental and theoretical investigation of two-cell stimulated-Brillouin-scattering systems, *J. Opt. Soc. Am. B* **8**, 2282–2288 (1991).

13. R. Mays, Jr. and R. J. Lysiak, Observations of wavefront reproduction by stimulated Brillouin and Raman scattering as a function of pump power and waveguide dimensions, *Opt. Commun.* **32**, 334–338 (1980).

14. L. P. Schelonka, The fidelity of stimulated Brillouin scattering with weak aberrations, *Opt. Commun.* **64**, 293–296 (1987).

15. L. P. Schelonka and C. M. Clayton, Effect of focal intensity on stimulated-Brillouin-scattering reflectivity and fidelity, *Opt. Lett.* **13**, 42–44 (1988).

16. V. E. Yashin, V. I. Kryzhanovskii, and V. A. Serebryakov, Reversal of the wavefront of nanosecond and subnanosecond light pulses in stimulated Brillouin scattering, *Sov. J. Quantum Electron.* **12**, 1086–1088 (1982).
17. A. A. Betin, A. F. Vasil'ev, O. V. Kulagin, V. G. Manishin, and V. E. Yashin, Phase conjugation in nonstationary stimulated Brillouin scattering of focused beams, *Sov. Phys. JETP* **62**, 468–476 (1985).
18. R. H. Lehmberg, Numerical study of phase conjugation by stimulated backscatter with pump depletion, *Opt. Commun.* **43**, 369–374 (1982).
19. I. Yu. Anikeev, I. G. Zubarev, and S. I. Mikhailov, Influence of saturation on the quality of wavefront reversal in the case of stimulated scattering of spatially inhomogeneous pump radiation, *Sov. J. Quantum Electron.* **16**, 88–91 (1986).
20. J. J. Ottusch and D. A. Rockwell, Stimulated Brillouin scattering phase-conjugation fidelity fluctuations, *Opt. Lett.* **16**, 369–371 (1991).
21. C. B. Dane, W. A. Neuman, and L. A. Hackel, Pulse-shape dependence of stimulated-Brillouin-scattering phase-conjugation fidelity for high input energies, *Opt. Lett.* **17**, 1271–1273 (1992).
22. A. M. Dudov, S. B. Kormer, S. M. Kulikov, Vik. D. Nikolaev, V. V. Portnyagin, and S. A. Sukharev, Competition between nonlinear processes in gaseous SF_6 as a result of pumping by 2-ns pulses, *JETP Lett.* **33**, 347–351 (1981).
23. Personal communication, D. A. Rockwell, Hughes Research Laboratories, Malibu, CA.
24. R. St. Pierre, H. Injeyan, and J. Berg, Investigation of SBS phase-conjugation fidelity fluctuations in Freon-113, in *Conference on Lasers and Electro-Optics, 1992*, Vol. 12, OSA Technical Digest Series (Optical Society of America, Washington, DC, 1992), p. 180.
25. H. Yoshida, V. Kmetik, H. Fujita, M. Nakatsuka, T. Yamanaka, and K. Yoshida, Heavy fluorocarbon liquids for a phase-conjugated stimulated Brillouin scattering mirror, *Appl. Opt.* **36**, 3739–3744 (1997).
26. J. M. Eggleston, T. J. Kane, K. Kuhn, J. Unternahrer, and R. L. Byer, The slab geometry laser—Part 1: Theory, *IEEE J. Quantum Electron.* **20**, 289–301 (1984).
27. T. J. Kane, J. M. Eggleston, and R. L. Byer, The slab geometry laser—Part II: Thermal Effects in a Finite Slab, *IEEE J. Quantum Electron.* **21**, 1195–1210 (1985).
28. L. A. Hackel and C. B. Dane, High power, high beam quality regenerative amplifier, U.S. Patent No. 5,239,408, August 24, 1993.
29. J. Miller, L. Hackel, C. Dane, and L. Zapata, High power regenerative laser amplifier, U.S. Patent No. 5,285,310, February 8, 1994.
30. L. F. Weaver, C. S. Petty, and D. Eimerl, Multikilowatt Pockels cell for high average power laser systems, *J. Appl. Phys.* **68**, 2589–2598 (1990).
31. I. M. Thomas, High laser damage threshold porous silica antireflective coating, *Appl. Opt.* **25**, 1481–1483 (1986).
32. C. B. Dane and L. A. Hackel, Long-pulse-width narrow-bandwidth solid-state illuminator laser, U.S. Patent No. 5,689,363, November 18, 1997.
33. G. C. Valley, A review of stimulated Brillouin scattering excited with a broad-band pump laser, *IEEE J. Quantum Electron.* **22**, 704–712 (1986).
34. A. A. Filippo and M. R. Perrone, Experimental study of stimulated Brillouin scattering by broad-band pumping, *IEEE J. Quantum Electron.* **28**, 1859–1863 (1992).

35. J. M. Auerbach and R. L. Schmitt, Diode-pumped monolithic Nd:YLF 1.053 m mini-laser and its application to injection seeding, *Solid State Lasers, Proc. SPIE* **1223**, 133–141 (1990).

36. D. C. Hanna, B. Luther-Davies, and R. C. Smith, Single longitudinal mode selection of high power actively Q-switched lasers, *Opto-electron.* **4**, 249–256 (1972).

37. Y. K. Park and R. L. Byer, Electronic linewidth narrowing method for single axial mode operation of Q-switched Nd:YAG lasers, *Opt. Commun.* **37**, 411–416 (1981).

38. W. L. Smith, Nonlinear refractive index, in *Handbook of Laser Science and Technology*, Vol. 3, M. J. Weber (ed.) CRC Press, Boca Raton, FL (1986), pp. 259–264.

39. L. M. Frantz and J. S. Nodvik, Theory of pulse propagation in a laser amplifier, *J. Appl. Phys.* **34**, 2346–2349 (1963).

40. J. M. Eggleston, L. M. Frantz, and H. Injeyan, Derivation of the Frantz–Nodvik equation for zig-zag optical path, slab geometry laser amplifiers, *IEEE J. Quantum Electron.* **25**, 1855–1862 (1989).

41. D. Eimerl, High average power harmonic generation, *IEEE J. Quantum Electron.* **23**, 575–592 (1987).

42. R. V. Lovberg, Eric R. Wooding, and Michael L. Yeoman, Pulse stretching and shape control by compound feedback in a Q-switched ruby laser, *IEEE J. Quantum Electron.* **11**, 17–21 (1975).

43. W. E. Schmid, Pulse stretching in a Q-switched Nd:YAG laser, *IEEE J. Quantum Electron.* **16**, 790–794 (1980).

44. D. C. Jones and D. A. Rockwell, Single-frequency, 500-ns laser pulses generated by a passively Q-switched Nd laser, *Appl. Opt.* **32**, 1547–1550 (1993).

45. J. Harrison, G. A. Rines, and P. F. Moulton, Stable-relaxation-oscillation Nd lasers for long-pulse generation, *IEEE J. Quantum Electron.* **24**, 1181–1187 (1988).

46. J. P. Roberts, K. W. Hosack, A. J. Taylor, J. Weston, and R. N. Ettelbrick, Efficient frequency-doubled long-pulse generation with a Nd:glass/Nd:YAG oscillator–amplifier system, *Opt. Lett.* **18**, 429–431 (1993).

47. A. M. Scott, W. T. Whitney, and M. T. Duignan, Stimulated Brillouin scattering and loop threshold reduction with a 2.1-μm Cr, Tm, Ho-YAG laser, *J. Opt. Soc. Am. B* **11**, 2079–2088 (1994).

48. D. G. Voelz, K. A. Bush, and P. S. Idell, Illumination coherence effects in laser-speckle imaging: Modeling and experimental demonstration, *Appl. Opt.* **36**, 1781–1788 (1997).

49. M. S. Mangir, J. J. Ottusch, D. C. Jones, and D. A. Rockwell, Time-resolved measurements of stimulated-Brillouin-scattering phase jumps, *Phys. Rev. Lett.* **68**, 1702–1705 (1992).

50. K. D. Ridley and A. M. Scott, Phase-locked phase conjugation using a Brillouin loop scheme to eliminate phase fluctuations, *J. Opt. Soc. Am. B* **13**, 900–907 (1996).

51. N. G. Basov, V. F. Efimkov, I. G. Zubarev, A. V. Kotov, S. I. Mikhailov, and M. G. Smirnov, Inversion of wavefront in SMBS of a depolarized pump, *JETP Lett.* **28**, 197–201 (1978).

52. D. S. Sumida, D. C. Jones, and D. A. Rockwell, An 8.2J phase-conjugate solid-state laser coherently combining eight parallel amplifiers, *IEEE J. Quantum Electron.* **11**, 2617–2627 (1994).
53. C. B. Dane and L. A. Hackel, Coherent beam combiner for a high power laser, U.S. Patent No. 6,385,228, May 7, 2002.

CHAPTER 6

Advanced Stimulated Brillouin Scattering for Phase Conjugate Mirror Using LAP, DLAP Crystals, and Silica Glass

HIDETSUGU YOSHIDA and MASAHIRO NAKATSUKA

Institute of Laser Engineering, Osaka University, Osaka 565-0871 Japan

6.1 INTRODUCTION

The stimulated Brillouin scattering (SBS) effect in various liquids and gases can be applied to a phase conjugator [1, 2], as described in previous chapters. In SBS liquid media such as heavy fluorocarbons [3–5] and tetrachlorides, it is necessary to eliminate the dissolved impurities and solid microparticles, and these media must be carefully handled due to their toxic properties. Quartz was used as a nonlinear medium in the first practical demonstration of SBS in 1964 [6]. The first observation of SBS in an optical fiber indicated that fiber provides highly reflective SBS due to its long interaction length [7]. Long path lengths can produce SBS even at a low incident power [8–11]. However, the incident pulse energy must be limited to a few millijoules to avoid surface damage of the fiber.

Organic L-arginine phosphate monohydrate (LAP) was examined in 1983 at Shangdon University, China as a nonlinear optical material [12]. LAP appeared to be a promising material for laser fusion experiments as a replacement for potassium dihydrogen phosphate (KDP) crystal [13, 14]. LAP and its deuterated analog, DLAP, have some attractive properties for frequency conversion, such as a highly nonlinear optical coefficient $d_{\text{eff}} = 1.9 \ d_{\text{eff}}$ [KDP], a short-wavelength cutoff of 260 nm, and a less hygroscopic nature than that of KDP [15]. In addition, it has a possibility of growing to a large dimension, greater than $100 \times 100 \times 100$ mm^3.

In a previous research the internal laser damage thresholds in LAP and DLAP were measured using a 1053-nm Nd:YLF laser with 1- and 25-ns pulse widths and were found to be much higher than for KDP and fused-silica glass [16]. The authors

---

Phase Conjugate Laser Optics, edited by Arnaud Brignon and Jean-Pierre Huignard
ISBN 0-471-43957-6 Copyright © 2004 John Wiley & Sons, Inc.

could not explain this extremely high damage threshold at that time. Then they observed intense stimulated Brillouin scattering (SBS) in LAP. This strong scattering prevents the focus from reaching a high energy density and suppresses the damage in the crystal. It also implies potential applications of SBS in phase conjugate mirrors (PCM). The absolute Brillouin steady-state gain of DLAP was reported to be the highest for the measured solid-state materials reported in the literature [17]. Therefore we investigate the possibility that LAP and DLAP can be used as a solid-state phase conjugator for high-power lasers.

In this chapter, we present the high SBS reflectivity in LAP and DLAP, using a Nd:YAG laser, and we discuss hereafter their respective properties. Then we show that these crystals can be used as phase conjugate mirrors in laser. We demonstrate the improvement of wavefront distortion induced by thermal effects in solid-state laser materials. We also present the SBS properties of bulk fused-quartz glass at high pulse energy operation.

6.2 CRYSTAL STRUCTURE OF LAP AND DLAP

In this section we review the crystal structures of LAP and DLAP based on several earlier papers [15, 17, 18]. These crystals are negative biaxial crystals that belong to the monoclinic class, space group $P2_1$, with two formula units per cell. The cell dimensions of LAP are $a = 10.75$ Å, $b = 7.91$ Å, $c = 7.32$ Å, and $\beta = 98.0°$, where β is the angle between the a and c axes and b is the twofold axis. The LAP chemical formula is shown in Fig. 6.1. L-Arginine is a linear-chain molecule with amino acid and guanadyl groups at each opposite end. The guanadyl and amino groups are protonated and hence have positive charges, which balance the negative charges of the carboxylate and dihydrogen phosphate ions. The crystal structure consists of alternate layers of phosphate groups and arginine molecules stacked along the a axis and held together by hydrogen bonds. Thus the bonding along the a axis is weaker than in other directions, and the crystal has a cleavage plane normal to (100), which is coincident with the $a \sin \beta$ direction. LAP has considerable absorption at 1064 nm because of overtones of molecular vibration of the hydrophilic functional groups marked with a boldface H in Fig. 6.1. Deuterizing these functional groups can reduce the strong absorption at the Nd laser wavelength. At 1064 nm the absorption coefficients of LAP and DLAP are 0.09 and 0.02 cm^{-1}, respectively.

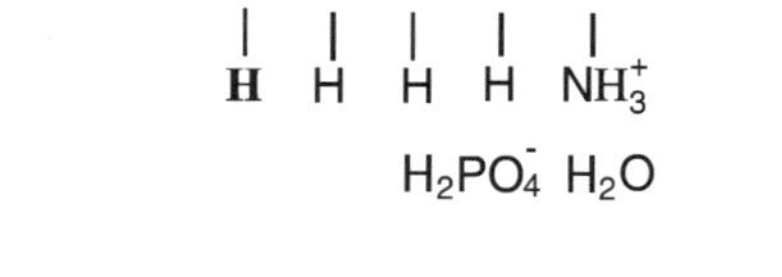

$$H_2N^+=\overset{\overset{\displaystyle H_2N}{|}}{C}-N-\overset{\overset{\displaystyle H}{|}}{\underset{\underset{\displaystyle H}{|}}{C}}-\overset{\overset{\displaystyle H}{|}}{\underset{\underset{\displaystyle H}{|}}{C}}-\overset{\overset{\displaystyle H}{|}}{\underset{\underset{\displaystyle H}{|}}{C}}-\overset{\overset{\displaystyle H}{|}}{\underset{\underset{\displaystyle NH_3^+}{|}}{C}}-COO^-$$

$$H_2PO_4^- \quad H_2O$$

Figure 6.1. LAP chemical formula. Functional groups marked with a boldface H are the origins of absorption at 1064 nm.

6.3 BASIC CHARACTERISTICS FOR STIMULATED BRILLOUIN SCATTERING

6.3.1 Damage threshold

The damage thresholds of LAP and fused silica measured with laser pulses at 1064 nm and 1.1-ns or 18-ns pulse durations are summarized in Table 6.1. The measured damage threshold of fused-silica glass at 1.1-ns pulse duration was 28 ± 3 J/cm^2, and no SBS reflectance was observed. It is shown that the damage threshold of LAP depends on laser polarization in either direction. The damage thresholds of LAP were equal (polarization $//$ b) and three times higher (polarization $//$ c) than for fused silica. The focal spot diameters inside the LAP crystal and fused silica were similar according to the measurements of the far-field pattern using a relay imaging method. The damage threshold measurement for a pulse duration shorter than the acoustic relaxation time can be used to demonstrate the real damage threshold. It is not yet understood why organic LAP has such a high damage threshold compared to fused-silica glass.

The damage thresholds defined by the incident energy at 18-ns pulse duration were not real values because of SBS reflection. Thus, when the SBS reflectivity is much higher, damage results from the transmitted focusable power through the reflecting point by SBS. The damage threshold fluence increases as the square root of the pulse duration with so-called empirical laws [19]. When LAP is irradiated with parallel b and c polarizations with an empirical law, it is reasonable to expect approximately 120 J/cm^2 and 360 J/cm^2 damage threshold, respectively.

When the polarization direction of the incident laser was parallel to the c axis, the bulk damage for LAP was not observed at the focused fluence of 1200 J/cm^2 based on the assumption that the entire beam energy reaches its focal point. The real damage threshold was estimated to be greater than 280 J/cm^2 when we took into consideration the effects of SBS and absorption loss. When the polarization direction of the incident laser was parallel to the b axis, the bulk damage was estimated to be 375 J/cm^2 based on the assumption that the entire beam reaches its focal point. The real focusable fluence was estimated to be 130 J/cm^2. The damage threshold obtained from the scaling law of the thermal process agreed comparatively with that of focusable fluence when we took into consideration the absorption loss. The damage in this case may be induced by a thermal effect due to the high absorption of LAP at 1064 nm.

6.3.2 Physical properties of SBS

The pumping pulse duration was more than 10 times longer than the relaxation time of the SBS medium, indicating that the medium reaches steady-state conditions early in the pulse. The gain coefficient g_B of several solids could be calculated from Refs. 20 and 21. The Brillouin linewidth $\Delta \nu$ is proportional to λ^{-2} (where λ is the pulse wavelength), so that $\Delta \nu$ at 1064 nm is one-fourth of that at 532 nm. The

TABLE 6.1. Laser-Induced Damage Threshold in LAP and Fused-Silica Glass[a]

Materials	Pump polarization	1.1-ns pulse duration			18-ns pulse duration			
		Incident fluence (J/cm^2)	SBS reflectance (%)	Damage threshold (J/cm^2)	Incident fluence (J/cm^2)	SBS reflectance (%)	Damage threshold (J/cm^2)	Sealed threshold (J/cm^2)
LAP crystal[b]	c	85	0	85	1200	70	>280	360
	b	28	0	28	375 Eimerl et al. [17]	55	130	120
Fused-silica glass		28	0	9.8–13.4 28	125	5	120	120

[a]Focus position inside sample: 1.1-ns pumping, 5 mm (absorption loss 4.5%); 18-ns pumping, 25 mm (absorption loss 20%).
[b]Laser condition of LAP crystal: (1) experiments; pump direction a plane; (2) Eimerl et al. [17]; no information.

linewidths for fused silica and DLAP at 1064 nm were thus estimated to be 41 MHz and 20 MHz, respectively, based on the reported values at 532 nm [22].

The experimental results of the SBS threshold energy and the estimated gain coefficients for LAP, fused silica, and DLAP are summarized in Table 6.2 [23]. The experimental value of g_B was deduced from measurements of SBS threshold energy at both 18- and 38-ns pulse durations. The steady-state gain coefficient g_B can be obtained by the SBS threshold energy E_{th} from a transient equation [24]. The acoustic relaxation times for fused silica and LAP at 1064 nm were estimated to be approximately 4 and 8 ns, respectively. The SBS gain coefficients for fused silica and LAP from measurements of SBS threshold energy at 18- and 38-ns pulse durations were 3.5 and 3.8 cm/GW and 9.5 and 11 cm/GW, respectively.

On the other hand, the gain coefficient for DLAP for two polarization directions (// b and // c) of the incident laser at 532 nm was reported to be 28–30 cm/GW [22]. The measured SBS gain coefficient for LAP was about one-third of the calculated and reported values for DLAP. The high SBS gain of DLAP crystal compared with LAP may be caused by the differences of electrostrictive coupling coefficient due to deuterization of crystal and the experimental combination of pumping polarization and crystal axis.

6.3.3 SBS reflectivity

The experiments were carried out with two Q-switched Nd:YAG laser systems: One was a single-shot laser, whereas the other was a 10-Hz-repetition-rate laser. The single-shot laser was convenient for use in measuring the SBS reflectivity, in determining the optimal direction of incident laser. The Nd:YAG laser operated at 10-Hz repetition was used to evaluate DLAP as a phase conjugate mirror for practical applications [25].

The following samples were used for SBS measurement: DLAP (13 mm × 20 mm × 20 mm), LAP (14 mm × 20 mm × 20 mm), and fused quartz (SiO_2) (40 mm × 25 mm × 25 mm) as a reference. The amount of deuterium substitution in the DLAP sample was 98%. Polished (100), (010), and (001) planes of DLAP and (100) plane of LAP without coating were used as incident faces. Figure 6.2 shows the experimental setup for measurement of SBS reflectivity. A single-shot Q-switched Nd:YAG oscillator delivered a linearly polarized 13- to 15-ns quasi-Gaussian pulse at 1064 nm. The beam quality was 1.5 times diffraction-limited. After passing through a variable attenuator and a Faraday rotator, the incident light (up to 100 mJ) was focused to a 10-mm point inside the input surface by a lens with a focal length of 100 mm to avoid surface damage. Experiments were also conducted at 532 nm. In this case, second harmonic generation from a type II KDP crystal was used. Pulses at 532 nm with durations of 10–11 ns and an energy of 10 mJ were used.

We investigated the SBS reflectivity in DLAP in various directions, (100), (010), and (001). In the directions (010) and (001), damage easily occurred, and the reflectivity was much lower than that in the (100) direction. This result indicates that the vibration of atoms is easily stimulated in the direction with weak hydrogen

TABLE 6.2. SBS Threshold Energy and Gain Coefficients of LAP and Fused-Silica Glass

Materials	Laser condition		SBS threshold energy (mJ)		SBS gain coefficient (cm/GW)				
	Pump polarization	Pump direction	18 ns	40 ns	Calculated	Estimated		Estimated from transmitted intensity, 40 ns	
						18 ns	40 ns		
Fused-silica glass			15	25	3.7	3.5	3.8	2.9–4.3 (Ref. 22)	(Ref. 7)
LAP	c	(Cleaved surface)	4.5	12		9.5	11	9–12.9	
DLAP	c	a			28–30 (Ref. 22)				

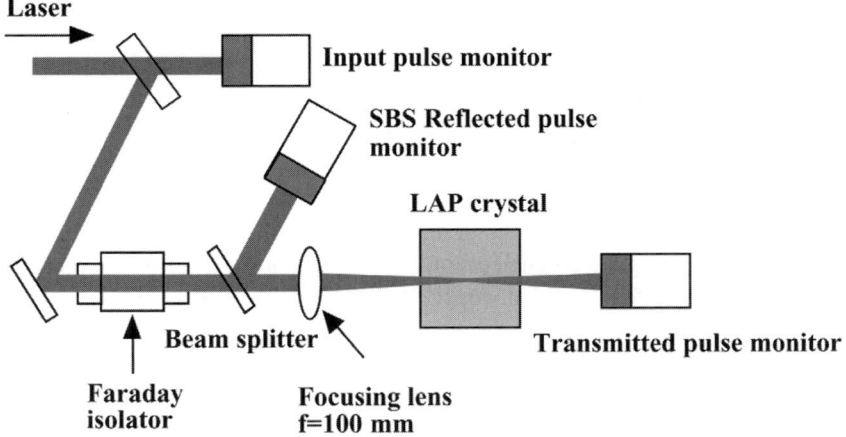

Figure 6.2. Experimental setup for measuring SBS reflectivity.

bonding and that this direction is effective for stimulated scattering. It is conceivable that the SBS that is due to the acoustic phonons built up along the (100) direction is quite sensitive to deuterization. From this result, therefore, we decided to carry out the experiments described below in the (100) incident direction.

Figure 6.3 shows the SBS reflectivities in fused-silica glass, LAP, and DLAP as a function of input energy at 1064-nm wavelength. These experiments were approximated to steady-state conditions. The SBS reflectivity was measured under the same focusing conditions, so the effective interaction length was kept constant. A 100-mm lens was used to focus a beam of diameter $2d = 8$ mm into crystal, so the effective interaction length could be estimated by a (Gaussian beam) confocal

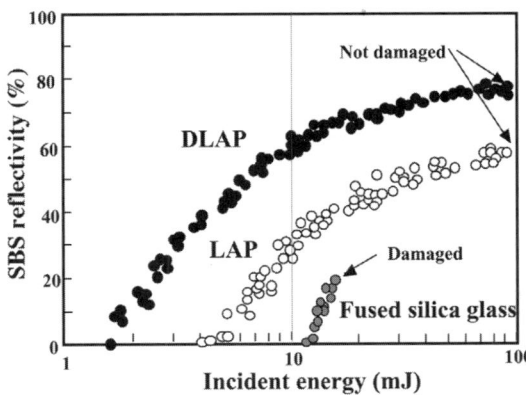

Figure 6.3. SBS reflectivity in fused-silca glass, LAP, and DLAP as a function of input pulse energy at 1064-nm wavelength.

parameter, $z_0 = f^2\lambda/d^2\pi$, of 0.2 mm. Thus a comparison of SBS gains could be roughly be made with reference to the SBS thresholds. The internal laser damage in fused-silica glass was observed at a SBS reflectivity of about 20%. The reflectivity in LAP and DLAP increased monotonically to maximum values of 59% and 78%, respectively, without damage. Figure 6.3 shows that the SBS thresholds in LAP and DLAP was 4.2 and 1.6 mJ, respectively. These input values are not corrected for surface and absorption losses. As mentioned above, LAP has strong absorption owing to overtones of molecular vibrations. To calculate the loss of the input energy, we measured the absorption coefficients of the crystals using a spectrophotometer. At 1064 nm, the absorption coefficients of the LAP and the DLAP samples were 0.11 and 0.01 cm^{-1}, respectively. After absorption compensation we calculated SBS thresholds of 3.6 mJ in LAP and 1.5 mJ in DLAP. LAP still showed a threshold that was more than two times higher than that in DLAP after absorption compensation.

Figure 6.4 shows the result of SBS thresholds as a function of focusing position from the input faces of the crystals. In the region of focusing below 1.2 mm, an increase of threshold was observed, which was related to the decrease of effective interaction length. It could be clearly confirmed that SBS threshold in LAP is essentially different from that in DLAP. That is, the SBS gain for DLAP is more than two times higher than that for LAP. DLAP has less absorption than LAP at 1064 nm.

To clarify the difference in SBS thresholds, we measured the SBS reflectivity at 532 nm, where LAP and DLAP have the same absorption coefficient of 0.01 cm^{-1}. Figure 6.5 shows the reflectivity dependence on input energy for LAP and DLAP at 532 nm. The reflectivity curves of LAP and DLAP are identical at 532 nm. After loss compensation, the thresholds decrease with wavelength and was as low as 0.4 mJ. A maximum reflectivity of 70% can be achieved for both crystals. It is significant that there is no difference in the SBS properties of LAP and DLAP at the frequency at which these crystals have the same absorption. This result shows that

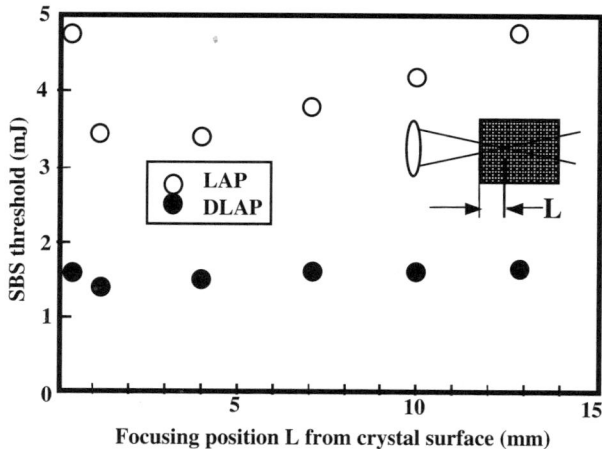

Figure 6.4. SBS threshold as a function of focusing position from the input surface.

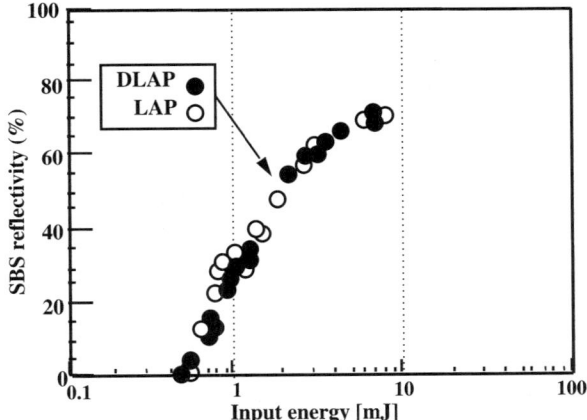

Figure 6.5. SBS reflectivity as a function of input energy at 532-nm wavelength.

the reason for the threshold difference at 1064 nm is the absorption, which increases SBS threshold in LAP. We believe that the heat caused by absorption in the focusing area of LAP permits the excitation of various noise modes of phonons, which resulted in the initiation of thermal scattering and which prevents a specific SBS mode from selectively reaching threshold.

For applications at 1064 nm, then, DLAP is obviously more efficient than LAP, whereas at 532 nm both crystals can be used.

6.4 APPLICATION OF SOLID-STATE SBS MIRRORS TO HIGH-POWER LASERS

6.4.1 Correction of aberrations

The SBS performances of DLAP was confirmed at a 10-Hz repetition rate [25]. In this case, phase conjugate reflectivity of 83% was demonstrated. If we consider the surface loss of noncoated DLAP, a maximum possible reflectivity of 90% is expected. Inasmuch as no sign of saturation is observed in this experiment, a DLAP SBS PCM could be achievable without serious loss and damage for an input average power of at least ~ 1 W.

To demonstrate the ability of the DLAP SBS PCM to correct aberrations, the experimental setup of Fig. 6.6 was used. The beam quality of this laser was 1.5 times diffraction-limited, and its power range was 0–500 mW (0–50 mJ). The beam was aberrated to a divergence of 6 mrad by being passed through a glass aberrator. The SBS phase conjugate pulses were evaluated by a cross check of the far-field spot images. Figure 6.7 shows the far-field images of the reflected pulses from an ordinary mirror and from DLAP SBS PCM. Figure 6.7a shows the image from the mirror without an aberrator. The spot diameter of Fig. 6.7a is 0.50 mm and is used as the standard for

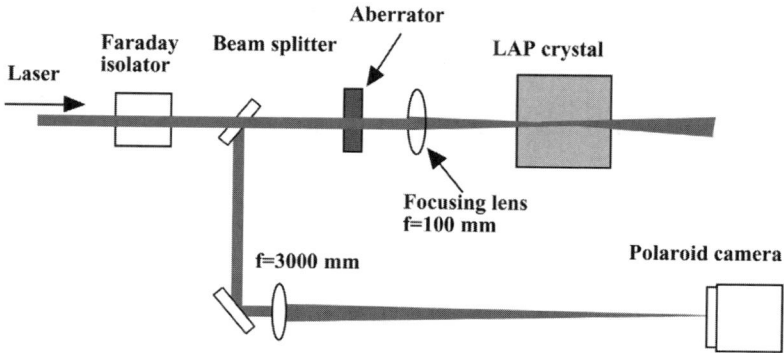

Figure 6.6. Experimental setup for demonstrating correction of aberration.

correcting the aberration. The image has two spots that are due to two surface reflections of the beam splitter. Figure 6.7b is the image of the aberrated beam after reflection on the ordinary mirror and double pass through the aberrator. In contrast, Fig. 6.7c, obtained from reflection by DLAP, indicates that an SBS mirror is useful to correct the aberration. The SBS threshold for the aberrated beam was extremely high because the spot area, the surface loss, and the interaction length were drastically different from those of a Gaussian beam. Figure 6.7c, which is the image just above the SBS threshold, needs 10 times higher energy than that of the Gaussian beam. A comparison of beam spot diameters roughly indicates the fidelity of phase conjugation. A diameter of 0.53 mm gives a fidelity of 0.94 in the case of Fig. 6.7c. In this experiment we obtained a fidelity of 0.7–1.0 with DLAP, regardless of the aberrator. Therefore we have found that a DLAP SBS PCM can clean up an aberrated beam and that the performance is stable for pulse energies below 50 mJ.

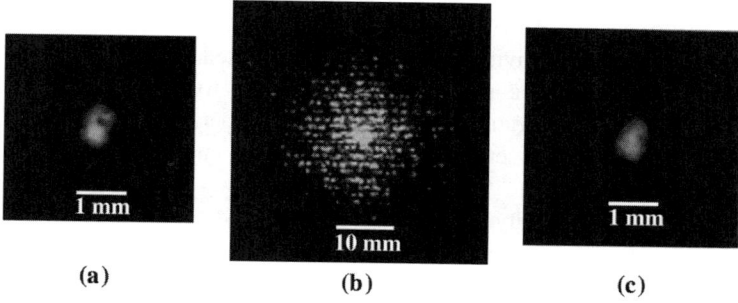

Figure 6.7. Far-field images of reflected pulses from a conventional mirror and from DLAP. (a, b) Images from a mirror without and with an aberrator, respectively. (b) Image showing that the aberrated beam cannot be focused. (c) Image from DLAP with aberrator. DLAP phase conjugate mirror can correct the aberration.

6.4.2 High-peak power laser system with LAP phase conjugate mirror

The double-pass Nd:YAG laser system with a SBS phase conjugate mirror using a LAP crystal is shown in Fig. 6.8 [26]. The laser oscillator was a linearly polarized, single-frequency, TEM$_{00}$ Nd:YAG Q-switched laser. The laser output pulse energy was amplified up to 130 mJ in a quasi-Gaussian pulse shape with an FWHM pulse duration of 18 ns. The laser system was operating at a repetition rate of 10 Hz. The polarization direction of the incident pulse was parallel to the LAP crystal c axis.

Figure 6.9 shows the performances of the laser system with the SBS mirror using a LAP crystal and a conventional high reflectivity (HR) mirror for comparison. The amplified output energy of 130 mJ was focused into the LAP crystal that exhibited a SBS reflectivity of 60%. However, due to the gain saturation, the output energy of the laser system after double pass through the amplifier was reduced by only 73% in comparison with the case of a conventional HR mirror was used in place of the LAP crystal. Figure 6.10 shows the compensation of the thermal lensing effect of the double-pass Nd:YAG amplifier by using a LAP SBS mirror. The focal length of the Nd:YAG rod with an HR mirror changed from 3000 to 1350 mm at a pumping energy of 50 J. The strong thermal lensing effect could be almost compensated in the double-pass amplifier by the LAP SBS mirror even at the pumping energy of 50 J.

The SBS fidelity of phase conjugation was measured by detecting the transmitted signal through a pinhole located at the focal point of lens. The fidelity using a LAP SBS mirror was constant and tightly bunched at 0.88 ± 0.8 up to an incident pulse energy of 165 mJ. This transmission corresponds to 1.3 times diffraction limited.

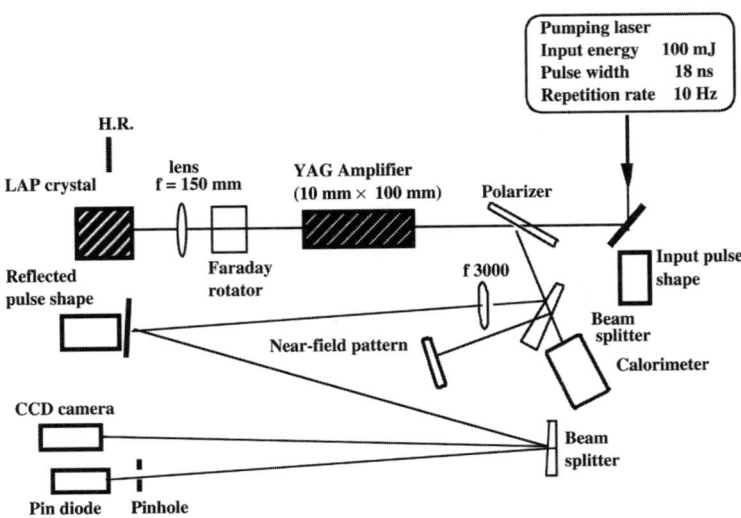

Figure 6.8. Experimental setup for the improvement of wavefront distortion of an Nd:YAG amplifier system by a LAP SBS mirror.

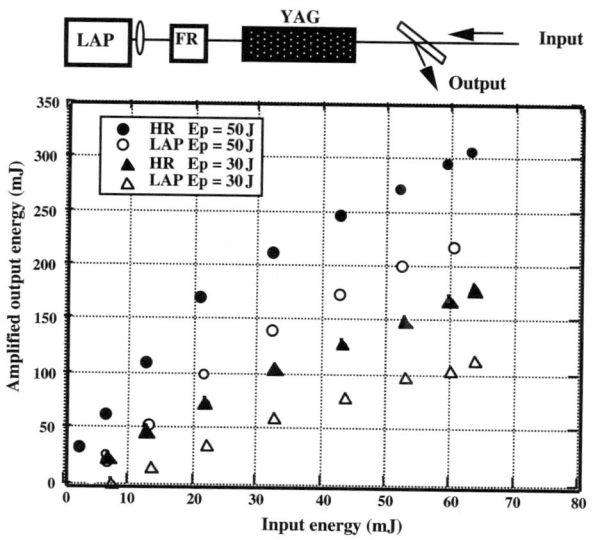

Figure 6.9. The output energy as a function of input energy with the double-pass Nd:YAG amplifier system.

The output beam quality using an HR mirror greatly decreased as the transmission was only of 0.3 at an incident energy of 150 mJ. The laser performances using a LAP PCM are almost comparable with those obtained with highly efficient SBS liquids. Damage-free operation could lead to the construction of more compact all-solid-state laser system.

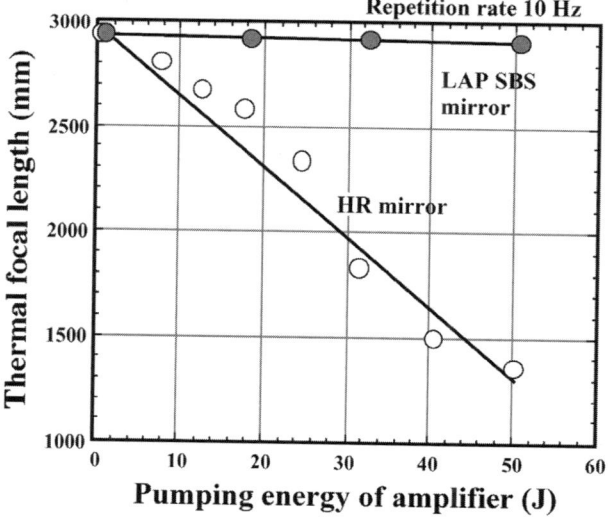

Figure 6.10. Compensation of the thermal lensing effect of the double-pass Nd:YAG amplifier with a LAP SBS mirror at a repetition rate of 10 Hz.

6.4.3 High-energy operation of Nd lasers with silica glass phase conjugate mirror

We show in this section that fused quartz glass can be used as an efficient SBS phase conjugator for applications that require a high incident energy of several joules. Fused quartz glass SBS PCM has been already used in a double-pass Nd:YAG laser system [27–31]. The laser beam was focused at 250–270 mm inside a 300-mm-long fused quartz glass rod by a lens of 500-mm focal length. The measured SBS reflectivity is shown in Fig. 6.11. It is defined as the ratio of output energy using an SBS mirror to that using an HR mirror of reflectivity 99.8% in place of the SBS mirror. A maximum intrinsic SBS reflectivity of 92% was obtained at an incident energy of 580 mJ with the compensation of the Fresnel loss at the input surface that was not antireflection coated to prevent surface damage. This operation energy is the highest value reported so far for a SBS PCM in solid materials. The SBS reflectivity at a 10-Hz repetition rate was similar to that for the single-shot operation. No optical damage was observed during the experiment.

Figure 6.12 shows the double-pass amplified output power with an SBS PCM in fused-quartz glass. Under these operating conditions, large aberrations and depolarization were introduced by the amplifier. The double-pass amplifier arrangement could produce an average output power of 7.4 W. The small-signal gain experienced by the SBS-reflected pulse during the second pass decreased in comparison with that in a case of a conventional HR mirror is used in place of the SBS mirror. This depletion in the small-signal gain coefficient, $g_0 = 0.31$ cm^{-1}, is 6% due to the frequency shift of about 16 GHz in the fused quartz glass, as the gain

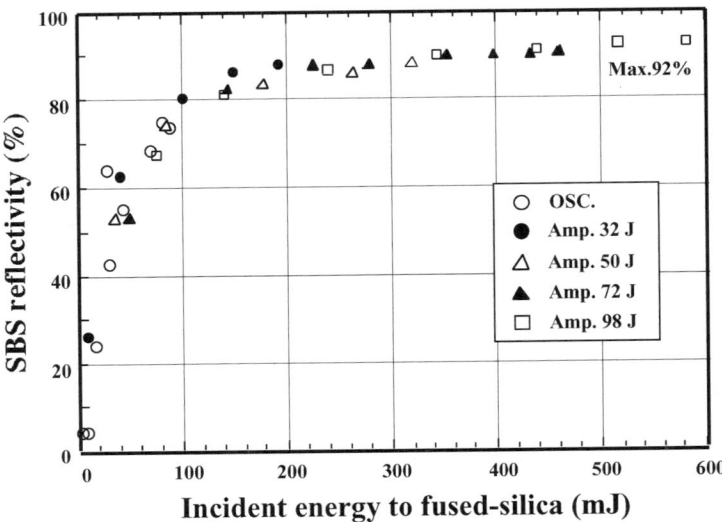

Figure 6.11. SBS reflectivity of a single-mode pump beam as a function of an incident energy focused into a fused-silica glass at a repetition rate of 10 Hz.

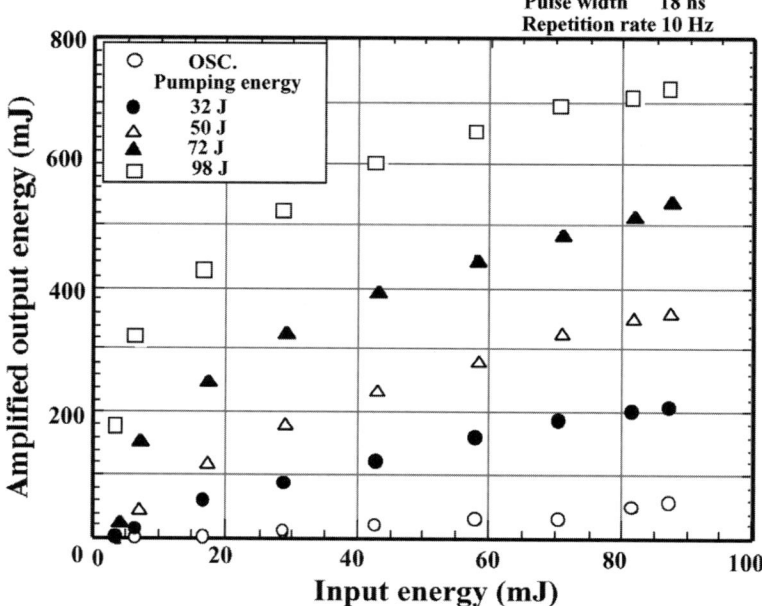

Figure 6.12. The output energy as a function of input energy with the double-pass Nd:YAG amplifier system. Pulse width, 18 ns; repetition rate, 10 Hz.

spectral width of Nd:YAG crystal is about 130 GHz. Thus, in the SBS PCM arrangement, the small-signal amplification factor, $G_0 = 23$, was reduced by about 25%. However, the output power using the HR mirror was limited to 5.5 W due to the laser-induced damage of the dielectric polarizer caused by the pattern deformation of the laser beam and also by the parasitic oscillation between the Nd:YAG rod and the conventional HR mirror.

The pulse duration changed from 18 to 14 ns due to the SBS threshold process. Figure 6.13 shows the typical oscilloscope trace of the incident and reflected pulse shapes. Transient pulse deformation was seen at the beginning of the pulse, because the reflection started after the onset of the SBS with an appropriate energy consumption to generate a density grating. The strong thermal lens effect could be almost compensated in the double-pass system by the SBS mirror made of fused quartz.

Figure 6.14 shows the SBS reflectivity of a bulk fused-quartz glass as a function of the incident pulse energy at 16-ns pulse duration using the focal length of 300 mm and 500 mm. As the incident energy increases, the SBS reflectivity increases monotonically. Using single transverse mode for the pump light, a fused-quartz glass can be used for higher pumping energy of few joules, without laser-induced damage for the $f = 500$ mm case. The SBS reflectivity reached the maximum value of 95% at the input energy of 2.3 J, which was approximately 153 times larger than the SBS threshold of about 15 mJ [32]. The incident energy was

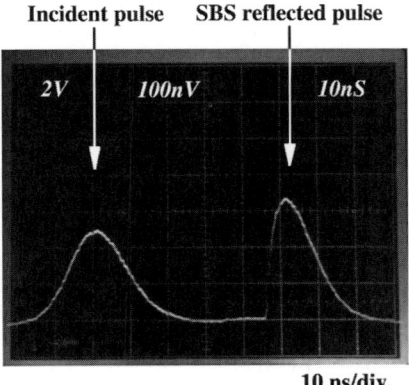

Figure 6.13. The typical oscillograms of the incident pulse and reflected pulse waveform from SBS phase conjugate mirror using a fused-silica glass.

limited by the input surface damage. As the incident energy increases, the temporal profile of the reflected high-energy pulse had a smooth and unmodulated envelope with a moderately steep leading edge. The damage threshold at a longer pulse operation depends on the SBS reflectivity. These results are consistent with the so-called empirical scaling law, where the damage threshold intensity is inversely proportional to the square root of the pulse duration. As the reflectivity grows above the threshold, the reflecting point of the incident laser is expected to move toward the source of the laser from the focus point. The bulk damage will be

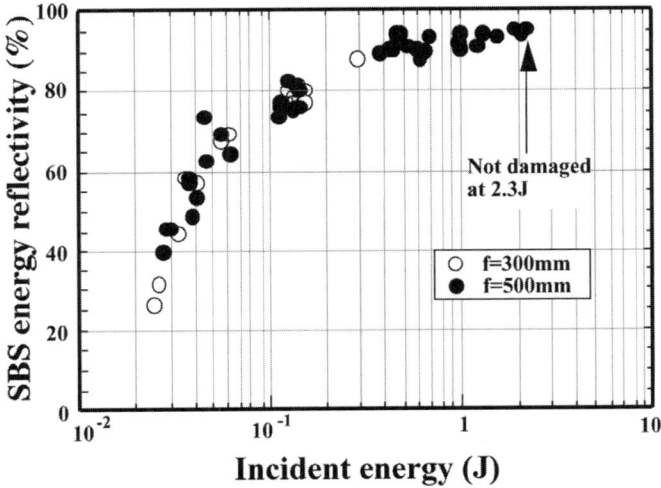

Figure 6.14. SBS reflectivity of a single-mode pump beam focused into the 300-mm-long fused-quartz glass with 300-mm (500-mm) focusing lenses.

induced by the considerable transmitted focusable power when the reflectivity by SBS process is low, for example, at a multimode operation. The SBS threshold can be reduced by use of a longer interaction length within the coherent length such as an optical fiber.

6.5 CONCLUSION

Among solid-state materials, LAP crystal exhibits one of the lowest SBS threshold at 1064-nm wavelength. We demonstrated that a LAP-SBS mirror can improve the wavefront distortion induced by thermal effects in solid-state laser materials. Deuterated LAP (DLAP) crystal is very interesting as a phase conjugate mirror for all-solid-state high-energy lasers because of high SBS gain coefficient and its low absorbance at 1064-nm wavelength.

Fused-silica glass is promising as a SBS phase conjugate mirror for all-solid-state high-energy lasers because of its high damage threshold and its low absorption. Thermal analysis shows that the fused-quartz SBS mirror is applicable for more than 100 W of input power if the SBS reflectivity is over 90%. Damage-free operation using fused silica could lead to the construction of a compact, laser-diode pumped, all-solid-state laser system with a high beam quality.

REFERENCES

1. B. Zeldovich, *Principle of Phase Conjugation*, Springer-Verlag, Berlin (1985).
2. D. A. Rockwell, *IEEE Quantum Electron.* **24**, 1124 (1988).
3. H. Yoshida, V. Kmetik, H. Fujita, T. Yamanaka, M. Nakatsuka, and K. Yoshida, *Appl. Opt.* **36**, 3739 (1997).
4. V. Kmetik, H. Fiedorowics, A. A. Andreev, K. J. Witte, H. Daido, H. Fujita, M. Nakatsuka, and T. Yamanaka, *Appl. Opt.* **37**, 7085 (1998).
5. H. Yoshida, A. Ohkubo, H. Fujita, and M. Nakatsuka, *Rev. Laser Eng.* **29**, 109 (2001) (in Japanese).
6. R. Y. Chiao, C. H. Townes, and B. P. Stoicheff, *Phys. Rev. Lett.* **12**, 592 (1964).
7. E. P. Ippen and R. H. Stoien, *Appl. Phys. Lett.* **21**, 539 (1972).
8. D. Cotter, *Electron. Lett.* **18**, 495 (1982).
9. E. A. Kuzin, M. P. Petrov, and B. E. Davydenko, *Opt. Quant. Electron.* **17**, 393 (1985).
10. A. Heuer and R. Menzel, *Opt. Lett.* **23**, 834 (1998).
11. H. J. Eichler, A. Haase, and O. Mehl, *Technical Digest of CLEO '98*, CThH2 (1998), p. 353.
12. D. Xu, M. Jiang, and Z. Tan, *Acta Chem. Sinica* **41**, 570 (1983).
13. S. B. Monaco, L. E. Davis, S. P. Velsko, F. T. Wang, and D. Eimerl, *J. Crystal Growth* **85**, 252 (1987).
14. A. Yokotani, T. Sasaki, K. Yoshida, and S. Nakai, *Appl. Phys. Lett.* **55**, 2692 (1989).

15. T. Sasaki, A. Yokotani, K. Fujioka, T. Yamanaka, and S. Nakai, in *Springer Proceedings in Physics*, Vol. 36, T. Kobayashi (ed.), Springer-Verlag, Berlin (1989), pp. 206–209.
16. A. Yokotani, T. Sasaki, K. Fujioka, S. Nakai, and C. Yamanaka, *J. Cryst. Growth* **99**, 815 (1990).
17. D. Eimerl, S. Velsko, L. Davis, F. Wang, G. Loiacono, and G. Kennedy, *IEEE J. Quantum Electron.* **25**, 179 (1989).
18. K. Aoki, K. Nagano, and Y. Iitaka, *Acta Crystallogr. Sect. B* **27**, 11 (1971).
19. W. Koechner, *Solid-State Laser Engineering*, 2nd ed., in *Springer Series in Optical Science*, Vol. 1, Springer, Berlin (1988), pp. 540–558.
20. C. L. Tang, *J. Appl. Phys.* **37**, 2945 (1966).
21. W. Kaiser and M. Maier, Stimulated Rayleigh, Brillouin, and Raman spectroscopy, in *Laser Handbook*, Vol. 2, F. T. Arecchi and E. O. Schulz-Dobois (eds.), North-Holland, Amsterdam (1977), pp. 1077–1150.
22. G. W. Faris, L. E. Jusinski, and A. P. Hickman, *J. Opt. Soc. Am. B* **10**, 587 (1993).
23. H. Yoshida, M. Nakatsuka, H. Fujita, T. Sasaki, and K. Yoshida, *Appl. Opt.* **36**, 7783 (1997).
24. H. J. Eichler, R. Menzel, R. Sander, M. Schulzke, and J. Schwartz, *Opt. Commun.* **121**, 49 (1995).
25. M. Yoshimura, Y. Mori, T. Sasaki, H. Yoshida, and M. Nakatsuka, *J. Opt. Soc. Am. B*, **15**, 446 (1998).
26. H. Yoshida, H. Fujita, M. Nakatsuka, and M. Yoshimura, T. Sasaki, and K. Yoshida, *Rev. Laser Eng.* **25**, 232 (1997) (in Japanese).
27. H. Yoshida, H. Fujita, M. Nakatsuka, and K. Yoshida, *Opt. Eng.* **36**, 2557 (1997).
28. H. Yoshida, H. Fujita, M. Nakatsuka, and K. Yoshida, *Rev. Laser Eng.* **26**, 138(1998) (in Japanese).
29. H. Yoshida, H. Fujita, M. Nakatsuka, and K. Yoshida, *Rev. Laser Eng.* **27**, 495 (1999) (in Japanese).
30. H. Yoshida, H. Fujita, M. Nakatsuka, and K. Yoshida, *Jpn. J. Appl. Phys.*, **38**, L521 (1999).
31. H. Yoshida, H. Fujita, M. Nakatsuka, and K. Yoshida, *Proc. SPIE*, **3889**, 812 (2000).
32. H. Yoshida, H. Fujita, M. Nakatsuka, M. Fujinoki, and K. Yoshida, *Opt. Commun.* **222**, 257 (2003).

CHAPTER 7

Stimulated Brillouin Scattering Pulse Compression and Its Application in Lasers

G. A. PASMANIK, E. I. SHKLOVSKY, and A. A. SHILOV

Passat, Toronto, Ontario, Canada M3J 3H9

7.1 INTRODUCTION

Temporal laser pulse compression through stimulated scattering has a long history, and its first observations [1–3] predate pioneering experiments on optical phase conjugation [4]. Despite early history of optical pulse compression, study of Brillouin compression began in earnest only after the well-known work of D. T. Hon [5]. It was as a part of this work that the SBS method was first applied to an Nd:YAG laser with the stated aim of shortening the pulse width in a double-pass power amplifier with phase conjugate (PC) Brillouin mirror. This classical work had the effect of inspiring further research of optical schemes for pulse compression as well as its applications ranging from generation of new wavelengths through nonlinear optical processes to laser-plasma-driven soft X-ray sources. Even though the main characteristics of SBS pulse compression were established some 20 years ago [6–20], SBS pulse compression-related research and literature remained intensely active until the early 1990s, after which a noticeable drop in the level of SBS pulse compression-related research was seen. To some degree, this decline in activity was a natural outcome of the work of the early 1990s in which the physics of optical pulse compression was well-defined. As well, the decline could be partially attributed to the nearly routine application of the phenomenon in laser systems and applications that followed. But beyond these factors, the rapid development of diode-pump technology has also played a role, resulting in the appearance of miniature solid-state lasers such as microchip lasers, capable of generating pulses with durations of 1 ns and less [21–23]. For powerful diode-pumped and excimer laser systems, however, SBS pulse compression still provides a viable, effective option for short laser pulse generation.

Phase Conjugate Laser Optics, edited by Arnaud Brignon and Jean-Pierre Huignard
ISBN 0-471-43957-6   Copyright © 2004 John Wiley & Sons, Inc.

It was shown almost two decades ago that a nanosecond laser pulse can be compressed to a few hundreds of picoseconds [24]. Given that the typical pulse width of diode-pumped Nd:YAG lasers is some nanoseconds in duration, Brillouin scattering provides a very attractive way of compressing a backward Stokes pulse to the subphonon lifetime region [25]. This approach still awaits common applications in such systems.

During the years following the first work in Brillouin pulse compression, various passive pulse compression schemes have been studied, including truncated SBS [26–31], two-step Brillouin compression [32, 33] and combined SBS–SRS compression—the latter comprising cascaded Brillouin and Raman compressors [34–39].

In the following sections we describe key experimental and theoretical studies as well as results from numerical simulations of laser pulse compression accompanying backward SBS. Since in many cases the choice of SBS-active material is crucial for efficient pulse compression, we outline selection criteria and discuss competing physical–optical processes that should be taken into account when selecting the pulse compressor medium. We also describe experimental results of cascaded SBS–SRS compression as well as application of this technique to the generation of picosecond pulses.

While we have endeavored to provide a well-rounded discussion of this field, the space restrictions inevitably limit the length of the discussion in this chapter. The authors regret that much of the worthy work that has been advanced in this field cannot be included here, but they also feel that the sheer size of the literature argues for the abiding interest in this area of physical optics.

7.2 PHENOMENOLOGICAL DESCRIPTION OF BRILLOUIN COMPRESSION

SBS pulse compression is based on the ability of an SBS-active material to amplify the leading edge of a backward Stokes pulse in the field of forward-propagating pump beam. Of the conditions under which efficient pulse compression can take place, the appearance of the steep leading edge of the backward Stokes pulse is the most important. The steep leading edge is a general feature of stimulated emission processes, the "steepness" being the time derivative of the optical pulse power, $\Delta P/\Delta t_S$, being in its turn a function of the material response time, pump duration, and nonlinear gain. Figure 7.1 illustrates the appearance of the steep leading edge in a Stokes pulse.

Depending upon the combined effects of the response time, τ, of the material, the intensity, P/A, in the focal region, the Brillouin gain, g_B, and the pump leading-edge rise time, t_{Ple}, the leading edge rise time of the Stokes pulse, t_{Sle}, may vary from units to tens of nanoseconds. The Stokes pulse leading edge extracts the energy stored in a nonlinear medium by the incident pump. Under appropriate conditions the stored energy can be completely extracted during the interval corresponding to rise time t_{Sle}, thereby producing a short Stokes pulse followed by a relatively long, low-

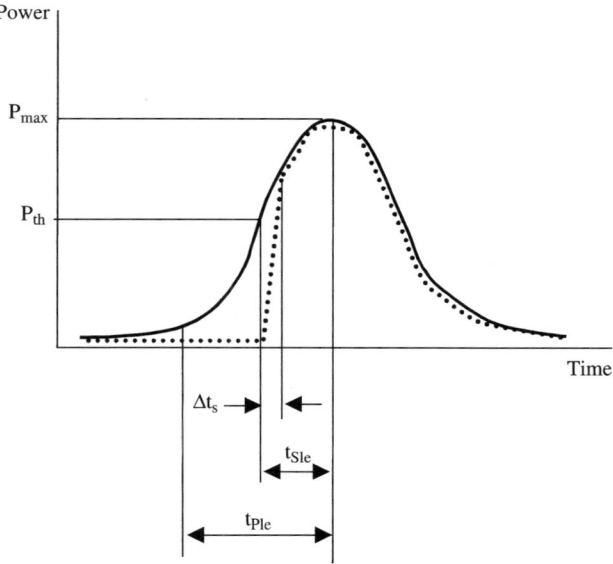

Figure 7.1. Representation of steep leading edge formation in a Stokes pulse (dashed).

intensity tail. Using a simple analogy with conventional laser amplification can be helpful in understanding still other conditions contributing to Brillouin pulse compression.

In the simplest, most popular configuration, backward Brillouin scattering is produced by focusing the pump into an SBS-active medium. The Stokes pulse originating in the focal area within the SBS medium moves back toward the lens and, due to the constructive interference with the remaining part of the pump, builds up a hypersonic grating. This, in turn, coherently reflects power from the pump beam, thus amplifying the Stokes pulse. The amplitudes of the pump and Stokes pulses should be matched in such a way that the hypersonic grating blocks the focal region from the oncoming pump. By careful matching and subsequently strong depletion of the pump, the peak amplitude of the Stokes pulse can be many times greater than that of the pump for a sufficiently long interaction length, as illustrated in Fig. 7.2.

Optimal conditions for the compression are different depending upon the pump pulse profile, the pump intensity, and the specific material used. In the case of a long, single-cell Brillouin compressor the maximum compression ration, K is approximately equal to

$$K \sim \int_L g_B \cdot I_P(z)\, dz \approx G_{th} \tag{1}$$

where I_P is the pump intensity, g_B is the stationary SBS gain coefficient, and $G_{th} = 30\text{--}40$ is the overall Brillouin gain needed to reach SBS threshold. It should

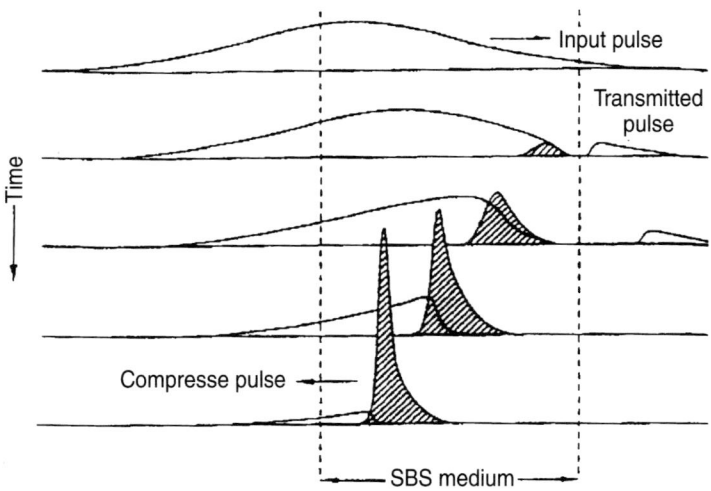

Figure 7.2. Formation of a compressed Stokes pulse (from Ref. 59).

be noted that for efficient SBS compression the spatial pulse length should match the interaction length, L, such that

$$L \geq \frac{ct_p}{2n} \quad (2)$$

where c is the light velocity and t_p is the effective pumping pulse width, which is defined as the interval between the initial SBS excitation and the end of the pump and Stokes pulses interaction.

Besides the compression ratio, there is another important characteristic in the pulse compression process. This is the energy conversion efficiency (or reflectivity) of the SBS compressor: $R = E_S/E_P$, where E_S and E_P are the Stokes and laser pulse energies, respectively. Since the Brillouin shift, Ω, equals $2n\omega_p v/c \sim 10^{-5}\omega_p$, where ω_p is the pump (laser) frequency and v is the speed of sound in the medium, it is possible to attain almost 100% conversion of the incident laser pulse into compressed Stokes pulse output. In practice, the energy conversion efficiency is somewhat less than 100% due to need for the intensity of the focused pulse to reach SBS threshold. During the time preceding the buildup of the Stokes pulse from noise, a portion of laser pulse with intensity below the threshold is transmitted through the cell.

There are two main configurations employed for optical pulse compression: (1) the simple SBS generator and (2) the cascaded SBS generator–amplifier configurations. In the SBS generator configuration, the buildup of the Stokes pulse from noise and its amplification in the field of ongoing laser pulse both take place within the same interaction volume. With the simple SBS generator configuration, it is

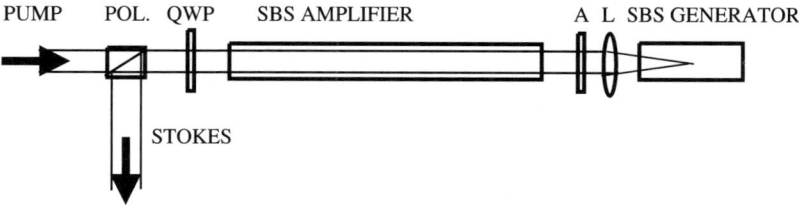

Figure 7.3. A two-cell optical configuration for SBS compression.

possible to efficiently compress laser pulses of relatively low peak power and energy, provided that detrimental and competing processes such as SRS, optical breakdown, and self-focusing are avoided. The parasitic influence of these processes can be observed even in carefully selected SBS-active materials. A configuration consisting of a tandem SBS generator and amplifier arrangement, first suggested for SBS phase conjugation of powerful laser beams [40], proved to be beneficial for SBS compression as well. Figure 7.3 shows general layout of two-cell arrangement for SBS pulse compression.

The seeding Stokes pulse, which in a single-cell compressor develops from spontaneous noise, is generated here by a separate cell (SBS generator) in a manner similar to conventional backward SBS. By insertion of an appropriate attenuator A and by maintaining intensity of the pump over the length of SBS amplifier slightly below SBS threshold, it is possible to efficiently compress the seed pulse from the SBS generator. It should be noted that due to linear attenuation of both the laser pump and Stokes beams, the conversion efficiency usually drops in this case. The operation of a two-stage SBS generator–amplifier compressor will be considered in somewhat greater detail later in this chapter.

7.3 THEORETICAL ANALYSIS OF BRILLOUIN PULSE COMPRESSION

In this section, analytical solutions of SBS compression are outlined. Of particular interest, the spherical wave approximation for transient scattering is used to derive a formula for the compressed pulse width. This expression is subsequently compared with the pulse-width expression derived from a plane wave approximation.

We recall that stimulated scattering is caused by nonlinear interaction of the intense laser pump beam with weak radiation resulting from spontaneous scattering of the pump beam from fluctuations in the refractive index. Solution of the equations describing the SBS process is especially simple under conditions of stationary-state scattering, in which there is no light attenuation and no pump depletion—that is, when $E_S \ll E_P$. In this case, one has

$$I_S(L) = I_S(0) \exp(gI_p L) \qquad (3)$$

where L is the interaction length.

The plane waves approximation has been used to describe pulse compression due to stimulated scattering ever since the first work devoted to the subject [41]. It can be shown [42] that if the laser pulse width, t_P, is less than the time needed for light to travel through the medium, then the logarithm of overall gain $G = \ln |E_S(L)/E_S(0)|^2$ can be expressed as

$$G = \left(\sqrt{1 + g_B I_P \frac{c}{n} \tau} - 1 \right) \frac{t_P}{\tau} \tag{4}$$

Here, $E_s(L)$ and $E_s(0)$ are complex amplitudes of laser electrical field at distance L from the entrance to nonlinear medium and at the entrance to the medium, respectively, and n is the refractive index of the medium. From Eq. (4) it follows that if $gI_P(c/n)\tau \gg 1$, the typical duration of the leading edge t_{Sle} of the scattered pulse is

$$t_{\text{Sle}} \approx \frac{\tau}{\sqrt{gI_P(c/n)\tau}} = \frac{t_P}{G} \tag{5}$$

For high values of transient SBS gain, backward scattering can occur from the nonlinear medium before the pump pulse reaches the focal region. This pre-scattering decreases SBS compression efficiency and results in lower compression ratios as well as multiple Stokes pulse generation. In order to avoid premature Stokes pulse generation, the SBS gain, G, over the cell's length should be less than $G_{\text{th}} = 30\text{--}40$. Thus, on the basis of the plane wave approximation a maximum compression ratio of 30–40 can be attained.

Let us now consider SBS pulse compression for a convergent laser beam (Fig. 7.4).

In the equations describing the complex amplitudes of the laser pump wave E_P and the Stokes wave E_S, it is convenient to make a transformation to spherical coordinates. Using these coordinates the laser wave is written in the form

$$E_P e^{ikz} = \frac{A_P}{r} \exp(ikr), \qquad E_S e^{-ikz} = \frac{A_S}{r} \exp(-ikr) \tag{6}$$

where r is the coordinate with respect to the focal point.

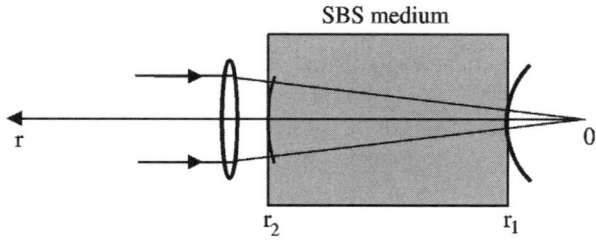

Figure 7.4. SBS pulse compression for a convergent laser beam.

The reduced equations for A_L and A_S can be written as follows:

$$\frac{n}{c}\frac{\partial A_P}{\partial t} + \frac{\partial A_P}{\partial r} = i\frac{k\Delta\rho}{2\varepsilon_0}A_S$$

$$\frac{n}{c}\frac{\partial A_S}{\partial t} - \frac{\partial A_S}{\partial r} = i\frac{k\Delta\rho^*}{2\varepsilon_0}A_P \quad (7)$$

$$\frac{\partial}{\partial t}\Delta\rho + \frac{\Delta\rho}{\tau} = i\beta A_P \frac{A_S^*}{r^2}$$

Equations (7) are valid when $r > l_F$, where l_F is one-half of the focal waist length.

Under the assumption that the pumping wave field is constant, we make the transformation to the traveling coordinate, $\xi = t - (nr/c)$, obtaining

$$\frac{2n}{c}\frac{\partial A_S}{\partial \xi} - \frac{\partial A_S}{\partial r} = i\frac{k\Delta\rho^*}{2\varepsilon_0}A_P$$

$$\frac{\partial}{\partial \xi}\Delta\rho + \frac{\Delta\rho}{\tau} = i\beta A_P \frac{A_S^*}{r^2} \quad (8)$$

For Eqs. (8) the initial conditions are as follows: $A_S(\xi = 0) = A_S(0, r) = $ const, and $\Delta\rho(\xi = 0) = 0$. Equations (8) were solved in Ref. 43 using Laplace transforms and assuming a rectangular laser pulse ($I_P = $ const) such that

$$A_S(p) = \int_0^\infty e^{ps}A_S(s)\,ds$$

$$\Delta\rho(p) = \int_0^\infty e^{ps}\Delta\rho(s)\,ds \quad (9)$$

We assume that there is no pump depletion in the above considerations. It can be shown that for exponential Stokes pulse growth the following equation must hold:

$$p(1+p\tau) = \frac{g_B|A_P|^2}{4rr_1} \cdot \frac{c}{n} \quad (10)$$

Here r_1 is the radius of wavefront curvature at the entrance of SBS cell (see Fig. 7.4).

In the general case we obtain the following equation for the overall SBS gain, G:

$$G = \left(\sqrt{1 + \frac{g_B|A_P|^2(c/n)\tau}{rr_1}} - 1\right)\frac{\xi}{\tau} = \left(\sqrt{1 + g_B I_P(r)\frac{c}{n}\tau\frac{r}{r_1}} - 1\right)\frac{\xi}{\tau} \quad (11)$$

where $I_P(r) = |A_P|^2/r^2$ is the laser pulse intensity at a spherical surface of radius r.

We can use Eq. (11) to find the duration of the Stokes pulse leading edge at the entrance, $r = r_2$:

$$t_{\text{Sle}}(r) = \frac{\tau}{\sqrt{1 + (g_B|A_P|^2(c/n)\tau)/(r_1 r_2)} - 1} = \frac{t_P}{G} \qquad (12)$$

When $(g_B|A_P|^2(c/n)\tau)/r_1 r_2 \approx 1$ the duration of the Stokes pulse leading edge is approximately equal to τ. Equation (12) shows that it is possible to shorten the Stokes pulse leading edge duration to $t_{\text{Sle}} \ll \tau$. However, to obtain pulses with duration much shorter than the medium's relaxation time, it is necessary that $(g|A_L|^2(c/n)\tau)/r_1 r_2 \gg 1$.

We now discuss the conditions required to generate a "compact" Stokes pulse— that is, a pulse with a reduced trailing edge. As mentioned earlier, the first condition is that the round-trip time for the light traveling through the nonlinear medium be greater than the effective length of the pump pulse [see Eq. (2)]. For example, if stimulated scattering occurs at a pump power level lower than its peak power, the effective output pulse duration (e.g., measured at the 0.1 peak power level) will significantly exceed the pulse duration FWHM. There is also a second condition. The thickness of a typical nonlinear "depletion" layer must be less than or equal to the length of the backward Stokes pulse being generated as Brillouin compression output. In this case the pump pulse passing through the nonlinear medium would be exhausted at the end of its interaction with the backward Stokes pulse; that is, the residual pump pulse energy passing through the "depletion" layer must be below the threshold leading to produce backward scattering, which would subsequently be added to the "tail" of the Stokes pulse. As a rule, the diameter of the input pump beam focused into nonlinear medium should not exceed the diameter of the focal waist by more than G_{th}. If this requirement is not satisfied, the backward Stokes beam emerging from the focal area widens and its energy density becomes too small to deplete the pump over its path from the front mirror to the focal region. This is illustrated in Figs. 7.5a and 7.5b for two pump beams with the same power and pulse shape but different convergence angles.

In Fig. 7.5a, a backward compressed "compact" Stokes pulse corresponding to a small pump convergence angle, is shown. It was formed while traveling from the focal region to the entrance of a cell filled with nonlinear medium. The length of the Stokes pulse is approximately equal to the length of the focal waist.

In Fig. 7.5b a configuration is shown depicting a greater angle of convergence for the pump beam. A "compact" Stokes pulse is also shown rising well before it reaches the front window of the cell with its length approximately equal to the length of the focal waist. Closer to the cell's front window, one sees that the Stokes pulse energy is increased, albeit through growth in its tail, rather than through increased peak power.

We now evaluate the dependence of duration of the exiting Stokes pulse on geometrical conditions. For the sake of clarity we consider a two-cell compressor configuration as shown in Fig. 7.6.

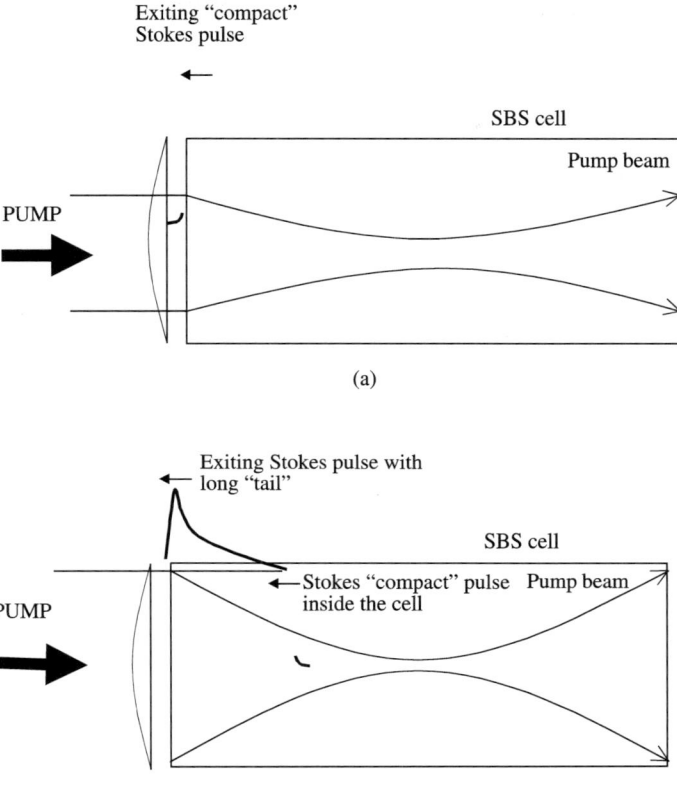

Figure 7.5. (a). Backward compressed "compact" Stokes pulse corresponding to a small pump convergence angle. (b) Same as for part (a) but with a greater angle of convergence for the pump beam.

In Fig. 7.6, D_1 is the diameter of the pump beam at the plane of lens L_1 and D_2 is the diameter of the pump beam at the exit of SBS cell 1. Let us assume that conditions for focusing the pump pulse within SBS cell 2 are optimal and that within this cell a short SBS pulse is excited, whose duration is approximately equal to the

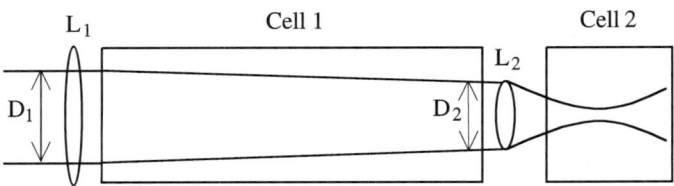

Figure 7.6. Two-cell compressor configuration.

light transit time through the focal waist. Moreover, we assume that the pump pulse duration is approximately equal to the light round-trip time through SBS cell 1 and that there is no excitation of backward SBS by the pump traveling through SBS cell 1. Finally, we assume that the focal distance of lens L_2 and the length of the focal waist in SBS cell 2 are substantially smaller than the length of SBS cell 1.

The width of Stokes pulse formed in SBS cell 1 depends on the rise time t_{Sle} of this pulse and its "tail" duration, t_{tail}. The rise time can be evaluated from Eq. (12): $t_{Sle} = t_P/G$, where G is the overall gain in SBS cell 1. In order to evaluate t_P in the transient regime, we will use the second condition stated above. We will make use the formula from Ref. 42 describing the interaction of two counterpropagating pulses in transient regime and apply this formula to "depletion" layer ΔL located near the entrance to SBS cell 1. The length ΔL of this layer is comparable or smaller than the length of interacting pulses:

$$G(\Delta L) = 2\sqrt{2g_B I_p \Delta L \frac{t_p(\Delta L)}{\tau}} \tag{13}$$

where $t_P(\Delta L) = 2(n\Delta L/c)$ is the width of the pump pulse, determined by the time of interaction between the pump and Stokes pulse incoming to the "depletion" layer.

Assuming that the interaction of counterpropagating pulses leads to the depletion of the pump pulse at $G(\Delta L) \geq 1$, we find

$$\Delta L \geq \frac{1}{4}\left(\frac{c\tau}{ng_B I_p}\right)^{1/2} \tag{14}$$

Taking into account Eq. (11), we obtain

$$\Delta L \geq \frac{ct_p}{4nG}\sqrt{\frac{D_1}{D_2}} \tag{15}$$

Furthermore, to find t_{tail} we have

$$t_{tail} = \frac{2n\Delta L}{c} = \frac{t_p}{2G}\sqrt{\frac{D_1}{D_2}} \tag{16}$$

Thus, for the full width of the Stokes pulse, there is the following approximate equation:

$$t_S = \frac{t_p}{G}\left[1 + \frac{1}{2}\left(\frac{D_1}{D_2}\right)^{1/2}\right] \tag{17}$$

It is obvious that the duration of the Stokes pulse will be minimal if D_1 is approximately equal to D_2. (The case where D_1 is less than D_2 requires special analysis because backward SBS might arise from thermal noise in cell 1 before the pulse escapes from the amplifier cell.)

Theoretical models of the compression process and methods for Stokes pulse width calculation were also suggested in Refs. 7, 20, and 69. In Ref. 67, the approach is based on the calculation of the temporal location of the leading and trailing edges of the Stokes pulse from threshold conditions for SBS. The model developed in Ref. 67 enables one to estimate the compression ratio for the Stokes pulse for monochromatic and nonmonochromatic pumping pulses as amplified from spontaneous noise. It was shown that in the case of nonmonochromatic incident laser pulses with optimal line widths, the compression ratio can be two to three times greater than G_{th} and attain values between 60 and 100 [68].

7.4 NUMERICAL SIMULATION

As noted above, theoretical analysis of Brillouin pulse compression normally makes simplifying assumptions concerning the transverse structure of laser beams. The relatively short lifetime of acoustical phonons (e.g., 1–2 ns for liquids) enables one to treat pulse compression on the basis of the following equations:

$$\frac{n}{c}\frac{\partial I_P}{\partial t} + \frac{\partial I_P}{\partial z} = -g_B I_P I_S$$
$$\frac{n}{c}\frac{\partial I_S}{\partial t} - \frac{\partial I_S}{\partial z} = g_B I_P I_S \quad (18)$$

Numerical solution of Eq. (18) reveals a temporal modulation of both transmitted and backward radiation, with a period of $2L/c$ [44, 45]. For this reason a number of authors have come to the conclusion that in the case of a collimated pump, only pulse shortening rather than true pulse compression is possible when the Stokes pulse originates from spontaneous noise [9, 46].

To define the limit of possible pulse compression, one has to take into account transient regime of SBS. Computational analysis of equations of type (18) under some simplified conditions shows that, given a sufficiently steep leading edge of the Stokes seed pulse, a quasi-soliton pulse of duration $t_S \approx 0.1\tau$ can be produced [11].

For the limit of pulse compression there are two expressions, which differ only by coefficients:

$$t_S \approx \frac{\lambda n}{2\gamma_e}\sqrt{\frac{\rho c}{I_P}} \quad \text{and} \quad t_S \approx \frac{1}{2\pi}\frac{\lambda n}{2\gamma_e}\sqrt{\frac{\rho c}{I_P}}$$

These expressions were derived in Refs. 47 and 15, respectively. Here γ_e is the electrostriction coefficient and ρ is the material density.

It should be emphasized that in the generation regime corresponding to the single-cell configuration and in order to attain the aforementioned duration of compressed Stokes pulses, the required energy density in the focal area is so great that it will inevitably exceed the threshold intensity of a number of detrimental effects. These effects include self-focusing, optical breakdown, and Raman scattering. However, this is not the case when the amplification regime corresponding to two-cell configuration is applied to SBS compression. The latter allows for efficient compression of powerful laser pulses.

Numerical simulation of SBS pulse compression using quasi-1D approximation was reported in Ref. 49. In the work described, carbon tetrachloride was used as the SBS-active medium. The pump was focused using a 50-cm focal length lens placed at the front window of the cell. It was found that with an increase in the convergence angle, $\beta \approx D/F$, at which a pump was focused into the cell, the Stokes pulse leading edge became increasingly steeper and less time was required for the pulse to reach maximum instantaneous power. Although SBS reflectivity also increases monotonically with the converging angle, a minimum Stokes pulse duration in the simulation could be obtained at an optimum beam diameter $D_{opt} = 2.25$ mm ($\beta_{opt} \approx 4.5$ mrad). For $\beta > \beta_{opt}$, the Stokes pulse leading edge does not change, but for larger angles the trailing edge elongates. The trailing edge tends to have the same shape as that found in the case of the tight focusing normally used in experiments on phase conjugation.

In the quasi-1D model the value of compression, K, strongly depends on the nonlinear medium's relaxation time, τ. For an 8-mJ pump with optimum converging angle, reduction of τ from 1 ns to 0.5 ns increases the compression K from 17.6 to 22.8 and increases the SBS reflectivity R from 0.72 to 0.82 [37]. In that work it was also shown that, given factors such as pulse width and energy, the compression, K, is more efficient for pumping pulses with leading edges having a steeper-than-Gaussian profile. In particular, for $P \sim t\exp(-t^2/1.5t_P^2)$ we obtain

$$K = 24.8 \quad \text{at } \tau = 0.5\,\text{ns} \quad \text{and} \quad E_L = 6\,\text{mJ}$$

A detailed study of the dynamics of SBS pulse reveals an alternate approach to SBS compression. Thermal density fluctuations lead to fluctuations in the amplitude and phase of the Stokes wave [50–53]. Earlier experimental observations of phase jumps and amplitude fluctuations in SBS were reported in [54–57]. Recently a comprehensive numerical study of phase modulations has been conducted and conditions found for generating a short Stokes pulse by catching a single oscillation on the Stokes pulse envelope [58]. The SBS process in the model was initiated by random noise obeying Gaussian statistics. No special pulse compression geometric conditions were imposed on the length of the Stokes–pump interaction. The noisy nature of the SBS phenomenon results in large-scale regular fluctuations in the Stokes pulse due to energy interchange between Stokes and pump. The model reveals threshold relaxation oscillations with damping time corresponding to phonon lifetime. The change in the relaxation oscillation dynamics is shown in Fig. 7.7. At this point we should note that the soliton-like behavior of the Stokes pulse in this case is not the result of pulse compression, but is due to the near-threshold conditions of the backward SBS excitation.

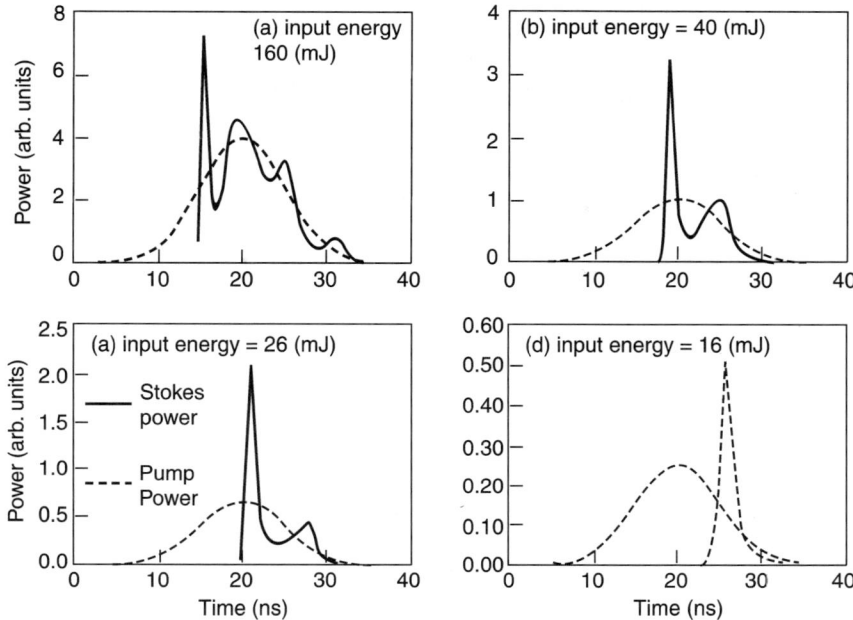

Figure 7.7. Change in Stokes profile with decrease of pump energy focused into an SBS cell [58].

In Ref. 25 a numerical model of Brillouin pulse compression for the tandem SBS oscillator–amplifier geometry was developed by using a generalization of the split-step method and fast Fourier transformations. The calculated Stokes pulse width as a function of pump energy for two pump durations is shown in Fig. 7.8.

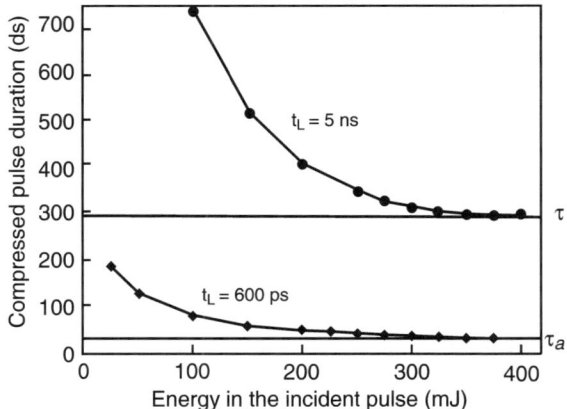

Figure 7.8. Calculated Stokes pulse width as a function of pump energy [25] (τ and τ_a are the phonon lifetime and period of hypersonic wave, respectively).

236 STIMULATED BRILLOUIN SCATTERING PULSE COMPRESSION

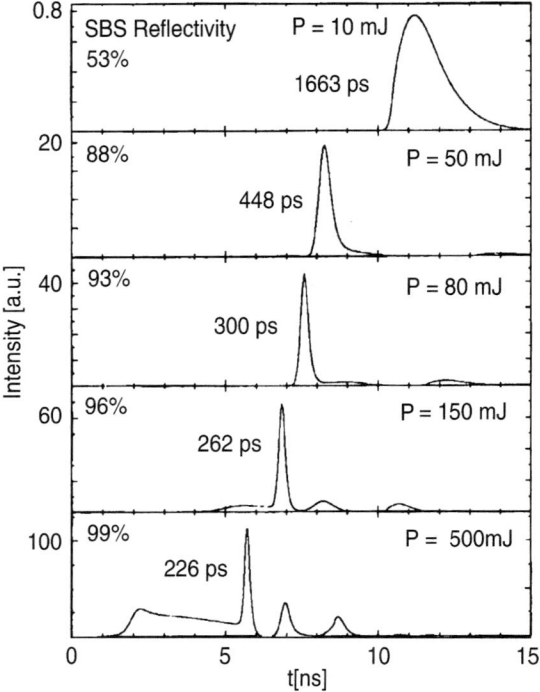

Figure 7.9. Numerically generated Stokes pulses in a two-cell methanol compressor. Laser pulse width, t_L, is 5.6 ns [60].

The above curves were simulated for a water SBS compressor pumped with laser pulse at 532 nm. In the transient regime of SBS compression, the model predicts Stokes pulses with duration shorter than phonon lifetime, τ. It can be seen that increasing the energy of pump results in decrease of the Stokes pulse duration.

Additional calculations of SBS pulse compression for particular experimental conditions can be found elsewhere [11, 15, 25, 60–62]. Some general characteristics are shared between results obtained from 1D numerical studies. As an example, Fig. 7.9 shows numerically generated Stokes pulses emitted from an SBS generator–amplifier compressor for an input laser pulse of 5.6 ns in width at different pulse energy.

In the 1D model it is assumed that the wavefronts of both the pump and Stokes pulse are plane. A 2D simulation can take into account a more realistic, nonuniform laser beam profile with correspondingly larger computational times. 3D calculations for Gaussian beams with axially symmetric intensity distributions [37] have shown the generally complex spatial and temporal structure of compressed Stokes pulses. As a whole, 3D models predict lower SBS pulse compression compared to 1D models, assuming similar operating conditions.

7.5 CHARACTERIZATION OF MATERIALS USED FOR SBS COMPRESSORS

Generally, the same nonlinear materials are used for laser pulse compression as are used for conventional SBS. Early experiments on pulse compression were conducted mainly using pressurized gases as the SBS media for SBS pulse compression of relatively long laser pulses (20- to 30-ns duration). Among them, argon and methane proved the most suitable for SBS compression due to their short phonon lifetime. Sulfur hexafluoride was found to have the advantage of high Brillouin gain even at relatively low operational pressures, as well as high resistance to optical breakdown.

Table 7.1 lists some characteristics of pressurized gases used in Brillouin pulse compressors. Nonlinear liquids offer better opportunities for SBS pulse compression than do pressurized gases due to their higher SBS gain and generally shorter phonon lifetime. Some nonlinear liquids, used for SBS PC and pulse compression, are listed in Table 7.2.

One of the key requirements for successful operation of a pulse compressor is the purity of the SBS medium used. An SBS mirror can effectively block the focal region from the powerful pump, and ideally only near-SBS-threshold energy leaks through the mirror into the focal waist during the buildup of refraction index grating. As a result, light intensity anywhere in the interaction volume does not reach the breakdown threshold even for an extremely powerful pump. Since optical breakdown dramatically depends upon the concentration of microparticles in the focal area, it is necessary to purify the material to the highest possible degree. Liquid fluorocarbons are a class of nonlinear materials used for SBS pulse compression that are most resistant to optical breakdown. For liquids of that class (e.g., FC-72 and FC-75) the breakdown threshold reaches 130 GW/cm² when irradiated by 1-ns pulses at 1064 nm [73]. Purified and filtered liquid tetrachloride also has a very high breakdown threshold, up to 100 GW/cm² at 1064 nm [42].

SBS compression in repetitively pulsed laser systems has some demonstrated peculiarities. At high pulse repetition rates even very low absorption within the SBS-active medium dramatically affects the pulse compression efficiency, both in terms of pump to Stokes energy conversion and the degree of compression. For example,

TABLE 7.1. Characteristics of Gases Used in SBS Pulse Compressors ($\lambda = 1064$ nm, $p = 10$ atm)

Parameters	Gases				
	N_2	SF_6	CH_4	Xe	Ar
Brillouin shift, $\Delta \nu$ (cm^{-1})	0.021	0.011	0.029	0.0088	0.023
Brillouin gain, g_B (cm/GW)	0.15	1.8	0.8	9	0.17
Phonon lifetime, τ (ns)	1.8	7.4	1.8	17	3.7

TABLE 7.2. Characteristics of Liquids Used for SBS–PC and Pulse Compression at 1 μm [42, 64, 73]

Liquid	$\Delta \nu$ (cm^{-1})	τ (ns)	g_B (cm/GW)
Benzene	0.141	3.0	18
Toluene	0.132	1.5	10
Carbon bisulfide	0.127	7	50
Acetone	0.102	4	18
Methanol	0.094	3.7	13
Water	0.125	3.4	4.8
n-Hexane	0.096	3.5	19
Tetrachlorides			
Carbon	0.096	1.3	6
Titanium	0.107	2	23
Silicon	0.074	2.2	10
Germanium	0.072	2.3	11
Tin	0.078	1.8	14
Fluorocarbons			
FC 72	0.041	0.95	6.2
FC 75	0.045	0.78	4.7
FC 77	0.047	0.72	1.5
Freon-113	0.059	0.85	6.5

acetone, methanol, and water can be used as effective SBS pulse compression media in the repetitively pulsed mode of operation, but only in the visible (mainly green) spectral region. In the near-infrared (NIR) region, where most solid-state laser fundamental emission wavelengths are found, the linear absorption of these liquids is too high ($\alpha \sim 0.01$–0.1 cm^{-1}), making impossible reliable compressor operation using these liquids at high (>100 Hz) repetition rates. The linear absorption in SBS medium results in thermal defocusing, which increases SBS threshold, reduces SBS reflectivity, makes the operation of the compressor unstable, and even terminates SBS.

Thermal defocusing for a single-mode pulsed light beam with diameter D and diffraction-limited divergence $\vartheta_d = 4/kD$, where $k = 2\pi/\lambda$, is characterized by the critical energy of thermal defocusing, E_{cr}. Actually, E_{cr} is the energy corresponding to formation of a thermolens, which subsequently increases beam divergence by a factor of two over the natural divergence [63]:

$$E_{\mathrm{cr}} = \frac{2\pi \rho c_p}{(\partial n/\partial T)_p \alpha k^2} \qquad (19)$$

Here c_p and α are the specific heat of the medium and the optical absorption coefficient, respectively. The importance of E_{cr} lies in the fact that for single-pulse energy lower than E_{cr}, thermal defocusing is negligible and does not affect performance of Brillouin mirror (compressor). However, even for such low pulse energy, thermal defocusing can be detrimental at high pulse repetition rates, owing to heat accumulation over a characteristic time, τ_d, the thermal diffusion time of the SBS medium. The thermal self-action is thus negligible if the total energy of laser pulses transmitted through the focal waist during time τ_d is no larger than critical energy, E_{cr}. This condition imposes very stringent requirement on the nonlinear media used. According to estimates conducted for liquid nonlinear materials in which laser pulses of duration t_P at repetition rate f are focused [64], the absorption coefficient of the material should satisfy the following condition:

$$\alpha \leq \frac{10^{-9} - 10^{-8}}{ft_P} \qquad (20)$$

Here α is in units of cm^{-1}, f is in Hz, and t_P is in seconds. The wide range of absorption coefficient values is due to variation in the SBS liquid parameter values. Unfortunately, even highly purified liquids in time lose their initial transparency and absorption increases by orders of magnitude due to the long-term effects of high-power pulses. The tandem cell schemes used in the SBS generator–amplifier mitigate the effect of induced aging of nonlinear liquids by focusing pulses with low threshold energies into the SBS generator. SBS compressor lifetime can be extended by using cells with large operational volumes to enhance liquids circulation as well as comprising a geometrical configuration that stops migration of microparticle impurities into the focal waist.

7.6 EXPERIMENTAL STUDY OF BRILLOUIN PULSE COMPRESSION

The light pulse compression phenomenon was first reported by Maier et al. in 1966 when they observed stimulated Raman scattering from sulfur dioxide [1]. Since that time it has been shown that SBS provides several potential advantages over compression systems based upon SRS. In particular, the production of second Stokes radiation does not restrict the efficiency of the compression, because SBS occurs actually in backward direction. Since the first Stokes pulse is of short duration and travels into an increasing area of the focused pumping beam, the second Stokes is amplified only in the transient regime, in a region with relatively low intensity. Consequently, the gain of the second Stokes remains below threshold. On the other hand, the peak power of the compressed pulse is considerably greater than the pump power, and SRS may achieve threshold. The effects related to SRS competition may be eliminated using atomic gases as an SBS-active media [12, 15, 20, 65] and liquids with weak Raman nonlinearity (freons, tetrachlorides).

The small frequency shift resultant in the SBS process contributes very little to the energy deposited in the SBS compressor, allowing operation at high pulse repetition rates, at least in principle. In addition, the small frequency shift enables the Stokes pulse to be further amplified in a laser amplifier.

Prior to the systematic experimental studies of the early 1980s, there were a number of works in which evidence of Brillouin pulse modulation and compression was presented [7, 55, 56, 66, 70]. Sequential SBS pulsing is possible if backward SBS develops before the leading edge of the incident pulse reaches the focal region and if the SBS pulse is strong compared to the pump intensity.

In Ref. 66, pulse formation by radiation transmitted through as well as backscattered from ethanol was reported. The investigators used 25-ns pulses at 532 nm as the pump. It was shown experimentally that due to the effect of SBS, amplitude modulation with very short rise time occurs with a modulation period equal to round-trip time in the nonlinear medium. The authors concluded that for a collimated or weakly converging pump beam, corresponding to a case when the pump intensity is subject to strong variations along the internal path of the nonlinear medium, a single pulse or train of pulses with power exceeding that of the pump might appear in backward SBS.

Pressurized gases were mainly used in early SBS pulse compressor experiments. At the time, these gases were virtually the only media that afforded very low absorption over relatively long path lengths (up to several meters) typical for SBS compressors used for compression of pulses with duration of tens of nanoseconds. Purified liquid nonlinear materials were later developed, investigated, and applied as part of research into SBS pulse compression and its application in laser systems.

Reference 5 describes an investigation in which a 200-mJ, 20-ns pulse from an Nd:YAG laser-based system was directed into a 1.3-m tapered waveguide filled with pressurized methane at 130 atm. The cell's length did not match the pulse duration in this situation. Consequently, the backreflected Stokes pulse had a sharp leading spike followed by a relatively long tail. This partially compressed pulse was directed back to the same cell after experiencing a corresponding delay. As a response of this input, a second SBS was observed with a pulse width of 2 ± 0.5 ns. In our opinion, this was not true second Stokes pulse compression, but rather its backscattering which made the 2-ns spike of the partially compressed first Stokes pulse more pronounced, through saturated amplification of the lasing medium. In Ref. 7 it was experimentally demonstrated that SBS in an argon cell with a length of $l > ct_P/2$ enables one to compress 20-ns laser pulses to pulses with duration of 1 ns with energy conversion efficiencies of more than 80%. It was found that there exist both an optimal pump energy, E_{opt}, and an optimal converging angle $\beta_{opt} = D_p/F$, where D_p is the pumping beam diameter at the entrance of the cell and F is the distance between the entrance and focal planes [10].

As a rule, SBS-active liquids possess substantially lower SBS thresholds than do pressurized gases. Unfortunately, the threshold for competing processes such as SRS is also lower in most nonlinear liquids than in gases. A configuration based upon nonlinear liquids and using a simple SBS compressor–generator geometry demonstrated subnanosecond pulse generation while utilizing a weak (less than tens of millijoules) pump. It is difficult to scale liquid SBS compressors beyond 1-J output energy level with a single-cell SBS compressor configuration, due to

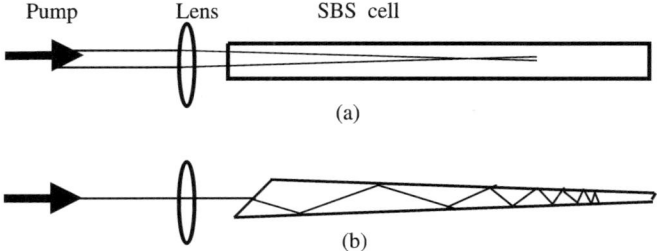

Figure 7.10. Two geometries of an SBS compressor: (a) long cell and (b) tapered waveguide.

competing breakdown and SRS within the extended focal waist. In the repetitively pulsed mode of operation, the situation gets even worse due to heat accumulation over the interaction length. As mentioned earlier, the issue driving this aspect of performance is the need for the Stokes seed to be accurately matched with an incident laser pump both in spatial domains and in time domains. This can be realized with relative ease in the absence of competing nonlinear processes such as SRS, optical breakdown, and self-focusing. In cases of high input intensity, these detrimental effects come into play, thereby degrading the pump to Stokes energy conversion, with subsequent low compression ratios and generation of multiple backward pulses.

There are basically two types of SBS compressors–generators used in pulse compression research. These are (1) compressors with long focal length geometry and (2) tapered waveguides (Fig. 7.10).

For weakly aberrated input beams, simple focusing into a cell containing a nonlinear medium can be used. For such beams it is possible to compress nanosecond pulses with energies as low as a few millijoules to fractions of a nanosecond. Reference 71 provides an example, in which a 6-mJ, 2.5-ns pulse of a Nd:YAG laser was shortened to 150 ps by SBS compressor of carbon tetrachloride. For laser pulses subjected to noticeable phase aberrations, it is necessary to apply a double-pass amplifier configuration with an SBS PC compressor based on the tapered waveguide filled with a nonlinear medium. Since the waveguide transforms the input beam into a multimode beam, tapered waveguide compressors require substantially higher input energy for excitation of SBS.

The need for short-pulse, high-power lasers for fields as diverse as controlled inertial confinement fusion and X-ray sources for lithography gave birth to the development of high-energy SBS pulse compressors. To scale Brillouin pulse compression up to the multi-joule level, the two-cell SBS generator–SBS amplifier configuration was used in Refs. 59, 62, 72, and 73. In its simplest design (Fig. 7.3) a laser beam is focused into a relatively short generator cell after having passed through an amplifier cell. The beam cross section within the amplifier cell should be adjusted so that the overall gain, $\int g_B I_P(z)\, dz$, is just below the threshold gain $G_{th} = 30$–40. Unfortunately, the configuration consisting of a short generator cell with an attenuator placed between the cells has limitations for energy scaling. If the pump energy is high enough to cause both breakdown and self-focusing in the oscillator, significant attenuation is necessary to avoid saturation of the SBS gain. On the other hand, unsaturated SBS gain results in poor compression ratios. It should also be pointed out that the pump-to-Stokes energy efficiency is strongly dependent upon whether or not the Stokes pulse escaping from the oscillator meets the incident pump in the amplifier well before the pump's peak power. In practice, it is difficult to match the Stokes and pump pulses and to simultaneously avoid parasitic effects of self-focusing and breakdown in the SBS oscillator incorporated in this application configuration.

In Ref. 59 a high-energy two-cell CCl_4-based SBS pulse compressor design was studied experimentally. The obtained results were in excellent agreement with numerical simulation predictions. The design incorporated independent excitation of the SBS generator, whose backward seed, after passing through an optical delay, was injected without further attenuation into the SBS amplifier. To block the SBS generator from the undepleted pump beam and simultaneously couple the seed to the amplifier, a polarizer was used instead of an attenuator. This SBS compressor arrangement demonstrated 80% energy efficiency with a 2.5-J laser pulse, while compressing the pulse width from 16 to 1.7 ns.

Simple although rather bulky, the two-cell SBS compressor architecture can be made quite efficient when SBS generator is used as a pulse pre- compressor. Such a compressor was used for compression of multi-joule laser pulses at the GEKKO-XII facility [73]. The layout of the compressor is shown in Fig. 7.11.

The component parameters used in the GEKKO-XII compressor were L1 focal distance 800 cm, L2 focal distance 100 cm, SBS amplifier length 260 cm, and SBS generator length 260 cm. SBS medium: Fluorinert FC 75 liquid of ultrahigh purity.

With this compressor it was possible to convert with 88% efficiency a 25-J, 25-ns pulse from an Nd:glass laser into 0.5-ns high-contrast pulse [74].

The generation of compressed Stokes pulses was demonstrated for visible and near-UV range of pumping wavelengths, mainly using either Nd:YAG laser harmonics or excimer laser radiation. Shorter wavelengths tend to produce better pulse compression because the phonon lifetime, τ, scales as λ^2. It should be noted, however, that many nonlinear liquids and pressurized gases experience two-photon absorption, which is yet another physical process competing with SBS UV pulse compression.

With the advancement of solid-state laser technology, commercial Nd:YAG lasers appeared, including injection-seeded models, offering a standard set of output

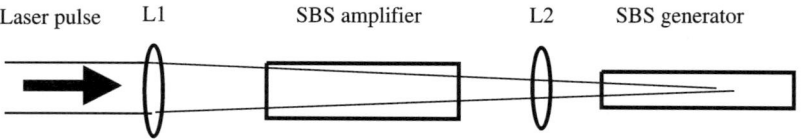

Figure 7.11. Two-cell arrangement with an SBS precompressor.

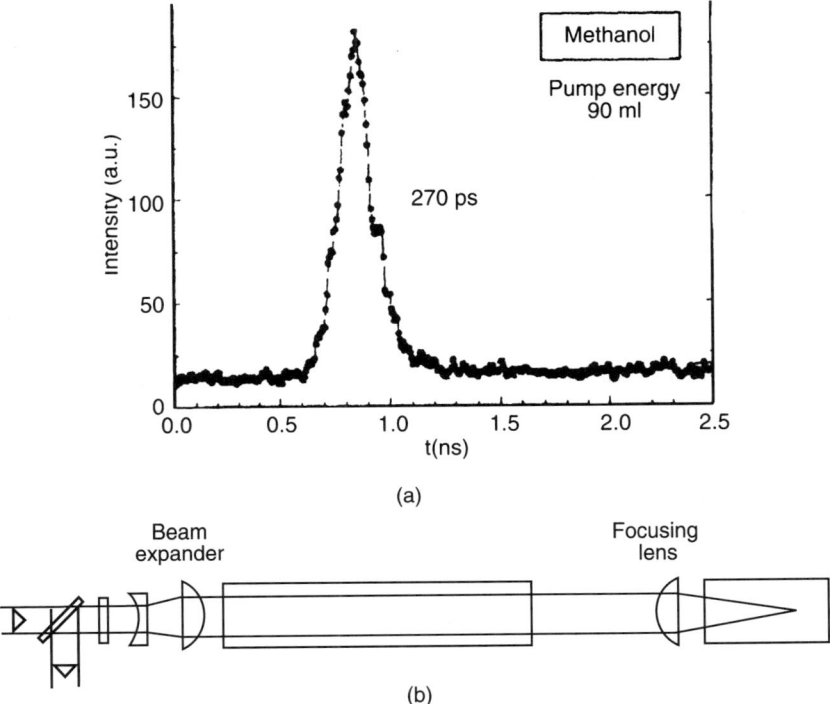

Figure 7.12. (a) Temporal profile of Stokes pulse from a compact methanol SBS compressor at 90-mJ pump energy at 532 nm. (Pump beam diameter 6.5 mm). (b) Layout of the compressor [60].

wavelengths corresponding to the second, third, and fourth harmonics of the fundamental 1064 nm output and each with pulse durations ranging between 5 and 10 ns. With such relatively short pulse widths it is possible to construct a compact pulse compressor ($\sim$ 100 cm long) by exploiting liquids as the SBS-active material. Several compact SBS compressors designed to compress pulses of 5–10 ns duration at 532 nm and 355 nm have been reported. In Ref. 60 it was shown that even the simplest SBS generator–amplifier configuration (Fig. 7.12) with no attenuator between the cell can compress a 5.6-ns, 532-nm pulse to 270 ps, as measured with a streak camera by integrating over a 2.5-mm aperture.

With an increase to the pump energy, the backreflected compressed pulse tends to be converted into multiple-pulse trains, consisting of a primary Stokes pulse and several Stokes "satellites." As previously mentioned, this is a general feature of the SBS dynamics, pertaining to both the SBS generator and the simplest SBS generator–amplifier configurations. The required input energy is a trade-off between the compression ratio and pump—to Stokes pulse energy conversion efficiency. Taking into consideration this trade-off, the term "optimum energy" was previously introduced to describe the characterization of pulse compression for a fixed pumping beam diameter [60]. In the latter publication, a useful minimum value $t_S^{\min}$, corresponding to the duration of the primary Stokes pulse, and a useful reflectivity R^* of the compressor, corresponding to the fraction of the primary pulse energy with respect to the pumping pulse, were introduced on the basis of numerical temporal intensity profiles calculations. The input energy was varied in the model between 80 and 500 mJ, resulting in changes to the useful reflectivity R^* from 80% to 23%. The corresponding change in the useful minimum $t_S^{\min}$ was from 300 ps to 226 ps. With methanol compressor (Fig. 7.12), it was possible to obtain pulses as short as 600 ps using pump energy up to 300 mJ. In the latter case it was necessary to expand the diameter of the beam traveling through the amplifier up to 9.5 mm in order to remain slightly below SBS threshold.

SBS pulse compression at 532 and 355 nm in various liquids was recently published in several reports. A compact variant of a single-cell generator–amplifier compressor was studied in Ref. 69. In this work, the pumping beam passed through a 110-cm-long cell filled with the liquid under study and was focused back into the liquid by a spherical mirror positioned at the rear end of the cell. Table 7.3, reproduced here from that publication, shows the obtained results.

Using several SBS stages enables one to obtain even higher compression ratios. However, this comes at the expense of energy efficiency. In Ref. 25 a frequency-doubled (532 nm) 300-ps pulse, generated by a two-cell SBS generator–amplifier compressor, was focused into a second-stage water-filled compressor, where the pulse was further compressed. The pulse width of that twofold compressed pulse, measured by streak camera, proved to be 160 ps. This value is shorter than the phonon lifetime in water, $\tau = 295$ ps. It was observed, however, that the second SBS compression was competing by SRS and optical breakdown, consequently limiting the energy in the second compressor to 5 mJ.

In the traditional SBS compressor schemes briefly considered above, the Stokes pulse seed originates from the spontaneous Brillouin noise. The so-called short-time phase modulation scheme, suggested and studied in Ref. 17, is another possible approach to pulse compression. The central principle in this approach is that at a

TABLE 7.3. Compression of ~5- ns Pulses for Pumping Beam Diameter of 6.6 mm [69]

Liquid	E_{th} (mJ)	E_{opt} (mJ)	R (%)	τ_t (ps)
		SBS at 532 nm		
Water	1.75	180 (10)	57.9	325
Methanol	0.62	84 (5)	54.0	366
Ethanol	0.63	67 (5)	56.7	—
CCl$_4$	1.0	65 (5)	56.9	200
		SBS at 355 nm		
Water	1.9	87 (5)	43.5	200
Methanol	0.5	39 (3)	41.9	—

moment corresponding to some point in the pulse leading edge, the laser frequency is shifted by the Brillouin frequency, Ω, within a short time interval (~ 1 ns). In the publication referred to, this was done by an electro-optical modulator. The pulse is then amplified to the required energy and directed into the SBS cell. A mirror located at the rear end of the cell reflects the pulse back through the cell. The frequency-shifted radiation serves as the Stokes seed, which is amplified by the remaining part of the pump. In this particular experiment [17] the 30-ns, 532-nm pulse with input energy fluence varying between 0.5 and 2 J/cm² was directed through a 4-m-long cell filled with 30 atm of argon. The duration of the compressed Stokes pulse was 0.5 ns, and its peak power was 15 times higher than that of the pump. Short-time phase modulation is a powerful tool for further pulse compressor scaling. SBS compression was further reported in Ref. 78 using four-wave mixing for producing the Stokes seed. This method was demonstrated to provide efficient compression for incident pulse energies up to $100 E_{opt}$.

Study of powerful UV SBS laser pulse compression was conducted mainly using excimer lasers. High-intensity, short-duration, broad-bandwidth excimer laser pulses are attractive for high-power repetitively pulsed UV and plasma X-ray sources, as well as for driving soft X-ray lasers [68]. However, the typical pulse duration of excimer lasers is tens of nanoseconds, which is far from the optimal duration for these applications.

The combination of high gain and relatively small saturation energy in excimer lasers makes it possible to amplify the first Stokes pulse rather than the complete Stokes signal, thereby making true SBS compression unnecessary but providing generation only of the steep leading edge. In Ref. 76, Stokes pulse shaping was made by a two-step SBS (Fig. 7.13). First a cell filled with pressurized SF_6 produced a Stokes pulse under action of a 76-mJ, 20-ns pump, generated by a KrF laser at 248 nm. The Stokes pulse produced in the first stage was then converted into a 90-ps backward pulse by a second cell filled with a liquid fluorocarbon. Saturated amplification of the second pulse resulted in a 54-ps output pulse.

That experiment is an example of using truncated SBS for laser pulse compression. SBS truncation can be realized using different processes: the aforementioned saturated amplification of the leading edge of the Stokes pulse

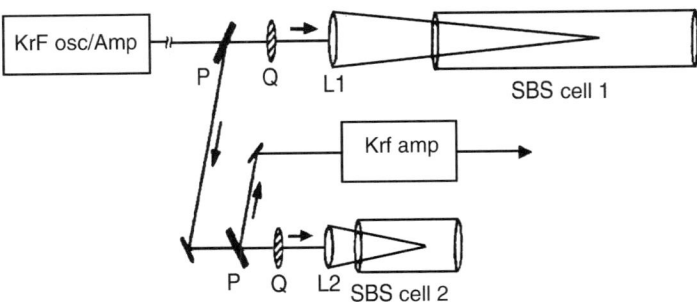

Figure 7.13. The scheme of two-step generation of short SBS pulses in a KrF facility [76].

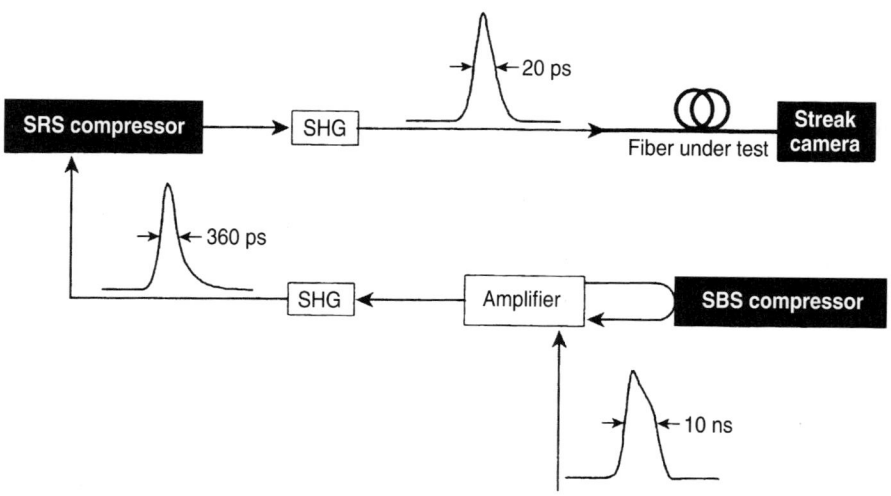

Figure 7.14. Cascaded pulse compression used in Ref. 38 for fiber testing.

[76], the combined effect of optical breakdown and SBS in liquid [28, 30], and the truncated SBS near a liquid surface [26]. It should be noted that stable generation of sub-100-ps laser pulses with truncated SBS is possible only under conditions of saturated amplification.

Cascaded compressors based on stimulated backward scattering enable one to attain light pulses with controllable durations of between 10 ns and 2 ps. Using short pulses from a laser oscillator ($t_L \leq 3$ ns), it is possible to generate pulses as short as 10 ps using only a two-step compressor [36]. The pulse shortening cascade scheme (see Fig. 7.14) was used at the University of Rochester Omega facility to produce picosecond pulses at 337-nm wavelength for fiber testing [38].

Normally, a cascaded compressor consists of a Brillouin cell, serving as a first-cascade compressor, followed by an SRS compressor. Incorporation of the second SBS compressor is reasonable only if the first SBS cell does not compress the Stokes pulse to durations shorter than 3–4 ns—that is, when the starting laser pulse is long. In Ref. 37, SBS compression of a 10-ns pulse at a 10-Hz repetition rate to ~ 360 ps with more than 50% energy efficiency was reported through use of proper focusing geometry and input energy. The output pulse was then frequency-doubled and further compressed to ~ 20 ps using a Raman cell filled with pressurized hydrogen.

In systems with cascaded compressors, double-pass laser amplifiers with SBS compressors are used to compensate for radiation phase distortions and to boost the energy delivered to the SRS cascade [32, 35, 39, 77]. Incorporation of SRS cascades into an optical parametric oscillator–amplifier (OPO–OPA) system or employing laser media with a wide bandwidth allows one to generate compressed pulses tunable within a wide spectral range. In Ref. 77 a triply cascaded compressor was studied experimentally, in which output picosecond pulses were amplified by a Ti:sapphire amplifier. With SRS compressors based on pressurized hydrogen, a 532-

nm pulse from amplified and frequency-doubled output from an SBS compressor was shifted to 629 nm by the first SRS stage and then converted to an 780-nm pulse by the second SRS compressor stage. When the Ti:sapphire was pumped by a 40-mJ pulse at 532 nm, single-pass amplification factors of between 10 and 15 were observed with resultant 2- to 3-mJ/picosecond pulse outputs.

7.7 APPLICATION OF SBS PULSE COMPRESSION TO DIODE-PUMPED SOLID-STATE LASERS WITH HIGH PULSE REPETITION RATE

For Brillouin optical compression in laser systems running at relatively high pulse repetition rates, competing and detrimental effects such as SRS and optical breakdown in nonlinear liquids degrade output performances and can obstruct reliable operation of the system. This is on account of the surplus energy and resultant convective flows in the nonlinear media that both of these effects cause.

Optical breakdown of SBS-active liquids is usually associated with microparticles in the focal waist area. At low pulse repetition rates these occasional breakdowns do not significantly affect the SBS phase conjugation/pulse compression processes. The probability of breakdown occurrence, however, not only is a function of microparticle concentration in the liquid per se, but also increases nonlinearly with pulse repetition rate. Increasing the pulse repetition rate with its attendant increase in optical breakdown leads to heat accumulation in the focal region and hence increased convective flow. As a result, more particles per unit time are drawn through the focal region by the convective flow—including microparticles from the inner wall of the SBS cell—and the frequency of optical breakdown increases. In addition, this effect gives rise to the increase of the SBS threshold and subsequent phase conjugation degradation.

SRS may also be detrimental for SBS compressor operation at high pulse repetition rates. The adverse influence of SRS on reliable optical pulse compression is associated with additional energy absorption in nonlinear medium due to Stokes losses in the process of Raman conversion of laser radiation. Because a Raman frequency shift of 300–500 cm^{-1} is observed for most of liquids used as SBS media, even a few percent of laser energy converted into Raman scattering, generates heat, and therefore contributes to convective flows in the medium. Usually these effects can be observed under action of 2- to 3-ns pulses if the average laser power reaches approximately 10 W. Convective flows lead to SBS quality degradation as well as to increasing probability of optical breakdown, as noted above.

In pulse compression geometry, the interaction length should be equal to or longer than half of pump pulse length. Forward SRS then dominates over backward SRS if there is no Stokes–anti-Stokes interaction to suppress the forward SRS gain. In order to reduce the conversion efficiency of forward SRS, one can select a geometry that provides a strong Stokes–anti-Stokes interaction. It can be done if $l_F \delta k \ll \pi$, where δk is the wave number mismatch, determined by the SRS frequency shift and nonlinear medium dispersion of refractive index. Thus, for suppression of forward SRS a short focal waist length, l_F, and correspondingly short focal distance are required. This may conflict with the pulse compression geometry requirements in which the focal length is equal to or longer than half of pumping pulse length. We see that for a particular laser pulse duration and a particular nonlinear material, one must optimize the compressor arrangement.

Recently we have investigated the two-cell compression scheme, comprising either a collimated or convergent pumping beam, passing through SBS amplifier and focused into SBS generator. In this scheme the required SBS amplifier interaction length and optimized the focal distance for the SBS generator, as defined by conditions suppressing SRS, optical breakdown, and so on, were achieved simultaneously. In order to avoid SRS, we used different media—in particular, media with practically equal Brillouin shifts but distinct Raman shifts—in both the SBS generator and amplifier. Here we remind the reader that some tetrachlorides have Brillouin shifts that are close to each other, with their difference within Brillouin linewidth (see Table 7.2).

Brillouin pulse compression in the SBS generator–amplifier geometry was attained through the use of a diode-pumped Nd:YAG master oscillator–power amplifier (MOPA) configuration, delivering TEM_{00} output with 5-ns pulse duration and controllable pulse repetition rates varying between 10 Hz and 1 kHz. The pulse energy was 10 mJ at low pulse repetition rates and 2 mJ at high pulse repetition rate. A 15-cm-long cell filled with $SiCl_4$ served as the SBS generator, and two cells of 40-cm length each filled with $SiCl_4$ and $GeCl_4$, respectively, were used as the SBS amplifiers. For low pulse repetition rates (10 Hz) the conversion efficiency and Stokes pulses' durations were measured as functions of pump energy for two different pump beam diameters in SBS amplifier $D = 1.6$ mm and $D = 1.3$ mm. (A collimated pump beam was used in these experiments, therefore $D = D_1 = D_2$.)

Figure 7.15 shows SBS reflectivity for the backward Stokes wave as a function of pump energy normalized to SBS threshold energy; that is, $E_{th} = 0.7$ mJ in this experiment.

It follows from these graphs that both SBS threshold and the reflectivity do not depend on the efficiency of the pump–Stokes interaction in the SBS amplifier. On the other hand, it is evident that this interaction is necessary for effective pulse compression. This is illustrated by Fig. 7.16, which shows Stokes pulses' width versus normalized pump energy for the same experimental condition.

One can see from Fig. 7.16 that when there is no SBS amplifier, with pump power increase the Stokes pulse duration rapidly reaches the pump duration. With an SBS amplifier the Stokes pulse duration decreases with increase in pump energy, and for narrower pump beams the pulse compression gets higher. The latter can be accounted for by the increase of the total SBS gain in the amplifier [Eq. (17)]. However, the growth of the overall gain up to a value corresponding to SRS threshold gives rise to the increase of Stokes pulse duration.

A similar scheme for SBS pulse compression was also used in picosecond MOPA laser operated at high repetition rate [48]. A schematic configuration of laser experimental setup is presented in Fig. 7.17.

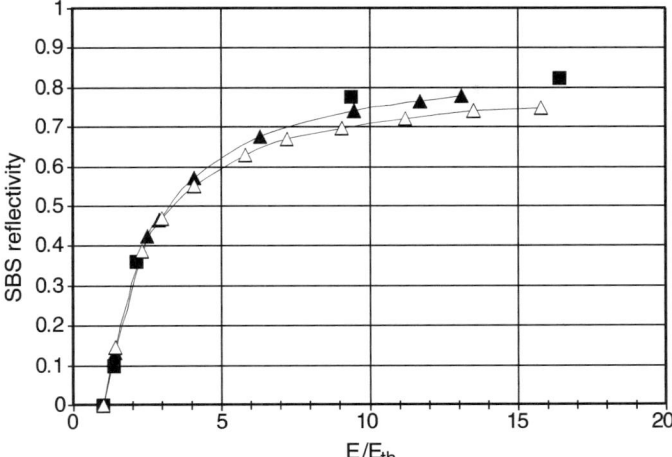

Figure 7.15. SBS reflectivity dependence on normalized pump energy E/E_{th}. ■—without SBS amplifier (no pulse compression); ▲—pump beam diameter in SBS amplifier $D = 1.6$ mm; △—pump beam diameter in SBS amplifier $D = 1.3$ mm.

The passively Q-switched diode-end-pumped master oscillator delivered pulses with 650-μJ energy in TEM$_{00}$ mode with 10- to 1000-Hz repetition rate, 3.2-ns pulse duration, and better-than-0.5% pulse-to-pulse energy in stability. After passing through the preamplifier, the energy of laser pulse increased to 2.4 mJ. Further, the radiation was directed to the pulse compressor based on SBS generator–amplifier scheme.

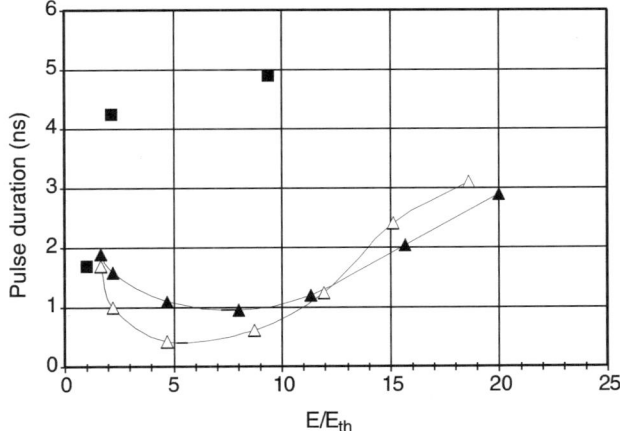

Figure 7.16. Stokes pulse duration versus normalized pump energy E/E_{th} for different experimental conditions. ■—without SBS amplifier (no pulse compression); ▲—pump beam diameter in SBS amplifier $D = 1.6$ mm; △—pump beam diameter in SBS amplifier $D = 1.3$ mm.

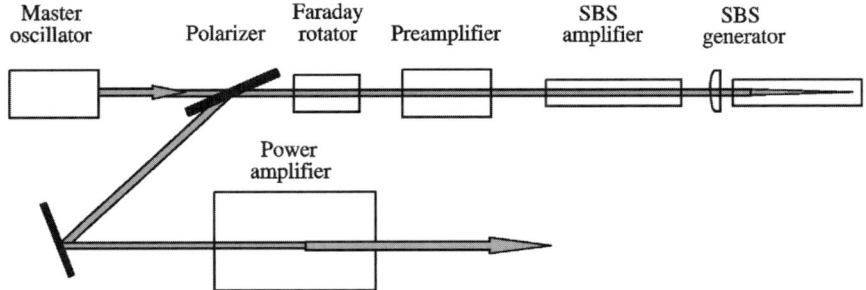

Figure 7.17. Experimental setup (see text).

We tested a few versions of SBS cells filled with highly purified liquid tetrachlorides such as $SiCl_4$, $SnCl_4$, and $GeCl_4$ with stainless steel or glass casings, which have different quality of cleaning of internal surface of SBS-cell casing. Although for low repetition rate both types of cells demonstrated the same performance, at high repetition frequency we could observe the drastic increase of optical breakdown frequency described above. For a stainless steel casing, used in the majority of our experiments at pulse repetition rate $f = 100$ Hz, the above effect was negligible. But with repetition rates in several hundreds of hertz, such negative effects became important. Replacing a stainless steel SBS casing with a glass one filled with the same liquid allowed an increase in the pulse repetition rate up to 1 kHz while the convection flow was still low.

The best results of SBS phase conjugation/pulse compression were achieved with an SBS-generator filled with $SiCl_4$, which has the highest SRS threshold and a very stable chemical composition, which keeps its properties even after the breakdowns, so the transparency of this liquid at 1064 nm did not change. A cell filled with $GeCl_4$, was used as the SBS amplifier. With an $SiCl_4$–SRS amplifier the optimal convergence angle found for a laser beam focused in a nonlinear medium was $\beta_{opt} = 1:50$. For $\beta < \beta_{opt}$, forward SRS process dominated over SBS. In the opposite case $\beta \gg \beta_{opt}$, we observed frequent laser optical breakdowns near the focal waist area. For optimal conditions, SBS reflectivity was 30% with compressed pulse width around 350 ps $\pm$ 25% (approximately 10-fold pulse compression). At the output of the double-pass preamplifier the laser pulse energy reached 2 mJ with 5% pulse-to-pulse energy instability, and the beam divergence was close to diffraction-limited divergence ($M^2 < 1.1$). Figure 7.18 exhibits pulse waveforms for the master oscillator output and after the preamplifier.

The compressed pulse was amplified in the power amplifier up to 35 mJ at pulse repetition rate of 100 Hz. Energy saturation in the amplifier was seen to give rise to changes in the pulse profile. In particular, the pulse trailing edge, being initially relatively long, got shorter, and the pulse shape became practically symmetric.

The same laser configuration was used with an SRS compression in a system developed for generation of picosecond pulses [39]. The layout of this system is shown in Fig. 7.19.

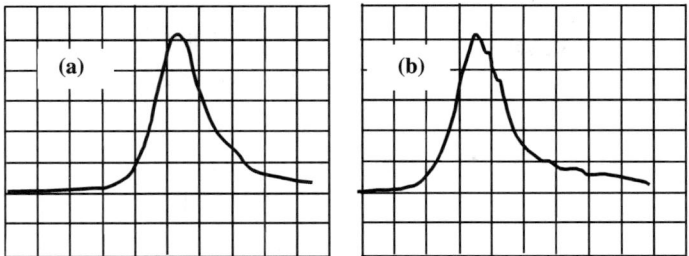

Figure 7.18. Oscilloscope traces of a laser pulse generated in the system. (a) Master oscillator output (2 ns/div). (b) Preamplifier output (200 ps/div).

In this system a fraction of the 2-mJ SBS-compressed pulse was focused into a calcite crystal-based SRS compressor, generating a 25-ps backward SRS pulse. This pulse was further amplified in a Raman amplifier to achieve output energies of 5 mJ and 2.5 mJ at 1202 nm and 601 nm, respectively.

Stimulated scattering of any kind is subject to fluctuations both in the intensity (energy) and temporal profiles due to the stochastic nature of the spontaneous seed. In the case of SBS compression, pulse-to-pulse stability can be significantly improved by operating in a high saturation amplification regime of both the SBS and lasing media. For SRS it is more difficult to make use of gain media saturation since the Raman shift exceeds the bandwidth of the majority of lasing materials, while employing of Raman gain is limited by relatively low pump fluence constraint, necessary in order to prevent self-excitation of SRS. Nevertheless, in the described system, it was possible to achieve 10% pulse-to-pulse energy in stability for 25-ps,

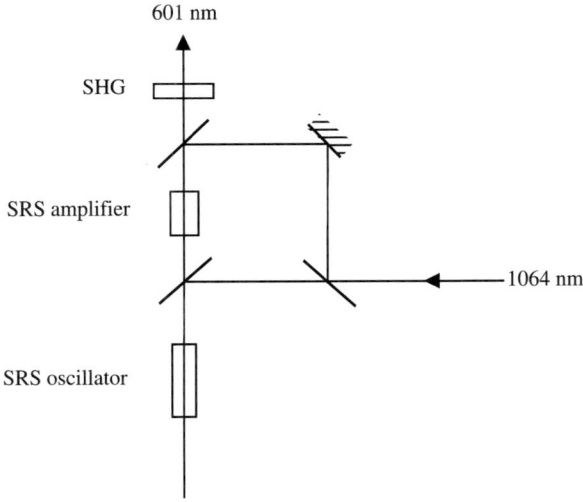

Figure 7.19. SRS compressor based on oscillator–amplifier scheme.

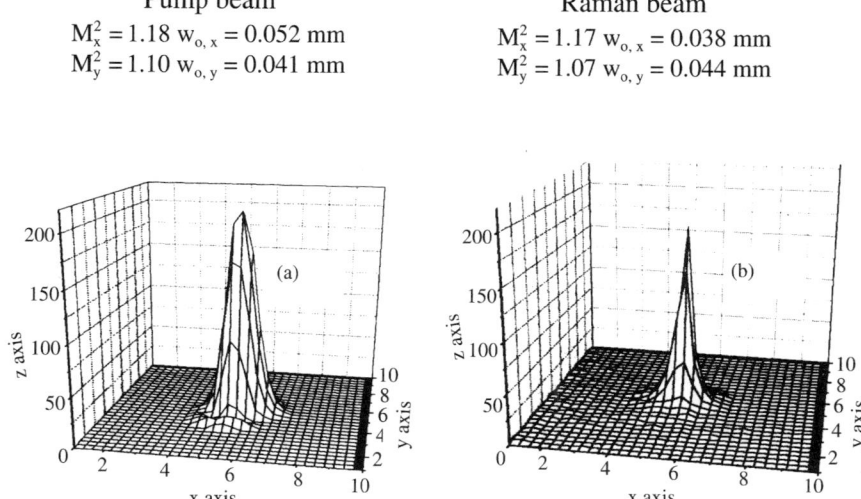

Figure 7.20. Energy distribution of the laser pulse at a focal point: for the beam at the input of an SRS compressor after SBS compression.

100-Hz output at 1202 nm with excellent spatial intensity distribution in the output beam (Fig. 7.20) by means of suitable time and spatial domain matching.

7.8 CONCLUSION

In this chapter we have provided an overview of some of the theoretical and experimental results relating to SBS pulse compression. We have identified the conditions required for compression of nanosecond pulses in nonlinear media, particularly in liquid tetrachlorides. The Brillouin compression method for generating short light pulses at repetition rate up to 1000 Hz has been demonstrated in Nd:YAG diode-pumped lasers. It was also shown that in order to generate a short backward pulse of duration t_S using the Brillouin scattering phenomenon, two conditions should be met. The first condition is to carefully match the pulse duration with the required length of the nonlinear medium. The second requirement is that the total gain integrated over length $\Delta L = ct_S/2n$ and interaction time Δt, determined by t_S, should be greater than unity in any region between the focal point and front window of the cell:

$$G(I, \Delta t, \Delta L) \geq 1 \tag{21}$$

Here $I = P(t - (nz/c))/A(z)$ is the pump intensity on the axis, where $A(z)$ is the cross-sectional area of the pumping beam.

Equation (21) is the defining relationship for the Stokes pulse width [see Eq. (17)]. At this power the Stokes wave excited in the focal region should exceed the fraction of Stokes pulse's power due to the pump radiation scattered by refractive index gratings distributed over the range between the focal region and the front window of the cell.

It should be emphasized that for production of short pulses (<1 ns) in solid-state systems operating at high pulse repetition rates (30–100 Hz), with output energies less than 1 J, microchip master oscillator technology is not applicable. This is because the power amplifiers required to boost the energy of the microchip master oscillator will inevitably distort the laser beam if there is no phase conjugate mirror in the system. Because conventional SBS mirrors cannot operate using a short (<1 ns) laser pump duration, a pulse of the master oscillator with duration of 4–5 ns should be amplified in a multipass master oscillator–power amplifier configuration with SBS pulse compressor playing the role as the phase conjugate mirror.

REFERENCES

1. M. Maier, W. Kaiser, and J. Giordmine, *Phys. Rev. Lett.* **17**, 1275 (1966).
2. M. Maier, W. Rother, and W. Kaiser, *Appl. Phys. Lett.* **10**, 80 (1967).
3. T. B. Stepanova, L. D. Khazov, and I. K. Nikitin, *Opt. Spekrosk.* **19**, 970 (1970) (in Russian).
4. B. Ya. Zel'dovich, V. I. Popovichev, and F. S. Faizulov, *Pis'ma ZhETF* **15**, 109 (1972) (in Russian).
5. D. T. Hon. *Opt. Lett.* **5**, 516 (1980).
6. V. S. Averbakh, A. I. Makarov, and A. K. Potyomkin, *Kvantovaya Elektron.* **6**, 2650 (1979) (in Russian).
7. S. B. Papernyi, V. F. Petrov, and V. R. Startsev, *Pis'ma ZhTF* **7**, 433 (1981) (in Russian).
8. S. B. Kormer, G. G. Kochemasov, S. M. Kulikov, V. D. Nikiloayev, and S. A. Sukharev, *ZhETF* **55**, 630 (1982).
9. V. A. Gorbunov, *ZhTEF* **52**, 2302 (1982) (in Russian).
10. S. B. Papernyi, V. F. Petrov, and V. R. Startsev, *Izv. ANSSSR, Ser. Fiz.* **46**, 1594 (1982) (in Russian).
11. D. T. Hon. *Opt. Eng.* **21**, 252 (1982).
12. V. A. Gorbunov, S. B. Papernyi, V. E. Petrov, and V. R. Startsev, *Kvantovaya Elektron.* **10**, 1386 (1983) (in Russian).
13. S. B. Papernyi, V. F. Petrov, V. A. Serebryakov, and V. R. Startsev, *Kvantovaya Elektron.* **10**, 502 (1983) (in Russian).
14. M. J. Damzen and H. R. Hutchinson, *Opt. Lett.* **8**, 313 (1983).
15. M. J. Damzen and H. R. Hutchinson, *IEEE J. Quantum Electron.* **QE- 19**, 7 (1983).
16. V. A. Gorbunov, *Opt. Spekrosk.* **55**, 1368 (1983) (in Russian).
17. S. B. Papernyi and V. R. Startsev, *Opt. Spekrosk.* **56**, 195 (1984) (in Russian).
18. V. A. Gorbunov, *Kvantovaya Elektron.* **11**, 1368 (1984) (in Russian).
19. S. B. Papernyi and V. R. Startsev, *ZhTEF* **54**, 1651 (1984) (in Russian).

20. S. B. Papernyi, V. B. Ivanov, S. B. Papernyi, and V. R. Startsev, *Izv. ANSSSR Ser. Fiz.* **48**, 1580 (1984) (in Russian).
21. J. J. Zayhowski and C. Dill III, *Opt. Lett.* **17**, 1201 (1992).
22. J. J. Zayhiwski, *Rev. Laser Eng.* **26**, 841 (1998).
23. Ye. Isyanova, J. G. Manni, and D. Welford, ASSL 2001, Report MD2 12 (2001).
24. B. I. Galagan, A. A. Manenkov, G. A. Matyushin, and V. S. Nechitailo. *Izv. ANSSSR, Ser. Fiz.* **48**, 1782 (1984).
25. I. Velchev, D. Neshev, W. Hogervorst, and W. Ubachs, *IEEE J. Quantum Electron.* **35**, 1812 (1999).
26. O. L. Bourne and A. J. Alcock, *Opt. Lett.* **9**, 411 (1984).
27. O. L. Bourne and A. J. Alcock, *Appl. Phys. B* **36**, 181 (1985).
28. Y. S. Huo, A. J. Alcock, and O. L. Bourne, *Appl. Phys. B* **38**, 125 (1985).
29. O. L. Bourne, A. J. Alcock, and Y. S. Huo. *Rev. Sci. Instrum.* **56**, 1736 (1985).
30. S. Yu. Natarov, P. P. Pashinin, E. I. Shklovsky, and I. A. Scherbakov, *Kvantovaya Elektron.* **14**, 477 (1987) (in Russian).
31. N. A. Kurnst and S. J. Thomas, *IEEE J. Quantum Electron.* **25**, 421 (1989).
32. R. R. Buzyalis, A. S. Dement'yev, and E. K. Kosenko, *Kvantovaya Elektron.* **12**, 2024 (1985).
33. K. Kuwahara, E. Takashi, Y. Matsumoto, S. Kato, and Y. Owadano, *J. Opt. Soc. Am. B* **17**, 1943 (2000).
34. R. R. Buzyalis, V. V. Girdauskas, and A. S. Dement'yev, *Kvantovaya Elektron.* **14**, 2267 (1987) (in Russian).
35. V. Kubichek, K. Hamal, I. Prochazka, P. Valach, R. Buzyalis, and A. Dement'yev. *Opt. Commun.* **73**, 251 (1989).
36. R. Buzyalis, A. Dement'yev, K. Hamal, V. Kubichek, I. Prohazka, and P. Valach, *Exp. Tech. Phys.* **39**, 327 (1991).
37. R. R. Buzyalis, V. V. Girdauskas, A. S. Dement'yev, E. K. Kosenko, R. Yu. Chegis, and M. S. Sheibakh, *Izv. ANSSSR Ser. Fiz.* **55**, 270 (1991) (in Russian).
38. LLE Review, *Quarterly Report*, Vol. 85, p. 29.
39. K. Deki, T. Arisawa, F. Matsuoka, A. Nishimura, T. Usami, Y. Shimobeppu, N. Hayasaka, I. Kubo, A. Shilov, E. Shklovsky, and G. Pasmanik, CLEO/Pacific Rim, 2001, Report P2-30.
40. N. F. Andreyev, V. I. Bespalov, M. A. Dvoretskiy, and G. A. Pasmanik, *Kvantovaya Elektron.* **11**, 1476 (1984) (in Russian).
41. G. I. Kachen and W. H. Lowdermilk, *Phys. Rev. A* **6**, 1657 (1977).
42. V. I. Bespalov and G. A. Pasmanik, *Nonlinear Optics and Adaptive Laser Systems*, Nauka, Moscow (1985), p. 23 (in Russian).
43. G. A. Pasmanik, Passat Ltd. Internal Report, p. 11 (1995).
44. B. N. Perry, P. Rabinivitz, and S. Newstein, *Phys. Rev. A* **27**, 1989 (1973).
45. R. V. Johnson and J. H. Marburger, *Phys. Rev. A* **4**, 1175 (1975).
46. M. N. Morozov, L. V. Piskunova, M. M. Suschik, and G. I. Freidman, *Kvantovaya Elektron.* **5**, 1005 (1978) (in Russian).
47. R. W. Hellwarth, *Optical Phase Conjugation*, Academic Press, New York, 1983, p. 169.
48. A. A. Shilov, G. A. Pasmanik, O. V. Kulagin, and K. Deki, *Opt. Lett.* **26**, 1565 (2001).

49. R. R. Buzyalis, V. V. Girdauskas, and A. S. Dement'yev, *Litovskiy Fiz. Sbornik* **26**, 713 (1986) (in Russian).
50. B. Ya. Zel'dovich, N. F. Pilipetskiy, and V. Shkunov, *Principles of Phase Conjugation*, Springer-Verlag, Berlin (1985).
51. E. M. Dianov, A. Ya. Karasik, A. V. Lutchnikov, and A. N. Pilipetskiy, *Kvantovaya Elektron.* **21**, 381 (1989) (in Russian).
52. R. W. Boyd, K. Rzazewski, and P. Naraum, *Phys. Rev. A* **42**, 5514 (1990).
53. A. L. Gaeta and R. W. Boyd, *Phys. Rev.* **A44**, 3205 (1991).
54. N. G. Basov, I. G. Zubarev, A. B. Mironov, S. I. Mikhailov, and A. Yu. Okulov, *Pis'ma ZhETF* **31**, 305 (1980) (in Russian).
55. M. V. Vasil'yev, A. L. Gyn'ameryan, A. V. Mamayev, V. V. Ragul'skii, P. M. Semyonov, and V. G. Sidorovitch, *Pis'ma ZhETF* **31**, 673 (1980) (in Russian).
56. V. I. Bespalov, A. A. Betin, G. A. Pasmanik, and A. A. Shilov, *Pis'ma ZhETF* **31**, 668 (1980) (in Russian).
57. E. Kuzin, V. Petrov, and A. Fotiadi, *Principles of Phase Conjugation*, Springer-Verlag, Berlin (1994).
58. S. Afshaavahid, V. Devrelis, and J. Munch, *Phys. Rev. A* **57**, 3961 (1998).
59. C. B. Dane, W. A. Neuman, and L. A. Hackel, *IEEE J. Quant. Electron.* **30**, 1907 (1994).
60. S. Schieman, W. Ubachs, and W. Hogervorst, *IEEE. J. Quant. Electron.* **33**, 358 (1997).
61. V. A. Gorbunov, S. B. Paperniy, V. F. Petrov, and V. R. Startsev, *Kvantovaya Elektron.* **13**, 900 (1983) (in Russian).
62. R. Fedosejevs and A. A. Offenberger, *IEEE J. Quantum Electron.* **21**, 1558 (1985).
63. E. L. Bubis, V. V. Drobotenko, O. V. Kulagin, G. A. Pasmanik, N. I. Stasyuk, and A. A. Shilov, *Soviet J. Quantum Electron.* **18**, 94 (1988).
64. N. F. Andreyev, A. A. Babin, E. A. Khazanov, S. B. Papernyi, and G. A. Pasmanik, *Laser Physics* **2**, 1 (1992).
65. J. R. Murray, J. Goldhar, D. Elmerl, and A. Szoke, *IEEE J. Quantum Electron.* **15**, 342 (1979).
66. N. F. Andreyev, V. I. Bespalov, A. M. Kiselev, A. M. Kubarev, and G. A. Pasmanik, *Kvantovaya Elektron.* **3**, 2248 (1976) (in Russian).
67. S. S. Gulidov, A. A. Mak, and S. B. Paperniy, *Pisma ZhETF* **47**, 393 (1988) (in Russian).
68. S. S. Gulidov, *Izv. ANSSSR, Ser. Fiz.* **52**, 294 (1988) (in Russian).
69. D. Neshev, I. Velchev, W. A. Majewski, W. Hogervorst, and W. Ubachs, *Appl. Phys.* **B68**, 671 (1999).
70. N. G. Basov, I. G. Zubarev, A. B. Mironov, S. I. Mikhailov, and A.Y. Okulov, *Pis'ma ZhETF* **31**, 685 (1980) (in Russian).
71. R. Buzyalis, A. S. Dement'yev, and E. K. Kosenko, *Kvantovaya Elektron.* **25**, 540 (1995) (in Russian).
72. A. A. Offenberger, D. C. Thomson, R. Fedosejevs, B. Harwood, J. Santiago, and H. R. Manjunath, *IEEE J. Quantum. Electron.* **29**, 207 (1993).
73. V. Kmetik, Investigation of Fluorocarbon Media for High Energy SBS Optical Phase Conjugation and Compression, Dissertation, University of Osaka, Japan (1998).
74. V. Kmetik, T. Kanabe, H. Fujita, M. Nakatsuka, and T. Yamanaka, *Rev. Laser Eng.* **26**, 322 (1998).

75. Y. Nagata, K. Midorikawa, S. Kubodera, M. Obara, H. Tashiro, and K. Toyoda, *Phys. Rev. Lett.* **71**, 3774 (1993).
76. K. Kuwahara, E. Takahashi, Yu. Matsumoto, S. Kato, and Y. Owadano, *J. Opt. Soc. Am. B* **17**, 1943 (2000).
77. R. R. Buzyalis, A. S. Dement'yev, and A. L. Deringas, *Kvantovaya Elektron.* **15**, 1660 (1988) (in Russian).
78. S. B Papernyi, V. F. Petrov, and V. R. Startsev, *Opt. Spekrosk.* **62**, 610 (1987) (in Russian).

■ CHAPTER 8

Principles and Optimization of BaTiO$_3$:Rh Phase Conjugators and Their Application to MOPA Lasers at 1.06 μm

NICOLAS HUOT, GILLES PAULIAT, JEAN-MICHEL JONATHAN, and GÉRALD ROOSEN
Laboratoire Charles Fabry, Institut d'Optique, 91403 Orsay, France

ARNAUD BRIGNON and JEAN-PIERRE HUIGNARD
Thales Research and Technology—France, 91404 Orsay, France

8.1 INTRODUCTION

High-power Nd:YAG lasers delivering a diffraction-limited TEM$_{00}$ mode are of interest for many applications such as pumping of optical parametric oscillator (OPO) or laser manufacturing. However, the laser beam quality is often affected by thermal aberrations induced in amplifier rods. Photorefractive self-pumped phase conjugation is one of the techniques that has been proposed to compensate for these aberrations. Since the early 1980s, it has been widely studied [1, 2] and has led to several applications such as optical gyroscope [3], linewidth narrowing of lasers [4], and laser diodes [5] or dynamic wavefront correction of lasers [6]. However, the development of applications at near-infrared wavelengths was limited by the spectral sensitivity of the photorefractive materials used in the phase conjugate mirrors. Barium titanate (BaTiO$_3$), one of the most interesting photorefractive crystals for phase conjugation, has large electro-optic coefficients and provides high reflectivity phase conjugation, but its spectral sensitivity was mainly limited to the visible range up to the 1990s. In 1994, intentional rhodium doping extended significantly its sensitivity to near-infrared wavelengths. Indeed, self-pumped phase conjugation was demonstrated in rhodium-doped barium titanate (BaTiO$_3$:Rh) in a total internal reflection geometry up to 0.99 μm in 1994 [7]. Most of the demonstrations were then made at laser diode wavelengths using BaTiO$_3$:Rh.

Phase Conjugate Laser Optics, edited by Arnaud Brignon and Jean-Pierre Huignard
ISBN 0-471-43957-6   Copyright © 2004 John Wiley & Sons, Inc.

Optical feedback from a phase conjugate mirror was used to narrow the linewidth of a laser diode emitting at 813 nm [8], double phase conjugation was performed at 800 nm [9], and beam clean-up was achieved with powerful laser diodes emitting at 860 nm [10]. The photorefractive response of $BaTiO_3$:Rh has also been investigated at 1.06 µm [7, 11]. The encouraging results obtained at this wavelength stimulated further research on self-pumped phase conjugation and made possible the dynamic wavefront correction of pulsed Nd:YAG laser sources.

This chapter is devoted to the results obtained with $BaTiO_3$:Rh at 1.06 µm. In the first section, photorefractive characterizations of $BaTiO_3$:Rh are presented, under both continuous-wave (CW) and nanosecond conditions of illumination. In the second part, self-pumped phase conjugation is described. The two geometries that proved to be successful at 1.06 µm are detailed and compared. Finally, the introduction of photorefractive self-pumped phase conjugate mirrors using $BaTiO_3$:Rh in master-oscillator power-amplifier (MOPA) laser sources is presented in the third section. In conclusion, the performances of the photorefractive self-pumped phase conjugate mirrors using $BaTiO_3$:Rh are compared to other possible techniques able to perform dynamic wavefront correction at 1.06 µm, such as photorefractive beam clean-up, stimulated Brillouin scattering, or adaptative optics.

8.2 OVERVIEW OF MATERIAL PROPERTIES

In 1993, near-infrared sensitivity of blue $BaTiO_3$ was reported in the literature [12]. Self-pumped phase conjugate reflectivities of 76% were measured between 860 nm and 1004 nm, but the reasons for the blue color of the crystal and its near-infrared sensitivity were unknown. In 1994, rhodium was identified as the photorefractive trap responsible for this near-infrared significant response, and intentionally doped $BaTiO_3$:Rh started to be investigated under CW illumination [7].

The following results have been obtained with $BaTiO_3$:Rh crystals from F.E.E. and Deltronics. Throughout the chapter, the samples are respectively designated by FEE-"crystal number" or Delt-"crystal number."

8.2.1 Characterization with CW illumination

Two kinds of experiments are conducted to investigate the photorefractive properties of $BaTiO_3$:Rh. Spectroscopic characterizations enable to determine the nature of the photorefractive traps. Photorefractive two-wave mixing experiments provide a measurement of the photorefractive gain. Additional experiments such as light-induced absorption measurements may discriminate between various photorefractive band transport models (one-center [13], two-center [14, 15], or three-charge state model [16, 17]).

Spectroscopic characterizations (electron spin resonance, absorption spectra, and pump/probe experiments) succeeded in indentifying the active impurities, lying in the bandgap of as-grown $BaTiO_3$:Rh. Two principal species have been indentified: iron under two valence states (Fe^{3+} and Fe^{4+}) and rhodium under three valence states (Rh^{3+}, Rh^{4+}, and Rh^{5+}) [18]. The thermal levels of these impurities have been found to be of 1 eV for $Rh^{3+/4+}$, 0.9 eV for $Fe^{3+/4+}$, and 0.7 eV for $Rh^{4+/5+}$ above the valence band. Optical transitions (corresponding to the absorption of a photon and creation of a free hole in the valence band) and thermal levels (corresponding to recombination of holes) are different because of the Franck–Condon shift that takes into account lattice relaxation. Indeed, following Ref. 18, the optical transitions are lower than 1.9 eV for $Rh^{3+/4+}$, lower than 2.8 eV for $Fe^{3+/4+}$, and lower than 1.6 eV for $Rh^{4+/5+}$. As a consequence, at 1.06 μm (energy of 1.2 eV), Fe^{4+} is very unlikely to be photoionized. The relevant band diagram for $BaTiO_3$:Rh illuminated at 1.06 μm therefore contains only one center under three charge states (Rh^{3+}, Rh^{4+}, Rh^{5+}). The corresponding bookkeeping model is given in Fig. 8.1.

The photorefractive band transport model used here is the three-charge state model, which was initially published in 1995 [16]. It accounts for light-induced absorption, which is observed in such crystals at 1.06 μm [19]. Initially, light-induced absorption was explained by a model with two different centers [14, 19] which appears later to be inappropriate for $BaTiO_3$:Rh at 1.06 μm [17].

In $BaTiO_3$:Rh, the dopant is substituted to Ti^{4+}. In the following equations, Rh^{4+}, which has the same charge state as Ti^{4+}, is considered as the neutral level of volume density N. Rh^{3+} appears as a hole acceptor of volume density N^-, while Rh^{5+} appears as a hole donor of volume density N^+. Only hole conductivity is considered [7]. The electrical neutrality of the crystal is achieved by shallow donor and acceptor densities N_A and N_D. The main results of this model are presented below [16, 17].

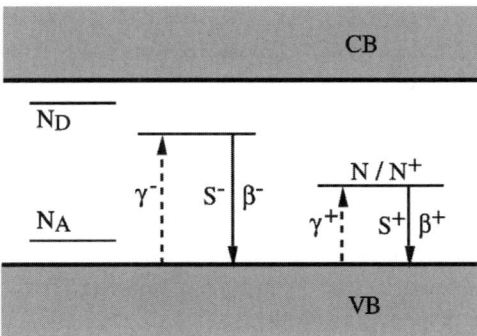

Figure 8.1. "Bookkeeping" diagram for the three-charge state model. N^-, N, and N^+ are the densities of Rh^{3+}, Rh^{4+}, and Rh^{5+}, respectively. $S^\pm$ are the photoexcitation cross sections, $\gamma^\pm$ are the recombination rates, and $\beta^\pm$ are the thermal excitation coefficients. N_A and N_D are optical inactive traps that ensure the electric neutrality of the crystal. Optical and thermal excitation as well as recombination of holes occur to and from the valence band.

Considering that the crystal is illuminated by a sinusoidal interference pattern of the form

$$I = I_0\left[1 + \frac{m}{2}\left(e^{ik_g x} + c.c.\right)\right] \tag{1}$$

where I_0 is the average illumination, m is the modulation of the system of fringes, and k_g is its wave number, one may assume that m is small enough to write any quantity involved in the equations as

$$Q = Q_0 + \frac{1}{2}\left(Q_1 \, e^{ik_g x} + c.c.\right) \tag{2}$$

The resolution of the usual rate equations, and Poisson equations at zero order leads to the determination of the light-induced absorption $\Delta\alpha$ which can be written in terms of the steady-state average density of Rh^{4+}, N_{0Stat}:

$$N_{0Stat} = \frac{1}{1-\kappa}\left\{\left[\kappa N_T^2 + \kappa(\kappa-1)(N_A - N_D)^2\right]^{1/2} - \kappa N_T\right\} \tag{3}$$

with

$$\kappa = \frac{S^+ I_0 + \beta^+ \, \gamma^-}{S^- I_0 + \beta^- \, 4\gamma^+} \tag{4}$$

where S^+, S^-, γ^+, γ^-, β^+, and β^- are defined in Fig. 8.1. N_T is the total density of dopants: $N_T = N^- + N + N^+$.

Assuming that the absorption of a photon by Rh^{4+} or Rh^{5+} leads to the excitation of a hole with a quantum yield of 1, the steady-state value $\Delta\alpha_{Stat}(I_0)$ of the induced absorption is a function of the illumination:

$$\Delta\alpha_{Stat}(I_0) = \alpha_{Stat}(I_0) - \alpha_{Stat}(0) = \frac{hc}{\lambda}(S^+ - 2S^-)\frac{N_{0Stat}(0) - N_{0Stat}(I_0)}{2} \tag{5}$$

hc/λ is the energy of an incident photon. As soon as $S^+/S^- \ll \beta^+/\beta^-$, the slope at the origin of the kinetics of the light-induced absorption (response to a step of illumination) can be expressed as follows:

$$\left(\frac{\partial \alpha}{\partial t}\right)_{t=0} \approx \frac{hc}{\lambda}(S^+ - 2S^-)S^- I_0 \frac{\gamma^+(N_T + N_A - N_D)^2}{\gamma^+(N_T + N_A - N_D) - \gamma^-(N_A - N_D)} \tag{6}$$

This will be a relevant number in the characterization of the crystals.

Solving the rate, continuity, and Poisson equations at first order yields the photorefractive spatially varying space-charge field:

$$E_1 = imk_g \frac{k_B T}{e} \frac{1}{1 + k_g^2/k_0^2(I_0)} \eta(I_0) \qquad (7)$$

with

$$k_0^2(I_0) = k_0^{-2} + k_0^{+2} \qquad (8)$$

and

$$\eta(I_0) = 1 - \left(\frac{\beta^-}{S^- I_0 + \beta^-} \frac{k_0^{-2}}{k_0^2} + \frac{\beta^+}{S^+ I_0 + \beta^+} \frac{k_0^{+2}}{k_0^2} \right) \qquad (9)$$

k_0^- and k_0^+ are the Debye wave numbers given by

$$k_0^{\pm 2}(I_0) = \frac{e^2}{k_B T \varepsilon_S} N_{\text{eff}}^{\pm}(I_0) \qquad (10)$$

with

$$N_{\text{eff}}^+(I_0) = N_0^+(I_0) \frac{(N_T + N_D - N_A)}{N_T}$$

and
$$\qquad (11)$$

$$N_{\text{eff}}^-(I_0) = N_0^-(I_0) \frac{(N_T - N_D + N_A)}{N_T}$$

$\eta(I_0)$ is a saturation factor that increases with I_0 to a value of 1 for sufficiently high illuminations. Similarly, the effective Debye wave number k_0 varies with I_0 and saturates at high intensities. Expression (7) is similar to the one obtained in the case of a two-site photorefractive material [14]. However, the behaviors for $k_0(I_0)$ and $\eta(I_0)$ are different, and the three-charge state model cannot be described as a subcase of the two-site model. In the three-charge state model, k_0^- and k_0^+ are not independent. There is a strong coupling between the populations of the three charge states, and k_0 may be defined in terms of the population N_0 of the intermediate state alone:

$$k_0^2(I_0) = \frac{e^2}{k_B T \varepsilon_S} \left[N_T - N_0(I_0) - \frac{(N_D - N_A)^2}{N_T} \right] = \frac{e^2}{k_B T \varepsilon_S} N_{\text{eff}} \qquad (12)$$

N_{eff} is the intensity-dependent effective density of traps.

A 0°-cut BaTiO$_3$:Rh crystal referred to as FEE-X16 was characterized by measuring the steady-state light-induced absorption (Fig. 8.2) together with its slope at $t = 0$ (Fig. 8.3). Additional two-wave mixing experiments were performed in the counterpropagating geometry [20] with ordinary polarized beams to determine N_{effsat}, the effective density of traps at saturation in intensity ($N_{\text{effsat}} = 5 \times 10^{16}$ cm^{-3} for sample FEE-X16). All these data are fitted simultaneously, and values of the internal parameters of the crystal with their error bars are deduced (Table 8.1). One can remark that the photoionization cross section from Rh$^{4+/5+}$ is 75 times higher than the one from Rh$^{3+/4+}$. This explains the light-induced absorption: Holes are photoionized from Rh$^{3+/4+}$ (Rh^{4+} is the main charge state present in the crystal). Some of these holes, retrapped in Rh$^{4+/5+}$, create Rh^{5+}. Then, Rh^{5+} is more likely to be photoionized than Rh^{4+}. Thus, absorption increases with illumination. As we will show later, the determination of the internal parameters of the crystal is a key point for further optimization of nonlinear functions like optical phase conjugation.

Various two-beam coupling gain values have been reported in the literature using BaTiO$_3$:Rh at 1.06 μm with CW illumination. Gains of 9.3 cm^{-1} and 11 cm^{-1} have been measured in 0°-cut samples [11, 21]. A gain of 23 cm^{-1} is obtained in a 45°-cut crystal referred to as FEE-Y32-B. In this latter experiment, the sensitivity of the material is deduced using the rise time τ_{ph} of the photorefractive effect for a given incident intensity I and the photoinduced index modulation Δn. A sensitivity of $S = \Delta n / I \tau_{\text{ph}} = 1.7 \times 10^{-7}$ cm^2 J^{-1} is reported [20], which is 1000 times smaller than those determined at 532 nm and 4 times smaller than those obtained at 670 nm with BaTiO$_3$:Fe or BaTiO$_3$:Co [22]. The relatively poor sensitivity of BaTiO$_3$:Rh at 1.06 μm is mainly due to the long photorefractive response time at 1.06 μm, as expected given the low value of the absorption coefficient (0.1 cm^{-1}).

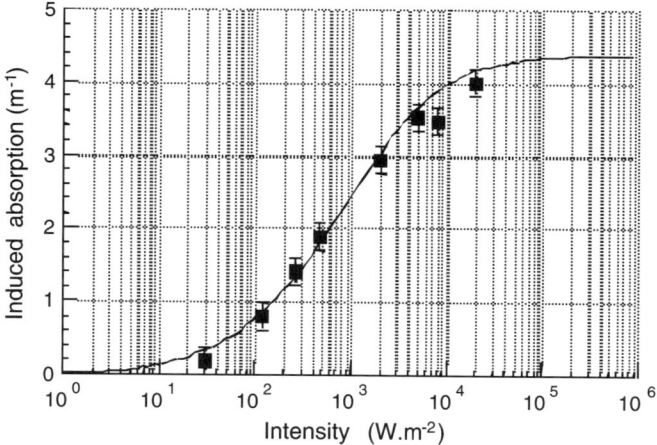

Figure 8.2. Steady-state light-induced absorption versus intensity for crystal FEE-X16. The full curve is a theoretical fit using Eq. (5) and parameters of Table 8.1.

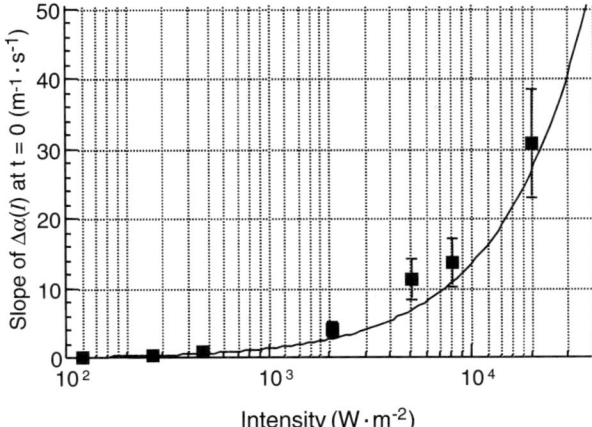

Figure 8.3. Slope at $t = 0$ of the light-induced absorption versus intensity for crystal FEE-X16. The full curve is a theoretical fit using Eq. (6) and parameters of Table 8.1.

For applications, the photorefractive characteristics of $BaTiO_3$:Rh crystals at 1.06 μm have to be reproducible. Even when they are cut from the same boule, photorefractive crystals may present significant variations in their performances due to inhomogeneities in the boule. The reproducibility is checked by investigating two $BaTiO_3$:Rh crystals referred to as FEE-X16-5/2 (0°-cut) and FEE-X16-45 (45°-cut) from the same boule grown with 1000 ppm of rhodium in the melt. As derived from the three-charge state model, the steady-state light-induced absorption, its slope at $t = 0$, and the effective density of traps at saturation in intensity are relevant parameters. Because the photorefractive gain Γ at saturation in intensity is related to N_{effsat}, comparing the different values of Γ measured in several samples in the same photorefractive configuration allows us to follow the eventual variations of N_{effsat}. In addition, three parameters are given: the photorefractive time constant τ_{ph} in the counterpropagating geometry for a given intensity, the erasure time in the dark τ_{dark}

TABLE 8.1. Parameters of Crystal FEE-X16 at 1.06 μm Deduced from the Simultaneous Fit of N_{effsat}, the Steady-State Light-Induced Absorption and Its Slope at the Origin, Using the Three-Charge State Model[a]

γ^-/γ^+	1.5 ± 0.3
S^-	$(1.2 \pm 0.5) \times 10^{-5}$ m² · J⁻¹
S^+	$(9.0 \pm 0.1) \times 10^{-4}$ m² · J⁻¹
β^-	$\beta^- \ll \beta^+$
β^+	2.3 ± 0.4 s⁻¹
N_T	$(3.3 \pm 0.6) \times 10^{23}$ m⁻³
$|N_A - N_D|$	$<4.4 \times 10^{21}$ m⁻³

[a]See caption of Fig. 8.1 for signification of parameters.

divided by τ_{ph}, and the absorption coefficient at saturation of the light-induced absorption. The experimental data given in Table 8.2 (ordinary polarizations and counterpropagating beams) show that the two samples have reproducible characteristics and that the homogeneity of the boule is correct. For sample FEE-X16-45, Γ and τ_{ph} are given for an angle of $\beta = 45°$ between the grating wave vector and the c axis. Consequently, these data cannot be directly compared to these obtained in 0°-cut crystals. Then, a second boule (FEE-X17), also with 1000 ppm of rhodium in the melt, has been grown on purpose in the same conditions as the previous one, in order to reproduce the properties of FEE-X16-5/2 and FEE-X16-45 crystals. Two 0°-cut crystals from this boule, FEE-X17-6/3 and FEE-X17-6/4, are investigated. The data reported in Table 8.2 indicate that these two crystals have similar properties and that the two boules are nearly equivalent. The only difference between the four crystals concerns the ratio τ_{dark}/τ_{ph}. Nevertheless, for the application to dynamic wavefront correction, the high value of this ratio for all the samples ensures that the photorefractive gain is well-saturated.

8.2.2 Performances of oxidized crystals

As detailed in the previous section, one major feature of BaTiO$_3$:Rh at 1.06 μm is a photoexcitation cross section 75 times higher for the Rh$^{4+/5+}$ than for the Rh$^{3+/4+}$ level. This indicates that the photorefractive rise time might be reduced by lowering the Fermi level. Indeed, it has been shown that the Fermi level lies around the Rh$^{3+/4+}$ level in as-grown BaTiO$_3$:Rh [18]. By oxidation, the Rh$^{3+/4+}$ level becomes full of holes. Consequently, Rh^{3+} vanishes and the density of Rh^{4+} rises. Moreover, Rh^{5+} is populated. This goes along with a decrease of the Fermi level, which drops from 1 eV (around the Rh$^{3+/4+}$ level) down to 0.7 eV (Rh$^{4+/5+}$ level). In oxidized crystals, it has also been proved in the same paper that an additional level is present (Fe$^{4+/5+}$), which comes from the oxidation of the Fe$^{3+/4+}$ level. But, because the optical transition of Fe$^{4+/5+}$ is larger than 2.5 eV, it is very unlikely photoionized at 1.06 μm [18]. So, at 1.06 μm in oxidized crystals, only one level remains with a relatively high photoexcitation cross section: Rh$^{4+/5+}$.

The influence of the oxidation of BaTiO$_3$:Rh on its photorefractive properties is predicted by the three-charge state model [23]. Indeed, during the process of oxidation, the total density of dopants remains constant:

$$[Rh^{3+}] + [Rh^{4+}] + [Rh^{5+}] = N_T \quad (13)$$

Moreover, the relative densities of Rh^{3+}, Rh^{4+}, and Rh^{5+} are controlled by the neutrality equation:

$$[Rh^{5+}] - [Rh^{3+}] \approx N_{AD} \quad (14)$$

where N_{AD} stands for $N_A - N_D$. Rh^{4+} does not appear in this equation because it is considered as the neutral level. Consequently, oxidizing the sample results in

TABLE 8.2. Relevant Parameters for BaTiO$_3$:Rh Crystals Characterized at 1.06 μm with CW Illumination (Ordinary Polarized Beams and Counterpropagating Geometry)

	X16	X16-45	X17 6/3	X17 6/4
α	(0.7 ± 0.02) cm^{-1}	(0.12 ± 0.02) cm^{-1}	(0.06 ± 0.02) cm^{-1}	(0.05 ± 0.02) cm^{-1}
$\Delta\alpha(1.1\ \mathrm{W\cdot cm^{-2}})$	(3.9 ± 0.2) m^{-1}	(5.7 ± 0.2) m^{-1}	(2.8 ± 0.2) m^{-1}	(3.6 ± 0.2) m^{-1}
$(\Delta\alpha/\partial t)_{t=0,\,I=1.1\,\mathrm{W\cdot cm^{-2}}}$	(14 ± 4) m^{-1}s^{-1}	(11 ± 3) m$^{-1}\cdot$s^{-1}	(11 ± 3) m^{-1}s^{-1}	(13 ± 4) m$^{-1}\cdot$s^{-1}
τ_{ph} at 6.7 W$\cdot$cm^{-2}	(8 ± 1)s	(2 ± 0.5)s $\beta = 45°$	(11 ± 1)s	(8 ± 1)s
τ_{dark}/τ_{ph}	240 ± 50	200 ± 40	90 ± 15	390 ± 70
Γ	2.1 cm^{-1}	0.6 cm^{-1} $\beta = 45°$	2.6 cm^{-1}	2.5 cm^{-1}

increasing N_{AD}. Equations (3) to (12) are still valid, but this time N_{AD} is the relevant parameter to be varied. The main results are the following: The absolute absorption increases with oxidation and the photorefractive rise time in a given photorefractive configuration decreases, which is suggested by the high photoexcitation cross section from $Rh^{4+/5+}$; light-induced absorption vanishes, which is consistent with a one-level two-charge state model; the dark decay time drops sharply with oxidation, which is consistent with a high thermal ionization rate from $Rh^{4+/5+}$ compared to $Rh^{3+/4+}$; and the two-wave mixing gain in a given photorefractive configuration remains nearly unchanged. The corresponding theoretical curves showing these evolutions are presented below, along with experimental results (Fig. 8.4). No free parameter is used in these simulations and the parameters of Table 8.1 are employed. It should be noted that an excessive oxidation would suppress any photorefractive effect, because the only charge state present in the crystal would be Rh^{5+}. No modulated space-charge field could be created. This is why the two-wave mixing gain drops for high values of N_{AD}.

Experimental verification of these statements have been conducted through a collaboration work between FEE (Germany), the University of Osnabrück (Germany), and the Laboratoire Charles Fabry de l'Institut d'Optique (France) [23]. Two 0°-cut crystals of $BaTiO_3$:Rh referred to as FEE-X14 and FEE-X16 and doped with the same quantity of rhodium (1000 ppm) have been characterized before and after oxidation. Absolute absorption, light-induced absorption, the photorefractive time constant τ_{ph} for an intensity of $6.7 \text{ W} \cdot \text{cm}^{-2}$, the ratio of the dark decay time τ_{dark} divided by τ_{ph}, and the photorefractive gain Γ in the counterpropagating geometry for ordinary polarizations are measured. The results are reported in Table 8.3. In Fig. 8.4, experimental data are plotted together with theoretical curves with no free parameter. One should note that $\tau_{dark}/\tau = \sigma/\sigma_{dark}$. The new photorefractive characterizations fulfill the predictions of the three-charge state model: a 2.8- to 5.2-fold decrease of the photorefractive rise time, along with a reduction by a factor of 30 of the ratio τ_{dark}/τ_{ph} and a sixfold increase of the permanent absorption. The light-induced absorption vanishes for sample FEE-X14 and is turned into light-induced transparency in FEE-X16, which may result from some shallow weakly absorbent level. Moreover, the photorefractive two-wave mixing gain remains constant. The estimated corresponding value of N_{AD} that is reached by oxidation is $8 \times 10^{16} \text{ cm}^{-3} < N_{AD} < 2 \times 10^{17} \text{ cm}^{-3}$.

These results show that the photorefractive rise time can be significantly reduced by oxidation at the expense of absorption. The strong increase of absorption will limit the use of oxidized $BaTiO_3$:Rh crystals for powerful nanosecond Nd:YAG wavefront correction because the damage thermal threshold will be reduced at least by a factor of 6 compared to as-grown crystals. Nevertheless, oxidized samples show the same behavior at 850 nm and at 1.06 μm. Therefore, oxidized $BaTiO_3$:Rh would be of great interest for applications at low-power laser diode wavelengths, where damage threshold is not a problem.

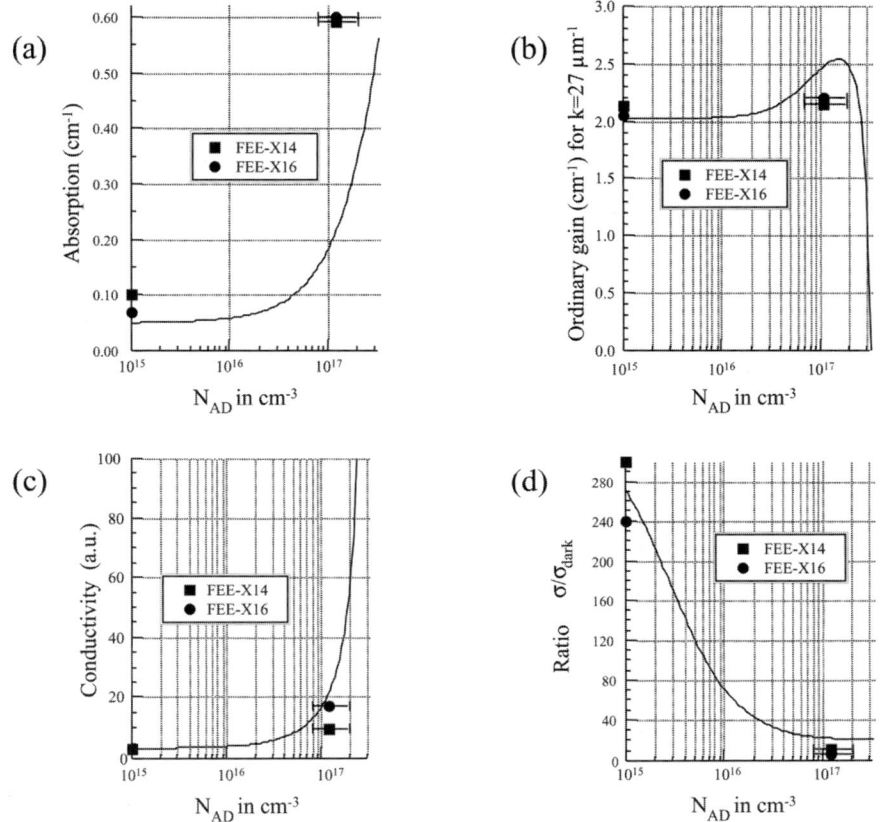

Figure 8.4. Theoretical predictions and experimental results of an oxidation of two BaTiO$_3$:Rh crystals (FEE-X14 and FEE-X16). (a) Absorption versus N_{AD}. Full line: with saturated light-induced absorption. Dotted line: without light-induced absorption. (b) Ordinary gain versus N_{AD} for counterpropagating beams ($k_g = 27$ µm^{-1}). (c) Photoconductivity at $I = 6, 7$ Wcm^{-2} versus N_{AD} normalized to its value for $N_{AD} = 10^{15}$ cm^{-3}. (d) Ratio σ/σ_{dark} versus N_{AD}.

8.2.3 Characterization with nanosecond illumination

In BaTiO$_3$, for pulse duration larger than 10 ns, the recombination time of free holes is shorter than the pulse duration. Consequently, the photorefractive effect builds up during the pulses. Nothing happens between the pulses, as long as thermal excitation can be neglected during the pulse interval.

Let us summarize the observed effects when BaTiO$_3$ is illuminated with nanosecond visible light. Under pulsed illumination, the saturation of the free carrier density was theoretically revealed [24]. This effect causes a nonconstant intensity × photorefractive rise-time product, contrary to the CW illumination at sufficiently high illumination (SI ≫ β). It was also shown that the density of free

TABLE 8.3. Relevant Parameters for BaTiO$_3$:Rh Crystals Before and After Oxidation (Ordinary Polarized Beams and Counterpropagating Geometry)

	X14		X16	
	As grown	Oxidized	As grown	Oxidized
α	(0.1 ± 0.02) cm^{-1}	(0.6 ± 0.02) cm^{-1}	(0.07 ± 0.02) cm^{-1}	(0.6 ± 0.02) cm^{-1}
$\Delta\alpha$ (1.2 W·cm^{-2})	(4.5 ± 0.2) m^{-1}	No light-induced absorption	(3.9 ± 0.2) m^{-1}	(-11 ± 2) m^{-1}
τ_{ph} at 6.7 W·cm^{-2}	6.3 s	2.2 s	8.3 s	1.6 s
τ_{dark}/τ_{ph}	300	11	240	6
Γ	2.1 cm^{-1}	2.1 cm^{-1}	2.0 cm^{-1}	2.2 cm^{-1}

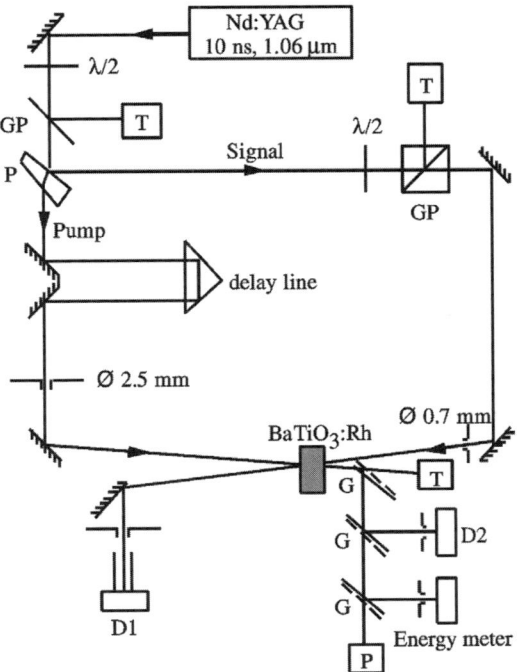

Figure 8.5. Experimental setup for two-wave mixing experiments in the counterpropagating geometry. GP, Glan polarizers; G, glass plates; T, light traps; D1 and D2, detectors; P, prismatic glass plate.

carrier could not be neglected compared to that of the traps [24]. As a consequence, the optimum grating spacing (highest photorefractive gain) is shifted. Moreover, an intensity-dependent electron–hole competition might occur for pulsed illumination [25, 26]. As a result, the photorefractive gain is widely intensity-dependent. It may decrease, vanish at the so-called compensation intensity, and even change its sign for sufficiently high illuminations. This intensity-dependent electron–hole competition under pulsed illumination has been observed in the visible range by two-wave mixing gain measurements [26, 27] and through photorefractive time constant measurements [28] for various pulse intensities. So, in the visible range, the photorefractive characteristics (gain and time constant) depend widely on the used pulsed intensity, which is unfavourable for applications. What happens with nanosecond illumination at 1.06 μm?

To answer this question, two-wave mixing experiments have been performed using a 10-ns Nd:YAG laser at a repetition rate of 10 Hz and the setup of Fig. 8.5 [29]. Counterpropagating extraordinary polarized beams and a 45°-cut $BaTiO_3$:Rh crystal referred to as FEE-Y32-B have been used. For intensities (average value on the pulse duration) up to 20 MW·cm^{-2}, the photorefractive gain keeps nearly constant (Fig. 8.6), whereas such experiments with visible illumination clearly show

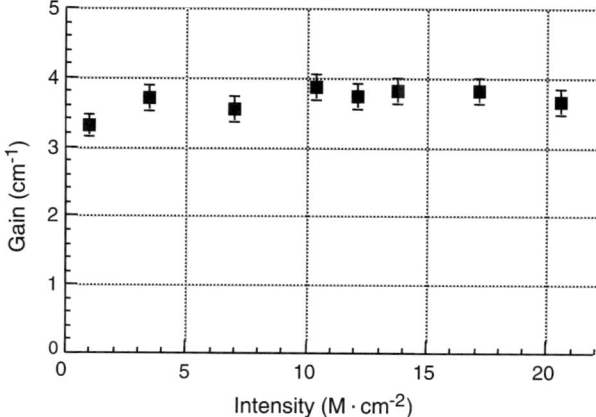

Figure 8.6. Photorefractive two-wave mixing gain measured versus intensity in the counterpropagating geometry with extraordinary polarized beams in sample FEE-Y32-B.

an inversion of the sign of the gain for this range of intensities [26, 27]. Therefore this crystal does not show electron–hole competition at this wavelength for this range of intensities. This can be simply explained by the fact that a photon does not carry enough energy to excite an electron to the conduction band. The same conclusions are also reached with another BaTiO$_3$:Rh crystal referred to as FEE-X14. Free carrier density saturation has been looked for by measuring the photorefractive time constant in the same photorefractive configuration for various pulsed intensities. For pulsed average intensities up to 20 MW · cm^{-2}, the inverse of the time constant is a linear function of the intensity. Because the time constant is inversely proportional to the free carrier density, Fig. 8.7 also indicates that no free carrier density saturation appears in this range of intensities. This conclusion is in agreement with the model of Ref. 26. Indeed, the authors introduce a saturation parameter $f_h = S_h I \tau_h$ which defines the relative number of photoionized photorefractive traps. S_h is the photoexcitation cross section, I is the intensity, and τ_h is the recombination time for holes. With typical values for these parameters [17], f_h remains smaller than 5% for $I < 20$ MW · cm^{-2} and saturation stays negligible. This might be a consequence of the low typical value of S_h at 1.06 μm, as suggested by the very low absorption at this wavelength.

The energies required to write a grating in the counterpropagating geometry are nearly equal in the CW and in the nanosecond regimes (100 J · cm^{-2}). This agrees with the conclusions of Ref. 24. In Y32-B (45° cut), the maximum two-wave mixing gain has been measured in the copropagating geometry for a grating spacing of 4 μm for extraordinary polarized beams. The measured gain, corrected from the erasing effect of the pump beam reflected by the rear face of the sample, is 16 cm^{-1}. As predicted by Ref. 26, this is smaller than the value obtained with CW illumination (23 cm^{-1}).

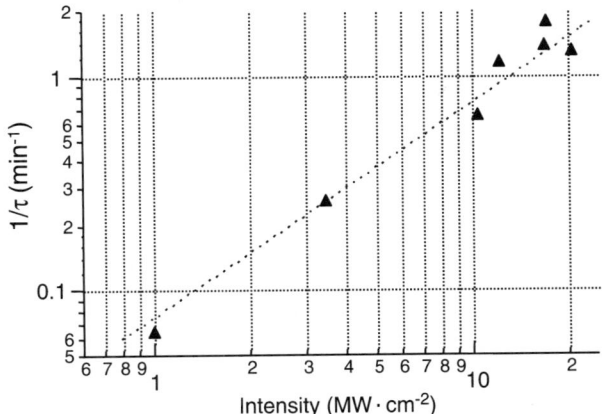

Figure 8.7. Inverse of the measured time constant (proportional to the density of free holes) versus pulsed intensity for sample FEE-Y32-B (repetition rate of 10 Hz). The dotted line is a guide for the eye.

As a conclusion, in the nanosecond illumination regime for pulsed average intensities up to 20 MW · cm^{-2} at 1.06 μm, the gain is intensity-independent, while the intensity × rise-time product keeps constant. The maximum gain is marginaly smaller than the one measured in the CW illumination regime. So, contrary to the results obtained in the visible range, nanosecond and CW illumination regimes are nearly equivalent for BaTiO$_3$:Rh at 1.06 μm for intensities up to 20 MW · cm^{-2}, which is favorable for applications.

BaTiO$_3$:Rh is a photorefractive material with high gain and low absorption at 1.06 μm. It is accurately described by a three-charge state model, which agrees with spectroscopic and photorefractive studies. Oxidation of these crystals reduces significantly the rise time of the photorefractive effect, but, because it also increases the permanent absorption, it is not suited for dynamic wavefront correction of powerful lasers. Nevertheless, oxidation might be very useful for low-power applications at laser diode wavelengths. For as-grown BaTiO$_3$:Rh at 1.06 μm, nanosecond and CW illuminations are nearly equivalent in terms of photorefractive gain and time constant for pulsed intensites up to 20 MW · cm^{-2}. This latter result, which differs from these obtained in the visible range, is a key point for applications to Q-switched Nd:YAG lasers. Such a high photorefractive gain and low absorption material is attractive for high reflectivity phase conjugation, as will be discussed in the next section.

8.3 SELF-PUMPED PHASE CONJUGATION

Phase conjugate mirrors fall into two main categories: phase conjugation by degenerate four-wave mixing and self-pumped phase conjugation. Degenerate four-wave mixing needs one signal beam and two externally provided pump beams. One major advantage of this configuration is that the phase conjugate reflectivity may be larger than unity if the signal intensity is small enough. This principle can be used in self-starting oscillators [30, 31]. The use of a phase conjugate mirror by degenerate four-wave mixing as the end mirror of a laser cavity has been studied theoretically [32] and experimentally [33], but this architecture needs an additional laser source to provide two pump beams, which is not favorable to applications. In self-pumped phase conjugation, this drawback is bypassed at the expense of a reflectivity limited to unity. Self-pumped phase conjugation has been widely studied since the early 1980s. In 1982, a self-pumped phase conjugator using $BaTiO_3$ was proposed [1]. In this self-starting geometry demonstrated at 514 nm with CW illumination, the two-pump beams were derived from the incident beam itself using beam-fanning and total internal reflection on the crystal faces. In the following sections, this geometry is called "internal loop geometry." In 1984, various self-pumped phase conjugate mirrors were proposed, modeled, and experimentally tested [2]. Among them, the "linear phase conjugate mirror" requires two additional mirrors that form a linear resonator providing the two counterpropagating pump beams [2]. This phase conjugate mirror, along with its semilinear version (only one additional mirror), has been used in laser cavities as end mirrors [6]. One major drawback of this geometry is that coherence length is required to provide efficient phase conjugation. In the case of a Q-switched Nd:YAG laser, it is about 1 cm. This is why linear phase conjugate mirrors are not used for the application discussed in this chapter. The "ring phase conjugate mirror" [2] uses two additional mirrors, which form a loop and feed the transmitted beam back to the photorefractive crystal. These two geometries (internal loop and ring) are often used for the improvement of laser sources such as laser diode injection and spectral narrowing [8] or self-corrected oscillators [34]. These two kinds of self-pumped phase conjugate mirrors with $BaTiO_3$:Rh are also currently developed for the application to dynamic wavefront correction of nanosecond Nd:YAG lasers at 1.06 μm [35, 36]. They are discussed in the following sections.

8.3.1 Internal loop self-pumped phase conjugate mirror

As said before, the internal loop phase conjugate mirror is widely used because the only required device is a photorefractive crystal. A theoretical description of such a phase conjugate mirror is difficult because the dimensions of the material and the incidence angle as well as its divergence affect the beam path in the photorefractive crystal; as a result, the numbers and positions of the interaction regions may vary from one case to another [37]. Moreover, the nature of the gratings (transmission or reflection gratings) involved in each interaction region is not easily determined. Experiments like low-coherence reflectometry are necessary [38]. Nevertheless, in all cases, this geometry is self-starting and the four-wave mixing develops from the beam-fanning. In a model considering plane waves interacting in two regions, the threshold in term of gain times interaction length ($\Gamma\ell$) product is shown to be $\Gamma\ell = 8$, which is high compared to other geometries mentioned above [2].

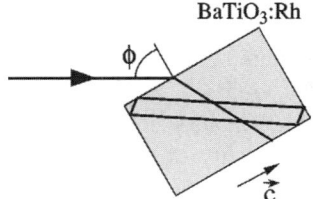

Figure 8.8. Arrangement of the internal-loop self-pumped phase conjugate mirror.

At 1.06 μm, this internal loop geometry is used with both CW and nanosecond illumination to perform phase conjugation in BaTiO$_3$:Rh. It is depicted in Fig. 8.8 [35]. When used with CW Nd:YAG lasers, steady-state instabilities may be removed by vibrating the crystal slightly in the vertical direction (amplitude 10 μm at a repetition rate of 200 Hz) so that competing parasitic gratings are washed out. The vibrating mechanical structure enables the phase conjugate reflectivity to reach 55% with CW illumination. This trick is no more necessary with nanosecond illumination as the coherence length of the source is short. As mentioned in the literature for other wavelengths [7, 37], it is observed that the reflectivity is strongly dependent on the incidence angle on the photorefractive crystal (Fig. 8.9). When the adequate angle is chosen, the maximum measured reflectivity is 32% with nanosecond illumination at 10 Hz (the crystal was not antireflection coated). As discussed in Section 8.2.1, the sensitivity of BaTiO$_3$:Rh at 1.06 μm is low compared to the usual values in the visible range. So, the rise time of the phase conjugate reflectivity at 1.06 μm has to be optimized in order to be competitive. Indeed, the phase conjugate mirror is unable to compensate for aberrations which vary on a time scale that is shorter than the reflectivity rise time. With the internal loop geometry, the reported rise time from zero to 90% of the maximum reflectivity is 20 min for an incident fluence of 230 mJ · cm^{-2}

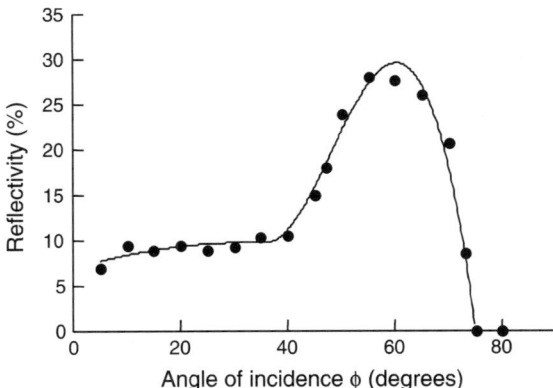

Figure 8.9. Steady-state reflectivity of the internal loop phase conjugate mirror versus incidence angle ϕ. The solid curve is a guide for the eye.

at a repetition rate of 10 Hz. This leads to a required fluence of $2700 \text{ J} \cdot \text{cm}^{-2}$. At a repetition rate of 30 Hz, the required fluence goes down to $2000 \text{ J} \cdot \text{cm}^{-2}$ (5 min for an input fluence of $230 \text{ mJ} \cdot \text{cm}^{-2}$) [35].

As depicted in Fig. 8.10, such a phase conjugate mirror is able to compensate for highly aberrated beams. The fidelity of phase conjugation is 90%, even when a phase plate aberrator is introduced in the beam path. It was measured by the power-in-the-bucket technique: The fidelity is defined as the ratio of the phase conjugate and incident pulsed energies that are transmitted through a pinhole in the focal

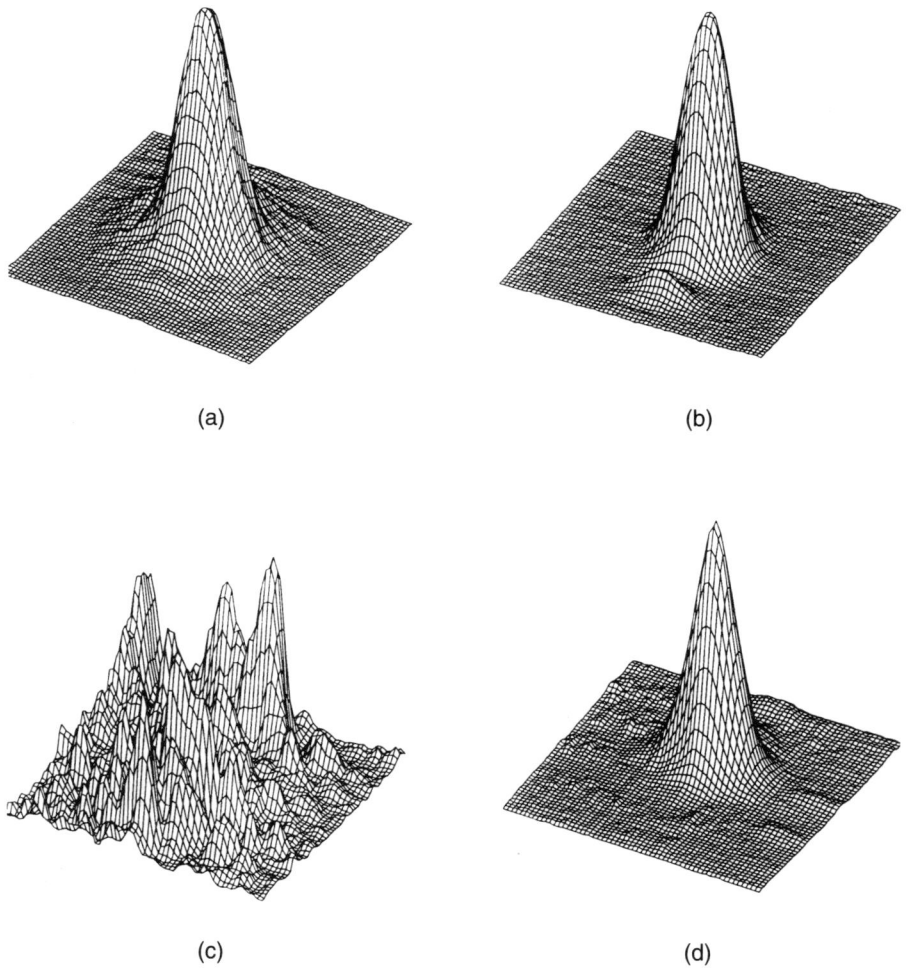

Figure 8.10. Near-field spatial beam profiles: (a) Incident beam from the Nd:YAG laser, (b) resulting phase conjugate return beam, (c) incident beam distorted by the introduction of a phase plate aberrator, and (d) phase conjugate beam corrected for aberration caused by the aberrator.

plane of a lens. This technique, easy to set up, is widely used [35, 39, 40]. Others, like shearing interferometry, enable us to observe phase-difference gradients between the reference beam and the phase conjugate beam [41]. When analyzed by a Fourier transform method [42], phase difference can be evaluated in the transverse plane. But, because phase conjugation is a nonlinear process, the fidelity may strongly depend on the nature of the incident beam (Gaussian, speckled, etc.). So, an extremely detailed analysis of the phase conjugate fidelity, only valid for a single kind of incident beams, seems useless. For the application to wavefront correction of laser sources, the focusing ratio is a relevant characteristics. This is why the power-in-the-bucket technique is employed.

8.3.2 Ring self-pumped phase conjugation

The ring self-pumped phase conjugate mirror consists of one photorefractive crystal and two additional mirrors that form a loop and feed the transmitted beams back to the crystal (Fig. 8.11). It is a self-starting phase conjugate mirror, and the four-wave mixing is initiated by beam-fanning. It has been studied theoretically in a plane-wave model at steady state with pump depletion [2]. The threshold in terms of $\Gamma\ell$ product is found to be $\Gamma\ell = 2$, which is four times lower than for the internal loop geometry. Moreover, the reflectivity increases sharply with $\Gamma\ell$, and high reflectivities are easily obtained. A $\Gamma\ell$ of 3 is enough to generate a reflectivity of 90% for a lossless loop [2]. As will be shown in this section, in this geometry, optical systems may be inserted in the loop to improve the fidelity of phase conjugation.

In the experiment of Fig. 8.11, a "roof cut" photorefractive crystal referred to as FEE-X16-45 is used. Indeed, under CW illumination, the total internal reflection on

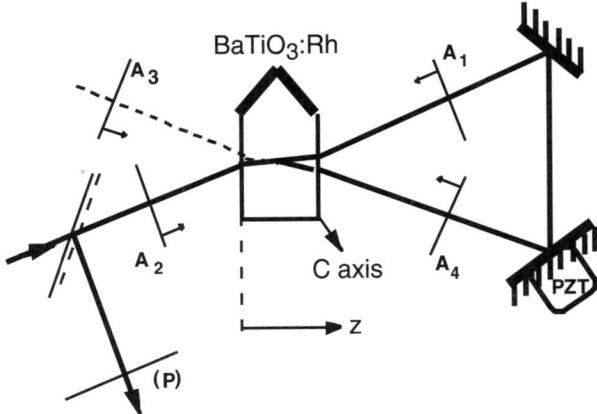

Figure 8.11. Schematic representation of the ring self-pumped phase conjugate mirror. Phase conjugate beams are observed in plane (P). A_i are the complex amplitudes of interacting beams in a plane wave model. PZT, piezo mirror used to wash out the reflection gratings.

the crystal faces allows the oscillation of spurious beams and prevents beam-fanning from rising. Thus, no phase conjugate beam can be observed. Thanks to this roof cut, antireflection coatings, and a 45° orientation of the c axis with respect to the input face, this oscillation is avoided as well as that of the beam-fanning between input and output faces. The coherence length of the usual CW sources allows both the desired transmission grating (between A_2 and A_3, or A_1 and A_4) and unwanted reflection gratings (between A_1 and A_2, or A_1 and A_3, or A_4 and A_2, or A_4 and A_3) to be recorded. This is why a vibrating mirror is used under CW illumination [43]: The response time of the photorefractive material is longer than the time necessary to shift the reflection grating which, therefore, cannot be recorded in the crystal. As a result, only the transmission grating remains. With nanosecond illumination, the coherence length of the source is too small to write reflection gratings and no vibrating mirror is needed.

As pointed out in Ref. 2, the reflectivity depends on the value of $\Gamma\ell$. As soon as $\Gamma\ell$ is high enough, the reflectivity saturates to its maximum value. The measured reflectivity with CW illumination is 70% for a transmission of the loop of 71%. With nanosecond illumination and other optical elements in the loop, a reflectivity of 79% is measured for a 81% loop transmission. Experimentally, the reflectivity proves to be limited to the transmission of the loop as soon as it reaches 5% (Fig. 8.12) [36]. A fit of experimental data by a plane-wave model at steady state including pump depletion leads to $\Gamma\ell = 8.2$ (Fig. 8.12). This is consistent with $\Gamma = 23$ cm^{-1} measured in Ref. 20 and a crystal thickness of 3.5 mm, as is the case in FEE-X16-45.

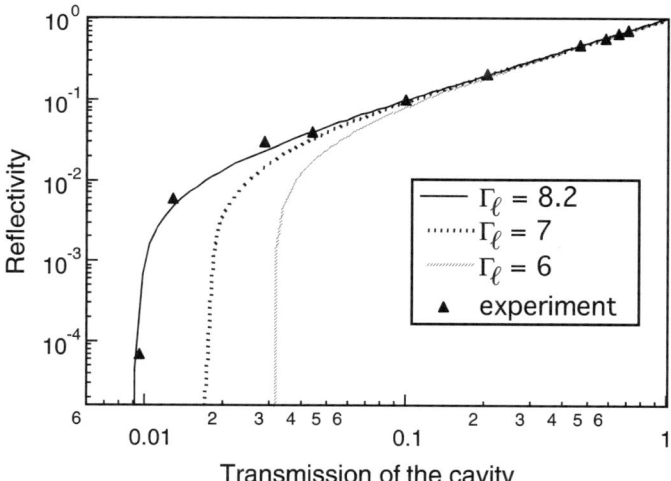

Figure 8.12. Reflectivity of the ring phase conjugate mirror versus transmission of the loop. Experimental data are fitted by theoretical curves obtained in a plane-wave steady-state model with pump depletion for different values of the $\Gamma\ell$ product.

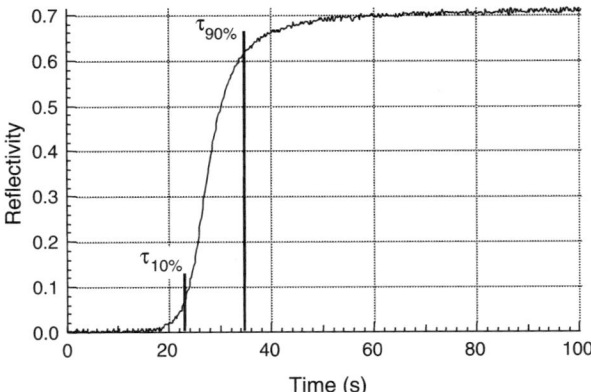

Figure 8.13. Experimental time evolution of the reflectivity of the ring self-pumped phase conjugate mirror for an incident CW intensity of 5 W · cm^{-2} at 1.06 μm.

The second important feature in photorefractive phase conjugation is the reflectivity rise time. A typical time evolution of the reflectivity is given in Fig. 8.13. It shows a sharp increase characterized by the time needed for the reflectivity to grow from 10% to 90% of its maximum value $\tau_{90\%} - \tau_{10\%} \equiv \tau_{pc}$. To optimize this feature, a plane-wave model is used. Absorption may be neglected in BaTiO$_3$:Rh at 1.06 μm ($\alpha = 0.1$ cm^{-1}). This model cannot take into account the spatial profiles of the interacting beams, like in three-dimensional numerical models [44], or the influence of beam-fanning [45]. But for $t > \tau_{10\%}$, one may reasonably consider that the desired transmission grating is the only one remaining in the material. Consequently, this simplified model is sufficient to optimize the reflectivity rise time, using coupled-wave equations for a transmission grating only [Eq. (15)], along with the time evolution of the space-charge field E_1 [Eq. (16)].

$$\frac{\partial A_1(z,t)}{\partial z} = i\frac{\Gamma}{4E_{sc}} E_1(z,t)A_4(z,t)$$

$$\frac{\partial A_2^*(z,t)}{\partial z} = i\frac{\Gamma}{4E_{sc}} E_1(z,t)A_3^*(z,t)$$

$$\frac{\partial A_3(z,t)}{\partial z} = -i\frac{\Gamma}{4E_{sc}} E_1(z,t)A_2(z,t)$$

$$\frac{\partial A_4^*(z,t)}{\partial z} = -i\frac{\Gamma}{4E_{sc}} E_1(z,t)A_1^*(z,t)$$

(15)

A_1, A_2, A_3, and A_4 are the complex amplitudes of the plane waves, and E_{sc} is the steady-state space-charge field.

As detailed in Section 8.2, the photorefractive effect in BaTiO$_3$:Rh at 1.06 μm finds its origin in the three charge states of rhodium Rh^{3+}, Rh^{4+}, and Rh^{5+}, and a three-charge state band transport model is suitable to account for experimental photorefractive characterizations of this crystal. However, to describe the time evolution of the space-charge field, the simpler single carrier single-site model without applied electric field is used here. The time evolution of the space charge field is then governed by

$$\frac{\partial E_1(z,t)}{\partial t} = -\frac{1}{\tau_{ph}}[E_1(z,t) - im(z,t)E_{sc}] \tag{16}$$

where τ_{ph} is the photorefractive time constant for a given grating spacing and for a given $I_0 = \sum_i |A_i|^2 \cdot m(z,t)$ is the modulation of the interference pattern defined by

$$m(z,t) = \frac{2(A_1 A_4^* + A_2^* A_3)}{I_0} \tag{17}$$

We assume that the initial seeding amplitude A_3 results from a scattering at the input surface: $A_3(z, t = 0) = \varepsilon$. This initial condition has no consequence on τ_{pc} as soon as ε is small enough. For a given I_0, the important features for the kinetics of the phase conjugation are τ_{ph}, $\Gamma\ell$, and the transmission T of the loop, as in photorefractive oscillators [46]. For a given incidence angle θ on the crystal (i.e., a given grating spacing Λ), τ_{pc} is deduced by calculating the corresponding values of τ_{ph} and $\Gamma\ell$ and by inserting these values in Eqs. (15) and (16) and solving numerically. A theoretical plot of τ_{ph}, $\Gamma\ell$, and τ_{pc} is given in Fig. 8.14 for an incident intensity on the crystal of $5\,W \cdot cm^{-2}$, an effective density of traps of $5 \times 10^{16}\,cm^{-3}$, extraordinary polarized beams, an antireflection-coated 45°-cut crystal whose thickness is 3.5 mm, and a lossless loop. The rise time τ_{pc} appears to be minimized for incidence angles between 20° and 50°, corresponding to $0.7\,\mu m \leq \Lambda \leq 1.6\,\mu m$. This result is checked

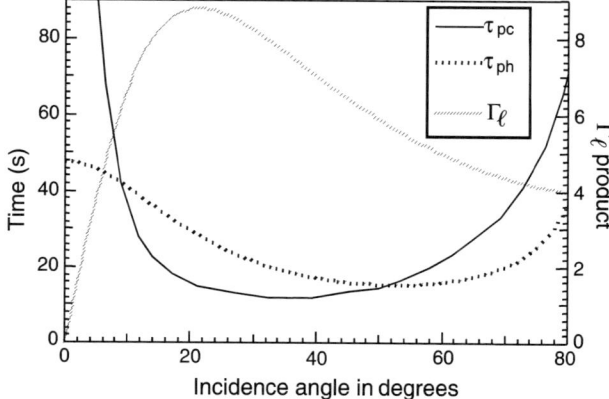

Figure 8.14. Theoretical variations of the $\Gamma\ell$ product, the photorefractive rise time τ_{ph}, and the reflectivity rise time τ_{pc} as a function of the incidence angle θ, for an incident intensity of $5\,W \cdot cm^{-2}$, extraordinary polarized beams, a 45°-cut antireflection-coated crystal whose thickness is 3.5 mm, an effective density of traps of $5 \times 10^{16}\,cm^{-3}$, and a lossless loop.

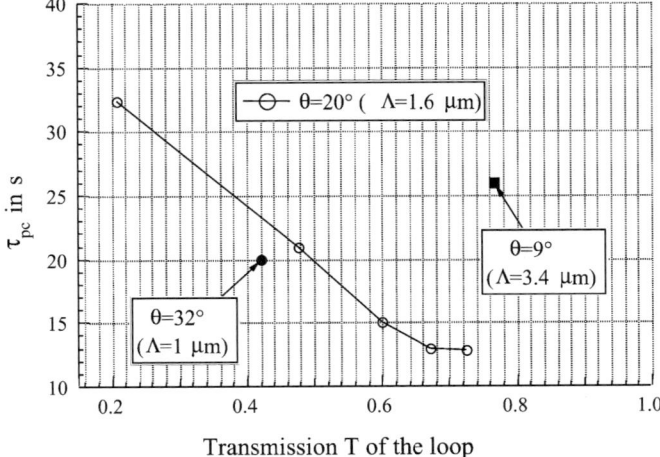

Figure 8.15. Experimental values of τ_{pc} for various incidence angles θ. As the transmission T of the loop varies with θ, the values of τ_{pc} for a given θ are to be compared with the value obtained at $\theta_{\text{reference}} = 20°$ for the same T.

experimentally (Fig. 8.15). However, as the transmission T of the loop varies with the incidence angle θ, the rise time τ_{pc} for a given θ has to be compared with τ_{pc} at a different $\theta_{\text{reference}}$ for the same T. Here the value of $\theta_{\text{reference}}$ is chosen to be $20°$. Experimental results are in agreement with the model: An incidence angle of $\theta = 32°$ leads to a value of τ_{pc} slightly smaller (20 s) than the one obtained at $\theta = 20°$ for the same T. At $\theta = 9°$, $\tau_{10\%-90\%}$ is much larger than the one obtained at $\theta = 20°$ for the same T. For reasons of comfort of work, an incidence angle of $20°$ is chosen for further experiments. In this case, for a transmission of the loop of 71%, the measured rise time for an incident CW intensity of $5 \text{ W} \cdot \text{cm}^{-2}$ is 12 s (Fig. 8.13), corresponding to an energy of $E_{10\%-90\%} = 60 \text{ J} \cdot \text{cm}^{-2}$. With nanosecond illumination, at $\theta = 20°$, the required energy is $E_{10\%-90\%} = 90 \text{ J} \cdot \text{cm}^{-2}$. The energy necessary to increase the reflectivity up to 90% of its maximum value is $300 \text{ J} \cdot \text{cm}^{-2}$. As the product intensity × rise time has been found to be constant, this can be extrapolated to give a response time of the phase conjugate mirror of 30 s for pulses of $10 \text{ mJ} \cdot \text{cm}^{-2}$ at a repetition rate of 1 kHz.

A further key point in phase conjugation is the fidelity of phase conjugation. This problem has been addressed theoretically and experimentally [44, 47]. Indeed, when the loop consists of two plane mirrors, the phase conjugate fidelity is poor. Two reasons may explain this.

First, the phase-matching condition

$$\vec{k}_2 - \vec{k}_3 = \vec{k}_4 - \vec{k}_1 = \vec{k}_g \tag{18}$$

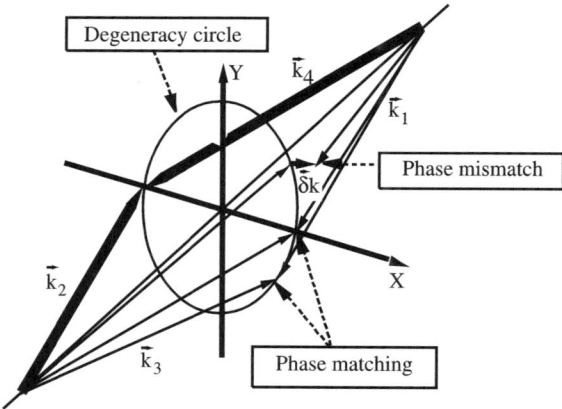

Figure 8.16. Representation of the wave vectors in the ring phase conjugate mirror. $\vec{k}_2$ (incident beam) and $\vec{k}_4$ (image of the incident beam through the loop) are fixed.

where $\vec{k}_i$ are the wave vectors of beams A_i and where $\vec{k}_g$ is the grating wave vector, may be fulfilled on a circle called the degeneracy circle whose diameter is k_g (Fig. 8.16). This circle is clearly visible in Fig. 8.17 obtained at 1.06 μm when a screen is inserted in the loop before the four-wave mixing process has reached its steady state.

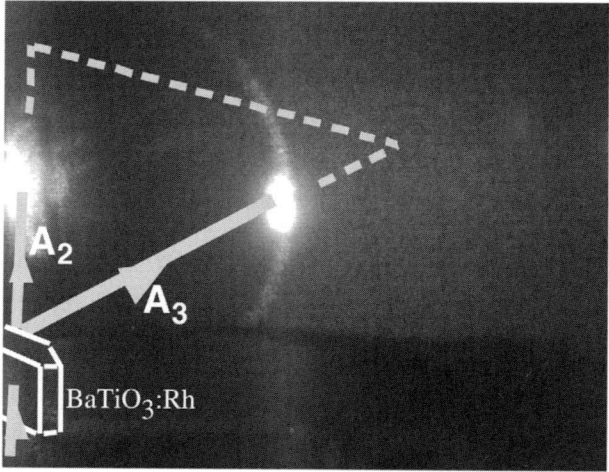

Figure 8.17. Observation of the degeneracy circle on beam A_3 at 1.06 μm on a screen inserted in the loop of the ring phase conjugate mirror, before the four-wave mixing has reached its steady-state. The position of the photorefractive crystal is represented by white lines. The beam path in the loop is also drawn in solid and dotted thick lines.

Second, the quasi-Bragg matching condition

$$|(\vec{k}_2 - \vec{k}_4 - \vec{k}_3 + \vec{k}_1)\overrightarrow{\Omega\Omega'}| = |\vec{\delta k} \cdot \overrightarrow{\Omega\Omega'}| < \pi \tag{19}$$

with Ω and Ω' two points in the interaction volume, allows wave vectors which point close to the degeneracy circle to exist (Fig. 8.16). This results in a roughly elliptical phase conjugate beam with a long axis along the Y axis [44, 47]. The insertion of optical elements in the loop improves the phase conjugate fidelity. With a three-prism system able to rotate the beam cross section by 90°, the phase conjugate beam is efficiently selected among the backscattered light. This device (Fig. 8.18), previously proposed [44] and tested in the visible range [47], is also used at 1.06 µm with nanosecond illumination and proves to be efficient [36]. Phase conjugation fidelities of 80% are measured by the power-in-the-bucket technique when lenses of focal length longer than 300 mm are inserted in the beam path. Because of size problems, the optical length of the loop cannot be made smaller than 12 cm. This prevents from correcting highly aberrated beams. Indeed, after one lap in the loop, the propagation of the diverging aberrated beam affects the overlap of the beams inside the BaTiO$_3$:Rh crystal.

An effort has been made in compacting the ring phase conjugate mirror. A ring self-pumped phase conjugate mirror using total internal reflection on the faces of a crystal cut in a triangle shape was already proposed [48] but offered no control of the phase conjugate beam and of the gratings involved. Another solution is to use one spherical mirror in the loop (Fig. 8.19). This configuration has been investigated theoretically. For doing this, an incident plane wave focusing near the crystal after one lap in the loop is considered (Fig. 8.19). It is also assumed that all waves are locally plane waves. The aberration considered is a tilt in the focusing beam ($\delta\alpha$ in the XZ plane, $\delta\phi$ in the YZ'' plane). In these conditions, δk expressed in (X'', Y, Z'') is

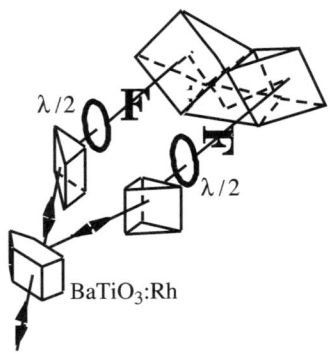

Figure 8.18. Ring self-pumped phase conjugate mirror with a three-prism system in the loop which performs a 90° rotation of the beam cross section. Two half-wave plates maintain the extraordinary polarization of interacting beams in the crystal, which is necessary to provide high photorefractive gains and high phase conjugate reflectivities. The loop length is 12 cm.

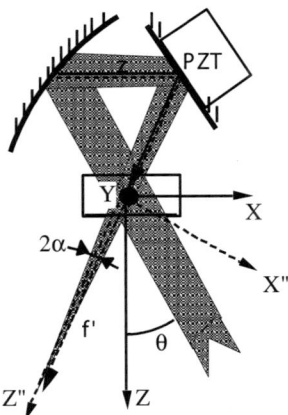

Figure 8.19. Compact ring phase conjugate mirror including one curved mirror in the loop.

given by

$$\delta \vec{k} = \vec{k}_2 - \vec{k}_4 - \vec{k}_3 + \vec{k}_1 = k \left| \begin{array}{c} \left[-1 - \cos(2\theta - \alpha) \dfrac{f' - z}{f'} \dfrac{1}{\cos^2 \alpha} \right] \delta\alpha \\ -\delta\alpha \, \sin(2\theta - \alpha) \dfrac{f' - z}{f'} \dfrac{1}{\cos^2 \alpha} \\ \delta\phi \left[\dfrac{f' - z}{f'} - 1 \right] \end{array} \right| \quad (20)$$

$2f'$ is the curvature of the spherical mirror. z, α and θ are defined in Fig. 8.19, $k = 2\pi/\lambda$. Introducing Eq. (20) into Eq. (19) leads to

$$\delta\alpha < \frac{\pi}{\ell k \left[1 + ((f' - z)/f')(1/\cos^2 \alpha)(\cos(2\theta - \alpha) + \sin(2\theta - \alpha)) \right]} \quad \text{if } \delta\phi = 0$$

$$\delta\phi < \frac{\pi}{\ell k \left[((f' - z)/f') - 1 \right]} \quad \text{if } \delta\alpha = 0 \quad (21)$$

ℓ is the typical size of the interaction region. It is chosen here to be 2 mm.

Equation (21) shows that the phase conjugate fidelity in the plane of incidence ($\delta\phi = 0$) is weakly dependent on the curvature $2f'$ of the spherical mirror. But in the Y direction ($\delta\alpha = 0$), the fidelity strongly depends on the ratio $(f' - z)/f'$. For a Gaussian incident beam, the "phase conjugate" beam is elliptic with a long axis along the Y direction if $(f' - z)/f' > 0.7$, whereas it remains Gaussian for $(f' - z)/f' < 0.7$ [49]. The ratio $(f' - z)/f'$ is also an expression for the change in the beam's diameter after one lap in the loop. The influence of this parameter on the phase conjugate fidelity is also pointed out in numerical simulations of Ref. 44. The

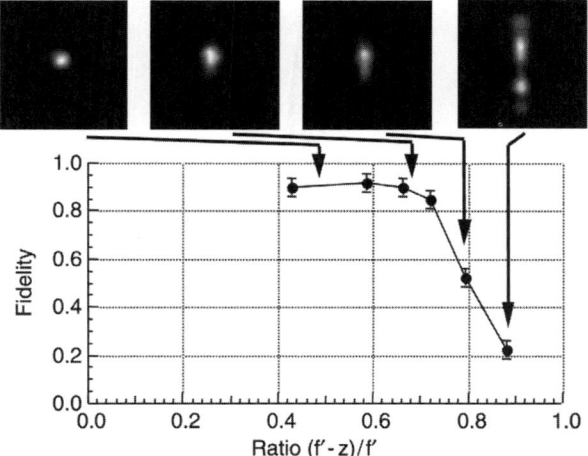

Figure 8.20. Phase conjugation fidelity of the compact ring mirror as a function of the ratio $(f' - z)/f'$ with nanosecond illumination at 1.06 μm. The corresponding far-field beam profiles are also given.

influence of $(f' - z)/f'$ has been tested experimentally by measuring the phase conjugate fidelity and monitoring the far-field profile of the phase conjugate beam. Even for small loop lengths close to the coherence length of the laser source, the vibrating plane mirror avoids reflection gratings in the crystal. The results reported in Fig. 8.20 are in agreement with the predictions. Using this simple configuration with a loop length of 6 cm, lenses of focal length from infinity down to 25 cm could be compensated for. Phase conjugation fidelities of 90% are obtained. Such a ring mirror may be adapted to any range of focal lengths of the aberrator—for instance, from 80 cm down to 15 cm—by increasing slightly the ratio $(f' - z)/f'$. The alignment of this device is not critical and the spherical mirror may even be replaced by a cylindrical mirror, because the fidelity in the incidence plane is weakly affected by the ratio $(f' - z)/f'$. With such a ring phase conjugate mirror, the ability to correct highly aberrated beams has also been demonstrated (Fig. 8.21) along with the ability to restore images, even when lenses or strong aberrations are inserted in the beam path. A compact ring phase conjugate mirror including a roof-cut BaTiO$_3$:Rh crystal and a vibrating cylindrical mirror in the loop is currently in a technology transfer procedure and is shown in Fig. 8.22. The typical dimensions of the device are 6 × 6 cm^2. This optimized photorefractive nonlinear mirror is as easy to use as a conventional mirror: The operator just needs to roughly adjust the orientation of the "black box-type" device so that the beam emerges, and a high fidelity phase conjugate beam is delivered in a few seconds.

Such a compact ring phase conjugate mirror seems to put together advantages of both internal loop and ring geometries presented before. It is able to correct highly aberrated beams, and the positioning is not critical. It thus appears to be a good

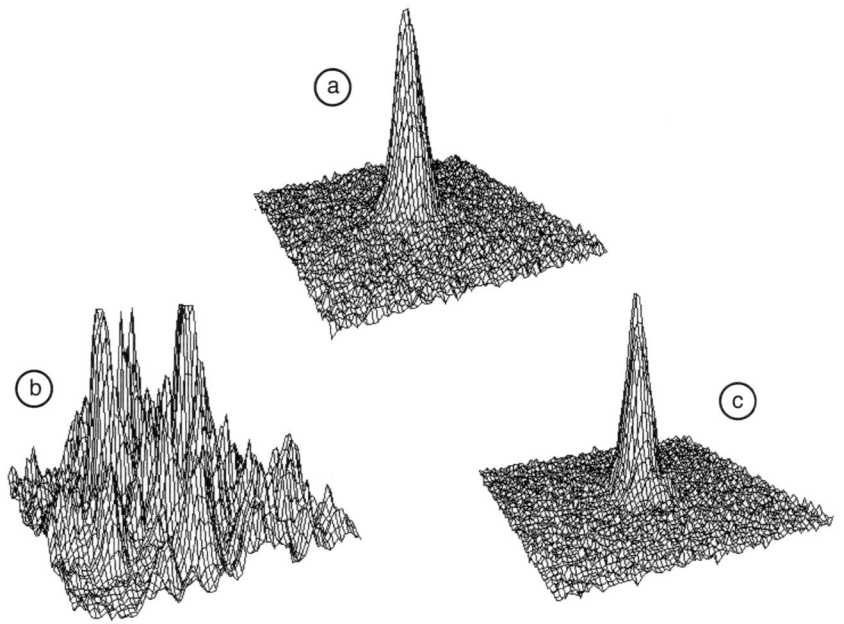

Figure 8.21. Correction of random aberrations at 1.06 μm with nanosecond illumination. (a) Incident beam. Far-field profile after a double pass through a random aberrator and (b) reflection on a dielectric mirror. (c) Reflection on the compact ring phase conjugate mirror.

Figure 8.22. Photograph of the packaged compact ring phase conjugate mirror including a roof-cut BaTiO$_3$:Rh and a vibrating cylindrical mirror.

candidate for all-solid-state dynamic wavefront correction of MOPA laser sources that will be presented in the next section.

8.4 DYNAMIC WAVEFRONT CORRECTION OF MOPA LASER SOURCES

$BaTiO_3$:Rh allows the phase conjugation of nanosecond pulses at 1.06 μm with high reflectivity and fidelity along with a reflectivity rise time that is compatible with laser manufacturer requirements. Moreover, a packaged ready-to-use ring phase conjugate mirror has been developed, showing that the application of this technique to dynamic wavefront correction of Nd:YAG lasers is realistic. We will now describe the origin of aberrations present in Nd:YAG amplifier rods and present MOPA architectures including a self-pumped phase conjugate mirror with photorefractive $BaTiO_3$:Rh. Experimental descriptions and performances will be given. The results obtained with a photorefractive phase conjugator will be compared to the performances of other existing techniques.

8.4.1 Origin of aberrations in Nd:YAG amplifier rods

As soon as the repetition rate of the amplifier rod gets higher than the inverse of the thermal relaxation time (a few hertz), permanent thermal effects occur in Nd:YAG rods [50]. These effects are caused by the pumping by flashlamps or laser diodes, because a significant amount of their energy is converted into heat. The rod is cooled at its periphery only, which leads to a significant radial gradient of temperature inside the material. This results in a nonuniform index variation. Indeed, three optical effects occur. First, the distribution of refractive index acts as a converging lens. Second, the nonuniform thermal expansion of the rod induces stresses in the material which, through the elasto-optic effect, cause double focusing (bifocusing) according to the local polarization of the propagating wave. Third, thermal stresses locally modify the index ellipsoid and induce birefringence with a local fast axis in the radial direction and a local slow axis along the tangential direction, resulting in a depolarization of the propagating wave across the beam [51].

Birefringence may be partially compensated by using two amplifier rods separated by a 90° polarization rotator [52] or by adding optical imaging systems along with the 90° polarization rotator between the two rods [53]. These devices, easy to set up, may only provide a partial correction of the depolarization if the rods are slightly different. Other tricks have been proposed to avoid the use of two identical amplifier rods. A single amplifier followed by a Faraday rotator and a phase conjugate mirror proved to be efficient, but bifocusing cannot be compensated [54]. This problem may be bypassed by splitting the depolarized beam into two orthogonal polarizations and performing a separate phase conjugation for each of these two polarizations and then passing through the amplifier again [55].

Thermal lensing in Nd:YAG rods has been widely studied, and a great effort has been developed to properly measure and evaluate theoretically this highly limiting effect. The thermal focal lens depends on the operating conditions of the amplifier, and especially on the absorbed average pump power [51, 56]. For a pulsed pumping, the higher the repetition rate, the shorter the thermal focal lens [56]. With typical values, the focal lens is about 50 cm for a flashlamp-pumped Nd:YAG amplifier at a repetition rate of 100 Hz. It has also been observed and explained that thermal lensing effects differ for intracavity and extracavity operations [57]. The thermal lens is hardly corrected by a compensative diverging lens having the same optical power as the laser rod, because transient effects occur in the rod—for instance, during the warm-up of the laser. Moreover, for a strong pumping, the thermal lens is not spherical but rather highly aberrated [58]. As an example, an experimental study was performed on a 1.064-μm longitudinally diode-pumped Nd:YVO$_4$ laser probed by a 1.074-μm Nd:YAP laser to avoid chromatic dispersion errors. It revealed that spherical aberration and astigmatism had a Zernike coefficient half as high as defocusing [59]. Consequently, adaptative devices are required. Optical phase conjugation is one of them. Section 8.4.2 is devoted to the use of photorefractive BaTiO$_3$:Rh crystals for the realization of MOPA with phase conjugate mirror. Section 8.4.3 compares this technique to other ways of performing dynamic wavefront correction: Photorefractive beam clean-up, intracavity active wavefront compensation, or, as widely discussed in other chapters of this book, stimulated Brillouin scattering, gain gratings, or thermal nonlinearity.

8.4.2 MOPA laser sources including a photorefractive self-pumped phase conjugate mirror

The first compensation of a passive aberration by double pass after reflection on a phase conjugate mirror (using stimulated Brillouin scattering) was performed in 1972 [60]. Less than one year later, a master-oscillator power-amplifier (MOPA) structure using a Ruby amplifier was demonstrated [61]. Many laser geometries have been tested with various nonlinear media in which phase conjugation was performed. The MOPA structures, depicted in Fig. 8.23 and demonstrated at 1.06 μm with either internal loop or ring BaTiO$_3$:Rh phase conjugate mirrors, have been evaluated in a collaborative work between Thales Research and Technology (France) and the Laboratoire Charles Fabry de l'Institut d'Optique (France) [62]. The laser beam experiences two (Fig. 8.23a) or four passes (Fig. 8.23b) in an Nd:YAG amplifier pumped by flashlamps at a repetition rate of 10 Hz. No care was taken to compensate for depolarization that is negligible at a repetition rate of 10 Hz. The oscillator is an intracavity-filtered Q-switched Nd:YAG laser delivering 20-ns pulses at a repetition rate of 10 Hz. The beam diameter at $1/e^2$ at the output of the oscillator is 2.5 mm. A first half-wave plate/Glan polarizer GP1 device is used to vary the incident energy in the amplifier rod. The device formed by GP1, the Faraday rotator FR1, the half-wave plate, and the Glan polarizer GP2 suppresses beams propagating back to the oscillator and performs the extraction of the output beam after a double pass in the amplifier rod (Fig. 8.23a). In the four-pass configuration, a Faraday rotator FR2 is inserted after the amplifier rod, so that the beam is directed to the phase conjugate mirror only after a double pass in the

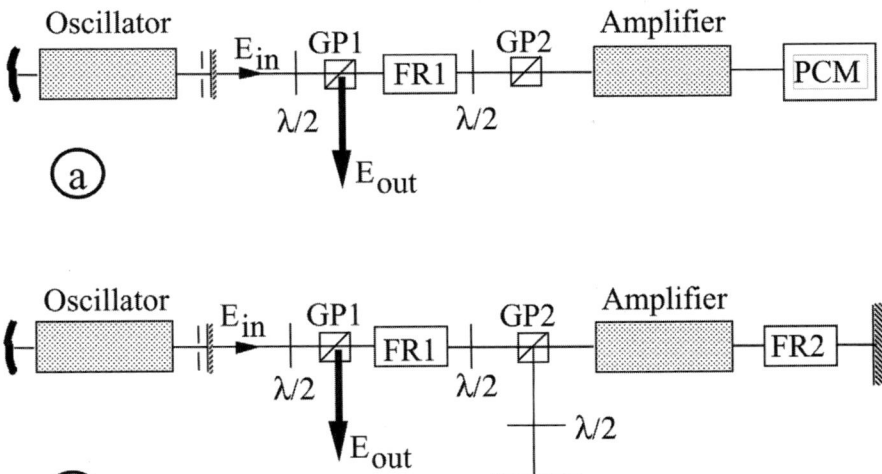

Figure 8.23. Nd:YAG master oscillator power amplifier architectures using BaTiO$_3$:Rh self-pumped phase conjugate mirrors. (a) Double-pass MOPA. (b) Four-pass MOPA. See text for a description of the beam path. PCM, phase conjugate mirror; GP, Glan polarizer; FR, Faraday rotator.

amplifier. A half-wave plate before the phase conjugate mirror changes the polarization from vertical to horizontal (extraordinary for the photorefractive crystal), which is necessary to provide high photorefractive gains in BaTiO$_3$:Rh. After a reflection on the phase conjugate mirror, the beam experiences a further double pass in the amplifier and its polarization is changed from vertical to horizontal by a double pass in FR2. The beam travels through polarizer GP2 and is extracted by polarizer GP1 after a rotation of its polarization induced by the half-wave plate and FR1 (Fig. 8.23b).

The energy per pulse at the output of the MOPAs is plotted in Fig. 8.24 versus energies per pulse at the output of the oscillator (after GP1) [62]. In the double-pass MOPA, the same output energy is obtained for an incident energy six times lower with the ring phase conjugate mirror than with the internal loop phase conjugate mirror. In the four-pass MOPA, the difference in the incident energy for the same output energy reaches a factor 40 in favor of the ring phase conjugate mirror. This is a consequence of the larger reflectivity available with the ring phase conjugate mirror, as pointed out in Section 8.3.

The two-pass-MOPA rise time, defined as the time needed to increase the output energy from 0 to 90% of its maximum value with a ring phase conjugate mirror, is given in Fig. 8.25 as a function of the incident energy per square centimeter on the photorefractive crystal. It is compared to the two-pass-MOPA rise time with an internal loop phase conjugate mirror. One observes that the internal loop phase

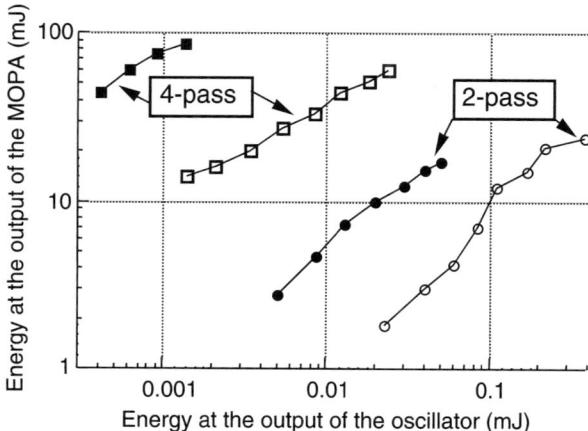

Figure 8.24. Output energy per pulse versus energy per pulse at the output of the oscillator for two- and four-pass MOPAs. Black markers: ring phase conjugate mirror. White markers: internal loop phase conjugate mirror.

conjugate MOPA is five times slower than the ring phase conjugate MOPA. This is consistent with the experiments performed on both phase conjugate mirrors alone (see Section 8.3.3).

For both phase conjugate mirrors, the output beam is a quasi-diffraction-limited Gaussian beam. Anyway, it should be kept in mind that the focal length of the thermal lens at a repetition rate of 10 Hz is larger than 2 m. Such aberrations are

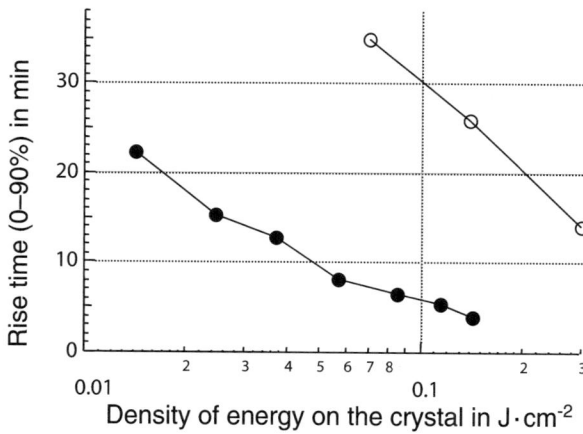

Figure 8.25. Rise time from 0 to 90% of the maximum output energy of the double-pass MOPA versus incident density of energy on the photorefractive crystal (repetition rate of 10 Hz). Black markers: ring phase conjugate mirror. White markers: internal loop phase conjugate mirror.

easily compensated by both phase conjugate mirrors. To demonstrate the ability of correction of strong aberrations, an additional random phase plate is inserted after the amplifier rod in the two-pass geometry; the output beam keeps TEM$_{00}$ at steady state as shown in Fig. 8.26.

The feasibility of nanosecond Nd:YAG MOPA laser sources with photorefractive self-pumped phase conjugate mirrors using BaTiO$_3$:Rh has been demonstrated with success at a repetition rate of 10 Hz. Satisfying results have also been obtained at a repetition rate of 30 Hz [62]. Such all-solid-state phase conjugate mirrors are well-suited for high-repetition-rate wavefront correction of diode-pumped MOPA, which exhibit a high average power but a low energy per pulse. Such a high-repetition-rate all-solid-state diode-pumped MOPA was demonstrated using an internal loop BaTiO$_3$:Rh phase conjugate mirror (Fig. 8.27) [63]. The master oscillator was a 7-mm-diameter, 10-mm-long Nd:YAG rod pumped by a CW fiber-coupled laser diode delivering 10 W at 808 nm. The acousto-optic Q-switch enabled the repetition rate of the laser to be varied from 10 Hz up to 1 kHz with a constant pulse duration (30 ns). Two water-cooled 5-mm-diameter, 95-mm-long Nd:YAG amplifiers completed the MOPA architecture. The pumping of each amplifier was performed

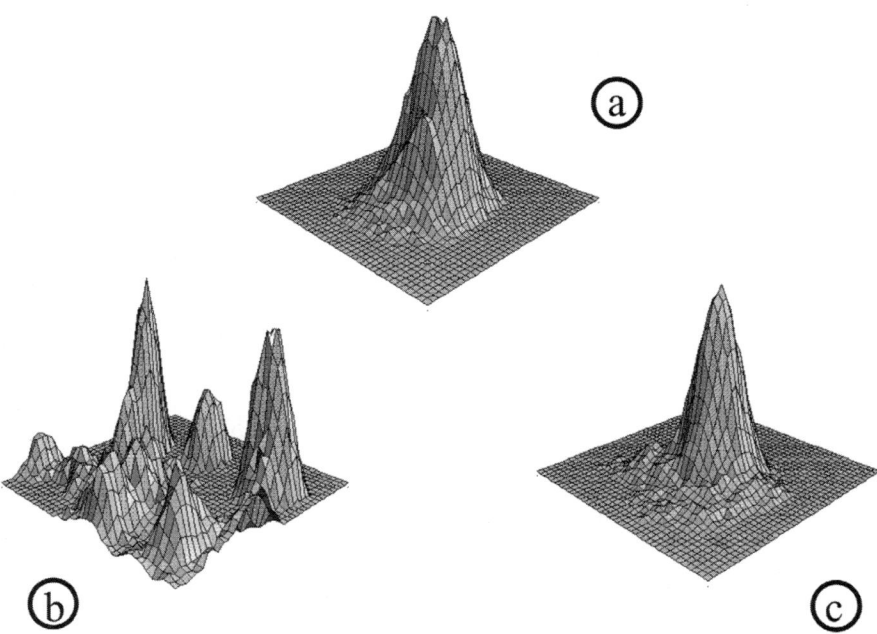

Figure 8.26. Compensation of strong aberration in a MOPA configuration. (a) Beam profile delivered by the oscillator. (b) Far-field output beam profile when a random phase plate is inserted after the amplifier and when the phase conjugate mirror is replaced by a dielectric mirror. (c) Far-field output beam profile with the random phase plate after the amplifier and with the phase conjugate mirror instead of the dielectric mirror.

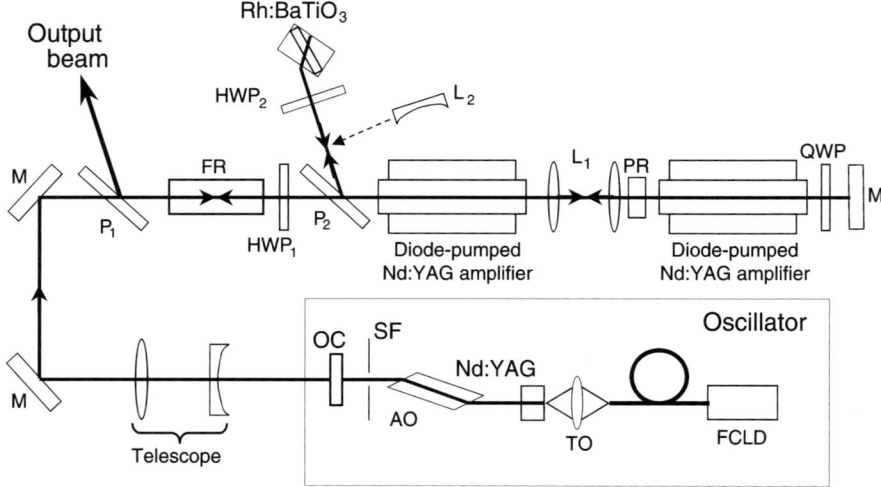

Figure 8.27. Schematic of the diode-pumped Nd:YAG MOPA laser with a BaTiO$_3$:Rh phase conjugate mirror. FCLD, fiber-coupled laser diode; TO, transfer optic; SF, spatial filter; AO, acousto-optic modulator; OC, output coupler: M and M′, highly reflecting plane mirrors; P$_i$, polarizer; FR, Faraday rotator; HWP$_i$, half-wave retardation plate; L1, focusing lens; L2, diverging lens; PR, polarizing rotator quartz plate; QWP, quarter-wave retardation plate.

by the placement around the active medium of 25 quasi-CW diode-laser arrays from Thales Laser Diodes, each delivering as much as 60 W in 200-µs pulses. A telescope and a half-wave plate inserted between the amplifiers ensured a birefringence compensation. An additional diverging lens in front of the phase conjugate mirror avoided focusing inside the BaTiO$_3$:Rh crystal. The architecture delivered an average power of 4 W at a repetition rate of 200 Hz in a diffraction-limited beam. The response time was 1 min. Both thermal lensing and high-order aberrations due the pumping geometry of the Nd:YAG rods were fully compensated with the BaTiO$_3$:Rh internal loop phase conjugate mirror.

These encouraging results stimulated research on high-average-power MOPA sources including a ring phase conjugate mirror, which is more efficient and with a shorter rise time than the internal loop phase conjugate mirror. Such an experiment was achieved in the Advanced Photon Research Center in Kyoto [64]. The MOPA source was made of a 100-Hz Q-switched oscillator seeding two zigzag slabs amplifiers that were imaged on each other by relay telescopes. Before and after phase conjugation, the beam experienced two passes in each amplifier. The beam was expanded at the output of the oscillator so that the incident intensity on the BaTiO$_3$:Rh crystal was maintained to $10 \text{ W} \cdot \text{cm}^{-2}$. As much as 36 W at a repetition rate of 100 Hz were obtained in a diffraction-limited beam with a response time of 20 s and a pulse duration of 36 ns.

The results obtained up to now suggest that the BaTiO$_3$:Rh phase conjugate mirror would lead to brilliant performances in kilohertz range high average power but relatively low energy-per-pulse MOPA sources.

8.4.3 Comparison of photorefractive self-pumped phase conjugation to other existing techniques

In the previous sections, we described wavefront correction by photorefractive self-pumped phase conjugation. Dynamic wavefront correction can also be achieved by two-wave mixing. This so-called beam clean-up technique (Fig. 8.28) consists of splitting the aberrated intense laser beam into two beams. One, the signal beam of very low power, is spatially filtered to provide a TEM$_{00}$ beam. The other is a powerful and aberrated pump beam. Both interfere in the photorefractive crystal. The $\pi/2$ phase shift between the interference pattern and the index grating in the material allows energy transfer from the pump beam to the signal beam, without phase transfer. As a result, the signal beam is amplified while keeping its TEM$_{00}$ structure. Beam clean-up has been studied with nanosecond illumination at 532 nm

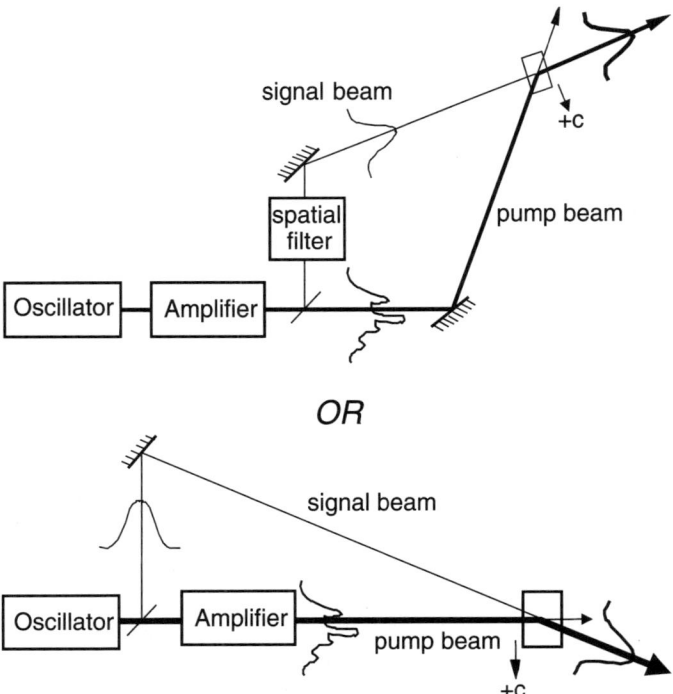

Figure 8.28. Principle of the beam clean-up: Energy is transferred from the powerful aberrated pump beam to the weak TEM$_{00}$ signal beam, without phase transfer. Two possible architectures are proposed here.

in BaTiO$_3$:Co, where 41% of the pump beam is transferred to the signal beam, which corresponds to 90% of the maximum energy transfer achievable, taking absorption into account [27]. At 1.06 μm in BaTiO$_3$:Rh, the overall efficiency, measured as the ratio of the output amplified signal power to the power of the beam delivered by the laser, is 28% [65]. This is a relatively poor efficiency compared to phase conjugation reflectivities at the same wavelength [36]. Indeed, it was shown that high phase conjugate reflectivities were achievable with relatively small gain time interaction length products [2]. High transfer efficiencies by two-wave mixing need higher $\Gamma\ell$ values. Moreover, beam-fanning is damageable in beam clean-up: In high gain crystals, part of the pump energy is transferred to beam-fanning and not to the signal beam. To get rid of this beam-fanning, the available photorefractive gain has to be reduced by incoherent illumination or by crystal tilting [27]. Moreover, to perform beam clean-up with sources of short coherence length, the path lengths of the two arms have to be adjusted carefully, so that the beams may interfere on the whole thickness of the crystal. In self-pumped phase conjugation, this is done automatically. These results show that beam clean-up seems less adapted to applications than phase conjugation. Anyway, beam clean-up has been recently used with success in BaTiO$_3$:Rh at 1.06 μm to convert a fiber-amplified multimode-transverse beam into a single-mode beam [66].

Photorefractive wave mixing is not the only way to perform wavefront correction. Indeed, one of the most-studied and well-developed techniques to perform phase conjugation is stimulated Brillouin scattering (SBS): A laser beam is focused on a fiber, a bulk material, a liquid, or gas cell (Chapters 2–7). As a summary, the stimulated scattering results from the coupling of two processes: (1) The input and scattered waves interfere and produce an acoustic wave by electrostriction, and (2) the input wave is Bragg-diffracted from the acoustic wave and produces the scattered wave. Above a given threshold in intensity, a large amount of the input wave is converted into the scattered wave. SBS is widely used and proves to be an efficient way to perform phase conjugation. One of the main advantages of stimulated Brillouin scattering compared to photorefractive self-pumped phase conjugation is that it performs a pulse-to-pulse phase conjugation. In other words, it is suitable for rapidly time-varying aberrations. But, unlike photorefractive materials, it is hardly used with millisecond illumination or high repetition rate (>1 kHz) nanosecond pulses, because the threshold in intensity is difficult to exceed.

As widely discussed in Chapter 11, the amplifier itself can also be used as the nonlinear medium in which gain gratings by way of gain saturation can be photoinduced. This idea, initially proposed in 1979 [67], leads to the development of loop resonators and single-mode self-starting self-Q-switched loops in which pulse-to-pulse phase conjugation is achieved inside the Nd:YAG amplifier rod itself [68, 69]. Nevertheless, it needs high-laser-gain amplifiers, which is not the case in high-repetition-rate operation regimes. This is why this structure seems well-suited for pulsed lasers in the 100-Hz operating range.

Other active techniques use adaptative optics. The active mirror can be a deformable mirror or an optically addressed light valve (OALV). The OALV is based on a spatial light modulator using liquid crystal technology. An incoherent intensity modulation modifies the orientation of the birefringent liquid crystal molecules, which provides a coherent phase modulation [70]. Such a system has been used to correct phase aberrations of pulsed nanosecond and femtosecond laser beams. Optically addressing is achieved by a mask imaged on the OALV and is made by electrically addressing a liquid crystal matrix [71]. A deformable mirror placed in the laser cavity may also control the divergence of the output beam and increase its brightness. With such a technique, an increase of the beam radiant intensity of 10 to 15 is reported without any drop in the output power in a 25-W Nd:YAlO$_3$ laser [72]. Various technologies and devices using adaptative mirrors may be found in Ref. 73.

8.5 CONCLUSION

In this chapter, we presented BaTiO$_3$:Rh, an infrared-sensitive photorefractive material. Its photorefractive properties at 1.06 μm are described under both CW and nanosecond illuminations. The main characteristics of this reproducible material are a high photorefractive gain (23 cm^{-1}) and a weak absorption (0.1 cm^{-1}). A three-charge state band transport model based on spectroscopic studies describes it accurately. This model also predicts a fourfold improvement of the time constant by an oxidation of the sample, along with an increase of absorption and of the dark conductivity. Contrary to the results obtained in BaTiO$_3$ with visible light, nanosecond and CW illuminations are equivalent for BaTiO$_3$:Rh at 1.06 μm in terms of gain and time constant. We also describe the performances of two self-pumped phase conjugate mirrors using BaTiO$_3$:Rh. Compact-ring phase conjugate mirrors present a higher reflectivity and a shorter rise time than internal-loop phase conjugate mirrors and may compensate as well for highly aberrated beams. Moreover, the control of the involved gratings enables us to obtain an accurate prediction of the performances of the mirror.

Both types of phase conjugate mirrors were introduced in MOPA architectures. As expected, the comparison of two 10-Hz flashlamp-pumped Nd:YAG MOPAs with ring and internal phase conjugate mirrors were in favor of the ring geometry. The efficiency was higher (a lower oscillator output energy leads to the same MOPA output energy), and the rise time was divided by a factor of 5, while maintaining a TEM$_{00}$ beam. A high-repetition-rate diode-pumped Nd:YAG MOPA was also demonstrated with an internal phase conjugate mirror leading to 4-W average power at 200 Hz in a diffraction-limited beam and with a rise time of 1 min. Recently, a ring phase conjugate mirror was introduced in a MOPA architecture using two zigzag slab amplifiers. The system delivered up to 36-W average power at 100 Hz. These results show that a photorefractive BaTiO$_3$:Rh phase conjugate mirror and especially the ring geometry seems very well suited for applications to all-solid-state dynamic wavefront correction of high-repetition-rate (>1 kHz) diode-pumped nanosecond Nd:YAG MOPAs that deliver a high average power with a relatively low energy per pulse.

One should also mention that other photorefractive materials are now infrared-sensitive and could be an alternative to BaTiO$_3$:Rh. For instance, photorefractive strontium and barium niobate (SBN) doped with cerium and calcium is sensitive up to 850 nm [74], but the low value of the gain at this wavelength cannot lead to high reflectivity phase conjugation. Nevertheless, by a preillumination with green light, SBN:Ce can be activated at 1.06 μm [75]. This material may be interesting for wavefront correction.

Tin hypothiodiphosphate (Sn$_2$P$_2$S$_6$) has been studied for a long time for its ferroelectric properties. It exhibits also interesting photorefractive properties at 1.06 μm. This material appears as an intermediate between the high gain but poor response time BaTiO$_3$:Rh and the short response time but poor gain semi-insulating materials [76]. At 10 W · cm^{-2}, a two-wave mixing gain of 7 cm^{-1} and a response time of 10 ms is reported [77]. Several problems still arise in this crystal: Sn$_2$P$_2$S$_6$ needs to be preilluminated with white light to be efficient at 1.06 μm. Moreover, parasitic thermal excitation of charges progressively decreases the gain. To partially eliminate this problem and obtain permanent high gains, low-temperature operations and moving grating techniques are used. This contributes to improve the performances of this promising infrared-sensitive material [78].

Efforts have been made to increase the sensitivity of potassium niobate (KNbO$_3$) in the near-infrared spectrum. Several dopants have been tested (Ce, Cu, Co, Fe, Rh, Mn, and Ni) and a significant response at 1.06 μm has been observed in Rh-, Fe-, Mn-, and Mn–Rh-doped KNbO$_3$ crystals [79]. The sensitivity of KNbO$_3$:Rh in the near infrared can be increased by a reduction of the crystal at high temperature, leading to a sensitivity of the same order as BaTiO$_3$:Rh [80]. The problem of the reproducibility of the performances of reduced samples remains, but such a crystal could be an alternative to BaTiO$_3$:Rh for various applications.

REFERENCES

1. J. Feinberg, Self-pumped, continuous-wave phase conjugator using internal reflection, *Opt. Lett.* **7**, 486–488 (1982).
2. M. Cronin-Golomb, B. Fischer, J. O. White, and A. Yariv, Theory and applications of four-wave mixing in photorefractive media, *IEEE J. Quant. Electron.* **QE-20**, 12–29 (1984).
3. B. Fischer and S. Sternklar, New optical gyroscope based on the ring passive phase conjugator, *Appl. Phys. Lett.* **47**, 1–3 (1985).
4. S. Weiss, M. Segev, and B. Fischer, Line narrowing and self-frequency scanning of laser diode arrays coupled to a photorefractive oscillator, *IEEE J. Quantum Electron.* **24**, 706–708 (1988).
5. R. A. McFarlane and D. G. Steel, Laser oscillator using resonator with self-pumped phase conjugate mirror, *Opt. Lett.* **8**, 208–210 (1983).
6. M. Cronin-Golomb, B. Fischer, J. Nilsen, J. O. White, and A. Yariv, Laser with dynamic intracavity distortion correction capability, *Appl. Phys. Lett.* **41**, 219–220 (1982).

7. B. A. Wechsler, M. B. Klein, C. C. Nelson, and R. N. Schwartz, Spectroscopic and photorefractive properties of infrared-sensitive rhodium-doped barium titanate, *Opt. Lett.* **19**, 536–538 (1994).
8. M. Lobel, P. M. Petersen, and P. M. Johansen, Suppressing self-induced frequency scanning of a phase conjugate diode laser array with using counterbalance dispersion, *Appl. Phys. Lett.* **72**, 1263–1265 (1998).
9. M. Kaczmarek, P. Hribek, and R. W. Eason, Near infrared incoherent coupling and photorefractive response time of "blue" Rh:BaTiO$_3$, *Opt. Commun.* **136**, 277–282 (1997).
10. S. MacCormack, G. D. Bacher, J. Feinberg, S. O'Brien, R. J. Lang, M. B. Klein, and B. A. Wechsler, Powerful, diffraction-limited semiconductor laser using photorefractive beam coupling, *Opt. Lett.* **22**, 227–229 (1997).
11. M. Kaczmarek and R. W. Eason, Very high gain single pass two-beam coupling in blue BaTiO$_3$:Rh, *Opt. Lett.* **20**, 1850–1852 (1995).
12. G. W. Ross, P. Hribek, R. W. Eason, M. H. Garrett, and D. Rytz, Impurity enhanced self-pumped phase conjugation in the near infrared in "blue" BaTiO$_3$, *Opt. Commun.* **101**, 60–64 (1993).
13. N. V. Kukhtarev, V. B. Markov, S. G. Odoulov, M. S. Soskin, and V. L. Vinetskii, Holographic storage in electrooptic crystals, *Ferroelectrics* **22**, 949–960 (1979).
14. P. Tayebati and P. Maghereftech, Theory of the photorefractive effect for Bi$_{12}$SiO$_{20}$ and BaTiO$_3$ with shallow traps, *J. Opt. Soc. Am. B* **8**, 1053–1064 (1991).
15. Ph. Delaye, L. A. de Montmorillon, I. Biaggio, J. C. Launay, and G. Roosen, Wavelength dependent effective trap density in CdTe: evidence for the presence of two photorefractive species, *Opt. Commun.* **134**, 580–590 (1997).
16. K. Buse and E. Krätzig, Three-valence charge-transport model for explanation of the photorefractive effect, *Appl. Phys. B* **61**, 27–32 (1995).
17. N. Huot, J. M. C. Jonathan, and G. Roosen, Validity of the three-charge-state model in photorefractive BaTiO$_3$:Rh at 1.06 μm in the cw regime, *Appl. Phys. B* **65**, 489–493 (1997).
18. H. Kröse, R. Scharfschwerdt, O. F. Schirmer, and H. Hesse, Light-induced charge transport in BaTiO$_3$ via three charge states of rhodium, *Appl. Phys. B* **61**, 1–7 (1995).
19. M. Kaczmarek, G. W. Ross, R. W. Eason, M. J. Damzen, R. Ramos-Garcia, and M. H. Garrett, Intensity-dependent absorption and its modelling in infrared sensitive rhodium-doped BaTiO$_3$, *Opt. Commun.* **126**, 175–184 (1996).
20. N. Huot, J. M. C. Jonathan, G. Pauliat, D. Rytz, and G. Roosen, Characterization of a photorefractive rhodium doped barium titanate at 1.06 μm, *Opt. Commun.* **135**, 133–137 (1997).
21. A. Brignon, D. Geffroy, J. P. Huignard, I. Mnushkina, and M. H. Garrett, Experimental investigations of the photorefractive properties of rhodium-doped BaTiO$_3$ at 1.06 μm, *Opt. Commun.* **137**, 311–316 (1997).
22. Ph. Delaye, J. M. C. Jonathan, G. Pauliat, and G. Roosen, Photorefractive materials: Specifications relevant to applications, *P. Appl. Opt.* **5**, 541–559 (1996).
23. N. Huot, G. Pauliat, J. M. C. Jonathan, G. Roosen, R. Scharfschwerdt, O. F. Schirmer, and D. Rytz, Fourfold improvement of the photorefractive time constant of BaTiO$_3$:Rh by oxidation, *Technical Digest CLEO Europe '98*, Paper CFE2, Glasgow, 14–18 September 1998.

24. G. C. Valley, Short pulse grating formation in photorefractive materials, *IEEE J. Quantum Electron.* **QE-19**, 1637–1645 (1983).
25. G. C. Valley, Simultaneous electron/hole transport in photorefractive materials, *J. Appl. Phys.* **59**, 3363–3366 (1986).
26. M. J. Damzen and N. Barry, Intensity dependent hole/electron competition and photocarrier saturation in $BaTiO_3$ when using intense laser pulses, *J. Opt. Soc. Am. B* **10**, 600–606 (1993).
27. L. Mager, G. Pauliat, D. Rytz, M. H. Garrett, and G. Roosen, Wavefront correction of nanosecond pulses at 532 nm by photorefractive two-wave mixing, *Nonlinear Opt.* **11**, 135–144 (1995).
28. N. Barry, L. Duffault, R. Troth, R. Ramos-Garcia, and M. J. Damzen, Comparison between continuous-wave and pulsed photorefraction in barium titanate, *J. Opt. Soc. Am. B* **11**, 1758–1763 (1994).
29. N. Huot, J. M. C. Jonathan, G. Roosen, and D. Rytz, Two-wave mixing in photorefractive $BaTiO_3$:Rh at 1.06 μm in the nanosecond regime, *Opt. Lett.* **22**, 976–978 (1997).
30. J. Feinberg and R. Hellwarth, Phase-conjugating mirror with continuous-wave gain, *Opt. Lett.* **5**, 519–521 (1980).
31. A. Desfarges, V. Kermène, B. Colombeau, M. Vampouille, and C. Froehly, Wave-front reconstruction with a Fourier hologram in a phase-conjugating mirror oscillator, *Opt. Lett.* **20**, 1940–1942 (1996).
32. P. A. Bélanger, A. Hardy, and A. E. Siegman, Resonant modes of optical cavities with phase-conjugate mirrors, *Appl. Opt.* **19**, 602–609 (1980).
33. R. C. Lind and D. G. Steel, Demonstration of the longitudinal modes and aberration-correction properties of a continuous-wave dye laser with phase-conjugate mirror, *Opt. Lett.* **6**, 554–556 (1981).
34. A. Litvinenko and S. G. Odoulov, Copper-vapor laser with self-starting $LiNbO_3$ nonlinear mirror, *Opt. Lett.* **9**, 68–70 (1984).
35. A. Brignon, J. P. Huignard, M. H. Garrett, and I. Mnushkina, Self-pumped phase conjugation in rhodium-doped $BaTiO_3$ with 1.06-μm nanosecond pulses, *Opt. Lett.* **22**, 215–517 (1997).
36. N. Huot, J. M. C. Jonathan, G. Roosen, and D. Rytz, Characterization and optimization of a ring self-pumped phase-conjugate mirror at 1.06 μm with $BaTiO_3$:Rh, *J. Opt. Soc. Am. B* **15**, 1992–1999 (1998).
37. A. V. Nowak, T. R. Moore, and R. A. Fisher, Observations of internal beam production in barium titanate phase conjugators, *J. Opt. Soc. Am. B* **5**, 1864–1877 (1988).
38. P. Lambelet, R. P. Salathé, M. H. Garrett, and D. Rytz, Characterization of a photorefractive phase conjugator by low-coherence reflectometry, *Appl. Phys. Lett.* **64**, 1079–1081 (1994).
39. W. T. Whitney, M. T. Duignan, and B. J. Feldman, Stimulated Brillouin scattering and phase conjugation of multiline hydrogen fluoride laser radiation, *J. Opt. Soc. Am. B* **7**, 2160–2168 (1990).
40. B. W. Liby, J. K. McIver, and D. Statman, Beam quality measurements of a unidirectional self-pumped phase conjugate mirror, *Opt. Commun.* **101**, 79–84 (1993).
41. B. Fleck, A. Kiessling, G. Notni, and L. Wenke, A simple interferometric method for measuring the fidelity of phase conjugated beams, *J. Mod. Opt.* **38**, 495–502 (1991).

42. M. Takeda, H. Ina, and S. Kobayashi, Fourier-transform method of fringe-pattern analysis for computer-based topography and interferometry, *J. Opt. Soc. Am. B* **72**, 156–160 (1982).
43. M. Cronin-Golomb, J. Paslaski, and A. Yariv, Vibration resistance, short coherence length operation, and mode-locked pumping in passive phase conjugators, *Appl. Phys. Lett.* **47**, 1131–1333 (1985).
44. V. T. Tikhonchuk and A. A. Zozulya, Structure of light beams in self-pumped four-wave mixing geometries for phase conjugation and mutual conjugation, *Prog. Quantum Electron.* **15**, 231–293 (1991).
45. N. V. Bogodaev, L. I. Ivleva, A. S. Korshunov, A. V. Mamaev, N. N. Poloskov, and A. A. Zozulya, Geometry of a self-pumped passive ring mirror in crystals with strong fanning, *J. Opt. Soc. Am. B* **10**, 1054–1059 (1993).
46. G. Pauliat, M. Ingold, and P. Günter, Analysis of the build-up of oscillations in self-induced photorefractive light-resonators, *IEEE J. Quantum Electron.* **25**, 201–207 (1989).
47. L. Mager, G. Pauliat, D. Rytz, and G. Roosen, Two- and four-wave mixing in photorefractive materials for dynamic correction of pulsed laser beams, in *Novel Optical Material & Applications*, I. A. Khoo, F. Simoni, and C. Umeton (eds.), Wiley, New York (1996), Chapter 6 pp. 149–174. See also L. Mager, C. Lacquarnoy, G. Pauliat, M. H. Garrett, D. Rytz, and G. Roosen, High quality self-pumped phase conjugation of nanosecond pulses at 532 nm using photorefractive $BaTiO_3$, *Opt. Lett.* **19**, 1508–1510 (1994).
48. M. Cronin-Golomb and C. D. Brandle, Ring self-pumped phase conjugator using total internal reflection in photorefractive strontium barium niobate, *Opt. Lett.* **14**, 462–464 (1989).
49. N. Huot, J. M. C. Jonathan, G. Pauliat, D. Rytz, and G. Roosen, Self-pumped phase conjugate $BaTiO_3$:Rh ring mirror at 1.06 μm: Optimization of reflectivity, rise time and fidelity, *SPIE* **3470**, 8–15 (1998).
50. W. Koechner, *Solid-State Laser Engineering*, Springer-Verlag, Berlin (1988).
51. J. D. Foster and L. M. Osterink, Thermal effects in Nd:YAG laser, *J. Appl. Phys.* **41**, 3656–3663 (1970).
52. W. C. Scott and M. de Witt, Birefringence compensation and TEM_{00} mode enhancement in a Nd:YAG laser, *Appl. Phys. Lett.* **18**, 3–4 (1971).
53. Q. Lü, N. Kügler, H. Weber, S. Dong, N. Müller, and U. Wittrock, A novel approach for compensation of birefringence in cylindrical Nd:YAG rods, *Opt. Quantum. Electron.* **28**, 57–69 (1996).
54. I. D. Carr and D. C. Hanna, Performance of a Nd:YAG oscillator/amplifier with phase-conjugation via stimulated Brillouin scattering, *Appl. Phys. B* **36**, 83–92 (1985).
55. N. G. Basov, V. F. Efimkov, I. G. Zubarev, A. V. Kotov, S. I. Mikhailov, and M. G. Smirnov, Inversion of wavefront in SMBS of a depolarized pump, *JETP Lett.* **28**, 197–201 (1978).
56. H. J. Eichler, A. Haase, R. Menzel, and A. Siemoneit, Thermal lensing and depolarization in a highly pumped Nd:YAG laser amplifier, *J. Phys. D* **26**, 1884–1891 (1993).
57. T. Y. Fan, Heat generation in Nd:YAG and Yb:YAG, *IEEE J. Quantum Electron.* **29**, 1457 (1993).

58. D. A. Rockwell, A review of phase-conjugate solid state lasers, *IEEE J. Quantum Electron.* **24**, 1124–1140 (1988).
59. L. Grossard, A. Desfarges-Berthelemot, B. Colombeau, and C. Froehly, Iterative reconstruction of thermally induced phase distorsion in a Nd^{3+}:YVO_4 laser, *J. Opt. A: Pure Appl. Opt.* **4**, 1–7 (2002).
60. B. Y. Zel'dovich, V. I. Popovichev, V. V. Ragul'skii, and F. S. D. Faizullov, Connection between the wave fronts of the reflected and exciting light in stimulated Mandel'shtam–Brillouin scattering, *JETP Lett.* **15**, 109–113 (1972).
61. O. Y. Nosach, V. I. Popovichev, V. V. Ragul'skii, and F.S.D. Faizullov, Cancellation of phase distortions in an amplifying medium with a "Brillouin mirror," *JETP Lett.* **16**, 435–438 (1972).
62. N. Huot, J. M. C. Jonathan, G. Pauliat, G. Roosen, A. Brignon, and J. P. Huignard, Nd:YAG oscillator power amplifier using $BaTiO_3$:Rh internal loop and ring self-pumped phase conjugate mirrors, *Technical Digest CLEO Europe '98*, Paper CWO2, Glasgow, 14–18 September 1998.
63. A. Brignon, S. Senac, J. L. Ayral, and J. P. Huignard, Rhodium-doped barium titanate phase-conjugate mirror for an all-solid-state, high-repetition-rate, diode-pumped Nd:YAG master-oscillator power amplifier laser, *Appl. Opt.* **37**, 3990–3995 (1998).
64. K. Tei, F. Matsuoka, M. Kato, Y. Maruyama, and T. Ariwasa, Nd:YAG oscillator–amplifier system with a passive ring self-pumped phase-conjugate mirror, *Opt. Lett.* **25**, 481–483 (2000).
65. A. Brignon, J. P. Huignard, I. Mnushkina, and M. H. Garrett, Spatial beam clean-up of a Nd:YAG laser operating at 1.06 μm with two-wave mixing in Rh:$BaTiO_3$, *Appl. Opt.* **36**, 7788–7793 (1997).
66. A. Brignon, Y. Louyer, J. P. Huignard, and E. Lallier, Beam clean-up of a multimode Yb-doped fiber amplifier with an infrared sensitive Rh:$BaTiO_3$ crystal, *Technical Digest CLEO 2001*, Paper CTuQ7, Baltimore, 6–10 May 2001.
67. A. Tomita, Phase conjugation using gain saturation of a Nd:YAG laser, *Appl. Phys. Lett.* **34**, 463–464 (1979).
68. P. Sillard, A. Brignon, and J. P. Huignard, Loop resonators with self-pumped phase conjugate mirrors in solid-state saturable amplifiers, *J. Opt. Soc. Am. B* **14**, 2049–2058 (1997).
69. K. S. Syed, G. J. Crots, and M. J. Damzen, Transient modelling of a self-starting holographic laser oscillator, *Opt. Commun.* **146**, 181–185 (1998).
70. P. Aubourg, J. P. Huignard, M. Hareng, and R. A. Mullen, Liquid crystal light valve using bulk monocrystalline $Bi_{12}SiO_{20}$ as the photoconductive material, *Appl. Opt.* **21**, 3706–3712 (1982).
71. J. C. Chanteloup, B. Loiseaux, J. P. Huignard, and H. Baldis, Detection and correction of the spatial phase of ultrashort laser pulses using an optically addressed light valve, *Technical Digest, CLEO '97*, Paper CThD8, Baltimore 1997.
72. S. A. Chetkin and G. V. Vdovin, Deformable mirror correction of a thermal lens induced in the active rod of a solid state laser, *Opt. Commun.* **100**, 159–165 (1993).
73. Special Issue *Opt. Eng.* **36** (1997). See also M. A. Vorontsov, G. W. Carhart, D. V. Pruidze, J. C. Ricklin, and D. G. Voelz, Adaptative imaging system for phase-distorded extended source and multiple-distance objects, *Appl. Opt.* **36**, 3319–3328 (1997).

74. R. A. Rakuljic, K. Sayano, A. Agranat, A. Yariv, and R. R. Neurgaonkar, Photorefractive properties of Ce- and Ca-doped SBN, *Appl. Phys. Lett.* **43**, 1465–1467 (1988).
75. A. Gerwens, M. Simon, K. Buse, and E. Krätzig, Activation of cerium-doped strontium-barium niobate for infrared holographic recording, *Opt. Commun.* **135**, 347–351 (1997).
76. S. G. Odoulov, A. N. Shumelyuk, U. Hellwig, R. A. Rupp, and A. A. Grabar, Photorefractive beam coupling in tin hypothiodiphosphate in the near infrared, *Opt. Lett.* **21**, 752–754 (1996).
77. S. G. Odoulov, A. N. Shumelyuk, U. Hellwig, R. A. Rupp, A. A. Grabar, and I. M. Stoyka, Photorefraction in tin hypothiodiphosphate in the near infrared, *J. Opt. Soc. Am. B* **13**, 2352–2360 (1996).
78. S. G. Odoulov, A. N. Shumelyuk, G. A. Brost, and K. M. Magde, Enhancement of beam coupling in the near infrared for tin hypothiodiphosphate, *Appl. Phys. Lett.* **69**, 3665–3667 (1996).
79. C. Medrano, M. Zgonik, I. Liakatas, and P. Günter, Infrared photorefractive effect in $KNbO_3$ crystals, *J. Opt. Soc. Am. B* **13**, 2657–2661 (1996).
80. M. Ewart, R. Ryf, C. Medrano, H. Wüest, M. Zgonik, and P. Günter, High photorefractive sensitivity at 860 nm in reduced rhodium-doped $KNbO_3$, *Opt. Lett.* **22**, 781–783 (1997).

CHAPTER 9

Spatial and Spectral Control of High-Power Diode Lasers Using Phase Conjugate Mirrors

PAUL M. PETERSEN, MARTIN LØBEL, and SUSSIE JUUL JENSEN
Optics and Fluid Dynamics Department, Risø National Laboratory,
DK-4000 Roskilde, Denmark

Laser diode arrays can produce impressive amounts of optical power, and they are attractive for their compactness and simplicity of operation. Unfortunately, these lasers tend to oscillate in multiple spatial and longitudinal modes, so their coherence properties are rather poor. We have developed new techniques for improving the temporal and spatial coherence of high-power diode lasers. Our techniques are based on frequency-selective phase conjugate feedback. In contrast to most other reported techniques for improving the coherence of high-power diode lasers, our techniques simultaneously improve both spatial and spectral coherence and allow the laser system to be operated far above its threshold.

In this chapter we show that optical phase conjugation leads to effective feedback that permits precise control of spatial and temporal coherence. We introduce the concept of frequency-selective phase conjugate feedback in two different feedback configurations.

In one configuration, a high-finesse etalon is placed in the external cavity. The etalon forces the high-power diode laser, which has low spatial and temporal coherence when running freely, to operate in a state with high temporal coherence and with the far-field very close to the diffraction limit.

In the other configuration, a grating is inserted into the phase conjugate cavity. In this case the laser also operates in a state of high spatial and temporal coherence, and its output is tunable over a broad wavelength range. A 100-h stability test showed that both the output power and the center wavelength are extremely stable.

Finally, we discuss how the use of phase conjugate feedback to improve the coherence properties of high-power diode lasers may be extended to yield new high-power blue laser sources using frequency doubling.

Phase Conjugate Laser Optics, edited by Arnaud Brignon and Jean-Pierre Huignard
ISBN 0-471-43957-6   Copyright © 2004 John Wiley & Sons, Inc.

9.1 INTRODUCTION

Semiconductor lasers are attractive for many applications because they are low in cost, highly reliable, and compact and have a long lifetime. Single-element semiconductor lasers provide spectrally and spatially coherent light at output powers up to 200 mW, a figure that is limited by the onset of optical damage to the output facet of the laser.

Many applications need higher output powers. One way to provide this is to form an array of laser elements, and laser diode arrays are now commercially available with output powers of more than 10 W. Unfortunately, laser diode arrays have a multimode non-diffraction-limited output that limits their usefulness in many applications. The high-power semiconductor lasers are divided into four categories: broad-area lasers, laser arrays, laser bars, and stacked arrays. They all suffer from poor spatial and temporal coherence properties.

There have been many attempts to improve the spatial and temporal coherence of high-power diode lasers. The literature includes references to gain tailoring and special design of the laser diode array [1–7].

Another technique, injection locking using external single-mode laser sources, has significantly increased the output power of single-mode semiconductors [8–14]. Injection locking of a single-mode master laser to a broad area laser has the advantages of producing a diffraction-limited, single-lobe, and single-frequency output. These systems, however, require optical isolation of the master laser, and they are difficult to operate since they need precise control of the temperatures of the master laser and diode laser array.

Other techniques for improving coherence are based on external cavities containing conventional optical components such as mirrors and gratings [15–30]. In applications requiring only high spatial coherence, this approach is robust, is easy to operate, and does not require an additional laser source. The output power from external-cavity systems, however, is still rather low.

An external-cavity design of particular interest has been suggested by Chang-Hasnain et al. [18]. This uses a thin stripe mirror on the back face of a quarter-pitch GRIN lens to select a single spatial mode of the diode laser array. This provides a simple, compact diode laser system with high spatial coherence.

Active feedback using nonlinear optical phase conjugation has also been proposed [31–42]. This kind of adaptive optical feedback provides very effective coupling of the individual diode elements, and may be used to improve the coherence properties of laser diode bars and stacked arrays. In 1993, MacCormack and Feinberg [34] applied phase conjugate feedback from a self-pumped $BaTiO_3$ crystal to obtain near-diffraction-limited output from a 20-element laser diode array. Narrow spectral operation of a GaAlAs diode laser has also been demonstrated using phase conjugate feedback [31]. However, in these cases the optical feedback led to self-induced frequency scanning of the lasing wavelength.

We have shown that prisms or dispersive gratings placed in the external cavity to oppose the material frequency dispersion of the phase conjugate $BaTiO_3$ crystal can eliminate frequency scanning and stabilize the output power [38]. More recently we have used phase conjugate feedback with a frequency-selective element in the external cavity. This feedback scheme eliminates frequency scanning and leads to an extremely stable laser output power with good spatial and temporal coherence properties [37, 39].

9.2 LASER DIODE ARRAYS WITH PHASE CONJUGATE FEEDBACK

The spatial modes of high-power gain-guided laser diode arrays have been described theoretically using coupled wave and nonlinear theories [43–48] and by Verdiell and Frey [49] using a perturbed broad area laser model. In the latter approach, the unperturbed problem is modeled in a manner similar to that of an infinite square-well potential: The field is confined in a square-well potential with a half-width x_0 corresponding to the lateral dimension of the junction. Outside the junction the model assumes infinite absorption, leading to perfect confinement. The spatial dependence of the electric field is given by [49]

$$\psi_m(x) = \frac{1}{x_0^{1/2}} \sin\left(\frac{m\pi x}{2x_0} + \frac{m\pi}{2}\right) \tag{1}$$

where m is the mode number of the higher-order transverse mode and where the x axis is parallel to the junction. The far field of the broad area laser modes is given by the Fourier transform of the near field in Eq. (1):

$$FT\{\psi\} = \frac{\sqrt{x_0}}{2\pi} \left[\exp\left(i\frac{(m-1)}{2}\pi\right) \mathrm{sinc}\left(\frac{m\pi}{2} - k_0 x_0 \sin\theta\right) \right.$$
$$\left. + \exp\left(-i\frac{(m-1)}{2}\pi\right) \mathrm{sinc}\left(\frac{m\pi}{2} + k_0 x_0 \sin\theta\right) \right] \tag{2}$$

where θ is the emission angle with respect to the normal of the laser output facet. The fundamental mode $m = 1$ of the far field given by Eq. (2) is a nearly Gaussian single lobe, while all the higher-order modes ($m > 1$) consist of two lobes in the far field. The radiation angles of the peak intensity of the two lobes are given by

$$\theta_m = \pm \frac{m\lambda}{2k_0 x_0} \tag{3}$$

Equation (3) shows that different laser modes have different radiation angles θ_m. Laser diode arrays operate in a number of spatial modes and can be modeled as a linear combination of the modes given in Eqs. (1) and (2). The full width at half-maximum of the fundamental mode $m = 1$ is obtained for a radiation angle:

$$\theta_{\mathrm{diff}} = \frac{1.189\lambda}{2x_0} \tag{4}$$

This angle is traditionally used to define the diffraction limit for laser diode arrays and broad area lasers.

Since the different spatial modes of a laser diode array radiate in different directions, it is possible to select a given mode using spatial filtering in an external cavity. This can then be used as the basis of a feedback system to improve the spatial coherence of the array.

Adaptive optical feedback from a phase conjugator forms a very effective feedback system, since the conjugator automatically feeds back light to the high-gain regions of the diodes and at the same time dynamically compensates for misalignment of the external cavity. Recently we have introduced the concept of frequency-selective phase conjugate feedback (FSPCF), which significantly improves both the spatial and temporal coherence of high-power laser diode arrays, as well as providing very stable output [42].

Figure 9.1 shows the configuration of the FSPCF. The phase conjugator provides feedback to every laser diode element in the array. Between the diode array and the phase conjugator, a frequency filter and a spatial filter select, respectively, a narrow frequency range and a single spatial mode from the output of the array. The phase conjugator is mounted off-axis, with an angle θ between the axis of the laser diode array and the phase conjugator. This asymmetric arrangement combines high output coupling efficiency with significant enhancement of both the spatial and temporal coherence of the laser output.

The phase conjugate wave from the phase conjugator retraces the output wave from the laser diode array, such that at all points the wavefronts of the phase conjugate wave coincide with those from the laser diode array. This autoretracing property is key to the use of phase conjugate feedback to improve the spatial and temporal coherence of laser diode arrays.

Since the same principle may also be applied to broad area lasers, laser bars, and stacked arrays, phase conjugate feedback can be used with diode laser systems at

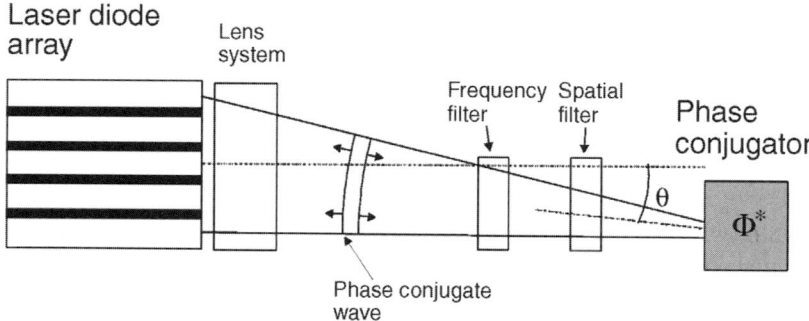

Figure 9.1. The FSPCF system. The phase conjugator provides feedback to the laser diode array. The spatial and temporal coherence are controlled by a frequency filter and a spatial filter in the external cavity.

output powers up to 100 W. External-cavity diode lasers using conventional optics for feedback, in contrast, are limited to a few watts.

Phase conjugate waves are usually generated using four-wave mixing in a nonlinear medium, with one incident probe beam and two external pump beams [50]. Semiconductors [51–62] and photorefractive [63–67] materials are the most promising candidates for phase conjugators used to improve the coherence of high-power diode lasers.

In photorefractive materials, a dynamic index grating is induced by spatial variation of light inside the medium. This photorefractive effect can be used to generate phase conjugate waves. The time constant for formation of the grating structure is typically between 1 ms and several seconds. Without illumination, the index grating decays with a time constant that varies from 1 ms to months, depending on the material. Several photorefractive materials exhibit optical phase conjugation without the need for external pump beams; instead, the pump beams are derived directly from the input beam. This kind of self-pumped phase conjugation is attractive for phase locking of laser diode arrays since it makes the configuration in the external cavity simple and self-starting (Fig. 9.2a).

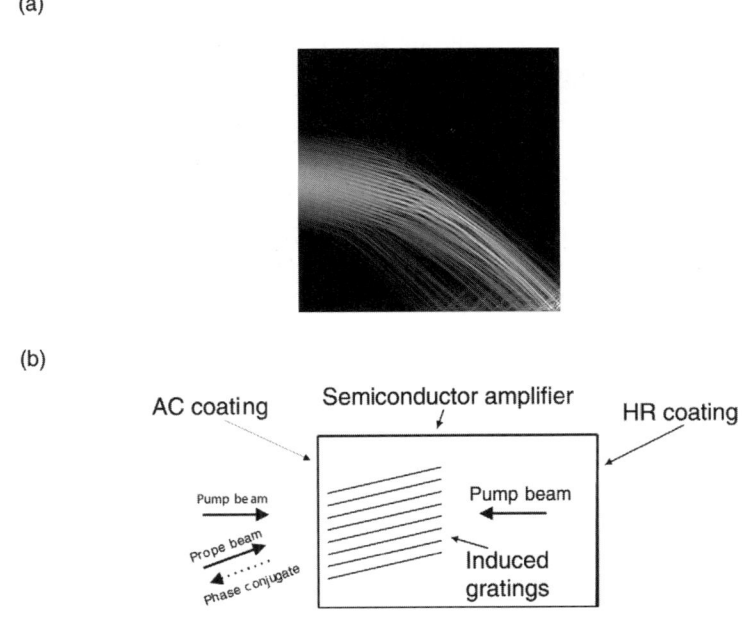

Figure 9.2. (a) Self-pumped phase conjugation in a $BaTiO_3$ crystal. No external pump beams are needed. The pump beams are derived from the probe beam. (From Ref. 40.) (b) Optical phase conjugation by four-wave mixing in a semiconductor amplifier. The backward pump beam is generated by reflection of the forward pump beam from the back facet with high reflective (HR) coating.

Much recent research has concentrated on designing new self-pumped photorefractive materials with high phase conjugate reflectivities that are sensitive at infrared wavelengths. $BaTiO_3$, for instance, gives reflectivities of greater than 70% in the visible and near-infrared parts of the spectrum [63, 64]. Reflectivities of up to 25% have been reported for self-pumped $KNbO_3$ at wavelengths of 500–1000 nm [65], and in the same wavelength range $Sr_{0.6}Ba_{0.4}Nb_2O_6$ (SBN) and $(K_{0.5}Na_{0.5})_{0.2}$-$(Sr_{0.75}Ba_{0.25})_{0.9}Nb_2O_6$ (KNSBN) also show high reflectivities when used as self-pumped phase conjugators [66]. In the infrared, at wavelengths greater than 1100 nm, the semiconductor material InP doped with iron exhibits self-pumped phase conjugation, though the reflectivity is low—up to 10%—and a square-wave voltage has to be applied to the material [67].

Optical phase conjugation may also be done within the semiconductor medium itself. Optical phase conjugation based on four-wave mixing in a semiconductor offers several advantages. The semiconductor has a fast response, and the material can easily be integrated with the laser diode array. Four-wave mixing in single-stripe diode lasers has been investigated in a configuration where the pump beams and the probe beam were injected collinearly into the narrow semiconductor stripe [52, 59].

Broad area lasers with large stripe widths permit optical phase conjugation with a spatially nondegenerate mixing geometry [55, 61]. In this geometry the probe beam is injected at a small angle with respect to the pump beam, and the backward pump beam is generated by reflection from the back facet of the semiconductor (Fig. 9.2b). Using this approach, phase conjugate reflectivities as high as 165% have been obtained [61]. Optical phase conjugate feedback from a broad area laser has been used to stabilize a laser diode [62], reducing its linewidth from 5 MHz to 25 kHz.

9.3 FREQUENCY-SELECTIVE PHASE CONJUGATE FEEDBACK WITH AN ETALON IN THE EXTERNAL CAVITY

In this section we demonstrate a powerful technique that forces a laser diode array, operated at a drive current far above threshold, to oscillate in a single spatial mode with high temporal coherence. The coherence length of the phase-locked output is proven to be increased significantly. To discriminate the longitudinal modes, a Fabry–Perot etalon is included in the external cavity, whereby only a limited number of modes are allowed to interact with the adaptive phase conjugate mirror (PCM). The etalon is the key component in this novel frequency-selective phase conjugate feedback (FSPCF) system. It eliminates frequency scanning and stabilizes the output. When operated with the FSPCF system, all the radiated energy from the freely running array is present in one mode. Furthermore, in an off-axis configuration the far field becomes almost diffraction-limited.

9.3.1 Experimental setup

The principle of operation is shown in Fig. 9.3. The laser diode array is a GaAlAs 10-stripe gain-guided device with a 100-μm-wide emitting junction. It is

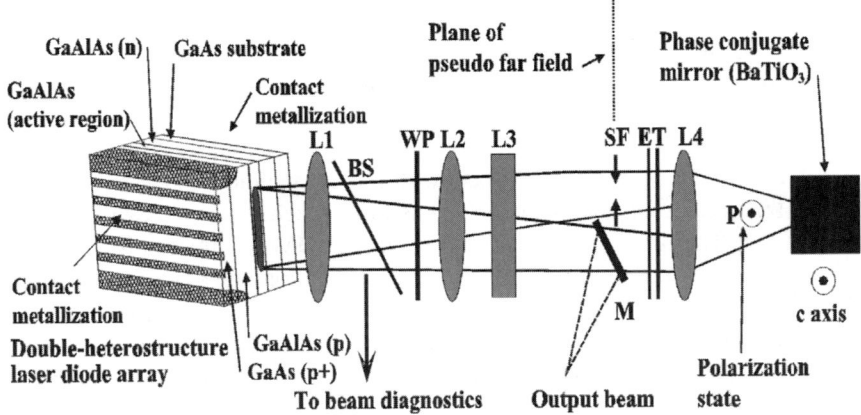

Figure 9.3. Experimental setup. A GaAlAs laser diode array is coupled to a PCM. L1: NA = 0.55, $f = 4.5$ mm; L2: lens $f = 76.2$ mm; L3: cylindrical lens $f = 150$ mm; L4: lens $f = 150$ mm; BS: beam splitter (2° wedge); WP: $\lambda/2$ wave plate (zero-order); ET: Fabry–Perot etalon; SF: spatial filter (two razor blades mounted on translation stages); M: removable mirror. (From Ref. 39.)

temperature-controlled by a peltier element. The array has a threshold of $i_{th} = 0.28$ amp and a maximum output power of 0.5 W at $3.2 i_{th} = 0.9$ amp. The center wavelength of the spectrum at 25° is 815 nm. The output beam of the array is collimated with a lens (L1) that has an effective focal length of 4.5 mm and a numerical aperture of 0.55. A spherical singlet with a focal length of 76.2 mm is used to generate a pseudo far field (image of Fourier plane of lens L1) at a distance of 475 mm from the array. In the plane of the pseudo far field, spatial filtering is applied. The spatial filter consists of two parallel razor blades mounted on two translation stages, and only radiation in-between the razor blades is allowed to pass the filter. A cylindrical lens with a focal length of 150 mm is used to collimate the beam in the transverse direction of the array. A spherical singlet with a focal length of 150 mm is used to focus the light and generates a 2-mm beam spot size at the face of the phase conjugator made up of a rhodium-doped $BaTiO_3$ crystal. The crystal is arranged in the self-pumped Cat configuration. Both a 45°- and a 0°-cut crystal have been used in the experiments. All lenses have a broadband antireflection coating ($R < 1\%$) in order to minimize the loss of the external cavity. A coupling loss of approximately 20% occurred between the array and lens L1, and an additional 10% is lost in the beam splitter, the wave plate, and the lenses (L2 and L3). The external cavity has a length of 775 mm. To obtain the largest photorefractive response, the high coherence axis, the polarization, and the c axis of the crystal must lie in the same plane, a requirement that can be accomplished if we rotate the beam polarization 90° while keeping the coherence axis fixed. This is done by rotating the array 90° around its optical axis and inserting a half-wavelength wave plate in the path in between the array and the PCM. In between the array and the PCM a solid etalon is placed. The etalon is a spectral filter that allows only certain resonance frequencies to pass and, subsequently, to interact with the adaptive PCM. Lens L3 is included to ensure the highest collimation upon passing the etalon. Two different etalons are tested in our experiments. The first etalon (etalon I) has a thickness of 300 μm with a finesse of approximately 17. The free spectral range is 0.75 nm at 814 nm (or 350 GHz). The FWHM bandwidth of the etalon is 0.04 nm at 814 nm (or 20 GHz). The second etalon (etalon II) has a free spectral range of 225 GHz and a finesse of 2.6. The full width at half-maximum (FWHM) bandwidth is 0.19 nm at 814 nm (or 86 GHz). Etalon I is used in the configuration unless otherwise stated.

We consider two different feedback configurations: (i) the on-axis configuration and (ii) the off-axis configuration. In the off-axis configuration a removable mirror (M) is inserted at the position of the generated pseudo far field, and it couples out one-half of the radiated far-field pattern. The beam reflected from the mirror is the output beam of the system. In the on-axis configuration the mirror and the spatial filter (SF) are removed and all the energy of the array is directed toward the PCM.

9.3.2 Characteristics of the on-axis configuration

First we consider the on-axis (twin-lobe) configuration. The etalon is a frequency filter and will only pass a limited number of frequencies that can subsequently interact with the phase conjugator.

When the crystal is illuminated, the reflectivity of the phase conjugator increases and the spectrum is narrowed down significantly. Figures 9.4a and 9.4b display the spectrum of the laser array when it is phase-locked due to the frequency-selective phase conjugate feedback (FSPCF) and when it runs freely, respectively. The drive current is two times the threshold current ($2i_{th}$), and at this level it could take several minutes before the phase conjugator had built up and locking would be established. In Fig. 9.4a it is seen that the cooperative interaction among the PCM, the etalon, and the array forces the spectrum to narrow down substantially. The bandwidth is measured to be reduced from a full width at half-maximum (FWHM) of 0.7 nm to less than 0.02 nm (resolution-limited). The etalon used (etalon I) has a bandwidth of 0.04 nm. The mode spacing between two adjacent spatial array modes is on the order of 0.02 nm [68, 69], and even the closest spaced array modes therefore have sufficiently different transmission losses in the etalon. A single spatial array mode can consequently be selected if the maximum transmission of the etalon is tuned to a frequency that matches an array mode with high gain. The tuning can be achieved by tilting the etalon with respect to the optical axis.

Figure 9.5a shows the far field of the array in the on-axis feedback configuration using etalon I. When the array runs freely, as seen in Fig. 9.5b, the far field consists of several spatial modes and the radiation is almost uniform within $\theta = -2°$ to $\theta = 2°$. However, when the array operates in one single spatial mode, the far field changes to a twin lobe, as seen in Fig. 9.5a. Only spectral filtering is applied in the on-axis configuration. Since the bandwidth of etalon I is comparable with spatial mode spacing (the spectral filtering is very selective), single spatial mode operation can still be achieved using the phase conjugate frequency-selective feedback.

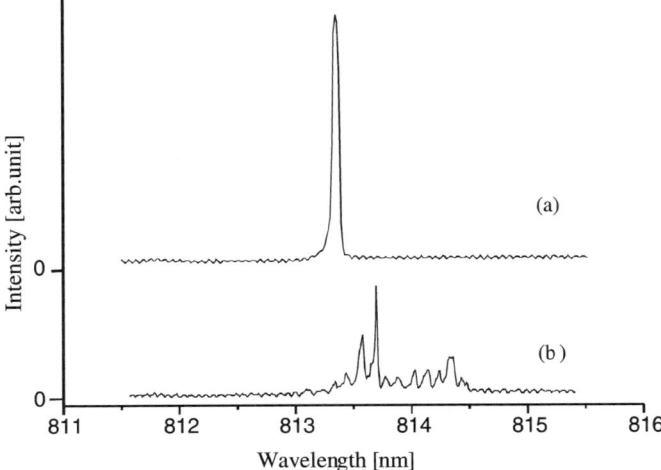

Figure 9.4. Spectrum from laser diode array. (a) Laser array is exposed to FSPCF at a drive current of $2i_{th}$ (on-axis configuration with etalon I). (b) Laser array runs freely with no feedback at a drive current of $2i_{th}$. (From Ref. 39.)

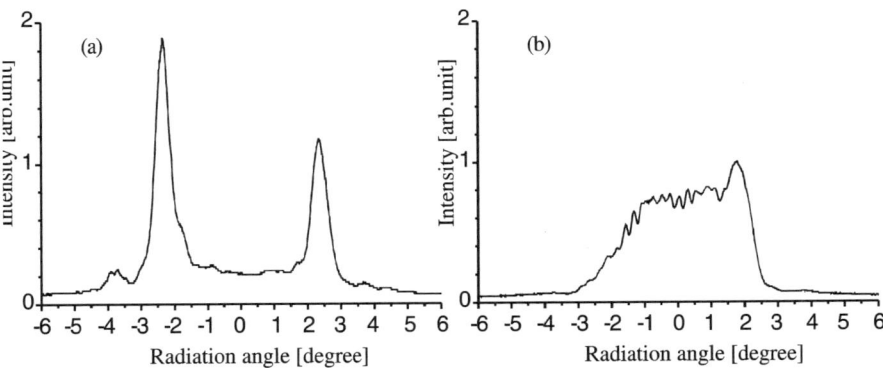

Figure 9.5. Far-field pattern of the laser diode array in the lateral direction. (a) Laser array is exposed to FSPCF at a drive current of $2i_{th}$ (the on-axis configuration with etalon I). (b) Laser array runs freely with no feedback at a drive current of $2i_{th}$.

9.3.3 Far-field spatial characteristics in the off-axis configuration

In the off-axis (single-lobe) configuration the far-field pattern is narrowed down significantly. Feeding back only one of the two lobes is somewhat similar to the previously reported self-injection locking techniques where a photorefractive PCM is used [32, 34]. The highest degree of brightness of the output of the system is obtained with the spatial filter and the etalon and when only one single lobe is fed back to the laser diode array. From an experimental and practical point of view, this configuration is attractive since a large fraction of the radiated energy can be extracted from the laser system.

As the reflectivity of the PCM is built up, the output far-field lobe grows at the expense of the injection lobe. To achieve single-lobe operation, it is often necessary to open the spatial filter slightly so that more energy will reach the PCM in order to start the self-pumped conjugator. After a few minutes the distance between the two razor blades can then be diminished again. Figure 9.6a shows the far field of the laser array when it runs freely at a drive current of $2i_{th}$. Figures 9.6b and 9.6c display the far field for a drive current of $2i_{th}$ and $3i_{th}$, respectively, when the FSPCF is applied.

In the plane perpendicular to the junction the far field is close to the diffraction limit and has a Gaussian shape that is unaffected by the number of oscillating spatial array modes. It is the lobe at the negative angles (the injection lobe) that is fed back to the laser diode array. The peak of the output lobe is centered at 2.3° and 2.2° for $2i_{th}$ and $3i_{th}$, respectively. The most efficient performance of the system always takes place with an emission around 2.3°. The edges of the two razor blades that form the spatial filter are positioned at $-1.9°$ and $-2.5°$. The FWHM of the output beam is measured to 0.75° at $2i_{th}$ and 0.92° at i_{th}, corresponding to 1.4 and 1.7 times the diffraction limit, respectively. The total power radiated from the laser diode array is increased by approximately 5% with the phase conjugate feedback compared with the power of the freely running laser diode array.

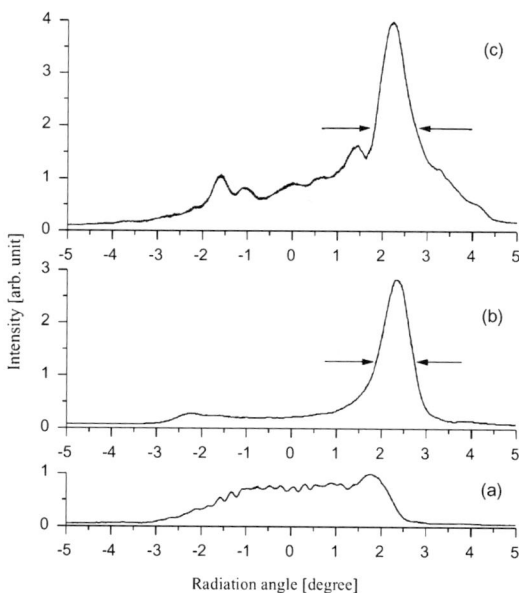

Figure 9.6. Far-field pattern of the laser diode array exposed to FSPCF in an of-axis self-injection locking configuration using an etalon in the external cavity. (a) Laser array runs freely at a drive current of $2i_{th}$. (b) Feedback applied at $2I_{th}$. (c) Feedback applied at $3I_{th}$.

The measured power of the output beam is 107 mW and 227 mW for a drive current of $2i_{th}$ and $3i_{th}$, respectively. When the laser diode array runs freely, the total power after the cylindrical lens (L3) is 147 mW and 320 mW for $2i_{th}$ and $3i_{th}$, respectively. At $2i_{th}$ and $3i_{th}$ and when the FSPCF is applied, the total radiated energy from the array (before lens L1) is 218 mW and 465 mW, respectively. More than 70% of the available energy after lens L3 is contained in the output beam, and 50% of the total radiated power is contained in the output beam. However, as can be seen from the profile in Fig. 9.6b, the output lobe (the positive lobe) holds more than 80% of the total radiated far-field power. Based on this profile, it is estimated that 80% of the radiated energy can be contained in the output beam, provided that transmission losses in the lenses and other optical components are eliminated.

The amount of feedback from the phase conjugator measured at the beam splitter is typically in the range of 0.4–1%. The highest feedback is obtained for low values of the current. The phase conjugate reflectivity of the PCM ranges from 12% to 15% for all drive currents. The amount of power that was fed back into the array is 0.5–1.4 mW (highest at lower current). At a drive current of $2i_{th}$ the total radiated output power is 218 mW, corresponding to an amplification of more than 21 dB of the power fed back into the array.

The far field shown in Fig. 9.6b can also be obtained if the etalon is removed from the external cavity. The lack of spectral filtering will, however, lead to the oscillations in several longitudinal modes.

9.3.4 The improvement of the spatial brightness

In many applications it is important that the laser beam can be focused to a small spot at the size of a wavelength. Due to the high number of spatial modes that are oscillating simultaneously in conventional high-power laser diode arrays, the laser beam cannot be focused to a small spot size. In this section it is tested how tight the output beam of the laser diode array with external phase conjugate feedback may be focused. The output beam has a size of 8×1 (mm)^2 (along the high and the low coherence axes of the laser array, respectively) measured at mirror M (see Fig. 9.3). The beam is collimated along the high coherence axis, but is slightly diverging along the low coherence axis. The output beam is expanded along the low coherence axis using two cylindrical lenses placed with a separation of 0.70 m. The two lenses have focal lengths of −60 mm and +800 mm, respectively. The output beam has a size of 10×10 mm^2 measured in the plane corresponding to the last cylindrical lens. Just after this cylindrical lens, the laser beam is focused using an achromat with a focal length of 40 mm. This is a typical focal length used when, for example, the output from diode arrays is applied to pump solid-state lasers. The measured spot size (full width at $1/e^2$) for the output beam is 11.7×11.8 (μm)^2. The measured spot is shown in Fig. 9.7.

The smallest spot size (full width at $1/e^2$) that can be achieved using a Gaussian beam is $4\lambda f/(\pi D)$, where λ is the wavelength, f is the focal length, and D is the beam spot size at a distance f before the lens. Inserting $D = 10$ mm and $f = 40$ mm gives a theoretical spot size of 4.1 μm. Comparison with the 11.7-μm size obtained in our experiments indicates that the output beam is of high quality yielding close to

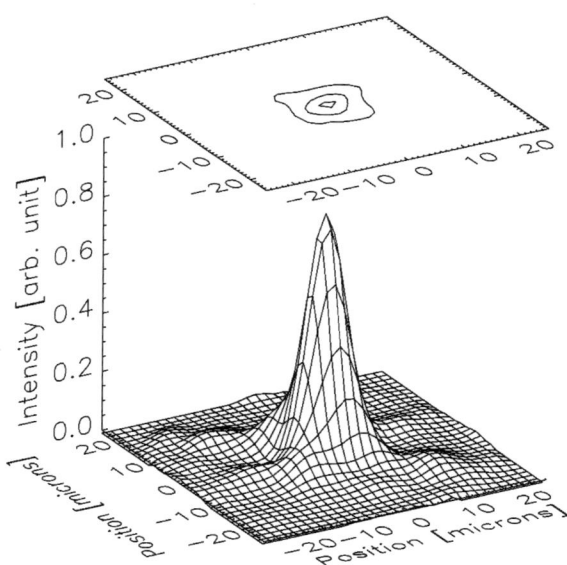

Figure 9.7. Minimum spot size of the output beam focused by a lens with a 40-mm focal length. The drive current is $2i_{th}$.

diffraction-limited performance. Moreover, the measured output spot size can be compared with the smallest spot size that can be achieved with conventional lenses, prisms, and so on, when attempts are made to focus the total radiation pattern of a freely running laser diode array. For the SDL-2432 laser diode array the smallest nearly circular spot size reported [70] was 90×100 $(\mu m)^2$ with a 40-mm focal length lens placed as the last lens in the configuration; this corresponds to an intensity in the focused beam spot that is more than 50 times smaller than the results presented here. Therefore, in comparison with the freely running diode laser, the intensity of the phase conjugate laser system in the focal point is increased significantly. The beam from the phase conjugate laser system may be focused to an even smaller spotsize using a lens with shorter focal length than 40 mm, and therefore the intensity in the focal point could be increased further.

9.3.5 Spectral characteristics of the laser system

In this section the characteristics of the spectrum and the coherence length obtained with the off-axis configuration are presented and discussed.

Figures 9.8a and 9.8b display a frequency-resolved near field of the output facet of the laser diode array when it runs freely and when the FSPCF is applied, respectively, for a drive current of $2i_{th}$. The near fields are obtained by imaging the near field of the array using the reflection from the beam splitter onto the input slit of a spectrum analyzer. The output slit of the spectrum analyzer is replaced with a two-dimensional photodetector array. The experimental setup is similar to the

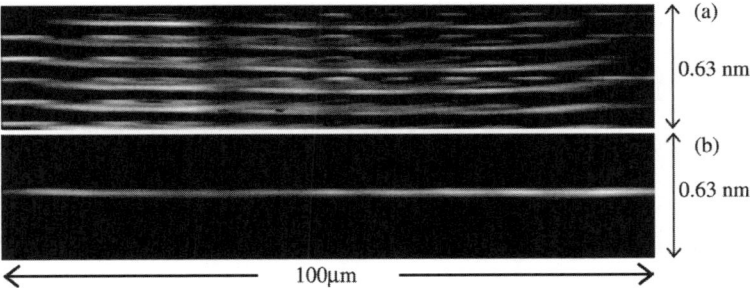

Figure 9.8. Spectrally resolved near-field of array. Spectrally resolved near-field of laser diode array exposed to off-axis phase conjugate feedback. Spectral and spatial filtering is applied in the external cavity. The drive current is $2I_{th}$. (a) Part of spectrum when laser array runs freely. Five clusters of array modes (five longitudinal modes) are identified. (b) The feedback causes the spectrum to narrow down to one single array mode (single-mode operation). The bandwidth (FWHM) is less than 0.02 nm. (From Ref. 39.)

configuration presented in Ref. 68. With this setup, all longitudinal and spatial modes of the laser array can be resolved. In Fig. 9.8a, five clusters of array modes are identified (five longitudinal modes). Some of the array modes do not extend over the full width of the junction. When the phase conjugator is turned on, the spectrum of the output narrows down to a single array mode. The bandwidth is measured to be less than 0.02 nm (resolution limited). For a drive current of $3i_{th}$ the spectrum broadens slightly and a few adjacent array modes start to oscillate. The bandwidth is measured to be $\Delta\lambda_{fwhm} = 0.1$ nm.

If the etalon with low finesse (etalon II) is used with the off-axis configuration, single-mode operation is still obtained at $2i_{th}$. This is due to the fact that for etalon II, only one longitudinal mode (a cluster of array modes) can be selected. The spatial filtering then selects one array mode from this cluster, as a result, single-mode operation can be obtained.

9.3.6 The improvement of the temporal coherence

The coherence length is measured by directing the output beam toward a standard Michelson interferometer that consists of two mirrors and a beam splitter. One of the arms is mounted on a translation stage. The reference arm has a fixed length of 2×110 mm. The visibility of approximately five fringes of the interference pattern is recorded by a 25-mm-wide photo array with 2048 pixels. Figure 9.9 presents the coherence degree versus the path difference of the two arms in the interferometer for different drive currents and when the FSPCF is not applied. The coherence degree is taken in the form $\gamma = ((I_{max} - I_{min})/(I_{max} + I_{min})) \times ((I_1 + I_2)/(2(I_1 I_2)^{1/2}))$, where I_{max} and I_{min} are the maximum and the minimum intensity in the interference pattern; and I_1 and I_2 are the intensity generated at the detector with the mirror in arms 2 and 1 blocked, respectively. By comparing the path differences for the case where the array runs freely and the case where the FSPCF is applied, it is

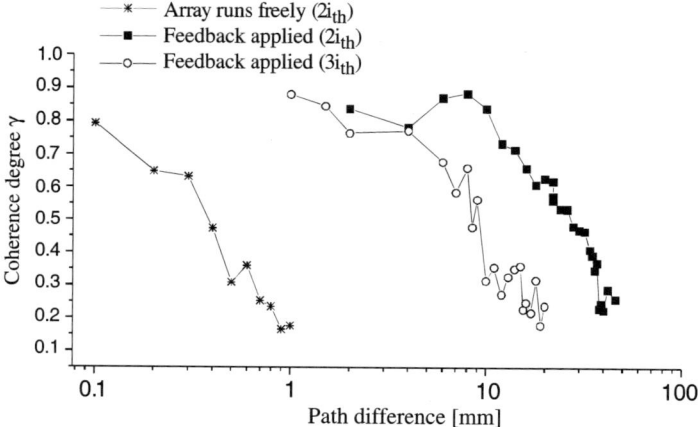

Figure 9.9. Coherence degree versus the path difference (etalon). The coherence degree versus the path difference of the arms in a standard Michelson interferometer. (From Ref. 39.)

observed that the coherence length is increased by a factor of 70. At a drive current of $3i_{th}$ the coherence length is increased by a factor of 40. As can be seen in the figure, the FSPCF increases the coherence length to at least 25 mm. However, if the coherence length is defined as $L_c = \lambda^2/\Delta\lambda_{fwhm}$, where $\Delta\lambda_{fwhm}$ is the FWHM bandwidth and λ is the lasing wavelength, the coherence length is calculated to be $L_c = 1$ mm when the laser diode array runs freely at a drive current of $2i_{th}$ (the bandwidth is 0.7 nm when the array runs freely); the phase conjugate feedback increases the coherence length to 70 mm. At a drive current of $3i_{th}$ the coherence length is increased by a factor of 40 to 22 mm. The coherence length may be further improved if we apply better antireflection coatings to the output facet of the laser diode array.

9.4 TUNABLE OUTPUT OF HIGH-POWER DIODE LASERS USING A GRATING IN THE EXTERNAL CAVITY

In this section we demonstrate how a laser diode array can be forced to oscillate in a single spatial mode with high temporal coherence. The laser diode array is driven with a drive current far above threshold, and the wavelength is widely tunable. Single longitudinal mode operation of a GaAlAs laser diode array operated with a drive current below threshold has been demonstrated by coupling it to an external cavity formed by a diffraction grating and a photorefractive phase conjugate mirror [71]. This system has proven to be widely tunable with respect to the frequency due to the presence of the diffractive grating. If operated far above threshold, this system is, in general, nonstable due to nonlinear processes in the photorefractive phase conjugate mirror (PCM), and self-induced frequency scanning can take place. In order to have a stable system, only one-half of the far field must be retroreflected back into the laser array.

Using the off-axis configuration, we have included a spatial filter in the external cavity that allows only a limited number of spatial modes to interact with the adaptive PCM. Before the output beam from the array reaches the PCM, it is diffracted in a grating. The angular dispersion of the grating makes the feedback system frequency selective. The PCM and the grating form the frequency-selective phase conjugate feedback (FSPCF) system that forces the laser array to oscillate in one single longitudinal mode. The interaction among the array, the dynamic gratings of the adaptive PCM, the spatial filter, and the diffractive grating enables a phase locking of the laser array and causes all the radiated energy to be transferred into one single mode. The coherence length of the phase-locked output is increased significantly. This technique allows a large fraction of the radiated energy to be extracted from the laser system with a single-lobe far-field pattern that is close to the diffraction limit. The center frequency of the output of the array can be tuned more than 5 nm by tilting the grating.

The experimental setup is shown in Fig. 9.10. It is almost similar to the setup with the etalon in the external cavity described previously. In the grating configuration, however, the frequency-selective element is a diffraction grating instead of an etalon. The laser diode array is a GaAlAs 10-stripe array described previously.

The output beam of the laser array is collimated with a lens (L1), and a second lens (L2) generates a pseudo far field at a distance of 400 mm from the array. The collimated light is focused with a lens (L3) that generates a 2-mm spot at the face of the PCM made up of a 0°-cut rhodium-doped (800 ppm) $BaTiO_3$ crystal in which a phase conjugation process takes place and returns the phase conjugate wavefront toward the array. The crystal has been cut along its crystallographic axes and measures $5.11 \times 5.53 \times 5.33$ mm^3 ($a \times a \times c$ axis). The crystal is arranged in a self-pumped Cat configuration with an angle of incidence of 60°. The high coherence axis is in the $a-c$ plane of the crystal in order to obtain the highest photorefractive response. A 2° wedge is used as beam splitter. The two reflections are used for beam diagnostics: One reflection is used for monitoring the pseudo far field generated with a lens, while the other one is directed to a spectrometer with a resolution of 0.02 nm. A coupling loss of approximately 20% occurs between the array and the collimating lens (L1), and a 10% loss takes place at the beam splitter, the wave plate, and the lenses. To avoid self-induced frequency scanning and to obtain a high output coupling efficiency, only one of the two far-field lobes is directed to the PCM; a mirror is placed at the position of the generated pseudo far field to pick out one-half of the radiated far-field pattern. The beam reflected off this mirror is the output beam of the system. The lobe that is directed to the PCM is diffracted in a 1200-line/mm ruled grating before it enters the $BaTiO_3$ crystal. The angular orientation of the grating is controlled by a piezoelement. The grating adds angular dispersion to the system and has strong influence on the dynamic behavior and the optical bandwidth of a multimode laser diode array when coupled to a self-pumped PCM. The interaction between the angular dispersion of the grating and the dynamic gratings formed in the $BaTiO_3$ crystal makes the external feedback system highly frequency-selective. The different array modes are discriminated by a spatial

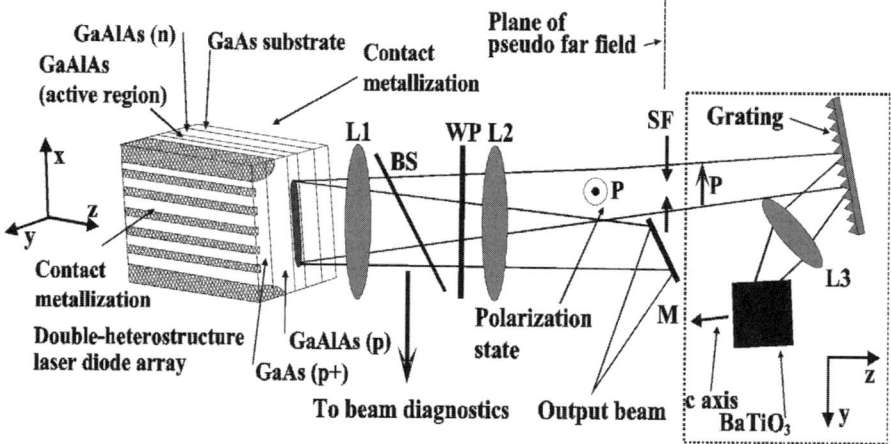

Figure 9.10. Experimental setup. A gain-guided GaAlAs laser diode array coupled to a PCM. L1: collimating lens, NA = 0.55, $f_1 = 4.5$ mm; L2: plano-convex cylindrical lens $f_2 = 60$ mm; L3: spherical singlet lens $f_3 = 100$ mm; BS: beam splitter (2° wedge); WP: half-wave plate (zero-order) at 815 nm; grating: 1200 lines/mm ruled grating with Blaze angle of 26.4° (750 nm); SF: spatial filter (two razor blades mounted on translation stages); P: indication of polarization (E-field). The components in the dotted box shall be rotated 90° around the z axis of the array. (From Ref. 37.)

filter placed in the plane of the generated pseudo far field. The spatial filter (SF) is formed by two razor blades mounted on translation stages, and by adjusting the position of them the number of spatial modes that can interact with the PCM can be controlled.

Spectral characteristics of the output beam Figure 9.11 shows the optical spectra of the output beam of the array under different conditions. Figure 9.11a displays the spectrum when the array runs freely. At a drive current of $2i_{th}$, several longitudinal modes are present and the FWHM bandwidth is 0.7 nm. Figures 9.11b–e present the spectra of the output beam at the same drive current when the FSPCF is applied, but for different tilts of the grating. Once the PCM has been built up and the phase locking has been established, the frequency can be tuned by tilting the grating. Between the recordings of Figs. 9.11b and 9.11e, the grating has been tilted 0.43°, which gives a sensitivity of the tunability of 12 nm/degree. A change in the tilt of the grating leads to a change in the angle of incidence at the air–crystal interface and, moreover, to a change of the position at the crystal surface at which the beam will enter. If the grating is tilted, the array will optimize the oscillating frequency for the best Bragg match to the existing gratings in the crystal and, thereby, again obtain high reflectivity from the PCM. By tilting the grating, the frequency is tuned by discrete steps corresponding to a longitudinal mode spacing of the array (0.11 nm); however, two longitudinal modes can oscillate simultaneously since the energy from

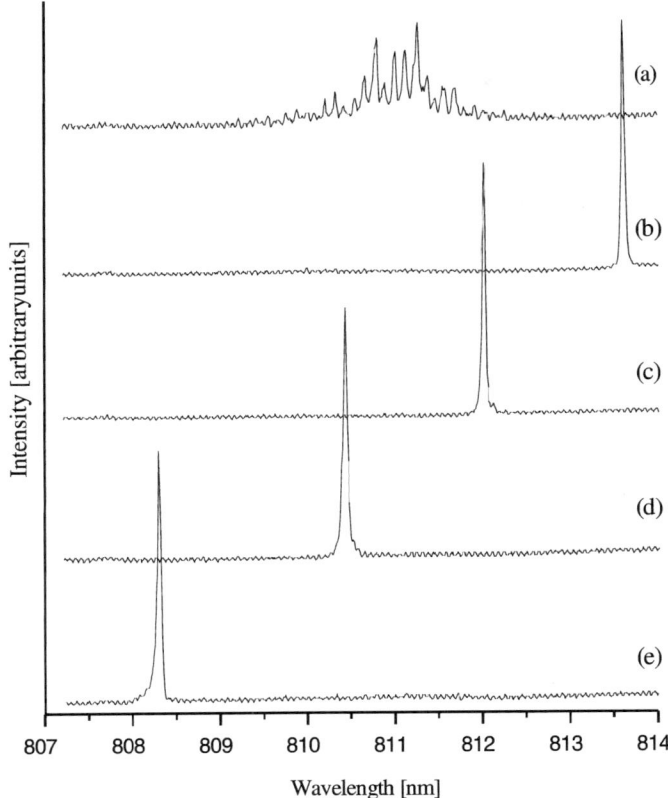

Figure 9.11. Laser diode array coupled with a grating and a PCM. Optical spectrum at a drive current of $2i_{th}$: (a) The array runs freely. (b) The FSPCF is applied. The angle of incidence at the grating is 19.73°. Bandwidth (FWHM) of 0.03 nm. (c) Same as (b), but angle of incidence is 19.60°. (d) Same as (c), but angle of incidence is 19.47°. (e) Same as (d), but angle of incidence is 19.30°. (From Ref. 37.)

one longitudinal mode is slowly transferred to the next mode as the grating is tilted. When the tilting of the grating stops, the adaptive PCM will adjust and optimize for the highest reflectivity, and within seconds the single-mode operation will again be obtained. For a fixed grating position the frequency can be tuned continuously over a range of 0.1 nm, corresponding to one longitudinal mode spacing by changing the temperature of the junction of the array less than 1°C. By adjusting the grating and the temperature of the junction, any wavelength within a 5-nm range can be achieved. The gain bandwidth of the GaAlAs array is huge, so the limited wavelength range may be explained as follows: The array and the PCM are only weakly coupled, and the PCM cannot suppress the oscillation of the natural modes of the array cavity if the spectrum is forced too far away by tilting the grating. The tuning range of 5 nm is similar to the observed range for the off-axis self-injection locking scheme where conventional feedback from a diffraction grating is used. This does indeed suggest that the limited tuning range has to do with the injection locking process and not the properties of the phase conjugator.

For a fixed grating tilt and for a drive current of $2i_{th}$ the FWHM bandwidth of the single-mode spectrum is measured to be less than 0.03 nm. By recording the frequency-resolved near-field pattern with a setup similar to the one in Ref. 68, it is verified that the array only operates in one array mode. As the drive current is increased to $3i_{th}$, however, additional array modes start to oscillate and increase the bandwidth to 0.1 nm.

9.5 STABILITY OF THE OUTPUT OF DIODE LASERS WITH EXTERNAL PHASE CONJUGATE FEEDBACK

In this section we discuss the stability of the output of the diode laser with external phase conjugate feedback. Both the long-term stability of the laser output and the sensitivity to feedback generated by external reflection of the output are examined. The output power and the center wavelength are found to be extremely stable in a 100-h experiment. External feedback of the output beam into the laser is seen to decrease both the spatial and the temporal coherence of the output significantly. Finally, in this section we outline an approach to obtain stable single-mode output when external feedback is present using spatial filtering in the path of the output beam.

The experimental setup shown in Fig. 9.12 is a gain-guided GaAlAs laser diode array implemented in a feedback system as described in Section 9.3. The laser is operated at two times threshold with a diode current of 0.45 A at 13°C, yielding a freely running output of 200 mW. In front of the laser there are a collimating lens pair (L_1 and L_2) and a half-wave plate ($\lambda/2$) to align the polarization axis with the high coherence axis. The feedback system consists of three parts: a phase

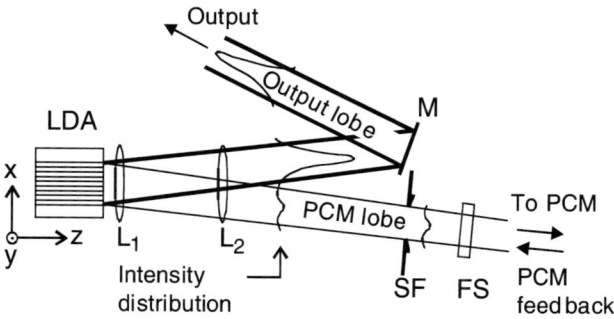

Figure 9.12. The weak intensity lobe is directed to the PCM (the PCM lobe), while the strong intensity lobe constitutes the output (the output lobe). LDA is the laser diode array, L_1 and L_2 are collimating lenses, SF is the spatial filter, FS is the frequency selective element, and M is a mirror. (From Ref. 42.)

conjugating mirror (PCM), a spatial filter (SF), and a frequency-selective element (FS). The PCM is a self-pumped photorefractive $BaTiO_3$-crystal doped with 800 ppm rhodium and measures $5.1 \times 5.5 \times 5.3$ mm^3 ($a \times a \times c$). The spatial filter is placed in the pseudo-far-field plane and consists of two razor blades that, when correctly adjusted, allow only a single spatial mode, (i.e., one array mode), to pass to be retroreflected by the PCM. The frequency-selective element is either a Fabry–Perot etalon with a finesse of 2.6 or a diffraction grating with 1200 lines/mm. Both elements narrow the spectrum of the laser output to less than 0.01 nm when inserted into the phase conjugate feedback path in the external cavity. The single-mode output lobe produced by this system is extracted by a mirror (M_1) in front of the spatial filter. The far-field radiation pattern is narrowed down to 1.5 times the diffraction limit. For operation at current levels higher than two times threshold, the improvement of the coherence is less pronounced. This problem can be avoided using antireflection coating of the front facet of the laser diode array, whereby the influence of the phase conjugate feedback at higher current levels is increased. Figure 9.13 shows the setup for external feedback measurements with spatial filtering in both the external feedback path and the phase conjugate feedback path.

9.5.1 Long-term stability of the phase conjugate laser system

The long-term stability of the output lobe was measured in a 100-hour "hands-off" experiment, where the output power and the wavelength were measured simultaneously every 20 seconds. In this experiment the etalon was used as the

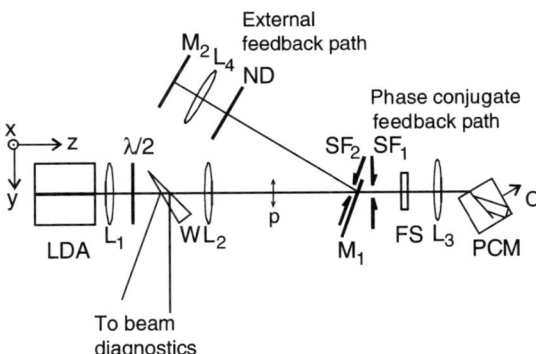

Figure 9.13. The setup for external feedback measurements with spatial filtering in both the external feedback path and the phase conjugate feedback path. LDA is the laser diode array; L_1, L_2, and L_3 are lenses of focal lengths 4.5 mm, 60 mm (cylindrical), and 100 mm, respectively; SF_1 and SF_2 are spatial filters in the phase conjugate feedback path and the external feedback path, respectively; FS is the frequency-selective element (a Fabry–Perot etalon of finesse 2.6, or a diffraction grating with 1200 lines/mm); PCM is the phase conjugating mirror; M_1 and M_2 are mirrors; ND is a neutral density filter; and p is the polarization. (From Ref. 42.)

frequency-selective element. The measurements are shown in Fig. 9.14. Figure 9.14a shows that the center wavelength has a standard deviation of less than 0.02 nm, which is less than one longitudinal mode spacing of about 0.11 nm; as a result, no mode hopping occurs within 100 hours of operation. Figure 9.14b demonstrates that the fluctuations of the output power are less than 1.4% of standard deviation. However, an increase in the fluctuation level is seen after approximately 50 hours of single-mode operation. This reflects an increase in the temperature fluctuations in the surrounding environment, which easily could be avoided by applying temperature stabilization to the etalon. The results in Fig. 9.14 reveal that both the output power and the wavelength of the phase conjugate laser diode array are extremely stable. The reason for such stable output is due to the fact that the phase conjugating crystal dynamically optimizes its internal grating structure to meet the condition of constructive interference in the external cavity.

9.5.2 The influence of external reflections of the output beam

When the sensitivity of the phase conjugate laser system to reflections back into the laser is measured, the output lobe is reflected back into the laser diode array by a mirror (M_2), as shown in Fig. 9.13. The intensity of the external feedback is varied with neutral density filters (ND). A lens (L_4) that focuses the beam onto the external mirror optimizes the signal that goes back into the laser diode array. The experiments are performed with and without a spatial filter (SF_2) in the external feedback path.

Figure 9.15 shows the far-field intensity distribution from the laser diode array (left-hand side) and its spectral distribution (right-hand side) when different amounts

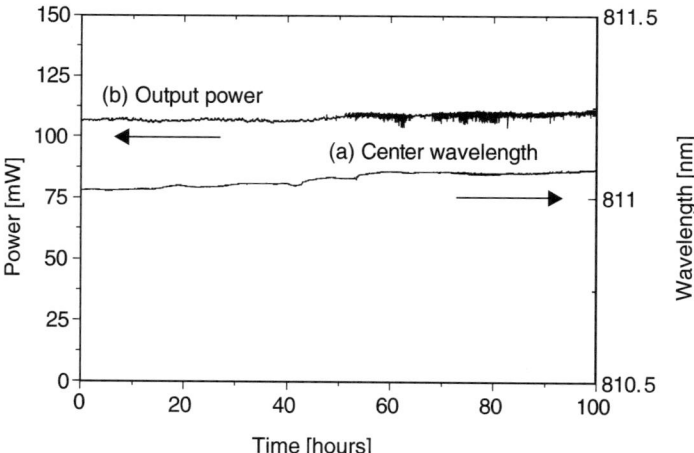

Figure 9.14. One hundred hours of "hands-off" measurements of the center wavelength (a) and the power of the output lobe (b). (From Ref. 42.)

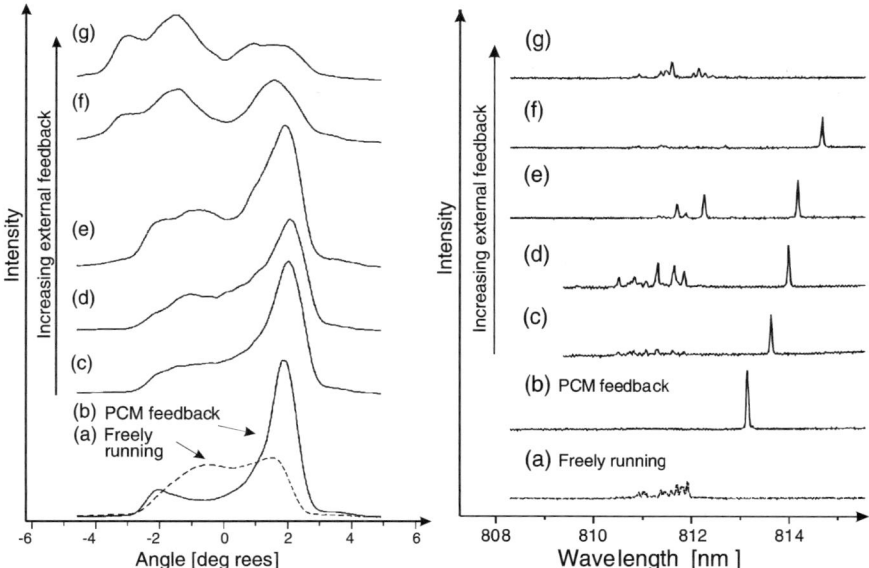

Figure 9.15. Far-field intensity profiles (left-hand side) and spectral distributions (right-hand side) for different levels of external feedback without any filtering in the external feedback path. The amount of external feedback is increased in the direction of the arrow by inserting ND filters of optical densities: (c) 1.0, (d) 0.7, (e) 0.5, (f) 0.04, and (g) no filter. (From Ref. 42.)

of feedback are applied without any filtering in the external feedback path; Figure 9.15a shows the multimode of the freely running laser; Figure 9.15b shows the single-mode operation of the laser when feedback is applied; Figures 9.15c–g reveal the deterioration of the spatial profile, the spectral broadening, and the mode hopping of the laser output as the amount of external feedback is increased. The spatial profile is changed from an almost diffraction-limited spot in Fig. 9.15a to a radiation pattern in Fig. 9.15g that is far from the diffraction limit, which corresponds to several spatial modes emitted from the laser diode array. The increase of the external feedback also leads to slow temporal alternations (on the order of minutes) of the output. At a given level of external feedback the laser diode array is forced out of single-mode operation. This occurs slowly over 10–15 minutes or immediately after the external feedback is applied, depending on the intensity and the feedback angle. In all the experiments the feedback has been optimized by adjusting the external mirror (M_2). However, due to diffraction loss, and so on, it is difficult to determine the absolute amount of reflected light that enters the laser diode array. The response to a certain amount of external feedback radiation is seen to vary from measurement to measurement. We believe that this is due to the fact that the feedback is coupled into different array modes. It is possible to alter the spatial profile by adjusting the angle of incidence of the external feedback, which to some extent confirms that different array modes are excited. The spatial profile is also seen to alter in time. This is in our opinion due to competition between different array modes. The amount of external feedback it takes to destroy the single-mode operation is found to be of the same order of magnitude as the phase conjugate feedback—that is, around 1% of the total output. Therefore, one must carefully avoid feedback of the output beam when the laser system is used for practical purposes.

An approach to prevent single-mode deterioration is to apply a spatial filter (SF_2) in the external feedback path, similar to the one in the phase conjugate feedback path (see Fig. 9.13). The second spatial filter (SF_2) is more simple than the spatial filter (SF_1) in the phase conjugate feedback path since we simply use the cut edge of the output extracting mirror. Using the mirror edge as a filter is advantageous since it is placed very near to the pseudo-far-field plane, where the spatial mode separation is at maximum. In contrast to the first spatial filter (SF_1), the second has only one filtering edge which, however, is found to be sufficient to secure that external feedback is coupled into one array mode only. Figure 9.16 shows the far-field intensity distribution (left-hand side) and the spectral distribution (right-hand side) of the laser diode array when different amounts of feedback are applied with spatially filtering in both the external feedback path and the phase conjugate feedback path; Figure 9.16a shows the freely running multimode output of the laser; Figure 9.16b shows the single-lobe but multimode frequency output when the spatially filtered external feedback is applied but no phase conjugate feedback is present (no FSPCF); Figure 9.16c shows the single-mode laser operation with the FSPCF applied but no external feedback; Figures 9.16d–g show the laser output with FSPCF applied and increasing amount of spatially filtered external feedback. It is clearly observed, from the curves in Fig. 9.16, that the configuration of double spatial filtering leads to spectral and temporal stability of the single-mode output even when the output is reflected back into the laser system. Furthermore, it is observed that the relative intensity of the two lobes in the spatial profile can at the same time be varied continuously by varying the amount of external feedback. A too-high external feedback intensity, however, destroys the single frequency but maintains the diffraction-limited output profile.

In summary, it is shown that the single-mode output from a laser diode array with phase conjugate feedback has stable long-time operation with respect to output power and center wavelength. However, the laser system is not robust against external feedback. External feedback of the same order of magnitude as the phase conjugate feedback (i.e., around 1% of the total output) is capable of destroying the single-mode operation of the laser diode array. The destruction of the single-mode operation is due to external feedback into several array modes leading to mode competition. In a configuration with spatial filtering in both the output beam and the phase-conjugate feedback path, it is shown that the system is robust against external feedback, and the output is stabilized with good spatial and temporal coherence even in the presence of external feedback up to approximately 1%.

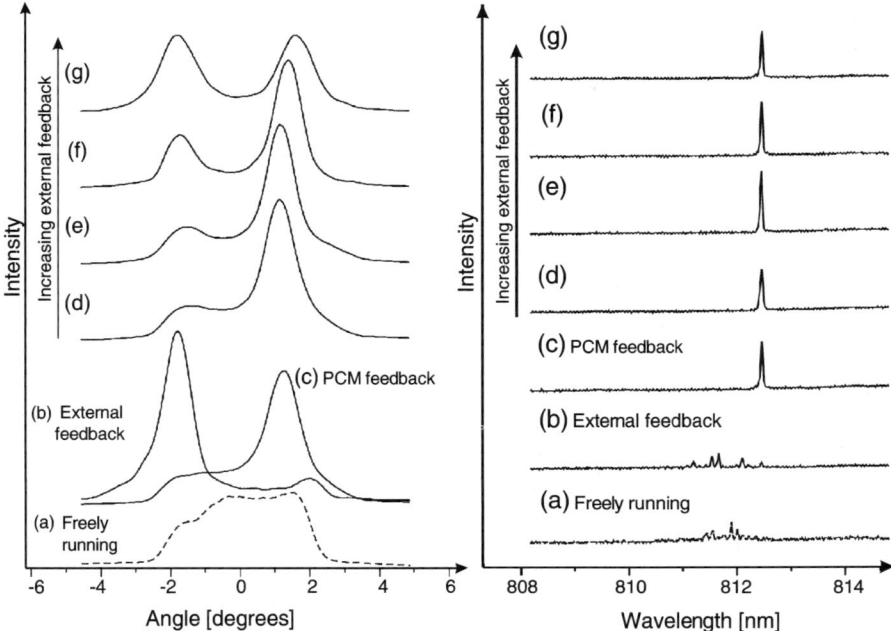

Figure 9.16. Far-field intensity profiles (left-hand side) and spectral distributions (right-hand side) for different levels of external feedback with spatially filtering in the external feedback path. The amount of external feedback is increased in the direction of the arrow by inserting ND filters of optical densities: (c) 1.0, (d) 0.7, (e) 0.5, (f) 0.04, and (g) 0.04. Curves (d)–(g) show that the intensity ratio between the two far-field lobes can be varied, while single frequency is preserved. (From Ref. 42.)

9.6 FREQUENCY DOUBLING OF HIGH-POWER LASER DIODE ARRAYS

Many new developments in medicine, optical storage, printing, and information technology require high-power laser light in the blue and violet parts of the spectrum. Blue and violet light is generally produced by frequency doubling of the output from longer-wavelength lasers, and to do this efficiently requires the high intensities normally available only from pulsed lasers or high-power continuous-wave (CW) lasers.

Until now, therefore, the main sources of blue and violet light have been Nd:YAG lasers using second-harmonic generation [72–74] and sum-frequency upconversion [75]. A high-power laser diode that could substitute the Nd:YAG laser would have considerable advantages.

Several research teams have investigated second-harmonic generation with diode lasers [76–78]. However, the low power available from a single infrared laser diode limits the usefulness of the resulting frequency-doubled output.

Laser diode arrays can provide much higher powers, but until now their low spatial and temporal coherence has meant that frequency doubling is limited to low power. Laser diode arrays with frequency-selective phase conjugate external feedback, though, yield output beams with much better spatial and temporal coherence and are therefore ideal candidates for high-power blue lasers based on frequency doubling. As well as providing much better coherence, frequency-selective phase conjugate feedback considerably increases the intensity of the output beam at the focal point.

The simplest way to produce blue light is to place a frequency-doubling crystal at the focal point of a lens system positioned in the output path of the laser diode array. A more efficient method of generating second harmonics is based on resonant external cavities. The advantage of this technique is that the intensity inside the cavity is much higher than outside the cavity. The frequency-selective phase conjugate feedback system described above uses an external resonant cavity containing a frequency-selective element such as an etalon. It is therefore quite straightforward to place the frequency-doubling element not just within the cavity, but inside the etalon itself. Since the beam intensity inside the etalon is much higher than outside, this approach will considerably increase the intensity of the frequency-doubled beam because the conversion efficiency is proportional to the square of the beam intensity entering the frequency doubler [79]. The presence of the frequency-doubling element within the etalon does not broaden the bandwidth of the output beam.

Figure 9.17 shows a laser diode array with external phase conjugate feedback and frequency doubling. The external cavity contains a collimating lens system, a nonlinear etalon, a diffraction grating, and a phase conjugator. The nonlinear etalon

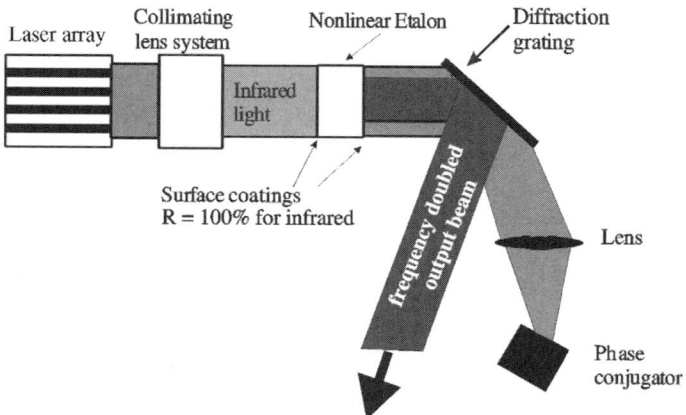

Figure 9.17. Frequency doubling of a high-power laser diode array. The external cavity consists of a collimating lens system, a nonlinear etalon, a diffraction grating, and a phase conjugator. The nonlinear etalon is a nonlinear crystal with highly reflective coatings applied to the endfaces of the crystal. The frequency-doubled beam is coupled out by the diffraction grating.

is a nonlinear frequency-doubling crystal with highly reflective coatings on each of the endfaces.

The beam inside the crystal is of high intensity, is highly collimated, and has a narrow bandwidth, making it ideal as a source for frequency doubling. The laser diode array acts as a gain medium with a conjugating mirror at one end and a conventional mirror at the other. The infrared laser light is maintained inside this cavity, and the frequency-doubled blue laser light is coupled out from the external cavity by the diffraction grating.

In conventional cavity-enhanced frequency doubling, the length of the external frequency-doubler cavity must be carefully controlled. This is usually done using an electric servo to tune the length of the cavity for maximum resonance. Adaptive phase conjugate feedback, on the other hand, automatically tunes the frequency-doubling cavity for resonance at the frequency of the source beam. This eliminates the need for a servo system.

9.7 CONCLUSIONS AND PERSPECTIVES

We have investigated several different architectures, all based on external frequency-selective phase conjugate feedback, that enhance the spatial and temporal coherence of high-power laser diode arrays.

With a high-finesse etalon in the external phase conjugate cavity it is possible to reduce the spectral bandwidth by several orders of magnitude and at the same time to increase the spatial coherence of the output beam close to the diffraction limit. This system can extract more than 80% of the power radiated by a freely running laser diode array.

In another configuration with a grating in the external cavity, the output is tunable over a broad range of wavelengths. Our experimental investigation showed that both the output power and the center wavelength of this system are very stable, with power fluctuations of less than 1.4% of standard deviation over a 100-hour test. External feedback of the output, however, leads to unstable operation, with oscillations in multiple spatial and temporal modes. To prevent these instabilities, we have introduced a special architecture with spatial filtering of the output beam.

In contrast to external-cavity diode lasers based on conventional optics, phase conjugate feedback can be used with diode bars and stacked arrays. It therefore has the potential to scale to output powers of hundreds of watts.

The significantly improved coherence of the output from a laser diode array with phase conjugate feedback may lead to many new applications of high-power laser diode arrays. In this chapter we have discussed how frequency doubling, for instance, can yield high-power blue and violet diode lasers. Other possible applications may be found in printing, material processing, medicine, optical sensors, pumping of solid-state lasers, and high-power single-mode fibers.

REFERENCES

1. E. Kapon, C. Lindsey, J. Katz, S. Margalit, and A. Yariv, Chirped array of diode lasers for supermode control, *Appl. Phys. Lett.* **45**, 200–202 (1984).
2. J. Salzman, R. Lang, S. Margalit, and A. Yariv, Tilted-mirror semiconductor lasers, *Appl. Phys. Lett.* **47**, 9–11 (1985).
3. C. Lindsey, P. Derry, and A. Yariv, Fundamental lateral mode oscillation via gain tailoring in broad area semiconductor lasers, *Appl. Phys. Lett.* **47**, 560–562 (1985).
4. M. Sakamoto and Y. Kato, High power (710 mW cw) single-lobe operation of broad area AlGaAs double heterostructure lasers grown by metalorganic chemical vapor deposition, *Appl. Phys. Lett.* **50**, 869–870 (1987).
5. M. Matsumoto, M. Taneya, S. Matsui, S. Yano, and T. Hijikata, Single-lobed far-field pattern operation in a phased array with an integrated phase shifter, *Appl. Phys. Lett.* **50**, 1541–1543 (1987).
6. D. Botez, L. Mawst, P. Hayashida, G. Peterson, and T. J. Roth, High-power, diffraction-limited-beam operation from phase-locked diode laser arrays of closely spaced "leaky" waveguides (antiguides), *Appl. Phys. Lett.* **53**, 464–467 (1988).
7. C. J. Chang-Hasnain, E. Kapon, and R. Brat, Spatial mode structure of broad area semiconductor quantum well lasers, *Appl. Phys. Lett.* **54**, 205–207 (1989).
8. L. Goldberg and M. K. Chun, Injection locking characteristics of a 1 W broad stripe laser diode, *Appl. Phys. Lett.* **53**, 1900–1902 (1988).
9. J. P. Hohimer, D. R. Myers, T. M. Brennan, and B. E. Hammons, Injection locking characteristics of a gain-guided diode laser arrays with an "on-chip" master laser, *Appl. Phys. Lett.* **56**, 1521–1523 (1990).
10. J.-M. Verdiell, R. Frey, and J.-P. Huignard, Analysis of injection locked gain guided diode laser arrays, *IEEE J. Quantum Electron* **27**, 396–401 (1991).
11. S. MacCormack and J. Feinberg, Injection locking of a laser-diode array with a phase-conjugate beam, *Opt. Lett.* **19**, 120–122 (1994).
12. M. W. Wright and J. G. McInerney, Injection locking semiconductor lasers with phase conjugate feedback, *Opt. Commun.* **110**, 589–698 (1994).
13. H. Li, T. L. Lucas, J. G. McInerney, M. W. Wright, and A. Morgan, Injection locking dynamics of vertical cavity semiconductor lasers under conventional and phase conjugate injection, *IEEE J. Quantum Electron.* **32**, 227–235 (1996).
14. A. Kamshilin, T. Jaaskelainen, V. V. Spirin, L. Y. Khriachtchev, R. Onodera, and Y. Ishii, Laser-diode injection locking with a double phase-conjugate mirror in photorefractive $Bi_{12}TiO_{20}$ fiberlike crystal under external alternating voltage, *J. Opt. Soc. Am. B* **14**, 2331–2338 (1997).
15. J. E. Epler, N. Holonyak, Jr., J. M. Brown, R. D. Burnham, W. Streifer, and T. L. Pauli, High-energy ($\lambda \leq 7300$ Å) 300 K operation of single- and multiple-stripe quantum-well heterostructure laser diodes in an external grating cavity, *J. Appl. Phys.* **56**, 670–675 (1984).
16. J. E. Epler, N. Holonyak, Jr., R. D. Burnham, T. L. Pauli, and W. Streifer, Far-field supermode patterns of a multiple-stripe quantum well heterostructure laser operated (7330 Å, 300 K) in an external grating cavity, *Appl. Phys. Lett.* **45**, 406–408 (1984).

17. J. Yaeli, W. Streifer, D. R. Scifres, P. S. Cross, R. L. Thornton, and R. D. Burnham, Array mode selection utilizing an external cavity configuration, *Appl. Phys. Lett.* **47**, 89–91 (1985).
18. C. Chang-Hasnain, D. F. Welch, D. R. Scifres, J. R. Whinnery, A. Dienes, and R. D. Burnham, Diffraction-limited emission from a diode laser array in an apertured graded-index external cavity, *Appl. Phys. Lett.* **49**, 614–616 (1986).
19. L. Goldberg and J. F. Weller, Single lope operation of a 40-element laser array in an external laser cavity, *Appl. Phys. Lett.* **51**, 871–873 (1987).
20. C. Chang-Hasnain, J. Berger, D. R. Scifres, W. Streifer, J. R. Whinnery, and A. Dienes, High power with high efficiency in a narrow single-loped beam from a diode laser array in an external cavity, *Appl. Phys. Lett.* **50**, 1465–1467 (1987).
21. H. Hemmati, Single longitudinal mode operation of semiconductor laser arrays with etalon feedback, *Appl. Phys. Lett.* **51**, 224–225 (1987).
22. C. Chang-Hasnain, A. Dienes, J. R. Whinnery, W. Streifer, and D. R. Scifres, Characteristics of the off-centered apertured mirror external cavity laser array, *Appl. Phys. Lett.* **54**, 484–486 (1989).
23. F. X. D'Amato, E. T. Siebert, and C. Roychoudhuri, Coherent operation of an array of diode lasers using a spatial filter in a Talbot cavity, *Appl. Phys. Lett.* **55**, 816–818 (1989).
24. A. C. Fey-den Boer, K. A. H. Van Leeuwen, H. C. W. Beijerinck, C. Fort, and F. S. Pavone, Grating feedback in a 810 nm broad-area diode laser, *Appl. Phys. B* **63**, 117–120 (1996).
25. D. Cassettari, E. Arimondo, and P. Verkerk, External-cavity broad-area laser diode operating on the D-1 line of cesium, *Opt. Lett.* **24**, 1135–1137 (1998).
26. V. V. Apollonov, S. I. Derzhavin, V. I. Kislov, V. V. Kuzminov, D. A. Mashkovsky, and A. M. Prokhorov, Phase locking of the 2D structures, *Opt. Express* **4**, 19–26 (1999).
27. A. Wakita and K. Sugiyama, Single-frequency external-cavity tapered diode laser in a double-ended cavity configuration, *Rev. Sci. Instrum.* **71**, 1–4 (2000).
28. V. Daneu, A. Sanchez, T. Y. Fan, H. K. Choi, G. W. Turner, and C. C. Cook, Spectral beam combining of a broad-stripe diode laser array in an external cavity, *Opt. Lett.* **25**, 405–407 (2000).
29. B. Chann, I. Nelson, and T. G. Walker, Frequency-narrowed external-cavity diode-laser-array bar, *Opt. Lett.* **25**, 1352–1354 (2000).
30. S. Mailhot, Y. Champagne, and N. McCarthy, Single-mode operation of a broad-area semiconductor laser with an anamorphic external cavity: Experimental and numerical results, *Appl. Opt.* **39**, 6806–6813 (2000).
31. M. Croning-Golomb and A. Yariv, Self-induced frequency scanning and distributed Bragg reflection in semiconductor lasers with phase-conjugate feedback, *Opt. Lett.* **11**, 455–457 (1986).
32. M. Segev, S. Weiss, and B. Fisher, Coupling of diode laser arrays with photorefractive passive phase conjugate mirrors, *Appl. Phys. Lett.* **50**, 1397–1399 (1987).
33. M. Segev, Y. Ophir, B. Fisher, and G. Eisenstein, Mode locking and frequency tuning of a laser diode array in an extended cavity with a photorefractive phase conjugate mirror, *Appl. Phys. Lett.* **57**, 2523–2525 (1990).
34. S. MacCormack and J. Feinberg, High-brightness output from a laser-diode array coupled to a phase-conjugate mirror, *Opt. Lett.* **18**, 211–213 (1993).

35. E. Milténi, M. O. Ziegler, M. Hofmann, J. Sacher, W. Elsässer, and E. O. Göbel, Long-term stable mode locking of a visible diode laser with phase-conjugate feedback, *Opt. Lett.* **20**, 734–736 (1995).
36. A. Shiratori and M. Obara, Wavelength-stable, narrow-spectral-width oscillation of an AlGaInP diode laser coupled to a $BaTiO_3$:Co stimulated photorefractive backscattering phase conjugator, *Appl. Phys. B* **65**, 329–333 (1997).
37. M. Løbel, P. M. Petersen, and P. M. Johansen, Tunable single mode operation of a high-power laser diode array using an external cavity with a grating and a photorefractive phase conjugate mirror, *J. Opt. Soc. Am. B.* **15**, 2000–2005 (1998).
38. M. Løbel, P. M. Petersen, and P. M. Johansen, Suppressing self-induced frequency scanning of a diode laser array with phase conjugate feedback using counterbalance dispersion, *Appl. Phys. Lett.* **72**, 1263–1265 (1998).
39. M. Løbel, P. M. Petersen, and P. M. Johansen, Single-mode operation of a laser diode array with frequency selective phase-conjugate feedback, *Opt. Lett.* **23**, 825–827 (1998).
40. M. Løbel, P. M. Petersen, and P. M. Johansen, The origin of laser frequency scanning induced by photorefractive phase conjugate feedback, *J. Opt. Soc. Am. B.* **16**, 219–227 (1999).
41. P. M. Petersen, S. J. Jensen, and P. M. Johansen, Phase locking of laser diode arrays using a photorefractive $Rh:BaTiO_3$ crystal, *Laser Resonators II*, Alexis V. Kudryashov (ed.), *SPIE Proceedings* **3611**, (1999), pp. 142–146.
42. S. Juul Jensen, M. Løbel, and P. M. Petersen, Stability of the output from a phase locked laser diode array, *Appl. Phys. Lett.* **76**, 535–537 (2000).
43. J. K. Butler, D. E. Ackley, and D. Botez, Coupled-mode analysis of phase-locked injection laser arrays, *Appl. Phys. Lett.* **44**, 293–295 (1984).
44. E. Kapon, J. Katz, and A. Yariv, Supermode analysis of phase locked arrays of semiconductor lasers, *Opt. Lett.* **10**, 125–127 (1984).
45. G. P. Agrawal, Lateral-mode analysis of gain-guided and index-guided semiconductor laser arrays, *J. Appl. Phys.* **58**, 2922–2931 (1985).
46. D. Mehuys, R. J. Lang, M. Mittelstein, J. Salzman, and A. Yariv, Self-stabilized nonlinear lateral modes of broad area lasers, *IEEE J. Quantum Electron.* **QE-23**, 1909–1920 (1987).
47. D. Mehuys and A. Yariv, Coupled-wave theory of multiple-stripe semiconductor injections lasers, *Opt. Lett.* **13**, 571–573 (1988).
48. R. J. Lang, A. G. Larsson, and J. G. Cody, Lateral modes of broad area semiconductor lasers: Theory and experiment, *IEEE J. Quantum Electron.* **QE-27**, 312–320 (1991).
49. J.-M. Verdiell and R. Frey, A broad-area mode coupling model for multiple stripe semiconductor lasers, *IEEE J. Quantum Electron.* **26**, 270–279 (1990).
50. P. M. Petersen, Theory of one grating nondegenerate four-wave mixing—and its application to a linear photorefractive oscillator, *J. Opt. Soc. Am. B* **8**, 1716–1722 (1991).
51. G. P. Agrawal, Four-wave mixing and phase conjugation in semiconductor laser media, *Opt. Lett.* **12**, 260–262 (1987).
52. Kyo Inoue, T. Mukai, and T. Saitoh, Nearly degenerate four-wave mixing in a traveling-wave semiconductor laser amplifier, *Appl. Phys. Lett.* **51**, 1051–1053 (1987).
53. G. P. Agrawal, Population pulsations and nondegenerate four-wave mixing in semiconductor lasers and amplifiers, *J. Opt. Soc. Am. B.* **5**, 147–159 (1988).

54. M. Lucente and G. M. Carter, Nonlinear mixing and phase conjugation in broad-area diode lasers, *Appl. Phys. Lett.* **53**, 467–469 (1988).
55. M. Lucente, J. G. Fujimoto, and G. M. Carter, Spatial and frequency dependence of four-wave mixing in broad-area lasers, *Appl. Phys. Lett.* **53**, 1897–1899 (1988).
56. R. Nietzke, P. Panknin, W. Elsässer, and E. O. Göbel, Four-wave-mixing in GaAs/ AlGaAs semiconductor lasers, *IEEE J. Quantum Electron.* **25**, 1399–1405 (1989).
57. T. Mukai and T. Saitoh, Detuning characteristics and conversion efficiency of nearly degenerate four-wave mixing in a 1.5-μm traveling-wave semiconductor laser amplifier, *IEEE J. Quantum Electron.* **26**, 865–874 (1990).
58. F. Favre and D. L. Guen, Four-wave mixing in traveling-wave semiconductor laser amplifiers, *IEEE J. Quantum Electron.* **26**, 858–864 (1990).
59. W. M. Yee and K. A. Shore, Enhanced uniform phase conjugation in two-section asymmetric laser diodes, *Opt. Lett.* **19**, 2128–2130 (1994).
60. L. Petersen, U. Gliese, and T. N. Nielsen, Phase noise reduction by self-phase locking in semiconductor lasers using phase conjugate feedback, *IEEE J. Quantum Electron.* **30**, 2526–2533 (1994).
61. P. Kürz, R. Nagar, and T. Mukai, Highly efficient phase conjugation using spatially nondegenerate four-wave mixing in a broad-area laser diode, *Appl. Phys. Lett.* **68**, 1180–1182 (1996).
62. P. Kürz and T. Mukai, Frequency stabilization of a semiconductor laser by external phase-conjugate feedback, *Opt. Lett.* **21**, 1369–1371 (1996).
63. G. J. Dunning, D. M. Pepper, and M. B. Klein, Control of self-pumped phase-conjugate reflectivity using incoherent erasure, *Opt. Lett.* **15**, 99–101 (1990).
64. G. W. Ross and R. W. Eason, Highly efficient self-pumped phase conjugation at near-infrared wavelengths by using nominally undoped $BaTiO_3$, *Opt. Lett.* **17**, 1104–1106 (1992).
65. D. Rytz and S. Shen, Self-pumped phase conjugation in potassium niobate ($KNbO_3$), *Appl. Phys. Lett.* **54**, 2625–2627 (1989).
66. G. Salamo, M. Miller, W. Clark, G. Wood, and E. Sharp, Strontium barium niobate as a self-pumped phase conjugator, *Opt. Commun.* **59**, 417–422 (1986).
67. R. Bylsma, A. Glass, and D. Olson, Self-pumped phase conjugation in InP:Fe, *Appl. Phys. Lett.* **54**, 1968–1970 (1989).
68. J. M. Verdiell, H. Rajbenbach, and J. P. Huignard, Array modes of multi-stripe diode lasers: A broad-area mode coupling approach, *J. Appl. Phys.* **66**, 1466–1468 (1989).
69. X. Tang, J. P. van der Ziel, and A. K. Chin, Characterization of the array modes of high-power gain-guided GaAs single-quantum-well laser arrays, *IEEE J. Quantum Electron.* **32**, 1417–1426 (1996).
70. T. Brabec, F. Krausz, E. Wintner, and A. J. Schmidt, Longitudinal pumping of lasers with multistripe laser diodes, *Appl. Opt.* **30**, 1450–1454 (1991).
71. S. Weiss, M. Segev, and B. Fisher, Line narrowing and self-frequency scanning of laser diode arrays coupled to a photorefractive oscillator, *IEEE J. Quantum Electron.* **24**, 706–708 (1988).
72. W. P. Risk and Lenth, Room-temperature, continuous-wave, 946 nm Nd-YAG laser pumped by laser-diode array and intracavity frequency doubling to 473 nm, *Opt. Lett.* **12**, 993–995 (1987).

73. G. J. Dixon, Z. M. Zhang, R. S. F. Chang, and N. Djeu, Efficient blue emission from an intracavity-doubled 946 nm Nd-YAG laser, *Opt. Lett.* **13**, 137–139 (1988).
74. D. H. Jundt, M. M. Fejer, R. L. Byer, R. G. Norwood, and P. F. Bordui, 69% Efficient continuous-wave second-harmonic generation in lithium-rich lithium niobate, *Opt. Lett.* **16**, 1856–1858 (1991).
75. W. P. Risk, J.-C. Baumert, G. C. Bjorklund, F. M. Schellenberg, and W. Lenth, Generation of blue-light by intracavity frequency mixing of the laser and pump radiation of a miniature neodymium–yttrium aluminum garnet laser, *Appl. Phys. Lett.* **52**, 85–87 (1988).
76. M. K. Chun, L. Goldberg, and J. F. Weller, Second-harmonic generation at 421 nm using injection-locked GaAlAs laser array and $KNbO_3$, *Appl. Phys. Lett.* **53**, 1170–1171 (1988).
77. A. Hemmerich, D. H. McIntyre, C. Zimmermann, and T. W. Hänsch, Second-harmonic generation and optical stabilization of a diode laser in an external ring resonator, *Opt. Lett.* **15**, 372–374 (1990).
78. W. J. Kozlovsky, W. P. Risk, W. Lenth, B. G. Kim, G. L. Bona, H. Jaeckel, and D. J. Webb, Blue light generation by resonator-enhanced frequency doubling of an extended-cavity diode laser, *Appl. Phys. Lett.* **65**, 525–527 (1994).
79. G. J. Dixon, C. E. Tanner, and C. E. Wieman, 432 nm source based on efficient second-harmonic generation of GaAlAs diode-laser radiation in a self-locking external resonant cavity, *Opt. Lett.* **14**, 731–733 (1989).

CHAPTER 10

Self-Pumped Phase Conjugation by Joint Stimulated Scatterings in Nematic Liquid Crystals and Its Application for Self-Starting Lasers

OLEG ANTIPOV

Institute of Applied Physics, Russian Academy of Science, 603950 Nizhny Novgorod, Russia

10.1 INTRODUCTION

Self-pumped phase conjugation (SPPC) of a laser beam can be accomplished by joint stimulated scattering of the self-intersecting optical beam in a nonlinear layer with a feedback loop. One of the most promising effects for the SPPC is stimulated scattering of Rayleigh type. It is well known that the stimulated Rayleigh scatterings (SRSs) can be caused by different origin of light-induced refractive index changes, such as thermal heating in absorbing media, molecule reorientation under electric field of the optical wave, photochromic effect, and so on. The SRS effects are easily observed in different media for a wide variety of optical wavelengths. For example, stimulated thermal scattering caused by the thermally induced nonlinearity was realized in gases, liquids, and solid states for optical beams with wavelengths ranging from UV to middle IR [1–3]. However, the increment of backward SRS is typically quite small due to fast relaxation of the small-period scattering grating. Therefore, a simple scheme of the backward Rayleigh-type scattering is scarcely applicable for the phase conjugation. On the other hand, the strong increment of near-forward SRS provides another attractive possibility for SPPC of laser beams with long pulse duration. The SPPC can be caused by joint near-forward SRS of two intersecting optical waves, when one strong wave, E_2, is formed from another wave, E_1, after one round trip through a feedback loop (Fig. 10.1). The geometry is similar to that for the SPPC in photorefractive media [4]. The principles of the SPPC by joint stimulated scattering will be discussed in more detail in Sections 10.2.1–10.2.3. One of the most promising nonlinear media for realization of SPPC by the

---

Phase Conjugate Laser Optics, edited by Arnaud Brignon and Jean-Pierre Huignard
ISBN 0-471-43957-6   Copyright © 2004 John Wiley & Sons, Inc.

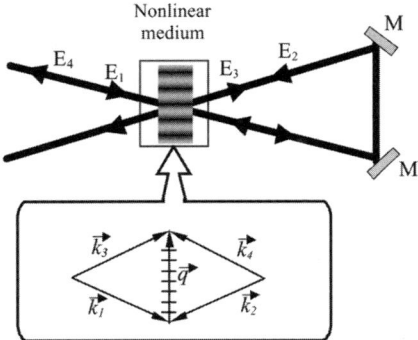

Figure 10.1. Schematic of the joint scattering of optical beams (an incident wave E_1 and a wave transmitted though the feedback loop E_2) in a nonlinear layer with a feedback loop (including mirrors, M) and wave-vector diagram of the interacting waves ($\vec{k}_1 - \vec{k}_4$ are the wave vectors of the initial and scattering waves $E_1 - E_4$).

joint stimulated scattering, nematic liquid crystals (NLCs), will be introduced herein.

It is well known that NLCs have extraordinarily large slow-response nonlinearities: orientational and thermal [5, 6]. The orientational nonlinearity is caused by refractive index changes due to collective rotation of a great number of mutually parallel nematic molecules under an electric field of the optical waves. The strong thermally induced changes of the NLC refractive index are determined by changes of the order parameter of the liquid-crystal phase and by the medium density. The order-parameter changes of the NLC refractive index predominate near the nematic–isotropic phase transition point. The both NLC nonlinearities allow for the effective degenerate two-wave and four-wave mixings of optical beams by induced dynamic refractive index gratings (RIGs) [7, 8]. Strong stimulated scatterings of Rayleigh type, the stimulated thermal scattering and stimulated orientational scattering, also take place in the NLC for wavelengths in the visible, near-, and middle-IR regions [5, 6, 9]. The nonlinear wave interactions have been used for several types of phase conjugation of laser beams. The SPPC effect in the NLC layer with a feedback loop will be discussed in Section 10.2, and lasers with the SPPC in the NLC layer will be addressed in Section 10.3.

Liquid crystals are not the only nonlinear medium for accomplishing phase conjugation by the joint scattering. This effect may be realized in different liquids with thermal nonlinearity; however, the threshold intensity for ordinary liquids is much more than in the NLC. Another interesting possibility for the SPPC by the joint scattering is offered by active laser media. Laser crystals, for example, also have a large number of optical nonlinearities. The optically induced changes of population inversion of a laser crystal amplifier can be accompanied not only by the gain saturation but by the refractive index changes as well [10, 11]. The scattering-like process of energy transfer from a strong laser beam to a weak one due to two-beam coupling in a flashlamp-pumped Nd:YAG amplifier, for example, occurs due to a dynamic RIG [10]. The joint stimulated scattering by the index and gain gratings induced by a self-intersecting optical beam in Nd:YAG amplifier also can provide the phase conjugation. The specificity of SPPC in laser crystals is beyond the scope of this chapter; however, its general principles may be understood by analogy with what will be discussed herein.

Nonlinear SPPC mirrors are known to ensure partial compensation of imperfections and phase aberrations after a backward pass through the active medium and other optical elements inside a laser resonator [12]. Therefore, the SPPC mirrors appear to be attractive for the creation of laser resonators of high-average-power lasers. The problem of laser resonators with phase conjugate mirrors has been intensively studied (see, for example, Refs. 12 and 13). In this chapter, the concept of the self-starting lasers with cavity completed by RIGs induced in nonlinear media by generating beams themselves will be discussed. The self-starting laser with cavity completed by SPPC mirrors based on joint stimulated scattering in NLCs will be described in Section 10.3.

10.2 SELF-PUMPED PHASE CONJUGATION BY JOINT STIMULATED SCATTERING

10.2.1 Geometrical features of joint stimulated scattering

The principle of joint stimulated scattering can be explained by simultaneous formation of the same scattering gratings $\delta n_{13} = \delta n_{24}$ by two pairs of optical waves: strong (pumping) waves $E_{1,2}$ and scattering waves $E_{3,4}$: $\delta n_{13} \sim E_1^* \cdot E_3$, $\delta n_{24} \sim E_2^* \cdot E_4$ (Fig. 10.1). To have a single grating, coupling all four optical beams, it is required that the wave vector synchronism be fulfilled (or almost fulfilled):

$$\vec{k}_3 - \vec{k}_1 = \vec{q} = \vec{k}_4 - \vec{k}_2 \tag{1}$$

where $\vec{k}_j$ is the wave vector of the optical beam E_j, and $\vec{q}$ is the grating vector.

Since directions of the pumping wave are fixed, ($\vec{k}_1$ and $\vec{k}_2$ are fixed), condition (1) determines two cones of scattering wave vectors with axes $\vec{k}_2 - \vec{k}_1$ and $\vec{k}_2 + \vec{k}_1$ [14]. However, the feedback loop separates the scattering wave propagating backward (or near backward) to the pumping waves: $\vec{k}_4 = -\vec{k}_1, \vec{k}_2 = -\vec{k}_3$. The phase conjugated scattering waves have minimal losses in the feedback loop and have a maximum (or near maximum) increment in the nonlinear layer.

In case of spatially inhomogeneous pumping beams $E_{1,2}(\vec{r})$, the requirement of the single grating formation determines the requirement for generation of the phase conjugated scattering beams. In other words, the condition

$$\delta n_{13} \sim E_1^*(\vec{r}) \cdot E_3(\vec{r}) = E_2^*(\vec{r}) \cdot E_4(\vec{r}) \sim \delta n_{24} \tag{2}$$

is fulfilled when

$$E_4(\vec{r}) \sim E_1^*(\vec{r}) \quad \text{and} \quad E_3(\vec{r}) \sim E_2^*(\vec{r}) \tag{3}$$

The phase conjugated beam (3) is the single solution of Eq. (2), when the pumping waves are spatially noncorrelated: $\langle E_2(\vec{r}) \cdot E_1^*(\vec{r}) \rangle = 0$ [14, 15].

The strong pumping waves can scatter independently in the nonlinear medium by separate gratings. However, the joint scattering of the two beams by a common grating can be much stronger than by the separate scattering. The parametric oscillation of the joint stimulated scattering with time-proportional increment takes place in the presence of positive feedback between an increase of the common grating and growing of the scattering waves.

One might say that the principle of the joint stimulated scattering, as explained here, is the same as that of a ring passive phase conjugate mirror based on the photorefractive medium [16–18]. Indeed, the analogy does can be drawn, but there is also an important difference between the joint SRS and the four-wave mixing in the photorefractive medium. The half-period spatial phase shift between the holographic RIG and the light interference pattern in the photorefractive medium provides unidirectional beam coupling. There is no spatial phase shift in non-photorefractive media, and the increment of the joint SRS is realized due to frequency detuning of the scattering and pumping waves leading to the movement of an index grating temporally shifted with respect to the light interference field.

From the point of view of formal analogies, there is another possibility to achieve SPPC in a similar optical scheme as the joint SRS. The backward propagating (phase conjugated) wave can be generated due to grating δn_{12} induced in the nonlinear medium by two pumping waves, $\delta n_{12} \sim E_1^* \cdot E_2$. In this case, however, it is necessary to use an additional laser amplifier inside the feedback loop, and the oscillation of the waves E_3 and E_4 occurs as a result of lasing in the cavity completed by the grating δn_{12} [19–21].

Finally, it should be noted that the SPPC by joint stimulated scattering was first achieved by backward stimulated Brillouin scattering (SBS) of laser beams in a nonlinear liquid with a loop scheme providing the parametric feedback [22]. The phase conjugation threshold in that scheme was measured to be less than for the ordinary backscattering. However, the SPPC by the near-forward SRS has a much lower oscillation threshold and can be used for long pulses of the optical beam and even for continuous-wave (CW) laser beams. The threshold of the SPPC by joint near-forward SRSs in one of the most promising nonlinear media (i.e., in NLCs), will be discussed in the next sections.

10.2.2 Theoretical description of phase conjugation by joint stimulated scattering in a nonlinear layer with feedback loop

10.2.2.1 Oscillation thresholds in the plane wave approximation
General conditions of generation of a phase conjugated wave by joint SRS in NLC with a feedback loop (in schematic of Fig. 10.1) can be illustrated by a simple

approximation of plane optical waves E_j and a universal model of NLC nonlinearity. Having passed through the NLC layer and the feedback loop, the incident wave E_1 propagates in the nonlinear layer again as a second pump wave E_2. In general, the feedback loop can include a nonreciprocal optical device providing an amplitude or phase nonreciprocity for counterpropagating waves.

The stimulated scattering occurs as a self-consistent process of simultaneous increasing both the grating and scattering waves. The intersecting strong waves $E_{1,2}$ can jointly scatter in the NLC layer by a common grating into scattering waves $E_{3,4}$ that propagate exactly backward or near-backward to the pumping waves [scattering in other directions that satisfy the requirement of the four-wave mixing synchronism (1) does not lead to the transformation of one scattering wave E_3 into another wave E_4 because of a finite view angle of the feedback loop]. In general, the joint scattering within the view angle of the feedback loop occurs even in the presence of small mismatch of the four-wave mixing synchronism [determined by Eq. (1)] for the interacting pumping and scattering waves (Fig. 10.2): $|\vec{\Delta k}|l = |\vec{k}_1 - \vec{k}_3 + \vec{k}_4 - \vec{k}_2|l \leq 1$, where l is the thickness of the nonlinear layer.

The SRS starts from spontaneous scattering of the pump waves on fluctuations and random inhomogeneities of the refractive index of the media. The interference fields of the mutually coherent pumping optical waves $E_{1,2}$ and the scattering waves $E_{3,4}$ induce gratings of refractive index $\delta n_{ij} = (\partial n/\partial Q) \cdot Q_{ij}$, where Q_{ij} is the temperature grating (for the thermal nonlinearity) or grating of director orientation (for the orientational nonlinearity) which for a medium with nonlinearity of diffuse type (such as orientational and thermal nonlinearity of NLC) can be given by the following equation [5–8]:

$$\frac{\partial Q_{ij}}{\partial t} + \chi \nabla^2 Q_{ij} = \gamma (E_i \cdot E_j^* \cdot \exp(i(\vec{k}_j - \vec{k}_i)\vec{r}) + c.c.) \tag{4}$$

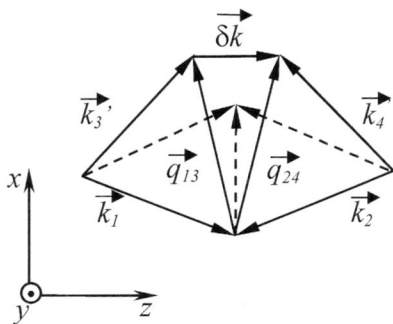

Figure 10.2. Wave-vector diagram of four-wave mixing in joint near-forward scattering, $\vec{k}_1$ and $\vec{k}_2$ are the wave-vectors of the pumping waves, $\vec{k}_3'$ and $\vec{k}_4'$ are the wave vectors of the scattering waves, $\vec{q}_{13}$ and $\vec{q}_{24}$ are grating vectors, and $\vec{\Delta k}$ is the detuning of the four-wave mixing synchronism.

where t is the time, χ is the coefficient of temperature diffusivity (for a thermal grating) or elasticity (for an orientational grating) of the medium, ∇^2 is the three-dimensional spatial Laplacian, γ is a coefficient determined by the mechanism for NLC nonlinearity, and $\vec{k}_i$ are wavevectors of the optical waves.

Generally, all intersecting optical waves E_1–E_4 interact by the gratings δn_{ij}. If the intersection angle of the waves in the nonlinear layer is small, $\theta_{in} \ll 1$, and when a transverse displacement of the beams in the nonlinear layer is much less than the beam diameter, the set of equations for the plane optical waves can be given in the one-dimensional approximation

$$\frac{1}{v}\frac{\partial E_1}{\partial t} + \frac{\partial E_1}{\partial z} = i\beta(T_0 E_1 + T_{12}^* E_2 + T_{13}^* E_3 + T_{14}^* E_4) - \frac{\alpha}{2} E_1$$

$$\frac{1}{v}\frac{\partial E_2}{\partial t} - \frac{\partial E_2}{\partial z} = i\beta(T_0 E_2 + T_{12} E_1 + T_{23}^* E_3 + T_{13}^* E_4 \exp(i\Delta kz)) - \frac{\alpha}{2} E_2$$

$$\frac{1}{v}\frac{\partial E_3}{\partial t} + \frac{\partial E_3}{\partial z} = i\beta(T_0 E_3 + T_{13} E_1 + T_{23} E_2 + T_{12}^* E_4 \exp(-i\Delta kz)) - \frac{\alpha}{2} E_3 \quad (5)$$

$$\frac{1}{v}\frac{\partial E_4}{\partial t} - \frac{\partial E_4}{\partial z} = i\beta(T_0 E_4 + T_{14} E_1 + T_{12} E_3 \exp(i\Delta kz) + T_{13} E_2 \exp(-i\Delta kz)) - \frac{\alpha}{2} E_4$$

where v is the light speed in the nonlinear layer, β is the nonlinear coefficient of the media, α is the absorption coefficient, $\Delta k = |\vec{\Delta k}|$, T_0 is the average-in-space perturbation of the refractive index, and $T_{12}, \ldots, T_{14}$ are the RIG complex amplitudes that can be determined (for the orientational or thermal nonlinearities of NLC) by the following equations [obtained from Eq. (4)]:

$$\frac{\partial T_0}{\partial t} + \frac{T_0}{\tau_0} = \gamma(E_1 E_1^* + E_2 E_2^* + E_3 E_3^* + E_4 E_4^*),$$

$$\frac{\partial T_{12}}{\partial t} + \frac{T_{12}}{\tau_{12}} = \gamma(E_1^* E_2 + E_3^* E_4 \exp(-i\Delta kz)),$$

$$\frac{\partial T_{13}}{\partial t} + \frac{T_{13}}{\tau_{13}} = \gamma(E_1^* E_3 + E_2^* E_4 \exp(i\Delta kz)), \quad (6)$$

$$\frac{\partial T_{14}}{\partial t} + \frac{T_{14}}{\tau_{14}} = \gamma E_1^* E_4, \qquad \frac{\partial T_{23}}{\partial t} + \frac{T_{23}}{\tau_{23}} = \gamma E_2^* E_3$$

where τ_{ji} and τ_0 are the relaxation time of refractive index grating and of the average-in-space perturbation, respectively: $\tau_{ji} = (\chi|\vec{q}_{ij}|^2)^{-1}$, $\vec{q}_{ij}$ is the grating's vector, $\vec{q}_{ij} = \vec{k}_i - \vec{k}_j$; $\tau_0 = r_0^2/(2\chi)$, r_0 is the optical beam radius. Equations (6) were derived by spatial averaging Eq. (4) and taking into account spatial structures of the gratings $Q_{ij} = T_{ij}\exp(i\vec{q}_{ij}\vec{r})$ and Gaussian transverse profile of the optical waves $\sim E_j \exp(-0.5 r^2/r_0^2)$, $r_0 \gg q_{ij}^{-1}$.

The boundary conditions for the waves are determined by optical characteristics of the feedback loop (losses or amplifications of the pumping and scattering waves

r_{ij} and their phase shifts φ_{ij})

$$E_2(z = l) = r_{21} \exp(i\varphi_{21})E_1(z = l),$$
$$E_4(z = l) = r_{43} \exp(i\varphi_{43})E_3(z = l) \quad (7)$$

and by initial noise for the scattering wave at the input boundary:

$$E_3(z = 0) = \varepsilon \exp(i\Omega t) \quad (8)$$

where Ω is the frequency detuning of the scattering waves ($\omega_3 = \omega_4$) with respect to frequency of the pumping waves ($\omega_1 = \omega_2$): $\Omega = \omega_3 - \omega_1$, $|\varepsilon| \langle\langle |E_1(z = 0)|$.

The thermal or orientational gratings with different period make different contributions to the scattering because their relaxation time and steady-state amplitudes are inversely proportional to square of the grating vector q_{ij} (if the grating period is much less than both the layer thickness and the radius of the optical beams). For optical pulses with duration τ_p much more than the walk-off time of the optical pulses through the scheme ($\tau_p \gg l/v$) and the relaxation time of small-scale reflection gratings (with period $\sim \lambda/2$), and comparable with the relaxation time of the transmitting grating (with the biggest period $\sim \lambda/\theta_{in}$), the set of Eqs. (6) can be simplified. In this case, the amplitudes of the small-scale reflecting gratings T_{12}, T_{14}, T_{23} induced by the interference field of the counterrunning (or near-counterrunning) optical waves are negligibly small in comparison with the amplitude of the large-scale transmitting grating T_{13} induced by the interference field of the pumping waves and the waves of near-forward scattering. Therefore, the small-scale reflecting gratings may be neglected and only equations for T_0 and T_{13} in the set (6) can be taken into consideration.

At the initial stage of the joint SRS (when the scattering wave intensities are much less than the pump-wave intensities) the set of Eqs. (5) can be transformed into the following set of equations:

$$\frac{\partial E_1}{\partial z} = i\beta T_0 E_1 - \frac{\alpha}{2} E_1,$$

$$-\frac{\partial E_2}{\partial z} = i\beta T_0 E_2 - \frac{\alpha}{2} E_2,$$

$$\frac{\partial E_3}{\partial z} = i\beta(T_0 E_3 + T_{13} E_1) - \frac{\alpha}{2} E_3,$$

$$-\frac{\partial E_4}{\partial z} = i\beta(T_0 E_4 + T_{13} E_2 \exp(-i\Delta k z)) - \frac{\alpha}{2} E_4 \quad (9)$$

In the case of four-wave mixing synchronism ($\Delta k l = 0$), the set of Eqs. (9) and the equation for T_{13} with boundary conditions (7) and (8) and zero initial condition for the grating ($T_{13}(t = 0) = 0$) has the following steady-state solution for the output wave:

$$E_4(z=0) = \left(\frac{r_{43}\exp(i\varphi_{43}) + r_{21}\exp(i\varphi_{21})}{\Gamma}\exp(\text{Int}(l,0)) - r_{21}\exp(i\varphi_{21})\right)$$
$$\times \varepsilon \exp\left(i\Omega t - \frac{\alpha l}{2} + i\beta \int_0^l T_0(z')dz'\right) \quad (10)$$

where

$$\Gamma = 1 - \frac{iG(r_{21}r_{43}\exp(i\Delta\varphi) + r_{21}^2)}{1+i\Omega\tau_{13}}\int_0^l dz\exp(\alpha(z-2l) + \text{Int}(l,z)),$$

$$\text{Int}(l,z) = \frac{iG}{1+i\Omega\tau_{13}}\int_z^l (\exp(-\alpha z') - r_{21}^2\exp(\alpha(z'-2l))\,dz')), \quad (11)$$

$$G = \beta\gamma\tau_{13}|E_1(z=0)|^2$$

The solution (10) shows the possibility of growing of the phase-conjugated wave $E_4(z=0)$ in the nonlinear layer for the positive real part of function $\text{Int}(l,0) - (\alpha/2)l$. This regime is an analogue of the ordinary stimulated scattering. However, a more interesting regime is observed when the denominator $\Gamma \to 0$. In this case the scattering wave increases infinitely. It means absolute instability of the scattering waves (i.e., the parametric generation of the phase conjugated wave) [23, 24]. The expressions (11) essentially determine the threshold of parametric generation for the increment G_{th} and the frequency detuning $\Omega\tau_{13}$. The scattering waves can strongly increase due to the joint SRS even below the oscillation threshold and can predominate over separate scattering of each wave. However, above the oscillation threshold of the joint SRS, the scattering wave amplitudes dramatically increase; therefore, this regime of "parametric oscillation" is most interesting for experimental realization of the self-pumped phase conjugation.

The parametric oscillation can be achieved by scattering of two pumping beams intersecting in a nonlinear layer in the presence of positive feedback. It is possible to say that this feedback is between the formation of the refractive-index gratings δn_{13} and δn_{24} at the opposite boundaries of the nonlinear layer. The feedback is established when the reflection of the optical beam E_1 by the "foreign" index grating δn_{24} amplifies specifically the wave E_3 created by scattering of the pumping wave by its "own" grating δn_{13}.

Transient analysis of the set of Eqs. (9) and (6) showed an exponential growth of the joint scattering wave near the oscillation threshold:

$$|E_4(z=0)|^2 \sim |\varepsilon|^2 \exp\left(2\left(\frac{G}{G_{\text{th}}} - 1\right)\frac{t}{\tau_{13}}\right) \quad (12)$$

where G_{th} is the threshold of the parametric oscillation which can be found (as well as the frequency detuning $\Omega\tau_{13}$) from expressions (11) when $\Gamma = 0$.

The threshold G_{th} and frequency detuning $\Omega\tau_{13}$ are functions of the absorption and reflection coefficient of the mirrors (Fig. 10.3). The minimum of the generation threshold $G_{th} \approx 16.5$ is achieved for $\alpha l = 1$.

For a small absorption in the nonlinear layer ($\alpha l \ll 1$) the generation threshold can be found analytically even in the general case $\Delta kl \neq 0$. The following transcendent equation determines the oscillation threshold in this case [23]:

$$\exp(p_2 - p_1) = \frac{p_2 + A_{11} + A_{12}r_{43}(\cos\Delta\varphi + i\sin\Delta\varphi)}{p_1 + A_{11} + A_{12}r_{43}(\cos\Delta\varphi + i\sin\Delta\varphi)} \quad (13)$$

where

$$p_{1,2} = -\frac{A_{11} + A_{22}}{2} \pm \frac{1}{2}\sqrt{(A_{11} - A_{22})^2 + 4A_{12}A_{21}},$$

$$A_{11} = i\Delta kl + \frac{iG_{th}}{1 + i\Omega\tau_{13}}, \quad A_{12} = \frac{iG_{th}}{1 + i\Omega\tau_{13}}r_{21}, \quad A_{21} = -A_{12}, \quad (14)$$

$$A_{22} = -i\Delta kl - \frac{iG_{th}}{1 + i\Omega\tau_{13}}r_{21}^2, \quad \Delta\varphi = \varphi_{34} - \varphi_{12}$$

Equation (13) with expressions (14) determine the oscillation threshold G_{th} and frequency detuning $\Omega\tau_{13}$ as a function of several parameters Δkl, $\Delta\varphi$, r_{21} and r_{43}.

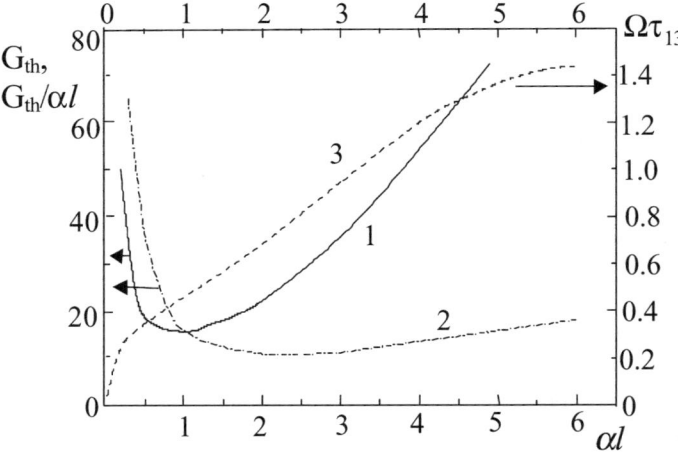

Figure 10.3. Dependencies of oscillation threshold of phase conjugate wave G_{th} (1), threshold normalized to absorption $G_{th}/\alpha l$ (2), and normalized frequency detuning $\Omega\tau_{13}$ of the wave at the oscillation threshold on the absorption αl.

For the phase conjugated waves (in the absence of the wave-vector detuning $\Delta kl = 0$), the expression of the oscillation threshold and the frequency detuning can be found explicitly:

$$G_{th} = \frac{(\ln B)^2 + (\arctan C + 2\pi N)^2}{(\arctan C + 2\pi N)(1 - r_{21}^2)},$$

$$\Omega = \frac{\ln B}{(\arctan C + 2\pi N)\tau_{13}}$$

(15)

where

$$B = \sqrt{\frac{1 + 2r_{21}r_{43}\cos\Delta\varphi + r_{21}^2 r_{43}^2}{2r_{21}r_{43}(1 + \cos\Delta\varphi)}}, \qquad C = \frac{(1 - r_{21}^2)\sin\Delta\varphi}{1 + r_{21}r_{43}\cos\Delta\varphi},$$

N is an integer

It can be shown that the solutions of equation $\Gamma = 0$, which was determined by Eqs. (11), transit to Eqs. (15) if $\alpha l \ll 1$.

The solutions (15) of the dispersion equation (13) include the branches with different values of N. The number N can be regarded as an analogue of the serial number of a longitudinal mode generated by the joint scattering in the nonlinear layer. It is possible to see from Eq. (15) that the oscillation threshold of the phase conjugated wave and its frequency are strongly dependent on the transmission coefficients of the feedback loop ($\Delta\varphi$, r_{21}, r_{43}) and the number N (Fig. 10.4). For the reciprocal feedback loop ($\Delta\varphi = 0$, $r_{21} = r_{43}$) the minimal generation threshold is achieved for the mode with $N = 1$ having nonzero frequency detuning Ω (Fig. 10.4). The detuning indicates a general peculiarity of SRS: The pump wave diffracts on the slow-moving grating into the scattering wave with a small frequency shift comparable to a scattering line width. Note that the threshold is infinite $G_{th} \to \infty$ for the equal intensity of the pumping waves $r_{21} = r_{43} \to 1$.

In the presence of the phase nonreciprocity of the feedback loop ($\Delta\varphi \neq 0$, $r_{21} = r_{43}$) the generation threshold can be considerably less than that for the reciprocal feedback. The smallest threshold can be achieved for $\pi/2$ phase nonreciprocity (compare curves 1 and 3 in Fig. 10.4). This fact can be explained as follows. The $\pi/2$ phase nonreciprocity provides an optimal phase shift between the optical interference field $E_1^* \cdot E_3$ (or $E_2^* \cdot E_4$) and the index grating induced by the interference field of another pair of the optical waves δn_{24} (or δn_{13}). As a result, the parametric interaction of two pairs of waves by mutual scattering on the grating is more effective than without any phase shift.

The threshold decrease in the scheme with the nonreciprocal feedback loop is particularly important in the case of equal-intensity pump waves ($r_{21} = r_{43} = 1$) when the reflection coefficient of the phase conjugate mirror based on the joint SRS can be close to unity [14, 25]. In this case the frequency of the phase conjugated

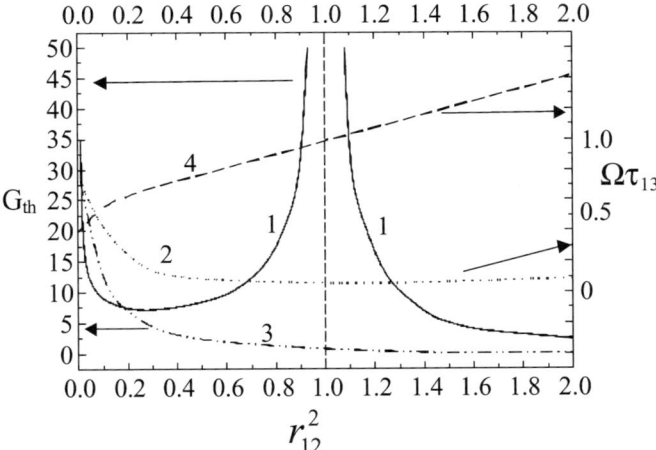

Figure 10.4. Oscillation threshold G_{th} of phase conjugate wave (curves 1 and 3) and its normalized frequency detuning $\Omega\tau_{13}$ (curves 2 and 4) versus transmission coefficient of the feedback loop $r_{21}^2 = r_{43}^2$ in the phase-reciprocal case, $\Delta\varphi = 0$ (curves 1 and 2), and nonreciprocal case, $\Delta\varphi = \pi/2$ (curves 3 and 4), for a poorly absorbing nonlinear layer ($\alpha l \ll 1$).

waves generated by the joint SRS coincides exactly with the line center of the common SRS ($\Omega\tau_{13} = 1$ for curve 4 in Fig. 10.4).

The reciprocal and nonreciprocal feedback systems differ also in the dependence of the parametric oscillation threshold on the mismatch of the spatial four-wave mixing synchronism ($\Delta kl = 2kl\cos\theta\sin(\delta\theta)$). The oscillation threshold in the scheme with the reciprocal loop has its minimum value for the scattering waves propagating at an angle $\delta\theta$ to the counterrunning pumping wave characterized by $\Delta kl \neq 0$ (curves 1 and 2 in Fig. 10.5). In other words, the oscillation threshold for the phase conjugated wave is not the smallest one. This fact takes place also for the parametric oscillation in a medium with local "photorefractive" nonlinearity [26–28]. If a scheme with the phase-nonreciprocal feedback ($\Delta\varphi = \pi/2$) is used, the parametric oscillation threshold has its minimum value for the scattering wave back-propagating to the pumps ($\delta\theta = 0$) even in the case of total reflectivity of the feedback mirrors $r_{21} = r_{43} = 1$ (curve 4 in Fig. 10.5). This means that in the scheme with the phase nonreciprocity, the phase conjugated wave can strongly predominate over nonconjugated component due to big difference of their increments.

To achieve SPPC by joint SRS in experiment, it is interesting to estimate the theoretical limit of the nonlinear reflection coefficient. The set of Eqs. (5) and (6) has several integrals that allow one to find simple relationships for the intensities of the interacting waves (for the short nonlinear layer ($l/v \ll \tau_p$), and the amplitude-reciprocal feedback $r_{21} = r_{43}$) [25]:

$$I_1(z) + I_3(z) = I_1(0)\exp(-\alpha z), \quad I_2(z) + I_4(z) = r_{12}^2 I_1(0)\exp(-\alpha(2l-z)) \quad (16)$$

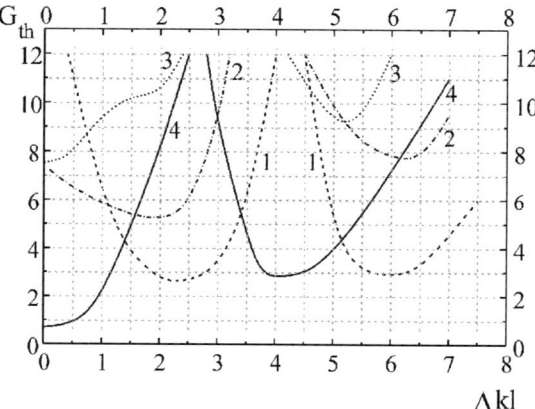

Figure 10.5. Generation thresholds of the scattering wave G_{th} versus detuning of four-wave synchronism Δkl for different transmission coefficients in the reciprocal loop (curves 1 correspond to $r_{21}^2 = r_{43}^2 = 0.75$, curves 2 correspond to $r_{21}^2 = r_{43}^2 = 0.1$, curves 3 correspond to $G_{th}(Q_{ef})$ for the scattering speckled beam with $\Theta/\Theta_d = 1$ and $r_{21}^2 = r_{43}^2 = 0.1$) and in the phase noreciprocal loop (curves 4 correspond to $\Delta\varphi = \pi/2$ and $r_{21}^2 = r_{43}^2 = 0.75$).

The theoretical limit of the nonlinear reflection of the SPPC mirror by Eqs. (16) can be estimated as follows:

$$\frac{I_4(z=0)}{I_1(z=0)} \leq r_{21}^2 \exp(-2\alpha l) \quad (17)$$

Expression (17) gives quite optimistic estimation for reflectivity of the SPPC mirror based on the joint SRS in a small-absorbing nonlinear layer ($\alpha l \ll 1$) and with a good-transmitting feedback loop ($r_{21} \to 1$). Numerical calculation of the set of Eqs. (5) and (6) also confirmed the possibility of high reflection coefficient of the nonlinear mirror [25].

This analytical analysis is valid only for the plane pumping waves. For real spatially finite beams (when the transverse beam displacement in the nonlinear layer is comparable to or higher than the beam diameter) the set of partial derivative equations is more complicated and has additional features.

10.2.2.2 Specific features of phase conjugation of speckle-inhomogeneous beams

The phase conjugation of a speckle-inhomogeneous beam is a traditional and most evident experimental test. However, as is well known, the theoretical description of the phase conjugation of the speckle beam is much more sophisticated than the theory in the plane-wave approximation even for conventional backscattering [5]. Similar (or even more complicated) problems appear in case of the joint scattering of two intersecting beams.

The equations for the joint scattering of the speckle-inhomogeneous beams need to be changed: A transverse Laplacian ($\nabla_\perp^2$) must be added to the right-hand side of the equation for electrical field of the optical waves. In a small absorbing medium ($\alpha l \ll 1$) the set of equations is as follows:

$$\frac{\partial E_1}{\partial z} + \frac{i}{2k}\nabla_\perp^2 E_1 = i\beta T_0 E_1,$$

$$-\frac{\partial E_2}{\partial z} + \frac{i}{2k}\nabla_\perp^2 E_2 = i\beta T_0 E_2,$$

$$\frac{\partial E_3}{\partial z} + \frac{i}{2k}\nabla_\perp^2 E_3 = i\beta(T_0 E_3 + T_{13} E_1),$$

$$-\frac{\partial E_4}{\partial z} + \frac{i}{2k}\nabla_\perp^2 E_4 = i\beta(T_0 E_4 + T_{13} E_2)$$

(18)

The steady-state approximation for the strongest (single) grating T_{13} and an assumption that the beam diameter is much more than the characteristic transverse dimension of speckles allow us to use a technique that has been developed for SBS of the speckle beams [5]. For a pumping beam with plane envelope, the solution of Eqs. (18) for complex amplitudes can be written using the Fourier transform:

$$E_{1,2} = \sum_m E_{1,2}^m \exp\left(i\vec{q}_m \vec{r} \pm \frac{iq_m^2 z}{2k} \pm i\beta <T_0> z\right) \quad (19)$$

The solution for the scattering waves can be found in a similar expression, taking into account the scattering wave increment and frequency detuning:

$$E_{3,4} = \sum_m E_{3,4}^m \exp\left(i\vec{q}_m \vec{r} \pm \frac{iq_m^2 z}{2k} + i\vec{q}\vec{r} + pz + i\Omega t \pm i\beta <T_0> z\right) \quad (20)$$

Using the set of equations (18) for wave amplitudes (19) and (20) and then averaging them over individual spatial harmonics and taking into account only constant (along the z axis) components in the nonlinear terms (in the approximation of small scattering increment at one characteristic length of speckle inhomogeneity z_{ch}) $E_{3,4}^m \sum_m |E_{1,2}^m|^2$ in $E_{3,4}|E_{1,2}|^2$ and $E_{2,1}^{m*} \sum_m E_{1,2}^m E_{4,3}^m$ in $E_{2,1}^* E_{1,2} E_{4,3}$, the relations for amplitudes of the scattering waves can be written as follows:

$$E_3^m = \frac{iG E_2^{m*} \sum_n E_1^n E_4^n}{(1+i\Omega\tau_{13})\langle|E_1|^2\rangle(pl - i(\vec{q}\vec{q}_m/k)l - iG/(1+i\Omega\tau_{13}))},$$

$$E_4^m = \frac{-iG E_1^{m*} \sum_n E_2^n E_3^n}{(1+i\Omega\tau_{13})\langle|E_1|^2\rangle(pl + i(\vec{q}\vec{q}_m/k)l + iGr_{21}^2/(1+i\Omega\tau_{13}))}$$

(21)

where $G = \gamma\beta\tau_{13}\langle|E_1|^2\rangle l$ is the steady-state increment of the common scattering of the wave E_1, and $\langle|E_1|^2\rangle$ is the average-in-space intensity.

Expressions (21) allow us to find two solutions for parameter p by expanding the denominators into series in terms of small parameter $\vec{q}\vec{q}_m l/kG \ll 1$. Then, the solution for correlation functions $\langle E_1 E_4 \rangle = \sum_n E_1^n E_4^n$ and $\langle E_2 E_3 \rangle = \sum_n E_2^n E_3^n$ shows the possibility of their infinite growth. The oscillation threshold is determined by solving the transcendent equation that can be written by analogy with Eq. (13):

$$\frac{A_1 + B}{A_2 - B} = \exp(iB) \tag{22}$$

where

$$A_1 = G_1(1 + 3r_{21}^2) - 2Q_{\text{ef}} + 2H(2r_{21}^2 + r_{21}^{-2})/G_1,$$
$$A_2 = G_1(1 + 3r_{21}^2) - 2Q_{\text{ef}} + 2H(r_{21}^{-4} + r_{21}^{-2} + r_{21}^2)/G_1,$$
$$B = \sqrt{(1 - r_{21}^2)^2 G_1^2 + 4Q_{\text{ef}}^2 - 4H(r_{21}^2 + r_{21}^{-2}) - 4G_1 Q_{\text{ef}}(1 + r_{21}^2)}, \tag{23}$$
$$Q_{\text{ef}} = \frac{q^2 l}{2k} + \sum_n \frac{\vec{q}\vec{q}_n |E_1^n|^2 l}{k\langle|E_1|^2\rangle} \approx \frac{q^2 l}{2k},$$
$$G_1 = \frac{G_{\text{th}}}{1 + i\Omega\tau_{13}}, \quad H = \sum_n \frac{(\vec{q}\vec{q}_n)^2 |E_1^n|^2 l^2}{k^2 \langle|E_1|^2\rangle}$$

Expressions (23) differ from similar expressions (14) by parameter Q_{ef}, used instead of $\Delta k l$, and a new parameter H. At $q = 0$, Eq. (22) with (23) transforms into Eq. (13) with (14). The parameter $H \approx (q^2 l/2k)(l/z_{\text{ch}}) \approx Q_{\text{ef}}(l/z_{\text{ch}})$ has an influence upon the dependence of the threshold G_{th} on the spatial "detuning" parameter Q_{ef}. The minimum of the dependence $G_{\text{th}}(Q_{\text{ef}})$ shifts to $Q_{\text{ef}} = 0$ already when $l/z_{\text{ch}} \geq 1$ (see curve 3 in Fig. 10.5). However, other minima of the dependence $G_{\text{th}}(Q_{\text{ef}})$ (other branches) can have approximately the same values as the first minimum $G_{\text{th}}(0)$. This fact means that the angular displacement of the scattering wave relative to the pumping wave at one speckle length leads to an increase in its oscillation threshold; however, the discrimination condition of these waves in comparison with the phase conjugated wave may not be satisfied. The phase conjugate condition can be expressed as a strong increase of the oscillation threshold for a scattering wave with displacement at an angle more than Θ_d with respect to the conjugated wave (where the angle Θ_d is the diffraction divergence of a single-mode beam with the same radius as that of envelope of the speckled beam E_1). Such displacement corresponds to $q_d = k\Theta_d$, and the phase conjugate condition can be written (by analogy with the theory of phase conjugation by the backward SBS [5]) as

$$G_{\text{th}}(q_d) \geq 2G_{\text{th}}(q = 0) \tag{24}$$

The parameter H can be expressed as $H = N_\Theta \, Q_d \, (l/z_{ch}) = N_\Theta \, (l/z_d)$ (where $N_\Theta = (\Theta/\Theta_d)^2$, Θ is the real divergence of the pumping beam, $z_d = (k\Theta\Theta_d)^{-1}$ is the diffraction length of the speckled beam). Analysis of the dependence $G_{th}(N_\Theta)$ for fixed (but different) values of the parameter (l/z_d) showed that the condition (24) (for predominance of the phase conjugated beam over angular displacement scattering waves) is fulfilled when the diffraction length of the speckled beam is comparable with the thickness of the nonlinear layer: $(l/z_d) \geq 1$ [29]. This means that for the nonlinear layer with a mirror at the boundary, the mechanism for selection of the phase conjugate wave is caused both by the diffraction in the nonlinear layer and by the nonlinear amplification (similar to the ordinal backward SBS of a focused beam [5]).

Strictly speaking, the analysis of the phase conjugate conditions is valid for plane envelope of the speckle-inhomogeneous beam. Another approach to investigation of generation of scattering beams by the joint SRS of speckle-heterogeneous laser beams with Gaussian envelope showed that the phase conjugation condition in the nonlinear layer with mirror boundary can be realized under similar condition, when $(l/z_d) \geq 1$ [15].

The SPPC condition obtained concerns only the nonlinear layer with a mirror-reflecting wall. In case of the feedback loop with a long pass, the phase conjugation can be achieved even in a layer with thickness less than the diffraction length of the pumping beam.

10.2.3 Experimental investigations of self-pumped phase conjugation of laser beams in nematic liquid-crystal layers

10.2.3.1 Self-pumped phase conjugation by thermal scattering
Phase conjugation of laser beams by joint SRS was investigated in NLC layers both with thermal and orientational nonlinearity. The SPPC of laser beam with pulse duration ranging from several microseconds to several milliseconds at different wavelengths (514.5 nm, 1064 nm, 10.6 μm) was realized in the NLC layer with a feedback loop [15, 23–25, 29–32].

A simple phase conjugate mirror was made using the NLC layer at the mirror surface (Fig. 10.6). The NLC layer with different thickness was used. The planar orientation of molecules in the NLC layer with thickness of 100–200 μm was achieved by creating a relief on the window surface coating. The homeotropic orientation of nematic molecules in the layer 0.5–2.0 mm in thickness was achieved by applying the ac electric field of 1–2 kV/cm at 50 Hz across the cell. The voltage was applied both to a thin electroconducting coating on the glass window and to a metallic mirror. Several mixtures of cyanobiphenyls (with intrinsic absorption of $\alpha \approx 0.1$–0.5 cm^{-1} at 1064 nm) or dye-doped cyanobiphenyls (with absorption of 4.0–10.0 cm^{-1}) were used as the NLC.

The laser beams were focused in the NLC layer. The focal waist length inside the NLC layer was chosen comparable with the layer thickness. The joint near-forward thermal scattering of the intersecting incident and mirror-reflected beams resulted in generation of a phase-conjugated beam (Figs. 10.7a and 10.7b). Both threshold of

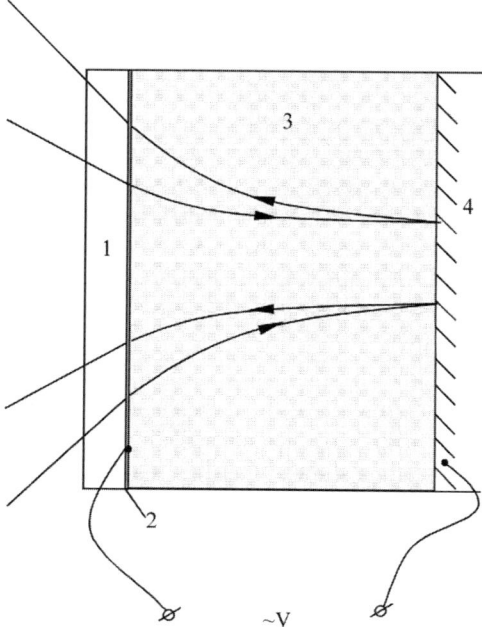

Figure 10.6. A nematic liquid crystal cell: (1) Entry window, (2) current-conducting layer, (3) NLC mixture, (4) totally reflecting mirror; "$\sim V$" is ac voltage for molecules orientation.

the SPPC-wave generation and the coefficient of energy transfer of the incident beam into the phase-conjugated beam were found to be dependent on the incidence angle, absorption, and temperature of the NLC. The threshold intensity of pumping wave was measured to be from several kW/cm² (the pumping power was several tens of mW) for the strongly absorbing layer with $\alpha \approx 4.0$ cm^{-1} to several tens of kW/cm² (the pumping power was several hundreds of mW).

The conditions of the phase conjugation was investigated both with a phase aberrator, which increases the divergence of the initial beam divergence by 10 times, and with an amplitude transparence "net." The phase conjugated beam was observed in both experiments when the initial beam was focused in the NLC layer at a small incident angle to normal to the mirror (Figs. 10.7c and 10.7d). In the case of the phase aberrator, power of the phase conjugated component increased when the confocal parameter of the beam was comparable to the NLC layer thickness (the phase conjugation takes place when $z_f = z_d \leq l$, in accordance with the theory described in Section 10.2.2). For a fixed confocal parameter, the power of the SPPC beam decreased when the incident angle of the input wave was much more than the geometry angle ($\theta_{in} > \theta_g$, where $\theta_g = r_0/f$, f is the focal length).

The temperature dependence of generation of the phase conjugated beam can be accounted for by the origin of the thermal nonlinearity of NLC. The fact is that the heat-induced changes of the NLC refractive index have two components caused both

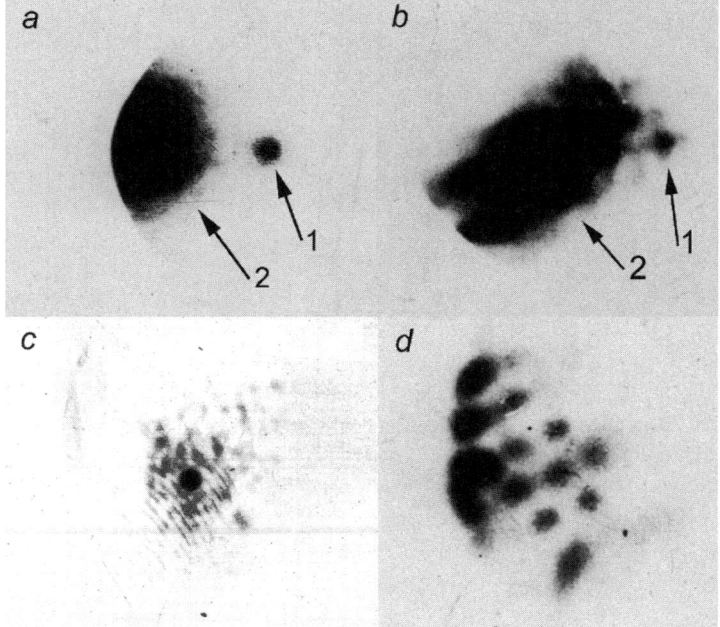

Figure 10.7. Space structures of the reflected beam from the NLC film with mirror boundary in the near field (a) and far field (b) (1, phase conjugate beam; 2, pump beam reflected by the mirror); the far field of the beam after two passes through a phase plate (c); and a phase conjugated image of the amplitude transparency "net" (d).

by density changes ρ and by changes of the NLC order parameter S [6]:

$$\frac{dn}{dT} = \left(\frac{\partial n}{\partial \rho}\right)\left(\frac{\partial \rho}{\partial T}\right) + \left(\frac{\partial n}{\partial S}\right)\left(\frac{\partial S}{\partial T}\right) \tag{25}$$

Near the phase transition point from nematic phase to an isotropic liquid, the thermal nonlinearity of NLC strongly increases primarily due to the strong temperature dependence of the order parameter which for a pure homogeneous NLC layer can be approximated by [33, 34]

$$S = S_0 + S_1(T_{\text{cr}} - T)^\eta \tag{26}$$

where S_0 and S_1 are normalization constants, T_{cr} is the temperature of the nematic–isotropic transition, and parameter η is determined by the type of NLC. Since the parameter $\eta \leq 1$ (for pentyl cyanobiphenyl 5CB $\eta \approx 0.16$ [34]), the temperature derivative of the order parameter strongly increases near the phase-transition temperature: $\partial S/\partial T \to \infty$, when $T \to T_{\text{cr}}$. Therefore, the coefficient of the NLC thermal nonlinearity near the phase transition point can be several orders of

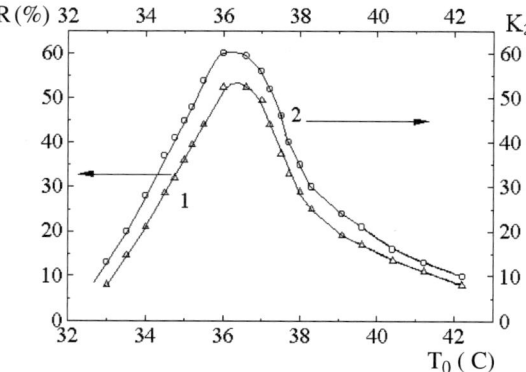

Figure 10.8. Dependencies of the reflection coefficient of the NLC mirror R (1) and of the power gain of a double-pass amplifier (2) on temperature of the cell walls in the NLC layer 5 mm in thickness with absorption coefficient of 0.5 cm^{-1} (pump intensity at the lens focus is 45 kW/cm^2, pump pulse duration is 0.6 ms, single-pass gain of amplifier is 11).

magnitude higher than that in an ordinary liquid. The NLC temperature is determined by the initial cell temperature and heating of the crystal in the beam channel during the laser pulse.

The SPPC reflectivity of the NLC mirror was optimized by appropriately choosing the initial temperature of the cell, the NLC absorption, and the pump intensity to achieve the temperature which is close to the phase-transition point in the area of the nonlinear interaction of the optical beams in the peak of their pulses [25, 35]. The power transfer of the pumping beam into the SPPC beam was achieved up to 60% (Fig. 10.8). However, if during the laser pulses heating of the nonlinear NLC layer in the beam channel destroyed the mesophase, the phase conjugation was suppressed (the nonlinear reflection coefficient and quality of the phase conjugated beam decrease due to strong opalescence in the phase transition point).

Another possibility of decreasing the SRS oscillation threshold and hence increasing the SPPC beam power was demonstrated in the scheme with a transmitting NLC layer and a nonreciprocal feedback loop (Fig. 10.9) [23]. The $\pi/2$ phase nonreciprocity of the feedback was achieved by using a quartz polarization rotator (rotating the beam polarization at 90°) and a quarter-wave plate. The cross-polarized pump and scattering beams had phase shifts in the quarter-wave plate which differed by $\pi/2$. The polarization orientation of the pump beam and director of the homeotropic NLC layer was chosen such as to avoid the orientational nonlinear effects (orientational SRS and lensing). The generation of the phase conjugated beam with polarization orthogonal to the initial beam polarization was obtained. The threshold of the SPPC beam generation in the scheme with the nonreciprocal feedback was measured to be less than in the case of the reciprocal feedback (Fig. 10.10).

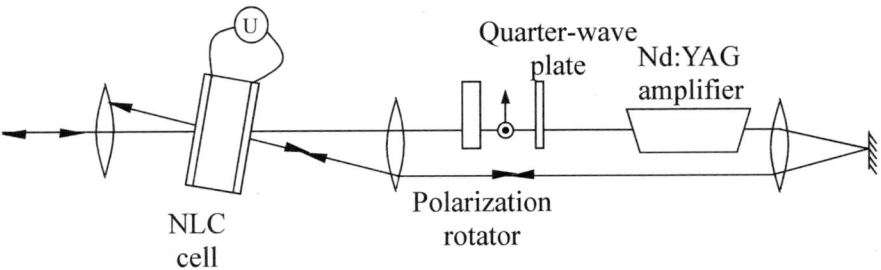

Figure 10.9. Schematic of experimental arrangement used in the study of self-pumped phase conjugation in a NLC layer with reciprocal or nonreciprocal feedback loop.

The thermal NLC mirror (the NLC layer on the mirror) with the optimized temperature was used to improve the beam quality in double-pass and four-pass Nd:YAG amplifiers at pulse duration 0.6–1.0 ms and repetition rate 10–20 Hz [25, 35]. The phase conjugation by near-forward SRS by the NLC mirror had the following advantages: small threshold of the pumping beam power (5–6 orders less than for backward SBS) and the possibility of using broadband radiation with high average power. The SPPC NLC mirror provided partial compensation of thermally induced aberrations in the laser amplifier and improvement of the beam quality after four passes through the Nd:YAG laser rod. The total amplification coefficient for the beam power of about 400 was achieved at the repetition rate of 10–20 Hz. The limit of fluence of the pumping beam due to overheating of the NLC layer during a laser

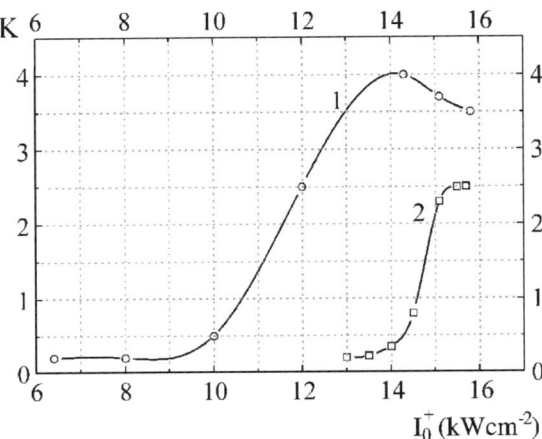

Figure 10.10. Experimental dependencies of the reflection coefficient K of a phase-conjugate mirror based on nematic liquid crystal on pump intensity for the reciprocal (1) and nonreciprocal feedback loop (2).

pulse was estimated as 0.6 kJ/cm^2 for absorption $\alpha \approx 0.1 \text{ cm}^{-1}$ and temperature interval of mesophase $T_{\text{cr}} - T \approx 40°\text{C}$. This limiting pump fluence can be increased by stabilizing the NLC temperature due to heat outflow from the thin nonlinear layer to good thermal-conducting walls.

10.2.3.2 Self-pumped phase conjugation by orientational scattering

The self-pumped phase conjugation of a laser beam in the NLC layer by joint orientational stimulated scattering was also investigated [24]. The origin of the orientational optical nonlinearity of NLC is caused by refractive-index changes due to rotation of a molecule under electric field of the laser beam [5, 6]. The nonlinear-optical effects of the orientational SRS and self-action have a longer decay time than the thermal effects and can be obtained, as a rule, by less intensive optical field.

The experiments were made using a quasi-CW Nd:YAG laser at 1064 nm with pulse duration of about 0.6–1.0 ms. The mutual orientation of pump-beam polarization and director of the homeotropic NLC layer was chosen in the experiment to maximize the orientational grating induced by the interference field of the cross-polarized pumping and scattering waves (Fig. 10.11) [24]. The generation of a phase-conjugated beam co-polarized to the pumping wave was achieved.

The possibility of decreasing the SPPC threshold by the $\pi/2$ phase noreciprocity of the feedback loop was also studied. The decrease of threshold and hence an increase in the phase conjugated beam power were achieved in case of the $\pi/2$ phase noreciprocity of the feedback loop. It should be noted that in the nonreciprocal scheme the quality of the phase conjugated beam also increased.

The SPPC mirror based on joint orientational scattering in the NLC layer with the feedback loop has showed potential for use in a double-pass amplifier for compensation of thermally induced aberrations in the laser crystal [24].

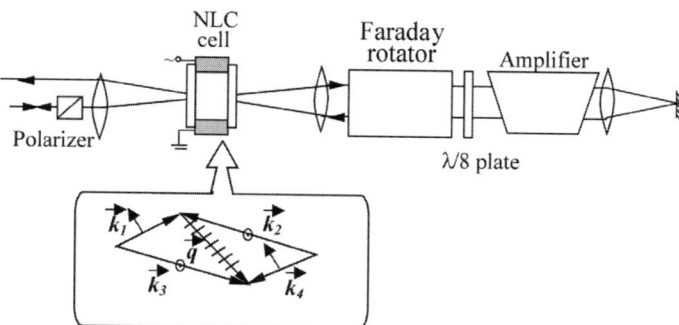

Figure 10.11. Schematic of experimental arrangement used in the study of phase conjugation by joint orientational scattering in a NLC layer and wave-vector diagram of the interacting waves.

10.3 SELF-STARTING LASERS WITH A NONLINEAR MIRROR BASED ON NEMATIC LIQUID CRYSTALS

10.3.1 Theoretical description of the principle of self-starting lasers

The small-threshold SPPC mirror based on the NLC layer with a feedback is attractive to use for compensation of different aberrations in a laser cavity. One solution of the laser is a self-starting laser oscillator with cavity completed by the NLC phase conjugated mirror.

The general idea of the laser oscillator with the cavity formed with participation of the nonlinear mirror can be illustrated by a scheme consisting of amplifiers and a NLC cell (Fig. 10.12). The optical wave E_1 starting initially from the level of spontaneous emission in the laser amplifier meets and interferes with waves E_2, E_3, and E_4 which are the result of propagation of the initial wave through the scheme. The interference fields of the optical waves induce in the NLC layer the gratings of refractive index Q_{ij} which were described by Eq. (4).

The stronger waves E_3 and E_4 can reflect from the RIG, providing energy transfer to the weaker waves E_1 and E_2. As has been shown in Section 10.2, for the NLC nonlinearity of diffuse type, it is only a moving RIG that can provide the nonlinear amplification of a weak wave (this fact results from the necessity to compensate for phase mismatch which occurs because of the wave mixing with participation of the RIG). Therefore, only the moving RIG can ensure positive feedback in the self-starting scheme. The moving RIG, in turn, can be effectively induced by the interference of optical waves with frequency detuning. The frequency shift of the optical waves reflected from the moving grating can be compensated in the cavity by movement of an ordinary mirror. Therefore, one may expect the self-starting oscillation in the presence of frequency detuning of the interacting waves ($\omega_i \neq \omega_j$) which can be caused by movement (or random mechanical vibrations) of the intracavity mirrors M_1 or M_2 and M_3.

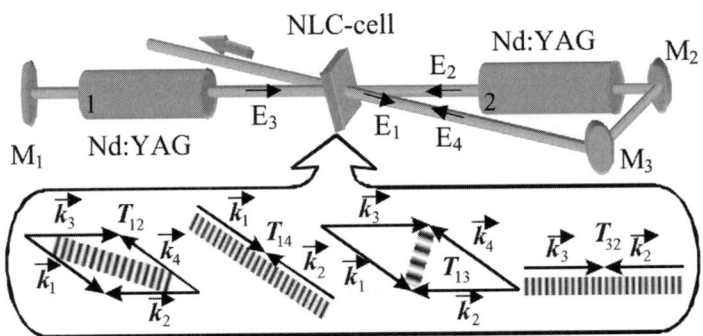

Figure 10.12. Model of the self-starting laser with holographic gratings in NLC and a wave-vector diagram: k_1-k_4 are wave vectors of generating beams, T_{ij} are gratings in NLC, and M_1-M_3 are linear mirrors.

The self-starting laser can be described in another way. The right-hand part of the scheme in Fig. 10.12, including an NLC cell, an Nd:YAG amplifier 2, and mirrors M_2 and M_3 is actually a self-pumped phase conjugator. In this interpretation, the laser oscillator consists of a linear mirror M_1, an Nd:YAG amplifier 1, and a self-pumped phase-conjugate mirror based on the NLC. The phase conjugate mirror can provide the adaptivity of the laser oscillator and the formation of a stable mode [12, 13].

A simple analytical solution of the self-starting laser can be obtained using the steady-state approximation for the strongest large-scale grating T_{13} and under an assumption that one weak wave E_1 gets amplified due to nonlinear energy transfer from three strong waves E_2, E_3, and E_4 by scattering. Equations (5) with the steady-state grating T_{13} can be used for the analysis. At determined intensities of the strong waves (only linear absorption was assumed for the strong waves), the solution for the wave E_3 can be found as follows:

$$E_1(l) = \frac{i\gamma\beta\tau_{13}}{1+i\Omega\tau_{13}} \int_0^l E_3(0)E_2(l)E_4^*(l)\exp(-\alpha(l-z'))$$

$$+ \frac{i\gamma\beta\tau_{13}}{1+i\Omega\tau_{13}} \int_{z'}^l |E_3(0)|^2 \exp(-\alpha z'')\,dz''\,dz' \quad (27)$$

where Ω is the frequency detuning of waves E_4 and E_2 with respect to E_3 and E_1. To obtain self-starting generation, it is necessary to take account of amplified spontaneous emission in intracavity amplifier. This can be done with the boundary condition for the wave E_2. Assuming that the absorption in the NLC cell is small ($\alpha l \ll 1$), the expression (27) can be transformed into

$$E_1(l) = \frac{\varepsilon r_{43}^*(\exp((i\gamma\beta\tau_{13}/1+i\Omega\tau_{13})|E_1(l)r_{32}r_{21} + \varepsilon r_{32}|^2) - 1)}{1 - r_{21}r_{43}^*(\exp((i\gamma\beta\tau_{13}/1+i\Omega\tau_{13})|E_1(l)r_{32}r_{21} + \varepsilon r_{32}|^2) - 1)} \quad (28)$$

where ε is the spontaneous emission amplitude at the boundary of amplifier 2, and r_{ij} is the reflection coefficient (including amplification) of the jth wave to the ith wave in the cavity.

Expression (28) determines the amplitude of the scattering wave E_1 as a function of the amplified spontaneous emission. This function has a region of multiple values (Fig. 10.13). Such S-like characteristic has an unstable branch with a critical point corresponding to the condition of the self-starting oscillation. The estimation of the appropriate level of the spontaneous emission for oscillations to occur is one of the most important points in the self-starting laser theory. However, in real experiments the time of generation development is much smaller than the relaxation time of the grating. Therefore, the steady-state approximation for the grating is invalid. Unfortunately, there is no transient analytical solution of the partial derivative equation, and one needs to use numerical computations when analyzing experimental conditions of the self-starting oscillation.

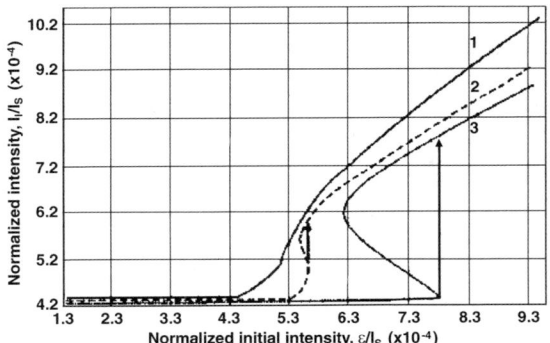

Figure 10.13. Dependencies of intensity of scattering wave E_1 normalized to amplifier saturation intensity I_S for different frequency detunings: $\Omega\tau_{13} = 1$ (curve 1), $\Omega\tau_{13} = 1.4$ (curve 2), and $\Omega\tau_{13} = 1.9$ (curve 3), $G(I_S) = \beta\gamma\tau_{13}I_S$, $r_{21} = r_{43} = 10$. Arrows indicate points of self-starting oscillations.

10.3.2 Numerical computation of the self-starting laser with an NLC mirror

To study the self-starting conditions of the laser oscillator, a set of Eqs. (5) for the complex amplitudes E_1–E_4 of the electric field of the optical waves was numerically calculated.

In our computations we took into consideration the amplification saturation by generating waves in amplifiers 1 and 2 (Fig. 10.12). The amplification of the waves in amplifiers was described by the following set of equations:

$$\mu_2 \frac{\partial E_1}{\partial t} - \frac{\partial E_1}{\partial z_2} = \sigma E_1 N_2, \quad \mu_2 \frac{\partial E_4}{\partial t} + \frac{\partial E_4}{\partial z_2} = \sigma E_4 N_2,$$

$$\mu_1 \frac{\partial E_2}{\partial t} - \frac{\partial E_2}{\partial z_1} = \sigma E_2 N_1, \quad \mu_1 \frac{\partial E_3}{\partial t} + \frac{\partial E_3}{\partial z_1} = \sigma E_3 N_1,$$
(29)

where σ are the unsaturated cross section of the laser transition inside the Nd:YAG amplifiers, $N_{1(2)}$ are the populations of the metastable level in the amplifier with index i ($i = 1$ or 2), $z_{1(2)}$ are coordinates inside the amplifiers; $\mu_{1(2)} = l_{1(2)}/(v\,\tau_R)$, $l_{1(2)}$ are lengths of the amplifier rods, τ_R is the relaxation time of the working transition, and v is the light velocity. The temporal behavior of the amplification of the amplifiers is described by equations for populations N_1 and N_2

$$\frac{\partial N_1}{\partial t} + \frac{N_1}{\tau_R} = N_e - \frac{N_1}{W_S}(E_2 E_2^* + E_3 E_3^*),$$

$$\frac{\partial N_2}{\partial t} + \frac{N_2}{\tau_R} = N_e - \frac{N_2}{W_S}(E_1 E_1^* + E_4 E_4^*)$$
(30)

where N_e is the pump velocity whose temporal profile was chosen close to that in experiment: $N_e(t) = [\exp(-t/t_{off})(1 - \exp(-t/t_{on}))]^2$ ($t_{off} = 300$ μs, $t_{on} = 200$ μs) and W_S is the saturation fluence.

The initial temporal conditions of the set of Eqs. (5), (29), and (30) were

$$E_1(z, t = 0) = E_2(z, t = 0) = E_3(z, t = 0) = E_4(z, t = 0) = 0,$$
$$T_0(z, t = 0) = T_{12}(z, t = 0) = T_{13}(z, t = 0) = T_{14}(z, t = 0)$$
$$= T_{23}(z, t = 0) = 0 \tag{31}$$

One of the most important problems in the theoretical description of the self-starting laser is the origin of the initial noise source which can provide the threshold condition of generation. Physically, it is the amplified spontaneous emission in the Nd:YAG amplifiers that is most likely to be the noise source for the initial excitation of oscillations in the system. For this reason, we considered the spontaneous emission by using a constant addition (ε) to the amplitude of the optical wave E_1 at the boundary of amplifier 2:

$$E_1(z = z_2^{out}, t) = E_1(z = l, t) r_{11} \exp(2ikL_{11} + i\Omega t) + \varepsilon \tag{32}$$

where Ω is the frequency shift due to vibration of the mirror M_1 which provides the frequency detuning of the interacting waves, L_{ij} are distances from the NLC boundary to the boundary of amplifiers 1 or 2, r_{ij} are the transmission coefficients of the optical waves in space, and $z_{1(2)}^{in(out)}$ are coordinates of the amplifier-rod boundaries.

From the viewpoint of the "self-starting" condition, the spontaneous emission (described by parameter ε) provides the necessary level of the optical wave intensity required to induce in the NLC cell the holographic gratings with an amplitude sufficient for positive feedback in the dynamic cavity. In a thin NLC layer (with thickness up to several millimeters) the optical waves with different frequencies within the fluorescence line of the Nd:YAG laser amplifier (linewidth of about 6 cm^{-1}) can effectively interact due to the common RIG (without any phase mismatch), similarly to monochromatic waves. Then the total fluorescence (within the whole frequency linewidth of the Nd:YAG amplifier) must be taken into account when analyzing the self-starting conditions. It is well known that the total intensity of spontaneous emission can be recalculated at the boundary of the laser amplifier and described by the following expression [36]:

$$|\varepsilon|^2 \approx \frac{I_S(d\Theta/4)(\exp(2\sigma N^0 l_2) - 1)^{1.5}}{(2\sigma N^0 l_2)^{0.5} \exp(3\sigma N^0 l_2)} \tag{33}$$

where I_S is the saturation intensity of the working laser transition, $d\Theta$ is the solid angle, N^0 is the density of the unsaturated population inversion.

Other boundary conditions for the NLC layer and the amplifiers were

$$\begin{align}
E_1(z=0,t) &= R_{ds}(E_2(z=0,t)+E_4(z=0,t)),\\
E_2(z=l,t) &= E_1(z=z_2^{in},t)r_{12}\exp(2ikL_{12}),\\
E_2(z=z_1^{out},t) &= E_2(z=0,t)r_{22}\exp(2ikL_{22}),\\
E_3(z=z_1^{in},t) &= E_2(z=z_1^{in},t)r_{23}\exp(2ikL_{23}),\\
E_3(z=0,t) &= E_3(z=z_1^{out},t)r_{33}\exp(2ikL_{33}),\\
E_4(z=z_2^{in},t) &= E_3(z=l,t)r_{34}\exp(2ikL_{34}),\\
E_4(z=l,t) &= E_4(z=z_2^{out},t)r_{44}\exp(2ikL_{44}+i\Omega t)
\end{align} \quad (34)$$

where R_{ds} is the diffuse scattering coefficient on the NLC boundary.

The set of Eqs. (5), (29), and (30) with the initial condition (31) and boundary conditions (32) and (34) describing the system that comprises the amplifiers and the nonlinear NLC cell was numerically calculated [37].

The numerical computations showed the presence of the nonlinear self-starting oscillations in the cavity even without any "linear" diffuse reflection on the rod boundary, $R_{ds}=0$ (Fig. 10.14). It was found that dynamics of the output beam is greatly different below and above the generation threshold. Below the generation threshold, there was only amplification of spontaneous emission, whereas above the threshold a pulse train was generated.

The generation threshold was found to be strongly dependent on different parameters of the scheme, such as the coefficient of the NLC nonlinearity ($\beta_{eff} = \beta \cdot \gamma$), the logarithmic gain $\alpha_{1,2}l_{1,2}$ of amplifiers, the intracavity frequency shift Ω, and the initial noise intensity (normalized to the saturation intensity of the resonant transition of Nd:YAG crystal) $|\varepsilon|^2/I_S$ (Fig. 10.15). The generation area was found in the plane of the frequency shift and the initial noise intensity at a fixed unsaturated

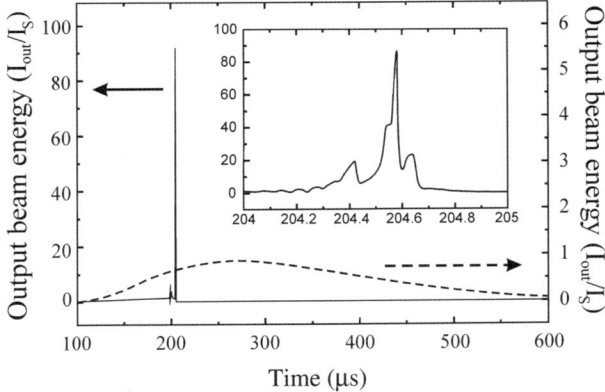

Figure 10.14. Numerically calculated oscillograms of generated pulse train beyond (solid curve) and below threshold (dashed curve).

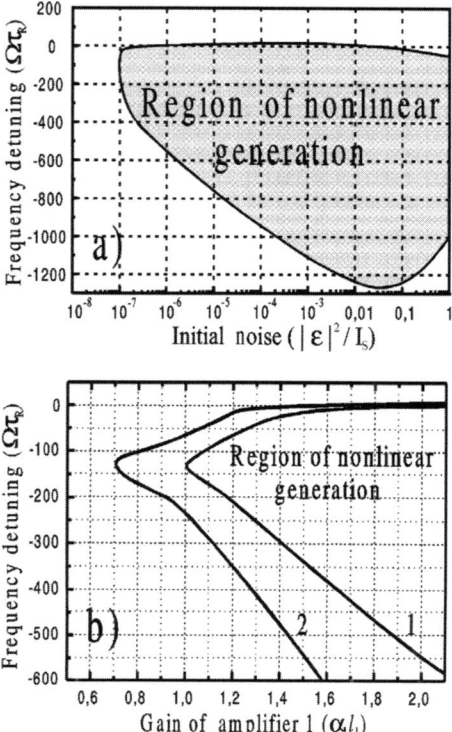

Figure 10.15. Numerically calculated regions of self-starting generation: (a) as a function of noise intensity (in the logarithmic scale) and frequency detuning at logarithmic field gains $\alpha_1 l_1 = 1.5$, $\alpha_2 l_2 = 2.95$ and NLC nonlinearity $\beta_{\text{eff}} = \beta \cdot \gamma = 18$; (b) as a function of logarithmic field gains $\alpha_1 l$ and frequency detuning at fixed noise intensity $|\varepsilon|^2/I_S = 10^{-7}$, $\alpha_2 l_2 = 2.95$ and different NLC nonlinearity $\beta_{\text{eff}} = 18$ (curve 1) and $\beta_{\text{eff}} = 36$ (curve 2).

gain ($\alpha_1 l_1 = 2\sigma N_1^0 l_1 = 1.5$) and constant NLC nonlinearity ($\beta_{\text{eff}} = 18$) (Fig. 10.15a). The dependence of the generation threshold on frequency detuning indicates the influence of intracavity mirror vibrations on the generation conditions. The resonance of the frequency detuning corresponds to the optimal nonlinear increment of the weakest waves due to two-wave mixing and four-wave mixing energy transfer from the stronger waves.

As mentioned above, the initial noise intensity, which determines the threshold of the self-starting generation, can be explained by the total spontaneous emission amplified in the Nd:YAG laser amplifiers. The estimation of the total fluorescence intensity, $|\varepsilon|^2/I_S$, using expression (33), yields $\approx 10^{-7}$ for the solid angle $d\Theta \approx 10^{-6}$ and the unsaturated amplifier gain $2\sigma N^0 l_2 \approx 2$. This value of the initial noise is in good agreement with numerically calculated threshold of self-starting generation (Fig. 10.15a).

The area of the self-starting generation was calculated also in the plane of the frequency detuning and the logarithmic gain in amplifier 1 at a constant noise intensity ($|\varepsilon|^2/I_S = 10^{-7}$) (Fig. 10.15b). This computation showed that there is an optimal frequency shift of the interacting waves at which the generation threshold in the self-starting laser is minimal. Such frequency shift can occur because of noise mechanical vibrations of the mirror in the cavity. Note that the calculated minimal threshold gain ($\alpha_1 l_1 = 1$) is in accordance with experimental results (all parameters of the calculated equations were chosen similar to those in the experiment with NLC cell with thermal nonlinearity).

The threshold logarithmic gain essentially decreased with an increase in the NLC nonlinearity β_{eff} (Fig. 10.15b, region 2). This result also agrees with experiment: Near the nematic–isotropic transition point the nonlinearity of NLC increases and the self-starting generation threshold decreases.

Thus, our numerical computation showed the presence of self-starting oscillations in the cavity completed by RIG induced in NLC by generating beams themselves.

10.3.3 Experimental investigation of the self-starting lasers

An experimental scheme of the laser oscillator with a dynamic holographic mirror written in the NLC cell by generation waves is shown in Fig. 10.16. The scheme consists of three flash lamp-pumped Nd:YAG amplifiers, a polarizer, ordinary mirrors, and a NLC cell. The pump pulse duration is 0.3 ms.

The origin of the gratings in the NLC completing the cavity was investigated by incorporating a quartz rotator (QR) and a Faraday rotator (FR) at different locations in the scheme. This arrangement allowed for four-wave interactions of a self-intersecting generated beam that started from spontaneous emission.

The laser system was investigated at a variable repetition rate of pumping, ranging from 1 Hz to 30 Hz. This variation allowed us to define the role of the RIGs in FWM due to the slow-response orientation nonlinearity or thermal nonlinearity of the NLC.

In our experiments, NLC cells with different thicknesses of nematic materials were used: a 75-μm layer of pure 5CB (nematic–isotropic transition temperature is 36°C) with the planar orientation of the NLC director, a 0.5-mm layer of pure 5CB

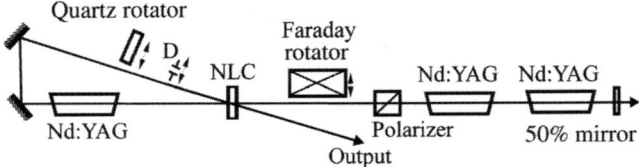

Figure 10.16. Optical scheme of a Nd:YAG laser with self-adaptive cavity based on FWM of generated waves in an NLC cell.

with the homeotropic orientation of the NLC director induced by the dc electric field applied to the walls, and a 5-mm layer of cyanobiphenyls mixture with the transition temperature of 59°C and the electrically induced homeotropic orientation. The different orientations of the director of the NLC cells, planar or homeotropic, allowed us to use different nonlinearities of NLC, either orientational or thermal.

10.3.3.1 Experiments with orientational dynamic gratings
In the first series of our experiments, we used a planar NLC cell with the nonlinear layer thickness of 75 μm. Different mutual orientations of the optical wave polarization and the NLC director were realized. In the absence of the QR the polarizations of the optical waves intersecting in the NLC cell were the same and were inclined at an angle (θ) to the orientation of the NLC director (Fig. 10.17).

The self-starting generation was observed when amplification exceeded some threshold level (Figs. 10.18 and 10.19). This generation was the nonlinear generation with participation of the NLC gratings. Indeed, threshold gain of the generation was measured to be considerably lower than that of generation without NLC or due to diffuse scattering without a second pass through the NLC layer.

It was observed that the threshold gain depends on the angle θ between the polarization and the director (Fig. 10.18). The minimum generation threshold was at optimal angle $\theta \approx 30°$. This is because the nonlinear coefficient of NLC and the magnitude of the dynamic grating are maximal for this angle in case of the orientational nonlinearity [5–7]. Thus, such angle dependence of the threshold gain confirms the main role of the orientational nonlinearity in the formation of dynamic gratings.

The generation was also observed when polarizations of almost counterrunning waves were orthogonal (in the presence of QR) (Fig. 10.19, circles). It may be explained by the interaction of the cross-polarized waves due to the orientational nonlinearity. The nonlinear generation was recorded in the scheme without any

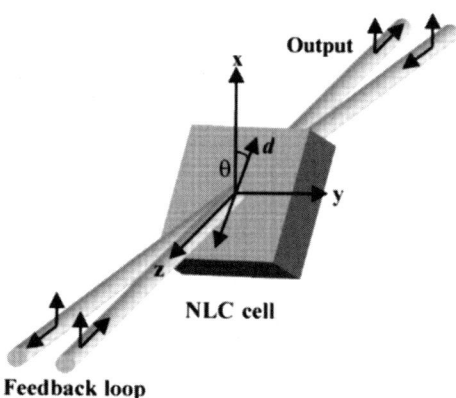

Figure 10.17. Schematic of variations of NLC-cell angle θ.

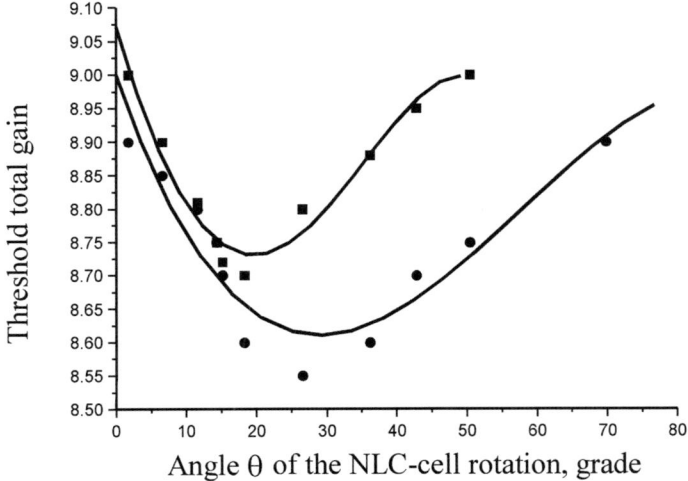

Figure 10.18. Dependencies of the threshold total gain αl_Σ ($l_\Sigma = l_1 + l_2$) of the self-starting scheme with a planar cell on rotation angle θ of an NLC cell at repetition rate 1 Hz (squares) and 30 Hz (circles).

Faraday rotator. If the Faraday rotator was included into the scheme, the generation was not observed because the wave-synchronism conditions were disobeyed.

The orientational mechanism for the formation of the large-scale reflecting RIG that completes the cavity is also demonstrated by the dependence of generation pulse

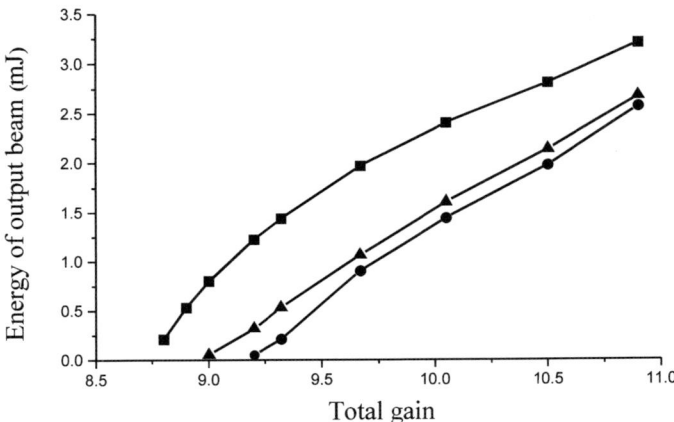

Figure 10.19. Dependencies of energy of output beam generated in the self-starting scheme with a planar cell on total gain αl_Σ at repetition rate 1 Hz. QR is located near the polarizer, so that $\theta = 90°$ (squares); the scheme without QR, the angle $\theta \approx 3°$ (triangles); QR is located in the feedback loop, so that generated beams are cross-polarized in the cell (circles).

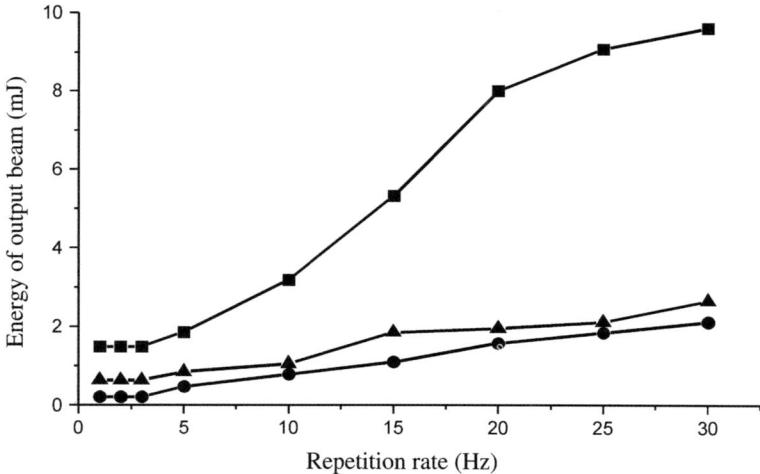

Figure 10.20. Energy of the output beam generated in the self-starting scheme with a planar cell versus repetition rate of pumping at total gain $\alpha l_\Sigma = 9.3$. QR is located near the polarizer, so that $\theta = 90°$ (squares). For the scheme without QR, the angle $\theta \approx 3°$ (triangles), QR is located in the feedback loop, so that the generated waves are cross-polarized in the cell (circles).

energy on repetition rate of pumping (Fig. 10.20). It was observed that the generated pulse energy considerably increases with increasing repetition rate. This increase may be explained as follows. The large-scale orientational grating had great relaxation time of about 100–150 ms for the intersection angle of the optical waves used in experiment (~ 0.1 rad). For a large pump-pulse period (repetition rate less than 10 Hz), the grating is able to relax between the pulses. If the repetition frequency exceeds 10 Hz, the grating will be maintained, increasing from pulse to pulse. This increase of the grating amplitude will provide an increase in the effective Q-factor of the cavity.

Thermal lenses with a focal distance of 200 cm were formed in the amplifiers at a pump pulse energy of 50 J and a pump repetition rate of 30 Hz. Under these conditions and when a diaphragm with an aperture of 2 mm was used, the laser displayed good direction stability and good quality of the output beam (Fig. 10.21a). Note that in the absence of the diaphragm the structure of the generation beam became irregular (Fig. 10.21b). An opposite situation was observed for a beam outgoing from the 50% mirror. Its structure was irregular when the diaphragm was incorporated into the scheme, and it was near Gaussian when the diaphragm was removed (Figs. 10.21c and 10.21d). However, energy of the beam outgoing from the 50% mirror was about 100 times smaller than that of the beam outgoing from the NLC. This value corresponds to the total amplification in all three amplifiers (K) in the presence of generation. Knowing the total amplification, we can estimate the reflectivity of the grating $R \approx 1/K \approx 0.01$.

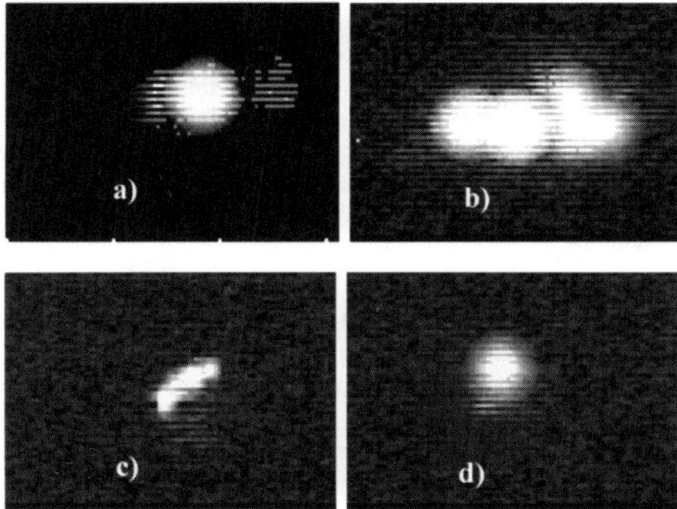

Figure 10.21. Transverse profiles of generated beams in different variants of the self-starting scheme with a planar cell at repetition rate 30 Hz with QR ($\theta = 90°$): (a) Output beam in the scheme with diaphragm; (b) output beam in the scheme without diaphragm; (c) generated beam outgoing from a 50% mirror of the scheme with diaphragm; (d) generated beam outgoing from a 50% mirror of the scheme without diaphragm.

The generation pulse train consisted of up to 30 pulses with pulse duration of 0.5–1.0 μs (Fig. 10.22). The pulse shape was temporally smooth, demonstrating single longitudinal mode operation (the bandwidth of the oscilloscope was about 50 MHz, a bandwidth large enough to visualize all possible intermode oscillations; however, these oscillations were not observed). This operation can be explained by

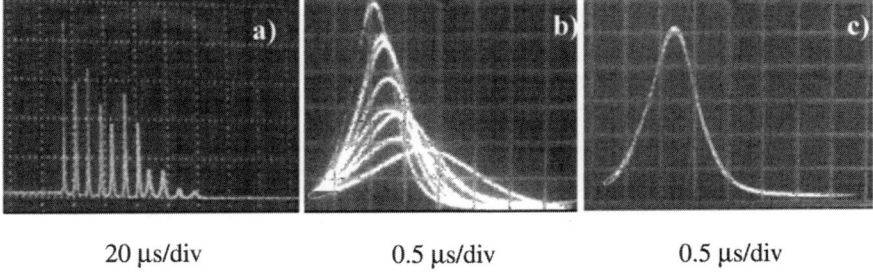

20 μs/div   0.5 μs/div   0.5 μs/div

Figure 10.22. Oscillograms of the beam pulse generated in the self-starting scheme at total gain $\alpha l_\Sigma = 10$ with planar cells on different time scales: (a) Generation pulse train; (b) 8 pulses from one generation pulse train; (c) one pulse from generation pulse train.

the necessity to have long coherence length of the self-intersecting wave for the dynamic gratings completing the cavity to be effectively induced.

Note that the single longitudinal mode operation of the scheme does not contradict the frequency shift of the interacting waves necessary for the self-starting oscillations. Indeed, the required frequency shift can be about $\tau_R^{-1} \sim 4$ kHz, which is much less than the estimated frequency bandwidth of the generated mode ($\sim$ MHz). In turn, the frequency shift of about 4 kHz is easily achieved by noised vibrations of the mirror mounting.

10.3.3.2 Experiments with thermal dynamic gratings in the NLC cell
In the second series of our experiments, NLC cells with the homeotropic orientation were used. This orientation of the NLC layer permits propagation of ordinary optical waves.

The self-starting generation was observed when amplification exceeded some threshold level (Fig. 10.23). Threshold gains of the nonlinear generation were considerably lower than those of the generation without NLC or due to diffuse scattering without a second pass through the NLC layer. The generation pulse energy did not depend on the repetition rate, so the relaxation time of the thermal grating was much less than 30 ms. Such temporal behavior corresponds to thermal nonlinearity of the NLC. The threshold of the self-starting generation decreased and the energy increased when the NLC was changed from 5CB with the nematic–isotropic transition temperature of about 36°C to a mixture of cyanobiphenyls with the nematic–isotropic transition temperature of about 59°C (Fig. 10.23). This fact

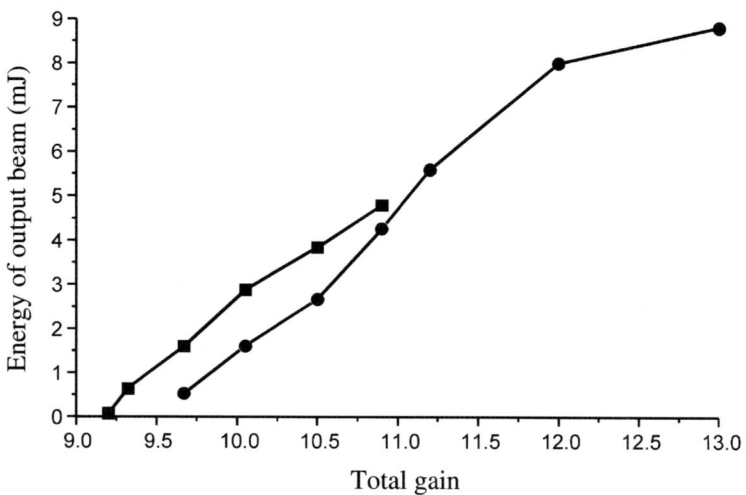

Figure 10.23. Energy of the output beam generated in the self-starting scheme with homeotropical cells versus total gain αl_Σ. Thickness of the cell is 0.5 mm (squares) and 5 mm (circles).

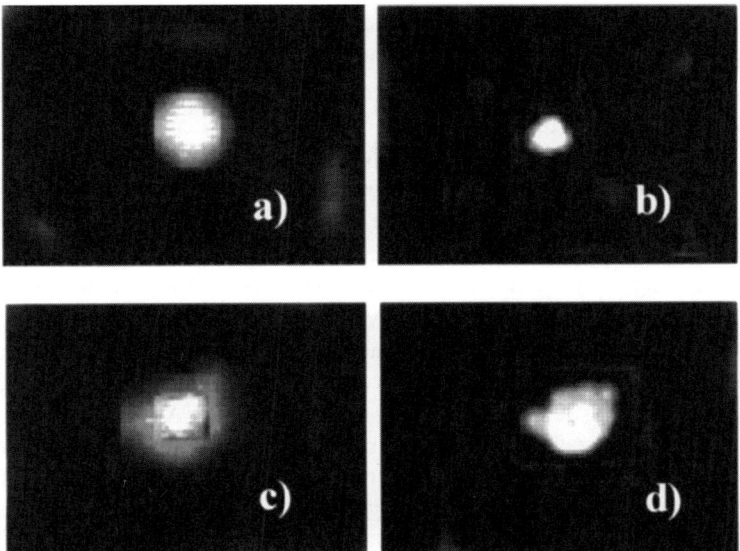

Figure 10.24. Transverse profiles of generated beams in different variants of the self-starting scheme at repetition rate 30 Hz without QR and diaphragm: (a) Output beam in the scheme with a homeotropical cell 0.5 mm in thickness; (b) output beam in the scheme with a homeotropical cell 5 mm in thickness; (c) generated beam outgoing from a 50% mirror in the scheme with a homeotropical cell 0.5 mm in thickness; (d) generated beam outgoing from a 50% mirror in the scheme with a homeotropical cell 5 mm in thickness.

can be explained by a strong increase of thermal nonlinearity near the phase transition point, which is different for different NLCs.

Our experiments performed at a high repetition rate of pumping showed good direction stability and high quality of the output beam in the schemes with both cells (Fig. 10.24). There was no need to use a diaphragm. The pulse shape and the pulse train structure were analogous to those in experiments with a planar cell: A single longitudinal mode generation with good beam quality was observed.

10.4 CONCLUSION

In this chapter, main principles of the self-pumped phase conjugation by joint SRS in a nonlinear layer with a feedback loop were discussed. It was demonstrated that the SPPC can be achieved in both the reciprocal scheme and the nonreciprocal scheme; however, the oscillation threshold in the feedback loop with $\pi/2$ phase nonreciprocity can be reduced in comparison with the reciprocal scheme. The described principles of the SPPC by joint near-forward stimulated scattering are sufficiently universal for different nonlinear media. This method of phase conjugation can be used at different optical wavelengths, being particularly promising at wavelengths where the well-known method of SPPC by stimulated Brillouin backscattering is difficult to accomplish, and for laser pulses with long durations (microsecond or more).

The SPPC mirror based on joint near-forward scattering in NLC can be used for compensation of aberrations in double-pass, four-pass, and even multipass laser amplifiers. A promising application of the phase-conjugate mirror is for the creation of self-starting lasers with good beam quality.

A self-starting laser oscillator with a cavity formed by the dynamic holographic grating induced in a nematic liquid crystal by the generating beam itself was investigated. It was found that the orientational nonlinearity in planar cells and the thermal nonlinearity in homeotropic cells can provide the formation of the RIG that completes the cavity of the self-starting laser. The Nd:YAG laser with a cavity formed by dynamic gratings in a liquid crystal is able to generate single transverse mode radiation with high beam quality in the pulse-repetitive regime. An increase in diffraction efficiency of the holographic grating at increasing repetition rate, when orientational nonlinearity of NLC is activated, is an attractive feature that opens up the possibility of using similar laser architectures in the creation of a more powerful CW generator with high quality of the output beam. On the other hand, comparison of self-starting lasers with the NLC nonlinear mirror and with the cavity completed by refractive-index gratings induced inside the laser crystal itself shows obvious advantages of the latter scheme in the pulse-repetitive regime at a repetition frequency of several tens of hertz.

ACKNOWLEDGMENTS

This work was supported in part by the Russian Foundation for Basic Research (through grants), INTAS (through grant I 97-12112), ISTC/EOARD (through grant no. 1913p), and the NATO "Science for Peace" foundation (through grant no. 974143). The authors would like to thank A. P. Zinoviev, A. S. Kuzhelev, D. V. Chausov, L. G. Kozina, and I. V. Yurasova, whose efforts have made this chapter possible.

REFERENCES

1. R. M. Herman, and M. A. Gray, Theoretical prediction of the stimulated Rayleigh scattering in liquids, *Phys. Rev. Lett.* **19**, 824–826 (1967).
2. G. I. Zaytsev, Yu. I. Kyzylasov, V. S. Starunov, and I. L. Fabelinskii, Stimulated thermal scattering of light in liquid, *JETP Lett.* **6**, 802–804 (1967) (in Russian).
3. H. J. Hoffman, Thermally induced phase conjugation by transient real-time holography: A review, *JOSA B* **3**(2), 253–273 (1986).
4. M. Croning-Golomb, B. Fisher, J. O. White, and A. Yariv, Passive (self-pumped) phase conjugate mirror, *Appl. Phys. Lett.* **41**, 689–691 (1982).

5. B. Ya. Zel'dovich and N. V. Tabiryan, The orientational optical nonlinearity of liquid crystals, *Sov. Phys. Uspekhy* **28**, 1059–1070 (1985).
6. Iam-Choon Khoo and Shin-Tson Wu, *Optics and Nonlinear Optics of Liquid Crystals*, World Scientific Publishing Co., Singapore (1993).
7. I. C. Khoo, Dynamics gratings and the associated self-diffraction and wavefront conjugation processes in nematic liquid crystals, *IEEE J. Quantum. Electron.* **22**, 1268–1276 (1986).
8. H. J. Eichler, P. Gunter, and D. W. Pohl, *Laser Induced Dynamic Gratings*, Springer-Verlag, Berlin (1986).
9. B. Ya. Zel'dovich and N. V. Tabiryan, *JETP Lett.* **30**, 510 (1979).
10. O. L. Antipov, S. I. Belyaev, A. S. Kuzhelev, and D. V. Chausov, Resonant two-wave mixing of optical beams by refractive index and gain gratings in inverted Nd:YAG, *J. Opt. Soc. Am. B* **15**, 2276–2281 (1998).
11. O. L. Antipov, A. S. Kuzhelev, D. V. Chausov, and A. P. Zinov'ev, Dynamics of refractive index changes in a Nd:YAG laser crystal under Nd^{3+}-ions excitation, *J. Opt. Soc. Am. B* **16**, 1072–1079 (1999).
12. A. E. Siegman, P. A. Belanger, and A. Hardy, Optical resonators using phase-conjugate mirrors, in *Optical Phase Conjugation*, R. A. Fisher (ed.), Academic Press, New York (1983), p. 465.
13. B. Ya. Zel'dovich, N. F. Pilipetsky, and V. V. Shkunov, *Principles of phase conjugation*, Springer, Berlin (1985).
14. O. L. Antipov, V. I. Bespalov, and G. A. Pasmanik, New possibilities of generation of pump-conjugate beams by stimulated scattering of opposed light waves, *Sov. Phys. JETP* **63**(5), 926–932 (1986).
15. O. L. Antipov, Mechanism of self-pumped phase conjugation by near-forward stimulated scattering of heterogeneous laser beams in nematic liquid crystal, *Opt. Commun.* **103**(5,6), 499–506 (1993).
16. M. Croning-Golomb, B. Fisher, J. O. White, and A. Yariv, Passive phase conjugate mirror based on self-induced oscillation in an optical ring cavity, *Appl. Phys. Lett.* **42**(11), 919–921 (1983).
17. M. Croning-Golomb, B. Fisher, J. O. White, and A. Yariv, Theory and applications of four-wave mixing in photorefractive media, *IEEE J. Quantum Electron.* **QE-20**(1), 15–30 (1984).
18. S. Odoulov, M. Soskin, and A. Khizniak, *Dynamic Grating Lasers*, Nauka, Moscow (1990).
19. I. M. Bel'dyugin, M. G. Galushkin, and E. M. Zemskov, Phase conjugation by use of a feedback for four-wave mixing, *Sov. J. Quantum Electron* **11**, 887–893 (1984) (in Russian).
20. A. A. Betin, E. V. Zhukov, and O. V. Mitropol'skii, Generation of illunination by four-wave mixing in scheme with feedback loop at 10.6 μm, *Sov. JTP Lett.* **12**(17), 1052–1056 (1986) (in Russian).
21. M. J. Damzen, R. P. M. Green, and G. J. Crofts, Reflectivity and oscillation conditions of a gain medium in a self-pumped loop geometry, *Opt. Lett.* **19**, 34–36 (1993).
22. V. I. Odintsov and L. F. Rogacheva, Effective phase conjugation in regime of parametric feedback, *Sov. JETP Lett.* **36**(8), 281–284 (1982) (in Russian).

23. O. L. Antipov and A. S. Kuzhelev, Self-pumped phase conjugation of laser beams in a nematic liquid-crystal layer with nonreciprocal feedback, *Quantum. Electron.* **25**(1), 49–52 (1995).
24. O.L. Antipov, S. I. Belyaev, and A. S. Kuzhelev, Orientational self-pumped phase conjugation in a layer of the nematic liquid crystal with a nonrecoprocal feedback, *Izv. Vuzov: Radiophysics*, **XXXVIII**(3–4), 304–311 (1995) (in Russian).
25. O. L. Antipov, A. S. Kuzhelev, and V. V. Turygin, Optimization of phase-conjugate mirrors made of nematic liquid crystals in a two-pass amplifier, *Quantum Electron.* **24**(5), 411–415 (1994).
26. Yu. A. Arutyunov, and A. I. Khizhniak, Generation conditions of phase conjugated waves in ring laser with local nonstationary nonlinearity, *Sov. Quantum. Electron.* **16**(4), 789–792 (1989).
27. A. P. Mazur, A. D. Novikov, S. G. Odulov, and M. S. Soskin, Parametric generation in media with local response in incompleted resonator, *JETP Lett.* **51**(10), 503–506 (1990).
28. A. V. Mamaev and V. V. Shkunov, Instability of near-contrunning waves in medium with local nonlinearity, *Sov. Quantum Electron.* **19**(3), 560–566 (1991).
29. O. L. Antipov, Parametric oscillation by joint stimulated scattering of near-counterruning laser beams, Ph.D. Thesis, Nizhny Novgorod State University (1992).
30. O. L. Antipov, N. A. Dvorjaninov, and V. Sheshkauskas, Parametric generation and self-pumped phase conjugation of intersecting laser beams in NLC layer doped by dye, *JETP Lett.* **53**, 610–612 (1991).
31. Iam-Choon Khoo, Hong Li, and Yu Liang, Self-starting optical phase conjugation in dyed nematic liquid crystals with stimulated thermal-scattering effect, *Opt. Lett.* **18**, 1490–1492 (1993).
32. P. Meindl, R. Macdonald, H. J. Eichler, and O. L. Antipov, Low threshold self-pumped phase conjugation of an Ar^+-laser beam in dye-doped nematic liquid crystals, *Mol. Cryst. Liq. Cryst.* **282**, 429–435 (1996).
33. M. A. Anisimov, *Critical Phenomena in Liquids and Liquid Crystals*, Nauka, Moscow (1987) (in Russian).
34. S. T. Wu, U. Efron, and L. D. Hess, *Appl. Phys. Lett.* **44**, 1033 (1984).
35. O. L. Antipov, S. I. Belyaev, H. A. Pasmanik, Four-pass laser amplifier with PC-NLC mirror, *Laser Phys.* **4**(6), 1185–1189 (1994).
36. G. J. Linford, E. R. Peressini, W. R. Sooy, and M. L. Spaer, Very long lasers, *Appl. Opt.* **13**, 379–390 (1974).
37. O. L. Antipov, D. V. Chausov, A. S. Kuzhelev, and A. P. Zinoviev, Self-starting laser oscillator with a nonlinear nematic liquid crystal mirror, *J. Opt. Soc. Am. B* **18**(1), 13–20, (2001).

CHAPTER 11

Self-Adaptive Loop Resonators with Gain Gratings

MICHAEL J. DAMZEN

The Blackett Laboratory, Imperial College, London SW7 2BW, United Kingdom

An overview is made of the use of dynamic gain gratings formed in an amplifying laser medium for optical wave-mixing, phase conjugation, and formation of self-adaptive loop resonators. Effects demonstrated in laser media include four-wave mixing (FWM), self-pumped phase conjugation (SPPC), and double phase conjugation (DPC). Application of gain gratings is presented for formation of novel self-adaptive solid-state laser systems with spatial, spectral, and temporal control of high-power radiation.

11.1 INTRODUCTION

The aim of this chapter is to give an overview of some of the key physics of gain-grating theory of nonlinear optical beam interaction by multiwave mixing in gain media and illustrate its application for spatial, spectral, and temporal control of laser radiation [1–3]. Some of the gain-grating effects demonstrated in laser amplifying media include phase conjugation by four-wave mixing [4–7], self-pumped phase conjugation [8–10], double phase conjugation [11, 12], and self-adaptive laser oscillators that can self-organize without any external optical input [13–15]. Most experiments have been conducted in pulsed laser systems where the gain is high [1–31]. More recently, continuous-wave operation has also been demonstrated [32] including continuous-wave diode-pumped solid-state operation [33, 34].

Nonlinear optics normally involves a nonlinear change in the polarization P of a material due to the presence of optical radiation field E. As with other nonlinear phenomena, to quantify the interaction of light field E with a resonant saturable

---

Phase Conjugate Laser Optics, edited by Arnaud Brignon and Jean-Pierre Huignard
ISBN 0-471-43957-6 Copyright © 2004 John Wiley & Sons, Inc.

gain material involves use of Maxwell's wave equation:

$$\nabla^2 E - \mu_0 \varepsilon_0 \frac{\partial^2 E}{\partial t^2} = \mu_0 \frac{\partial^2 P}{\partial t^2} \qquad (1)$$

where μ_0 and ε_0 are the permittivity and permeability, respectively, of free space and P is the medium polarization. In nonlinear optics with nonresonant media, the standard convention is to express the medium polarization as a power series expansion of the electric field, $P = \varepsilon_0(\chi^{(1)}E + \chi^{(2)}E^2 + \chi^{(3)}E^3 + \cdots)$, with the first term representing the linear optics regime and the higher-order terms representing the nonlinear optical behavior of the medium. The power series works successfully as a perturbation expansion with the assumption that the higher-order terms are small compared to the linear term. This condition holds well in nonresonant (or weakly resonant) media, except at extreme intensity conditions ($>10^{13}$ W/cm^2). The situation is quite different, however, for a strongly resonant medium interaction. In resonant materials, the optical radiation induces stimulated emission (or absorption), causing transfer of population between different energy states of the medium. This can lead to saturation of the amplification (or absorption) amplitude gain coefficient $\alpha(I) = \frac{1}{2}\sigma N(I)$, where σ is the stimulated emission cross section for an amplifying medium and $N(I)$ is the saturable (intensity-dependent) population inversion. The intensity-dependence of the gain coefficient causes corresponding changes in the resonant component of the susceptibility $\chi_R = i2\alpha/k$, where k is the magnitude of the wave vector of the optical radiation. Under steady-state conditions, the gain coefficient saturates with intensity I according to Eq. (2):

$$\alpha = \frac{\alpha_0}{1 + I/I_s} \qquad (2)$$

where I_s is the saturation intensity that characterizes the gain medium. In general, the medium will consist of both a nonresonant and resonant polarizability; for example, in a solid-state laser medium, a nonresonant solid-state host component P_{host} equals $\varepsilon_0 \chi_{\text{NR}} E$ and a resonant active lasing dopant component P_R is given by $\varepsilon_0 \chi_R E = \varepsilon_0 (i2\alpha/k) E$. Expanding the resonant polarization using Eq. (2) and field notation ($I/I_s = |E|^2/A_s^2$) gives

$$P_R = i\left(\frac{2\varepsilon_0}{k}\right)\alpha_0\left(E - |E|^2 E/A_s^2 + \cdots\right) \qquad (3)$$

The expansion contains terms with odd powers of E. The first term represents the small-signal gain of the inverted medium. The second term is similar to a third-order nonlinearity, because it is responsible for phase conjugation via four-wave mixing in a nonresonant FWM process and is the source of the multiwave beam coupling of FWM in a saturable gain medium. The power series expansion, however, breaks down (diverges) when I is equal to or greater than I_s, which may be at quite modest intensity values (e.g., $I_s = 2$ kW/cm^2 in Nd:YAG). A proper analysis of gain gratings and multiwave mixing requires the use of Eq. (2), and a perturbation expansion is not generally valid. The change in susceptibility is also seen to be imaginary, assuming line-center interaction of a simple resonance line profile. If operating off-line center, the gain coefficient is complex and α_0 becomes $\alpha_0/(1 + i\delta)$, where δ is the detuning normalized to linewidth of the resonance.

The interference of two (or more) coherent beams $I(x)$ in the laser amplifier can cause a spatial modulation of the population inversion $N(x)$. This modulation of the inversion is commonly known as spatial hole burning in lasers and leads to formation of a gain grating $\alpha(x) = \frac{1}{2}\sigma N(x)$, where σ is the stimulated emission cross section. The gain grating can act as a very efficient diffractive optical element exhibiting high diffraction efficiency that may be even greater than unity [16] and also exhibiting very high phase conjugate reflectivity using four-wave mixing (FWM) geometries [5–7].

Phase conjugation by FWM has been performed in several types of gain media— for example, dye lasers [5, 17], CO_2 lasers [4], excimer lasers [18], flashlamp-pumped solid-state Nd:YAG [2, 5–7], diode-pumped solid-state Nd:YVO$_4$ [19, 33, 34], and copper vapor lasers [20]. There is a very significantly large body of work on phase conjugation in CO_2 active media in the Russian and former Soviet literature. In CO_2 laser media, gain, refractive index, and thermal nonlinearities can be manifest and the precise physical process is generally more complex to describe than the pure gain process. More recently, experiments have focused on development of solid-state lasers based on the gain-grating mechanism. In this chapter, the theory developed applies broadly to most gain media, but we shall limit our illustration of experimental implementations to the solid-state laser systems. Indeed, the highest phase conjugate reflectivities of >250,000% have been demonstrated in flash-lamp-pumped Nd:YAG by FWM geometries [6, 7]; in self-pumped phase conjugate geometries, reflectivities of >1,000,000% [21] have been achieved. The simultaneous presence of refractive index and gain gratings can also occur, and this will be treated briefly. The formation of thermally induced refractive index gratings will also be treated separately at the end of the chapter as an alternative to gain gratings for phase conjugation and adaptive laser operation.

The key to high phase conjugate reflectivity is to have a laser medium with high small-signal gain since this is a key nonlinearity of the gain medium, providing both amplification of the interacting beams and available magnitude of the gain modulation induced. Table 11.1 lists some of the key features of a gain medium as a nonlinear phase conjugation device and gives a list of corresponding features of the stimulated Brillouin scattering (SBS) and photorefractive materials as a comparison. The table uses Nd:YAG as the example of a typical gain medium because this has provided the most solid-state laser material demonstrations of gain-grating phase conjugation and adaptive laser operation. As Table 11.1 shows, some of the main benefits of the gain medium as a phase conjugation device are its high power capability, fast response, and excellent material quality. Since there is no absorption required for the nonlinearity of gain saturation, there are no power dissipation issues (except for the inversion mechanism) allowing the potential for arbitrary scaling in power levels. Also, since the gain mechanism depends on gain saturation, there can be high-efficiency extraction of the energy stored in the amplifying medium. These qualities make phase conjugation by gain gratings very interesting for control of high-power laser radiation. Compared to saturable gain gratings, the photorefraction process is very slow, limited in power handling capability

TABLE 11.1. Comparison of Characteristics of Laser Gain, SBS, and Photorefraction as Processes for Optical Phase Conjugation Devices

	Laser gain	SBS	Photorefraction
Examples of dimensions and material types	$L = 100$ nm; rod diameter $= 6$ mm	$L = 200$ mm; cell diameter $= 10$ mm	$L \times w \times h = 5 \times 5 \times 5$ mm
	e.g., Nd:YAG (lamp- or diode-pumped) Ti:sapphire (laser-pumped)	Liquid or gas cell (carbon disulfide; methane) multimode optical fiber (silica) (core ~ 100 μm)	e.g., BaTiO$_3$, BSO, CdTe
Power capability	High power: multiwatt High energy: ~ joule per pulse	High energy: joule per pulse	Low power μW – W
Response time (typical)	ns (pulsed)–230 μs (CW)	ns	μs–minutes
Nonlinearity (gain)	Small-signal gain $\exp(gL) \sim e^1 - e^6$	Brillouin-active gain $\exp(g_B L) \sim e^{30}$	Two-beam coupling gain $\exp(\Gamma L) \sim e^{0.5} - e^{10}$
Wavelength	Nd:YAG (1064 nm) Ti:sapphire (~ 700–1100 nm)	Liquid (visible, near IR) Gas (UV–near IR) Fiber (near IR)	BaTiO$_3$ (visible) CdTe (near-IR)
Material properties	Excellent quality, large sizes, good availability Nonabsorbing allowing high power capability	Large sizes; cheap Weak absorption and competing nonlinearities limit power handling	Some variability of dopant characteristics Limited sample size Strong absorption limits power
Numerical aperture (N.A.)	Low to moderate NA	Moderate to high RA	High NA
	Moderate resolution	Moderate to high resolution	High resolution
Key application area	High power (energy) laser beam control	High average power (energy) laser beam control	Low power image/signal processing Interferometry

because it relies on absorption for the redistribution of charge in its mechanism; it also relies on limited sample size and variable dopant quality of some samples. SBS represents the main competing nonlinear optical mechanism for the control of high-power lasers since it can have low absorption for the radiation, when using high-purity SBS media, and has fast response time. A major limitation of SBS in bulk media is its high-power threshold requirement, making it mainly suitable in high-energy pulsed laser systems. At high peak power, competing nonlinearities can occur (e.g., stimulated Raman scattering, breakdown, thermal nonlinearity) limiting the range of operation. More recently, interest has been shown in using multimode optical fibers for reducing the power threshold of the SBS process and allowing the use of long pulse and even continuous-wave laser operation.

11.2 THEORY OF MULTIWAVE MIXING IN GAIN MEDIA

In this section we consider the appropriate equations for the evolution of the gain medium and optical fields in a four-wave mixing type interaction (see Fig. 11.1) where, for simplicity, we illustrate a single (transmission) gain grating being written. After summarizing the fundamental starting equations, we proceed to investigate the steady-state case, then the transient regime of operation (interaction time ≪ laser inverted state lifetime), and finally a full general time regime of interaction.

11.2.1 Rate equation for the laser gain coefficient

The appropriate equation for the evolution of the gain coefficient in a four-level laser system is given by

$$\frac{\partial \alpha(r,t)}{\partial t} = R(t) - \frac{1}{U_s} I_T(r,t) \cdot \alpha(r,t) - \frac{\alpha(r,t)}{\tau_\ell} \qquad (4)$$

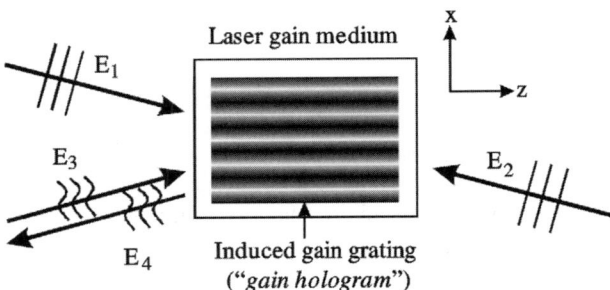

Figure 11.1. Gain FWM showing how two beams (E_1 and E_3) write a gain grating and a third beam (E_2) scatters off this grating with a π phase shift and is the phase conjugate of one of the writing beams ($E_4 \propto E_3^*$).

where $\alpha(t)$ is the amplitude gain coefficient, $I_T(t)$ is the total optical field intensity, $R(t)$ is a pumping term that generates the medium inversion, τ_ℓ is the gain (or upper-state) lifetime, $U_s = h\nu/\sigma$ is the saturation fluence of the lasing transition, h is Planck's constant, ν is the laser frequency, and σ is the stimulated emission cross section. On the right-hand side of Eq. (1), the first term is a rate of pumping the inversion, the second is the stimulated emission rate, and the third is the spontaneous emission rate. Equation (4) can be recast as

$$\tau_\ell \frac{\partial \alpha(r,t)}{\partial t} = \tau_\ell R(t) - \left(1 + \frac{I_T(r,t)}{I_s}\right) \cdot \alpha(r,t) \tag{5}$$

where I_s is the saturation intensity of the medium given by $I_s = U_s/\tau_\ell$. For example, Nd:YAG has values $U_s = 0.5$ J/cm^2, $\tau_\ell = 230$ μs, and $I_s = 2$ kW/cm^2; the corresponding values for Ti:sapphire are $U_s = 0.8$ J/cm^2, $\tau_\ell = 3$ μs, and $I_s = 250$ kW/cm^2.

11.2.2 The optical field equation

The total intensity in the medium is given by $I_T = 1/2nc\varepsilon_0 |E_T|^2$, where E_T is the total field in the medium, n is the refractive index, c is the speed of light, and ε_0 is the permittivity of free space. If four optical fields are present, as in the four-wave mixing (FWM) geometry of Fig. 11.1, E_T can be written as

$$E_T = \sum_{j=1}^{4} A_j e^{-ik_j \cdot r} \tag{6}$$

When the total field, together with the appropriate atomic polarization for the lasing species ($P_R = \varepsilon_0 \chi_R E = -i\varepsilon_0 (2\alpha/k) E$), is substituted into the nonlinear Maxwell wave equation (with plane wave and slowly varying envelope approximations) and we consider that pairs of fields are taken to be counterpropagating to each other. The following coupled wave equations are obtained for the optical field amplitudes

$$\left((-1)^{j+1}\frac{\partial}{\partial z} + \frac{n}{c}\frac{\partial}{\partial t}\right) A_j(r,t) = +(\alpha \cdot E_T \cdot e^{+ik_j \cdot r})_j, \quad j = 1 \to 4$$

$$= \sum_{k=1}^{4} \kappa_{jk} A_k \tag{7}$$

where selection of the appropriate phase-matched components in the $\alpha \cdot E_T$ source term leads to the FWM coupling coefficients κ_{jk}, and α is determined by solution of Eq. (5).

11.2.3 The intensity interference pattern

In the general case, there are four induced gain gratings due to interference of pairs of beams as shown in Fig. 11.2. For the sake of simplicity, we consider the single

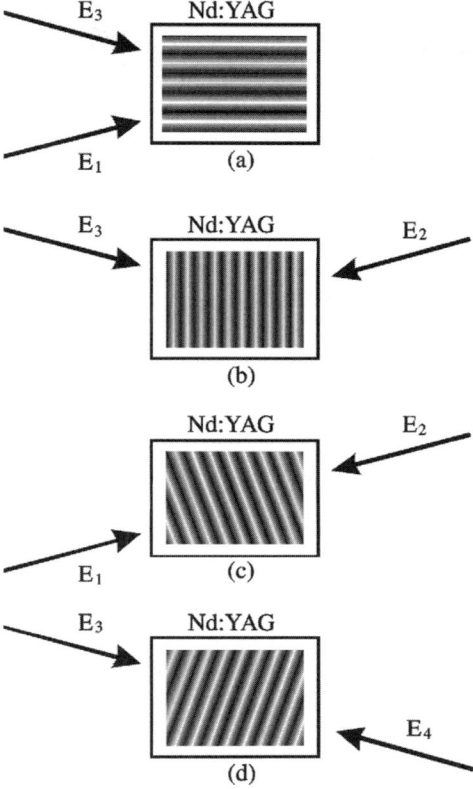

Figure 11.2. Grating formation due to FWM showing (a) a transmission grating written by beams E_1 and E_3, (b) a reflection grating written by beams E_2 and E_3, (c) a pump-pump standing-wave grating written by beams E_1 and E_2, (d) a probe-conjugate standing-wave grating written by beams E_3 and E_4.

transmission grating case when, for example, the interacting fields are such that A_1 and A_3 are co-polarized but with orthogonal polarization state to A_2 and A_4 which are also co-polarized. The resulting normalized intensity pattern is given by

$$\frac{I_T(r,t)}{I_s} = \sigma_i + |\tau_i|\cos(K_\tau x - \phi_\tau) = \sigma_i + \frac{1}{2}(\tau_i e^{-iK_\tau x} + \tau_i^* e^{+iK_\tau x}) \tag{8}$$

$$\sigma_i = \frac{1}{A_s^2}\sum_{j=1}^{4} A_j(r,t) \cdot A_j^*(r,t) \tag{9a}$$

$$\tau_i = |\tau_i|e^{+i\phi_\tau} = \frac{2}{A_s^2}\left[A_1 \cdot A_3^* + A_2^* \cdot A_4\right] \tag{9b}$$

where $K_\tau x = k \cdot (r_1 - r_3)$ is the fast spatially varying phase of the interference pattern (in the direction x, which is normal to the bisector of beams A_1 and A_3) and σ_i and τ_i are the normalized amplitudes of the average and modulated components, respectively, of the intensity pattern and where A_s is the saturation field amplitude ($I_s = (1/2) n c \varepsilon_0 A_s^2$).

11.3 THE STEADY-STATE REGIME

In the steady-state regime ($\partial/\partial t \to 0$) the solution to Eqs. (4) and (5) can be written as

$$\alpha = \frac{\tau_\ell R}{1 + I_T/I_s} = \frac{\alpha_0}{1 + I_T/I_s} \tag{10}$$

where α_0 is the (unsaturated) small-signal gain coefficient. When the interference pattern of Eq. (8) is substituted for I_T, a gain grating is written in the form

$$\alpha = \frac{\alpha_0}{1 + \sigma_i + |\tau_i|\cos(K_\tau x - \phi_\tau)} = \frac{\Gamma_0}{1 + M_\tau \cos(K_\tau x - \phi_\tau)} \tag{11}$$

where Γ_0 and M_τ are the incoherently saturated gain coefficient and a modulation parameter, respectively, and are given by

$$\Gamma_0(z) = \frac{\alpha_0}{1 + \sigma_i(z)} \tag{12a}$$

$$M_\tau(z) = \frac{|\tau_i(z)|}{1 + \sigma_i(z)} \tag{12b}$$

The periodically modulated gain coefficient of Eq. (11) can be expanded in a Fourier cosine series as follows:

$$\alpha(x, z) = \sum_{n=0}^{\infty} \alpha_\tau^{(n)}(z) \cos\left[n(K_\tau x - \phi_\tau)\right] = \frac{1}{2} \sum_{n=0}^{\infty} \alpha_\tau^{(n)}(z) e^{-in(K_\tau x - \phi_\tau)} + \text{c.c.} \tag{13}$$

with coefficients given by

$$\alpha^{(0)}(z) = +\frac{\Gamma_0}{\sqrt{1 - M_\tau}} \tag{14a}$$

$$\alpha^{(n)}(z) = (-1)^n \frac{2\Gamma_0}{\sqrt{1 - M_\tau}} \left(\frac{1 - \sqrt{1 - M_\tau}}{M_\tau}\right)^n \tag{14b}$$

The substitution of Eq. (13) into Eq. (7) and the selection of the appropriate phase-matched components lead to the following set of coupled equations describing the steady-state FWM interaction:

$$+\frac{dA_1}{dz} = \gamma A_1 + \kappa_\tau A_3 \tag{15a}$$

$$-\frac{dA_2}{dz} = \gamma A_2 + \kappa_\tau^* A_4 \tag{15b}$$

$$+\frac{dA_3}{dz} = \gamma A_3 + \kappa_\tau^* A_1 \tag{15c}$$

$$-\frac{dA_4}{dz} = \gamma A_4 + \kappa_\tau A_2 \tag{15d}$$

with coupling coefficients

$$\gamma(z) = \alpha_\tau^{(0)}(z) = +\frac{\Gamma_0}{\sqrt{1 - M_\tau}} \tag{16a}$$

$$\kappa_\tau(z) = \frac{1}{2}\alpha_\tau^{(1)}(z)e^{+i\phi_\tau} = -\frac{\Gamma_0}{M_\tau}\left(\frac{1}{\sqrt{1 - M_\tau}} - 1\right)e^{+i\phi_\tau} \tag{16b}$$

A further useful point to note is that manipulation of the system of Eqs. (15) yields the conserved quantity $A_1(z) \cdot A_2(z) + A_3(z) \cdot A_4(z) =$ constant, for the steady-state FWM interaction.

The first terms on the right-hand side of Eqs. (15) account for amplification of the fields in the medium, and the second terms provide coupling between the fields due to the gain grating. If we set $A_2 = A_4 = 0$, Eqs. (15a) and (15c) describe a two-beam coupling interaction. The sign of the coupling coefficient κ_τ is such that diffraction of one of the beams A_1 (or A_3) from the self-induced transmission grating into the other beam A_3 (or A_1) is destructive. There is an effective two-beam coupling loss as opposed to the case of photorefraction where, with a $\pi/2$ phase-shifted refractive grating, one beam sees a constructive interference leading to two-beam coupling gain. It is clear from this phasing of the gratings that gain grating and refractive index nonlinearities can lead to rather different physical interactions and must be considered separately.

The physical reason for the negative sign in the coupling equations for gain gratings is that the modulation of the gain is in antiphase to the interference writing it: High intensity leads to regions of low gain. This effect is shown graphically in Fig. 11.3, where the interference pattern [Eq. (8)] and corresponding gain modulation [Eq. (11)] are depicted for various intensities. In can also be seen that at highly saturating intensity levels the gain modulation can be very clearly nonsinusoidal, consisting of many higher harmonic components [Eqs. (13) and (14b)].

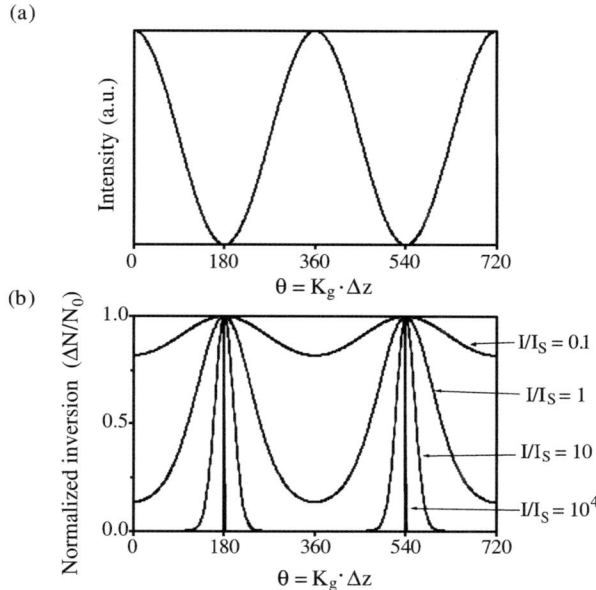

Figure 11.3. Plots of (a) the intensity pattern caused by two coherent interfering beams and (b) the corresponding modulation of the population inversion (gain grating), for four levels of saturation.

Despite the gain grating being in antiphase to the writing beams, the grating can have very strong diffraction efficiency for a Bragg-matched optical beam—for example, as in a FWM geometry. To illustrate the diffraction efficiency of a gain grating, we consider the case where $A_1 = A_3$ and take a weak A_2 (probe) beam with orthogonal polarization. The beams A_1 and A_3 form a transmission grating from which A_2 diffracts. Making the writing beams of equal intensity causes maximization of the modulation of the grating and hence maximization of the diffraction efficiency defined as $\eta = |A_4(0)/A_2(L)|^2$. In Fig. 11.4, the diffraction efficiency is plotted against normalized writing beam intensity (I/I_s) for various small-signal (amplitude) gain-length products $\alpha_0 L$ ($G_0 = \exp 2\alpha_0 L$). In all gain cases the diffraction efficiency rises with increasing writing intensity up to maximum value and then decreases with higher writing intensity. The maximum diffraction increases with higher gain and requires lower writing intensity to achieve this value. At small-signal gains $G_0 > 50$, the maximum diffraction efficiency can become greater than unity. The greater than unity value is possibly due to the stored energy in the amplifying medium.

When the counterpropagating (pump) beams A_1 and A_2 are made equal in strength and beam A_3 is a weak beam, we have the optimum case for high four-wave mixing phase conjugate reflection defined as $R = |A_4(0)/A_3(0)|^2$. Again we take A_2 orthogonally polarized to A_1 and A_3 so that a single (transmission) grating case is considered. Figure 11.5 shows the phase conjugate reflection R as a function of

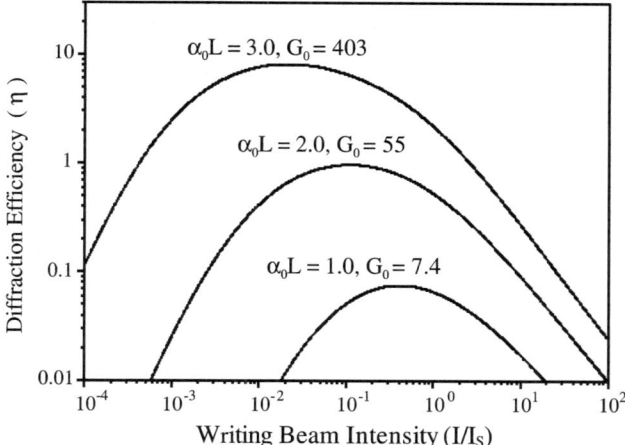

Figure 11.4. Steady-state diffraction efficiency of a single transmission gain grating versus normalized writing beam intensity for three values of small-signal gain, as indicated.

normalized pump beam intensity for various small-signal gain-length products. The plots are similar in shape and magnitude to the diffraction efficiency curves of Fig. 11.4, and greater-than-unity reflectivity is also observed at high gains $G_0 \sim 100$.

11.4 THE TRANSIENT REGIME

In the above analysis it has been assumed that we are dealing with continuous-wave interactions or interaction times much longer than the upper-state lifetime of the gain

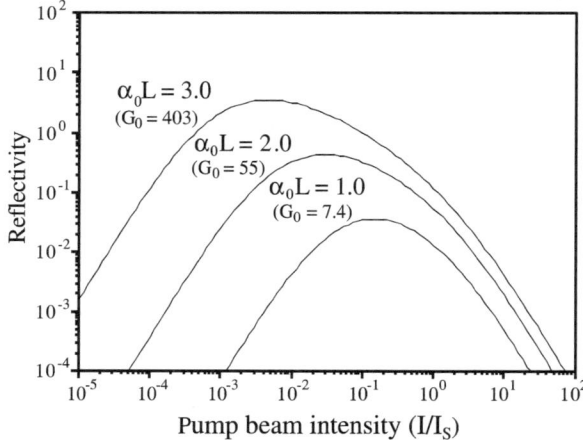

Figure 11.5. Steady-state FWM reflectivity for a weak probe beam from a single transmission gain grating versus normalized pump beam intensity for three values of small-signal gain, as indicated.

medium. For short pulse interactions that are much less than this lifetime (for Nd:YAG the upper-state laser lifetime is ~ 230 µs), the spontaneous decay of the inverted population can be neglected and in many cases so can the pumping term during the interaction. In this transient regime ($R \to 0$, $1/\tau_\ell \to 0$) the solution to Eqs. (4) or (5) can be written as

$$\alpha = \alpha_0 e^{-U_T/U_s} \tag{17}$$

where α_0 is the initial (unsaturated) small-signal gain coefficient and U_T is the time-integrated version of the interference pattern and is given by

$$\frac{U_T(r, t)}{U_s} = \frac{1}{\tau_\ell I_s} \int_0^t I_T(r, t') \, dt'$$

$$= \sigma + |\tau| \cos(K_\tau x - \phi_\tau)$$

$$= \sigma + \frac{1}{2}(\tau e^{-iK_\tau x} + \tau^* e^{+iK_\tau x}) \tag{18}$$

where the time-integrated normalized average and modulated fluences are given by

$$\sigma(r, t) = \frac{1}{\tau_\ell} \int_0^t \sigma_i(t') \, dt' = \frac{1}{\tau_\ell A_s^2} \int_0^t \sum_{j=1}^4 A_j \cdot A_j^* \, dt' \tag{19a}$$

$$\tau(r, t) = \frac{1}{\tau_\ell} \int_0^t \tau_i(t') \, dt' = \frac{2}{\tau_\ell A_s^2} \int_0^t [A_1 \cdot A_3^* + A_2^* \cdot A_4] \, dt' \tag{19b}$$

When the time-integrated interference pattern of Eq. (18) is substituted for U_T in Eq. (17a), gain grating is written in the form

$$\alpha = \alpha_0 e^{-(\sigma + |\tau|\cos(K_\tau x - \phi_\tau))} = \Gamma_0 e^{-|\tau|\cos(K_\tau x - \phi_\tau)} \tag{20}$$

where Γ_0 is the incoherently saturated gain coefficient given by

$$\Gamma_0(z, t) = \alpha_0 e^{-\sigma(z, t)} \tag{21}$$

The gain grating of Eq. (20) can be expanded in a Fourier cosine series as follows:

$$\alpha(x, z, t) = \sum_{n=0}^{\infty} \alpha_\tau^{(n)}(z, t) \cos[n(K_\tau x - \phi_\tau)]$$

$$= \frac{1}{2} \sum_{n=0}^{\infty} \alpha_\tau^{(n)}(z, t) e^{-in(K_\tau x - \phi_\tau)} + \text{c.c.} \tag{22}$$

with coefficients given by

$$\alpha^{(0)}(z, t) = +\Gamma_0 I_0(|\tau|) \tag{23a}$$

$$\alpha^{(n)}(z, t) = (-1)^n 2\Gamma_0 I_n(|\tau|) \tag{23b}$$

where $I_n(x)$ is modified Bessel function of order n and argument x. The substitution of Eq. (22) into Eq. (7) and the selection of the appropriate phase-matched components leads to the following set of coupled equations describing the highly transient FWM interaction:

$$\left(+\frac{\partial}{\partial z} + \frac{n}{c}\frac{\partial}{\partial t}\right) A_1(z, t) = \gamma A_1 + \kappa_\tau A_3 \tag{24a}$$

$$\left(-\frac{\partial}{\partial z} + \frac{n}{c}\frac{\partial}{\partial t}\right) A_2(z, t) = \gamma A_2 + \kappa_\tau^* A_4 \tag{24b}$$

$$\left(+\frac{\partial}{\partial z} + \frac{n}{c}\frac{\partial}{\partial t}\right) A_3(z, t) = \gamma A_3 + \kappa_\tau^* A_1 \tag{24c}$$

$$\left(-\frac{\partial}{\partial z} + \frac{n}{c}\frac{\partial}{\partial t}\right) A_4(z, t) = \gamma A_4 + \kappa_\tau A_2 \tag{24d}$$

with coupling coefficients

$$\gamma(z, t) = \alpha_\tau^{(0)}(z, t) = +\Gamma_0 I_0(|\tau|) \tag{25a}$$

$$\kappa_\tau(z, t) = \frac{1}{2}\alpha_\tau^{(1)}(z, t)e^{+i\phi_\tau} = -\Gamma_0 I_1(|\tau|)e^{+i\phi_\tau} \tag{25b}$$

Graphs of diffraction efficiency or FWM reflectivity as a function of normalized writing beam fluence U/U_s [10] are similar in behavior and magnitude to the steady-state plots of Figs. 11.4 and 11.5.

11.5 THE GENERAL TIME REGIME

When none of the above approximations can be made, a full transient analysis is required to describe this intermediate regime of operation. There are two approaches to solving this problem.

11.5.1 Instantaneous coupling coefficients

One approach to solving the problem is to Fourier decompose the gain grating into instantaneous time-dependent harmonics as follows:

$$\alpha(x, z, t) = \sum_{n=0}^{\infty} \alpha_\tau^{(n)}(z, t) \cos\left[nK_\tau x - \Phi_\tau^{(n)}\right]$$

$$= \frac{1}{2} \sum_{n=0}^{\infty} \alpha_\tau^{(n)}(z, t) e^{-i(nK_\tau x - \Phi_\tau^{(n)})} + c.c.$$

$$= \frac{1}{2} \sum_{n=0}^{\infty} \tilde{\alpha}_\tau^{(n)}(z, t) e^{-inK_\tau x} + c.c. \tag{26}$$

where $\tilde{\alpha}_\tau^{(n)}$ is now a complex Fourier amplitude containing phase information about the grating. This expansion can now be substituted into the gain Eq. (5) as below:

$$\tau_\ell \frac{\partial}{\partial t} \left[\frac{1}{2} \sum_{n=0}^{\infty} \tilde{\alpha}_\tau^{(n)} e^{-inK_\tau x} + c.c. \right] = \tau_\ell R(t) - \left(1 + \frac{I_T(r, t)}{I_s}\right)$$

$$\times \left[\frac{1}{2} \sum_{n=0}^{\infty} \tilde{\alpha}_\tau^{(n)} e^{-inK_\tau x} + c.c. \right] \tag{27}$$

The expansion of Eq. (26) is also inserted in the nonlinear wave Eq. (1). When the intensity interference pattern of Eq. (7) is substituted into Eq. (27) and phase-matched components are selected, a set of coupled field equations are obtained as follows:

$$\left(+\frac{\partial}{\partial z} + \frac{n}{c} \frac{\partial}{\partial t}\right) A_1(z, t) = \gamma A_1 + \kappa_\tau A_3 \tag{28a}$$

$$\left(-\frac{\partial}{\partial z} + \frac{n}{c} \frac{\partial}{\partial t}\right) A_2(z, t) = \gamma A_2 + \kappa_\tau^* A_4 \tag{28b}$$

$$\left(+\frac{\partial}{\partial z} + \frac{n}{c} \frac{\partial}{\partial t}\right) A_3(z, t) = \gamma A_3 + \kappa_\tau^* A_1 \tag{28c}$$

$$\left(-\frac{\partial}{\partial z} + \frac{n}{c} \frac{\partial}{\partial t}\right) A_4(z, t) = \gamma A_4 + \kappa_\tau A_2 \tag{28d}$$

with coupling coefficients

$$\gamma(z, t) = \alpha_\tau^{(0)}(z, t) \tag{29a}$$

$$\kappa_\tau(z, t) = \frac{1}{2}\tilde{\alpha}_\tau^{(1)}(z, t) \tag{29b}$$

and where the evolution of the complex grating harmonic amplitudes are described by

$$\tau_\ell \frac{\partial}{\partial t}\alpha_\tau^{(0)}(z, t) = \tau_\ell R(t) - (1 + \sigma_i)\alpha_\tau^{(0)} - \frac{1}{4}(\tau_i\tilde{\alpha}_\tau^{(1)*} + \tau_i^*\tilde{\alpha}_\tau^{(1)}) \tag{30a}$$

$$\tau_\ell \frac{\partial}{\partial t}\tilde{\alpha}_\tau^{(n)}(z, t) = -\frac{1}{2}\tau_i\tilde{\alpha}_\tau^{(n-1)} - (1 + \sigma_i)\tilde{\alpha}_\tau^{(n)} - \frac{1}{2}\tau_i^*\tilde{\alpha}^{(n+1)}, \quad n = 1 \to \infty \tag{30b}$$

11.5.2 Time-integrated coupling coefficients

A second approach to this problem is to use the time-integrated solution to Eq. (4) as given below

$$\alpha(t) = \left[\alpha_0 + \int_0^t R(t')\exp\left(\frac{U(t')}{U_s} + \frac{t'}{\tau_\ell}\right) dt'\right]\exp\left(-\frac{U(t)}{U_s} - \frac{t}{\tau_\ell}\right) \tag{31}$$

where α_0 is the initial small-signal gain (i.e., at $t = 0$). Using a similar Fourier decomposition to Eq. (26) the gain grating can be reexpressed as

$$\alpha(t) = \left[\alpha_0 + \int_0^t R(t')e^{+(\sigma+t'/\tau_\ell)}\sum_{n=-\infty}^{+\infty}I_n(|\tau|)e^{-in(K_\tau x - \phi_\tau)} dt'\right]e^{-(\sigma+t'/\tau_\ell)}$$

$$\times \sum_{n=-\infty}^{+\infty}(-1)^n I_n(|\tau|)e^{-in(K_\tau x - \phi_\tau)} \tag{32}$$

The substitution of this expansion into the nonlinear wave Eq. (7) and the selection of phase-matched, fluence-dependent coupling coefficients yields the following set of coupled field equations:

$$\left(+\frac{\partial}{\partial z} + \frac{n}{c}\frac{\partial}{\partial t}\right)A_1(z, t) = \gamma A_1 + \kappa_\tau A_3 \tag{33a}$$

$$\left(-\frac{\partial}{\partial z} + \frac{n}{c}\frac{\partial}{\partial t}\right)A_2(z, t) = \gamma A_2 + \kappa_\tau^* A_4 \tag{33b}$$

$$\left(+\frac{\partial}{\partial z} + \frac{n}{c}\frac{\partial}{\partial t}\right)A_3(z, t) = \gamma A_3 + \kappa_\tau^* A_1 \tag{33c}$$

$$\left(-\frac{\partial}{\partial z} + \frac{n}{c}\frac{\partial}{\partial t}\right)A_4(z, t) = \gamma A_4 + \kappa_\tau A_2 \tag{33d}$$

with coupling coefficients

$$\gamma(z, t) = \alpha_0 e^{-(\sigma+t/\tau_\ell)} I_0(|\tau|)$$

$$+ 2e^{-(\sigma+t/\tau_\ell)} \sum_{n=0}^{\infty} I_{2n}(|\tau|) \int_0^t R(t') e^{+(\sigma+t'/\tau_\ell)} I_{2n}(|\tau|) \, dt' \quad (34a)$$

$$\kappa_\tau(z, t) = \alpha_0 e^{-(\sigma+t/\tau_\ell)} I_1(|\tau|) e^{+i\phi_\tau}$$

$$+ e^{-(\sigma+t/\tau_\ell)} \sum_{n=0}^{\infty} \left[(-1)^n I_n(|\tau|) \int_0^t R(t') e^{+(\sigma+t'/\tau_\ell)} I_{n+1}(|\tau|) \, dt' \right.$$

$$\left. +(-1)^{n+1} I_{n+1}(|\tau|) \int_0^t R(t') e^{+(\sigma+t'/\tau_\ell)} I_n(|\tau|) \, dt' \right] e^{+i\phi_\tau} \quad (34b)$$

It should be noted that whereas Eqs. (30) are an infinite series of equations describing the real-time Fourier components of the gain grating, Eqs. (34) are only two equations for the time-integrated coupling coefficients but contain an infinite number of time-integrated quantities. In practical implementations of these two schemes, the maximum value of terms n can be limited to a finite value that accurately models the degree of gain saturation caused by the optical fields.

11.6 SELF-PUMPED PHASE CONJUGATION

The principle of producing self-pumped phase conjugation in a gain medium is illustrated by the schematic diagram shown in Fig. 11.6. The system consists of a laser amplifier (G_1) in a self-intersecting loop geometry. The system is injected with an input beam A_1, which passes through amplifier G_1, and then through any other loop elements, to become beam A_3. The self-intersection of beam A_3 with input beam A_1 writes a transmission gain grating in the saturable amplifier G_1 by spatial hole burning. If beam A_3 experiences phase distortions, such as can occur in a strongly pumped laser amplifier, this information is encoded in the volume gain grating. This grating acts as a diffractive optical element to form a ring cavity, which can go into self-oscillation in the backward direction, provided that a threshold condition is fulfilled. In this case a backward mode develops, consisting of beams A_2 and the diffracted component A_4. Beams A_2 and A_4 can form a self-consistent spatial mode if they are the phase conjugate of beams A_1 and A_3, respectively, and the system can thereby correct for phase distortions present in the loop and in the self-intersecting amplifier G_1, itself. For generality, the loop is considered to contain a second amplifier (G_2) and a nonreciprocal transmission element (NRTE) with forward (clockwise) and backward (anticlockwise) amplitude transmission factors equal to t_+ and t_-, respectively. The forward transmission factor t_+ of the NRTE can be selected to ensure that beams A_1 and A_3 are of comparable intensity to

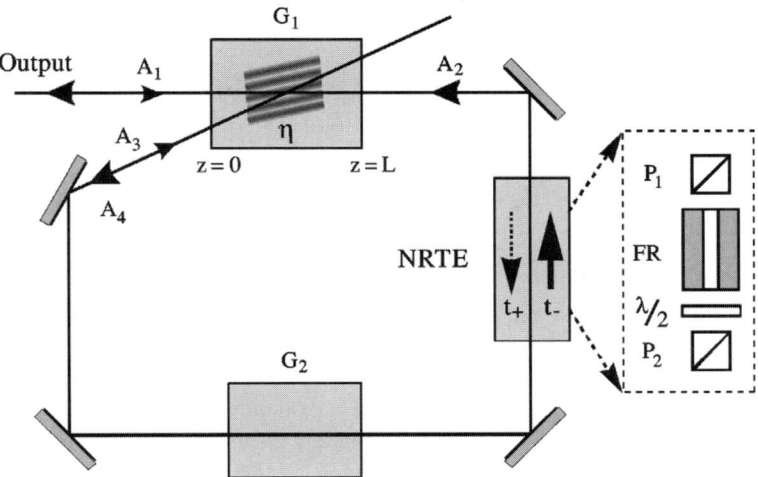

Figure 11.6. Simplified schematic of the self-conjugating loop laser, in which an amplifier G_1 is in a self-intersecting loop geometry with the loop incorporating a second amplifier G_2 and a nonreciprocal transmission element (NRTE) comprising a Faraday rotator (FR), a half-wave retardation plate ($\lambda/2$), and two polarizers (P1, P2). $A_1 \to A_4$ are the interacting optical fields.

write a deeply modulated gain grating and thereby optimize its diffraction efficiency. In experiments, the NRTE has consisted of a Faraday rotator (with 45° rotation angle) and half-wave plate positioned between two polarizers, as shown in Fig. 11.6. By rotating the waveplate, the forward and backward transmission factors can be changed.

The threshold condition for oscillation in the backward direction occurs if the round-trip loop gain of the anticlockwise ring cavity exceeds unity, that is,

$$t_- \eta g'_S \geq 1 \qquad (35)$$

where g'_S is the saturated amplitude gain of amplifier G_2 and $\eta = A_4(0)/A_2(L)$ is the amplitude diffraction efficiency of the gain grating in amplifier G_1. For the steady-state case the diffraction efficiency of the gain grating is given by [9]

$$\eta = \frac{A_4(0)}{A_2(L)} = -t_+^* \frac{I_{in}}{I_S} \exp(\alpha'_0 L') \exp(3\alpha_0 L) \sinh(\alpha_0 L) \qquad (36)$$

where $\alpha_0 L$ and $\alpha'_0 L'$ are the small-signal (amplitude) gain-length products of the FWM amplifier G_1 and loop amplifier G_2.

Following a similar derivation to the steady-state analysis [9], one can obtain an analytical expression for η for the transient case. In the transient analysis, account has to be taken of the nonideal temporal overlap of the self-intersecting input pulse, leading to an expression for the diffraction efficiency [21]:

$$\eta = \frac{A_4(0)}{A_2(L)} = -\frac{1}{2} t_+^* \sqrt{G'} G^2 \frac{U_{in}}{U_S} \exp\left[-\ln 2 \left(\frac{\tau_L}{\tau_P}\right)^2\right] \quad (37)$$

where G and G' are the small signal gains of amplifiers G_1 and G_2, respectively, τ_L is the round-trip time of the loop, and τ_P is the FWHM duration of the input pulse. The negative sign of η indicates a π phase shift induced on the diffraction of A_2 into A_4, since the modulation of the gain is in antiphase to the intensity pattern. Threshold is reached when the round-trip loop gain equals $+1$ ($t_- \eta \sqrt{G'} = +1$). Hence the normalized threshold input fluence (U_{th}/U_S), obtained from Eq. (37), which is required to achieve sufficient diffraction efficiency (η) to fulfill threshold Eq. (35) under weak saturation conditions is given by

$$\frac{U_{th}}{U_S} = \frac{2}{|t_+ t_-| G' G^2} \exp\left[\ln 2 \left(\frac{\tau_L}{\tau_P}\right)^2\right] \quad (38)$$

together with a resonance condition for the phase of the transmission factors:

$$\frac{t_+^* t_-}{|t_+ t_-|} = -1 = \exp[-i(2m-1)\pi] \quad (39)$$

where m is an arbitrary integer. The transmission factors t_+ and t_- are defined by $t_\pm = |t_\pm| \exp[-i(k_\pm L_C + \delta_\pm)]$, where $k_\pm$ is the modulus of the wave vector ($k_\pm = 2\pi n_\pm/c$) of the forward (k_+) or backward (k_-) traveling wave, L_C is the optical loop length, and $\delta_\pm$ is an additional phase shift induced inside the loop by the NRTE. This results in the following resonant frequency relationship between the backward and forward radiation:

$$\nu_- = \nu_+ + \left(m - \frac{1}{2} + \frac{\delta}{2\pi}\right)\frac{c}{L_C} \quad (40)$$

where δ is a differential phase shift between backward and forward running radiation ($\delta = \delta_+ - \delta_-$). From Eq. (40) it is seen that the backward resonant mode spacing c/L_C is independent of the differential phase shift δ. It is noted that with a differential phase shift of $\delta = \pi$, the forward and backward frequencies can be degenerate ($\nu_- = \nu_+$, for $m = 0$). With zero differential phase shift $\delta = 0$ the loop cavity is antiresonant to the input frequency and the nearest resonant backward modes are at $\pm c/2L_C$ (for $m = 0, +1$) from the input frequency.

11.6.1 Experimental setup

A diagram of an experimental self-adaptive laser system [21] is shown in Fig. 11.7. The two laser amplifiers (G_1 and G_2) are flashlamp-pumped Nd:YAG rods of length 115 mm and diameter 6 mm housed symmetrically in a pump chamber with a single central flashlamp and such that the small-signal gains of these amplifiers are equal. The injection beam A_1 was derived from a single-longitudinal-mode (SLM) Q-switched Nd:YAG laser oscillator operating on a wavelength of 1064 nm and producing pulses of 16-ns duration with a TEM_{00} spatial mode output at a repetition rate of 10 Hz. An optical isolator was placed between the injection laser and the self-adaptive system to prevent the backward oscillation causing feedback and damage to the injection laser.

The NRTE consisted of a Faraday 45° rotator and zero-order half-wave plate positioned between two Glan-air polarizers with parallel transmission axes. The forward amplitude transmission of this element could be varied by rotating the angle of the half-wave plate. A rotation of the waveplate by θ results in a 2θ rotation of the polarization state; and if we define $\theta = 0°$ as the orientation for $t_+ = 0$, then the following expressions for the forward and backward transmission of the loop are obtained:

$$t_+ = t_0 \exp[-ik_+L_C]\sin(2\theta) \tag{41a}$$
$$t_- = t_0 \exp[-ik_-L_C]\cos(2\theta) \tag{41b}$$

where t_0 describes the passive losses due to surface reflections, and so on. Expressions (41a) and (41b), when substituted into Eq. (39), illustrate that the

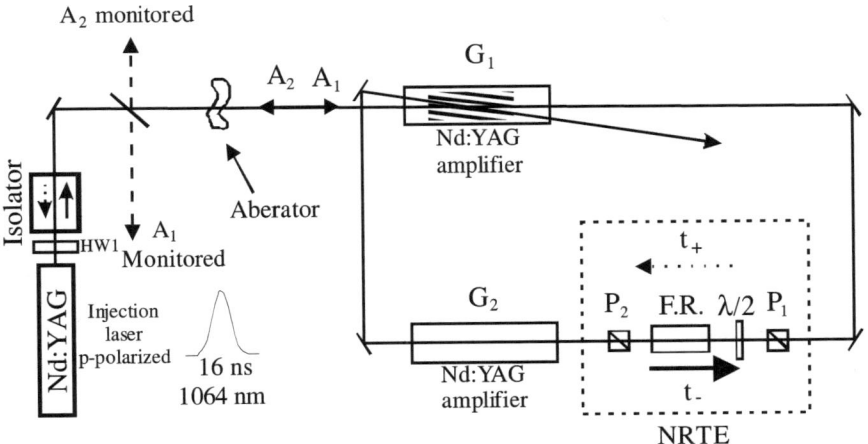

Figure 11.7. Experimental setup of the self-conjugating loop laser. The amplifiers G_1 and G_2 are Nd:YAG laser rods housed in a dual-rod amplifier head and pumped by a single flashlamp. Also shown are the injection laser, optical isolator, and components to monitor beams A_1 and A_2.

angular dependence of t_+ and t_- enables the NRTE to introduce a differential phase shift between the forward and backward running radiation of either $\delta = \pi$ for $-45° < \theta < 0°$ or $\delta = 0$ for $0° < \theta < 45°$.

11.6.2 Spatial and phase conjugation behavior

The spatial characteristics of the self-conjugating loop were investigated by monitoring the spatial profiles of the input and output beams via a low reflectivity beam splitter (see Fig. 11.6). A near spatially Gaussian injection beam was used, and the corresponding output beam was also found to exhibit a TEM$_{00}$ Gaussian spatial form but is somewhat reduced in size relative to the input due to a spatial aperturing of the beam by the finite size of the gain grating [22]. The phase conjugate corrective nature of the system was investigated by passing the injecting beam through a phase aberrator and then observing the output beam after passage back through the same aberrator. After passing back through the phase distorter, the output beam had a Gaussian spatial profile and demonstrated good correction of the phase distortions. Similar behavior was shown when the phase distorter was placed inside the loop [22].

11.6.3 Energy and temporal behavior

Figure 11.8a shows the variation of output energy (E_{out}) and reflectivity (E_{out}/E_{in}) as a function of input energy (E_{in}) for a small signal gain in the amplifiers of $G_0 = 50$ and with the NRTE adjusted to give a forward transmission factor $T_+ = |t_+|^2 \approx 0.0015$. Figure 11.8a also displays the buildup time, defined as the time delay between peak of input and peak of output radiation. Figure 11.8b contains the numerical results of the equivalent modeled system using the transient system of Eq. (24) together with boundary conditions of the loop geometry of Fig. 11.6. Numerical fluence units (U/U_s) are normalized to saturation fluence ($U_s = 550$ mJ/cm^2 in Nd:YAG [6, 23]) and can be compared with experimental data by using the relation $U_{in} = E_{in}/\pi w_{in}^2$, where the input beam radius $w_{in} = 1.9$ mm in this example. Examples of the temporal pulse traces are shown in Fig. 11.9a for a range of input energy and corresponding numerically computed pulses are shown in Fig. 11.9b.

From the graph of Fig. 11.8a it is seen that there is a threshold input energy of 75 µJ required to initiate the oscillator and that the output energy rises with increasing input energy. At an input energy ~ 4 mJ, the output maximizes at ~ 330 mJ corresponding to $\sim 30\%$ optical energy extraction efficiency of the 1.1-J stored energy in both rods for the small signal gain of 50. A maximum reflectivity of ~ 700 is reached at an input energy of ~ 0.17 mJ. When the amplifier gains were raised to $G = 125$, the output reflectivity reached values greater than 10,000 times [21].

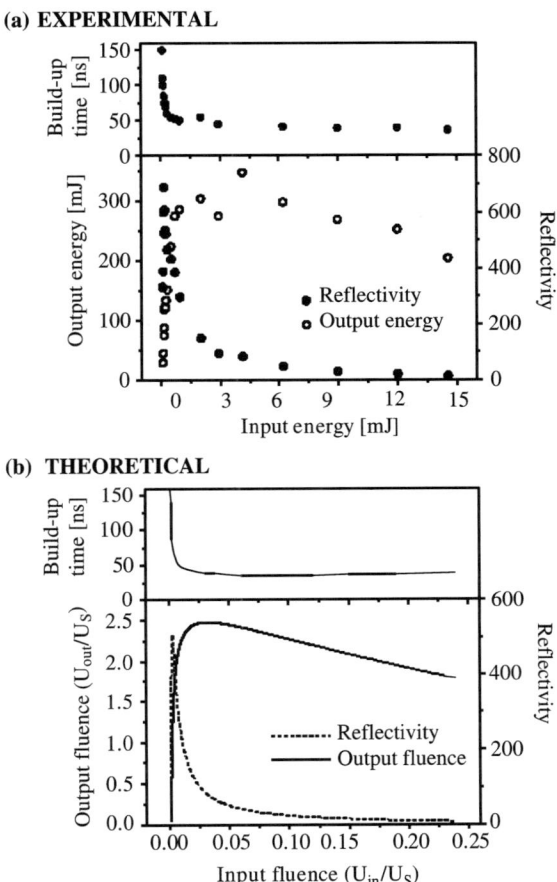

Figure 11.8. (a) Experimental output energy (E_{out}) (left axis), reflectivity (E_{out}/E_{in}) (right axis), and buildup time as a function of input energy (E_{in}) for a small signal gain of $G_0 = 50$ and forward transmission of $T_+ = 0.0015$. (b) Numerically simulated output fluence (U_{out}), reflectivity (U_{out}/U_{in}), and buildup time as a function of input fluence (U_{in}).

11.7 DOUBLE PHASE CONJUGATION

When two coherent pump beams are required to write a grating in the nonlinear medium and each beam is separately phase conjugated, we have what is referred to as double phase conjugation (DPC), which has been studied extensively in photorefractive crystals. It was predicted theoretically by Cronin-Golomb et al. [24] and demonstrated experimentally by Weiss et al. [25]. In a gain medium, since there is no equivalent of two-beam coupling gain as occurs in photorefractives, the nonlinear geometry for gain DPC and the physical process is completely different from photorefractive DPC.

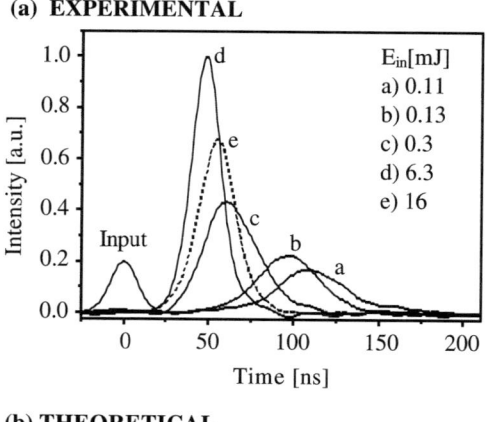
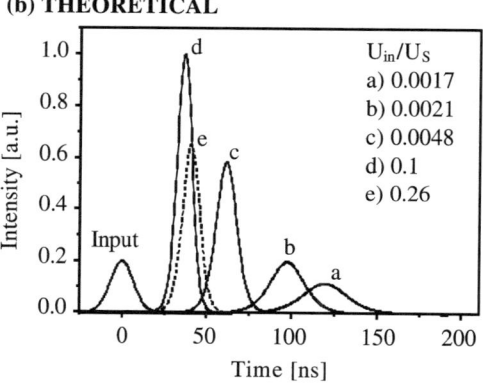

Figure 11.9. (a) Experimentally recorded and (b) numerically simulated output pulses from the resonator for different (a) input energy and (b) fluence, with $G_0 = 50$ and $T_+ = 0.0015$. Also shown is the input pulse (scale not comparable) to indicate the relative buildup time of the output pulses with respect to the arrival of the input at amplifier G_1.

A schematic of the gain DPC system is shown in Fig. 11.10, where G and G' are laser amplifiers and the NRTEs are nonreciprocal transmission elements. The NRTEs are attenuating in the clockwise direction, and they have near-unity transmission in the anticlockwise direction as illustrated by the arrows. The clockwise and anticlockwise amplitude transmission factors are denoted by t_+, t'_+ and t_-, t'_-, respectively. In Fig. 11.10 the arm 1 input field is denoted by A_1, and the arm 2 input field is denoted by A'_1. Input beam A_1 passes through amplifier G and is then attenuated by the NRTE (on the right-hand side of Fig. 11.10) and becomes A'_3, as noted on Fig. 11.10. Beam A'_3 then writes a transmission gain grating by interference with input beam A'_1. Similarly, beam A'_1 passes through amplifier G', and then it is attenuated by NRTE' (left-hand side) and becomes A_3 to write a transmission grating with beam A_1 in amplifier G. As for the SPPC, the NRTEs optimize the system by ensuring that the writing beams are of comparable intensity to produce a high modulation depth interference for formation of an optimum gain grating. The gain gratings in amplifiers G and G' have amplitude diffraction efficiencies denoted by η and η', respectively, where

$$\eta = \frac{A_4(0)}{A_2(L)} \quad \text{and} \quad \eta' = \frac{A'_4(0)}{A'_2(L)} \tag{42}$$

Above a certain injected fluence the diffraction efficiency of the gratings give rise to oscillations in the anticlockwise direction, where the condition for oscillation is

$$t_- \eta \cdot t'_- \eta' \geq +1 \tag{43}$$

Under this condition, spontaneous emission, counterpropagating to the injected beams (A_1 and A'_1) experiences a net loop gain greater than unity, which leads to the buildup of an anticlockwise mode (A_2, A'_2 and A_4, A'_4). As with the SPPC case, the anticlockwise mode is expected to be the phase conjugate of the input beams in order to produce a self-replicating mode on each loop transit. This can be understood since, if A_2 is the complex conjugate of A_1, then from four-wave mixing theory A_4 is conjugate to A_3, and hence A'_2 is conjugate to A'_1, and so on.

In an experimental Nd:YAG version of the DPC system [12], the layout was similar to the schematic diagram of Fig. 11.10 except, to reduce the threshold requirement of Eq. (43), a double pass of the writing beams through the amplifiers was made, this adds an extra pass of gain in both amplifiers and considerably lowers the threshold of the system. This system used amplifiers having small-signal gains of 100 and 35 and injection by two 16-ns pulses with TEM$_{00}$ spatial mode (see Fig. 11.11a) into arms 1 and 2. Phase conjugate reflectivity as high as 160 was observed [12]. The spatial quality of output of one of the arms of DPC system is shown in Fig. 11.11(b) exhibiting high-quality TEM$_{00}$ operation, although somewhat larger beam size to the input. When an aberrator was inserted in the DPC loop system, the output quality was almost unaffected as shown in Fig. 11.11c. The strength of the aberration

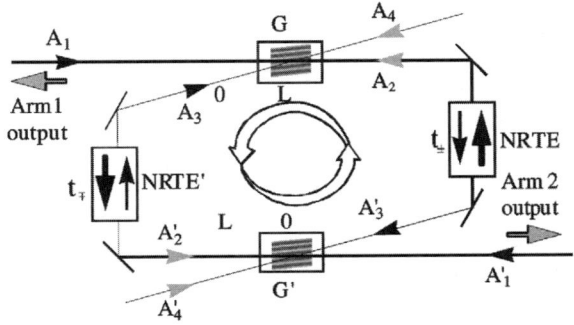

Figure 11.10. Geometry of the double phase conjugation system in gain media (G and G').

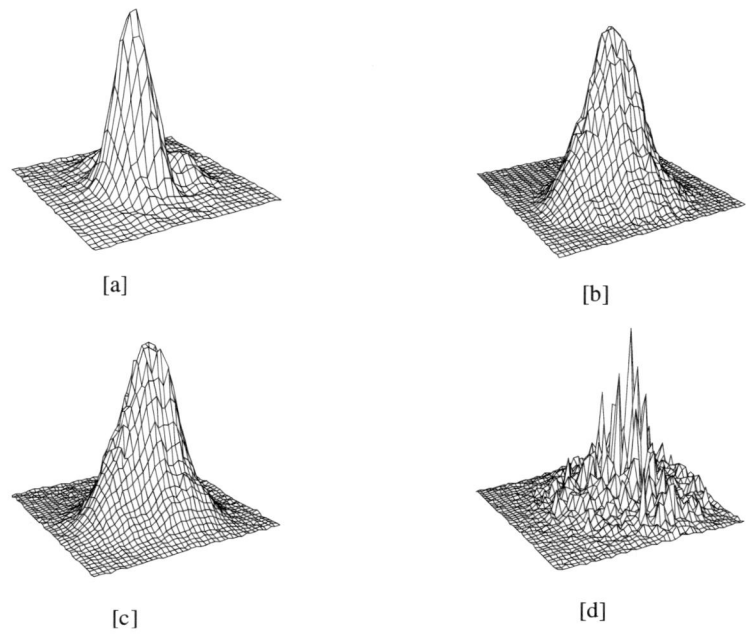

Figure 11.11. Spatial output of a gain double phase conjugator: (a) Input beam, (b) output beam from arm 1, (c) output beam when phase aberrator plate placed in loop, and (d) forward pass of the input beam through the aberrator plate.

in the forward pass through the distorter is shown in Fig. 11.11d, thereby confirming the aberration correction ability of the DPC system.

11.8 SELF-STARTING ADAPTIVE GAIN-GRATING LASERS

In the previous section it was shown that wave-mixing of beams in gain media can produce high-reflectivity phase conjugation by four-wave mixing FWM, as well as efficient self-pumped and double phase conjugation. In all cases (FWM, SPPC, and DPC) the phase conjugate reflectivity of the system could be much higher than unity. This feature allows the possibility of incorporating them in combination with a partial reflector to form a phase conjugate laser system. These systems (with the exception of the FWM mirror) are self-starting without requirement of any external input beam, and they are spatially self-adaptive due to the phase conjugate quality of the gain grating hologram formation. Self-starting adaptive lasers with no input beam have been demonstrated using a FWM gain mirror geometry [7], using a SPPC mirror [14, 26, 27] and a DPC mirror [12]. All these self-starting laser systems used a solid-state gain medium and operated in a pulsed mode with the exception of Ref. 28, where a copper vapor laser was used. More recently, a continuous-wave diode-pumped solid-state version has been operated [32].

A key interest in such adaptive laser systems arises in high average power solid-state lasers where maintenance of high beam quality is problematic due to thermally induced phase and polarization distortions. A schematic of a self-starting adaptive laser oscillator is shown in Fig. 11.12. The system shows the gain SPPC loop scheme in combination with a partial reflector as an output coupler. In this type of system, initiation of laser action is due to amplified spontaneous emission from the laser amplifiers. The spontaneous emission has low spatial and spectral coherence but has been shown to be sufficient to form a weak gain diffraction grating that can lead to the growth of the intracavity flux and the selection of increasing coherence of the radiation. Operation on a TEM_{00} spatial and single-longitudinal mode has been demonstrated [14, 27]. In one system based on flashlamp-pumped Nd:YAG laser rods [14], the transient onset and spectral selectivity of the gain gratings produced an output with energy of 600 mJ in a 10-ns single-longitudinal mode pulse. Self-adaptation of the spatial quality was demonstrated by correction of an intracavity phase plate distorter.

The diagram in Fig. 11.12 shows the self-adaptive laser with additional polarization elements (quarter-wave retardation plates QW1, QW2) that allow the system to correct polarization distortions in the loop section of the resonator [15]. The use of two quarter-wave plates to achieve counterpropagating orthogonally circular polarized beams in laser resonators is a common technique to prevent spatial hole burning due to the counterpropagating beams and is associated with mode hopping of the laser frequency. In the FWM beam configuration, shown by the shaded interaction region of Fig. 11.12, it also provides the necessary conditions to achieve vector phase conjugation (VPC) [29] and thereby correct for depolarization distortions as well as phase distortions in the laser amplifiers. The action of VPC by FWM in a laser amplifier has been investigated experimentally in an Nd:YAG amplifier [30] and analyzed theoretically [31].

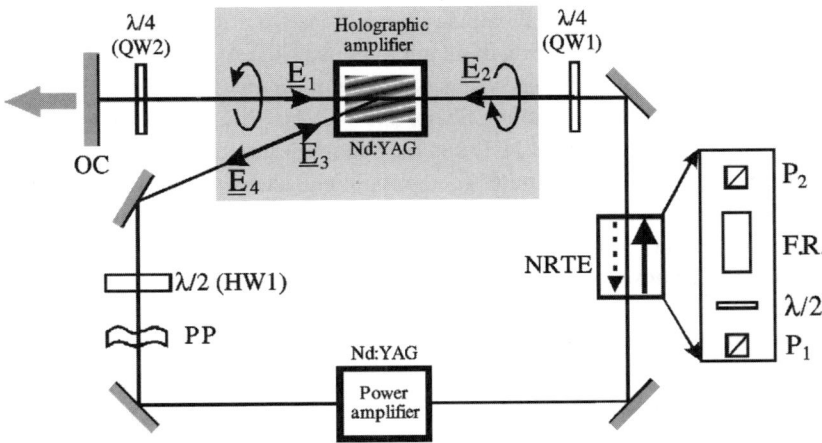

Figure 11.12. Schematic of a self-starting, self-adaptive laser oscillator incor-porating a design to produce vector phase conjugate (VPC). OC, output coupler; QW1 and QW2, quarter-wave plates; HW1, half-wave plate; FR, Faraday rotator; P1 and P2, polarizers; NRTE, nonreciprocal transmission element; PP, phase plate.

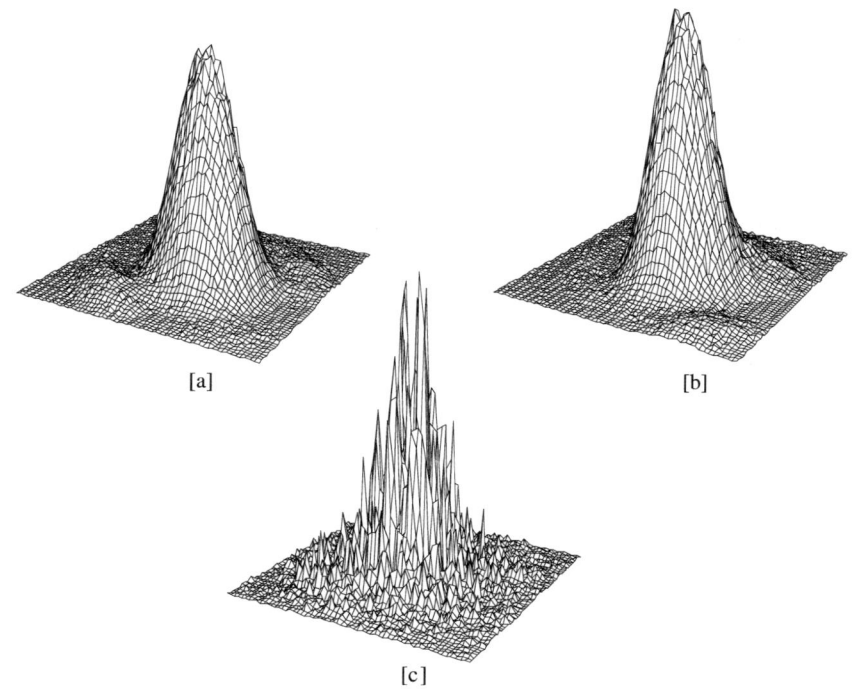

Figure 11.13. Spatial output of the self-starting adaptive laser with VPC: (a) Output without added distortion, (b) output when phase distortions are added to loop, and (c) spatial profile of beam after forward pass through phase distortions.

Figure 11.13 shows the spatial output of the VPC self-starting adaptive laser system shown in Fig. 11.12 and is based on flashlamp-pumped Nd:YAG amplifiers [15]. Figure 11.13a is the spatial output of the system without intracavity distortion (except for the thermally induced effects in the amplifier), and Fig. 11.13b is the output with phase distorter placed in loop. Figure 11.13c is the spatial distorted input beam due to the phase plate. The system demonstrates good phase conjugate correction of the distortion. A variable polarization element was placed in the loop, and the output was insensitive to the polarization state of the intracavity radiation [15].

The demonstrations of the self-starting adaptive lasers based on gain gratings show that this technology offers promise for improving the performance of laser systems. Advantages include the adaptive correction of both phase and polarization distortions in the laser system, as well as self-Q-switching for short high-peak powers and single-longitudinal mode selection.

11.9 SELF-ADAPTIVE LOOP RESONATORS USING A THERMAL GRATING HOLOGRAM

The previous sections of this chapter have shown the effectiveness of using nonlinear dynamic gain gratings for formation of adaptive laser resonators that can compensate for intracavity distortions. A few papers [35, 36] have also shown that thermally induced refractive index gratings formed by optical beams in a nonsaturable absorbing medium can also be an effective nonlinear mechanism for formation of adaptive laser resonators. The optically induced thermal grating mechanism has been shown to be capable of producing injected adaptive resonators [35] and a self-starting adaptive resonator [36] initiated by spontaneous emission.

The dynamics of thermal gratings induced by absorption of optical radiation is governed by the heat diffusion equation:

$$\frac{\partial T(r, t)}{\partial t} = \frac{\kappa}{\rho c_p} \nabla^2 T(r, t) + \frac{\alpha}{\rho c_p} I(r, t) \tag{44}$$

where T is the temperature, κ is the thermal conductivity, ρ is the density, c_p is the specific heat capacity, α is the absorption coefficient, and I is the optical intensity.

We consider the case of a cosinusoidally modulated intensity pattern, such as can be produced by the interference of a pair of coherent beams, and write it in the form

$$I(r, t) = I_0(t) + \Delta I(t)\cos(q \cdot r + \phi) \tag{45}$$

where I_0 is the mean intensity, ΔI is the modulation amplitude, q is the wave vector of the interference modulation, and ϕ is a phase constant. This intensity distribution acts as a driving term [in Eq. (44)] to produce a temperature distribution of the form

$$T(r, t) = T_b + \Delta T_0(t) + \Delta T(t)\cos(q \cdot r + \phi) \tag{46}$$

and thereby results in a corresponding modulation of the refractive index (due to its temperature dependence)

$$n(r, t) = n_b + \Delta n_0(t) + \Delta n(t)\cos(q \cdot r + \phi) \tag{47}$$

where T_b and n_b are the unperturbed temperature and refractive index of the medium, respectively, ΔT_0 and Δn_0 ($=(dn/dT)\Delta T_0$) are the mean change in temperature and refractive index, respectively, and ΔT and Δn ($=(dn/dT)\Delta T$) are the corresponding amplitudes of the cosinusoidally modulated components of the temperature and refractive index, respectively.

For the formation of the intensity pattern given by Eq. (45), we consider the interaction of four optical fields with directions and polarizations as shown in Fig. 11.14. This is a four-wave mixing (FWM) configuration commonly used in nonlinear optics for generation of phase conjugation. We write fields in the form

$$E_i(r, t) = \frac{1}{2} A_i \exp[i(\omega t - k_i \cdot r)] + \text{c.c.} \tag{48}$$

where A_i is the field amplitude, ω is the angular optical frequency, and k_i is the optical wave vector. The intensity pattern of this field and polarization combination is given by

$$I(r, t) = I_0 + \left[\Delta \tilde{I} \exp(iq \cdot r) + \text{c.c.}\right] \tag{49}$$

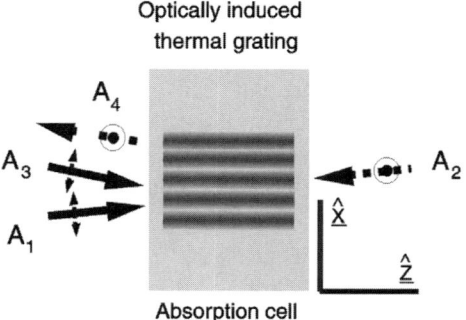

Figure 11.14. Formation of a thermally induced refractive index grating by four-wave mixing in an absorbing cell. Polarization of interacting beams are adjusted for formation of single (transmission) grating.

where $I_0(t) = \sum_{i=1}^{4} |A_i(t)|^2$ is the mean intensity and $\Delta \tilde{I}(t) = (A_1 \cdot A_3^* + A_2^* \cdot A_4)$ is the (complex) amplitude of the intensity modulation with wave vector $q = (k_1 - k_3) = (k_4 - k_2)$. The wave vector has magnitude $q = 2k \sin \theta$ and is in the $+x$ direction, where $k = |k_i|$ is the magnitude of the optical wave vector and θ is the half-angle between beams A_1 and A_3. The polarization combination has been chosen such that a single transmission grating is written.

Using Maxwell's wave equation, along with the heat diffusion equation for this four-wave mixing beam geometry, and assuming transient conditions (interaction time $\ll$ relaxation time of thermal grating $\tau = (\rho c_p / \kappa q^2)$), we derive coupled wave equations for the four-wave mixing optical fields in the following form:

$$\frac{\partial A_1}{\partial z} = \left(-\frac{\alpha}{2} + i\beta\sigma(t)\right) A_1 + i\beta\tau(t) A_3 \tag{50a}$$

$$-\frac{\partial A_2}{\partial z} = \left(-\frac{\alpha}{2} + i\beta\sigma(t)\right) A_2 + i\beta\tau^*(t) A_4 \tag{50b}$$

$$\frac{\partial A_3}{\partial z} = \left(-\frac{\alpha}{2} + i\beta\sigma(t)\right) A_3 + i\beta\tau^*(t) A_1 \tag{50c}$$

$$-\frac{\partial A_4}{\partial z} = \left(-\frac{\alpha}{2} + i\beta\sigma(t)\right) A_4 + i\beta\tau(t) A_2 \tag{50d}$$

where α is the absorption coefficient, $\beta = (2\pi/\lambda) \cdot (\alpha/\rho c_p) \cdot (dn/dT)$ incorporates the thermal parameters of the medium, and σ and τ are the mean and modulated components of the optical fluences, respectively, and are given by

$$\sigma(t) = \int^t \sum_{i=1}^{4} |A_i|^2 \, dt = \int^t I_0(t) \, dt \tag{51a}$$

$$\tau(t) = \int^t (A_1 \cdot A_3^* + A_2^* \cdot A_4) \, dt = \int^t \Delta\tilde{I}(t) \, dt \tag{51b}$$

Equations (50a)–(50d) contain terms that describe (a) the absorption of each beam (α), and (b) phase changes in each beam ($i\beta\sigma = ik\Delta n_0$) due to the mean intensity-induced change in the refractive index (Δn_0), and final terms on the right-hand side of the equations involve diffraction from the cosinusoidal modulation in the refractive index (Δn) that exchanges energy between pairs of beams ($A_1{:}A_3$ and $A_2{:}A_4$) that are Bragg-matched to the thermally induced refractive index grating.

For general solution of Eq. (50), with arbitrary input strengths and temporal pulse shapes, it is convenient to use a numerical computational method. Analytical solutions can, however, be derived for some important special cases. One such case is when $A_1, A_3 \gg A_2, A_4$ and the optical driving terms σ and τ can be considered functions of A_1 and A_3 only, and beam A_2 as a weak probe beam. When beam A_4 is initially zero at the input to the medium, this beam will be created in the medium by diffraction of beam A_2 from the refractive index grating induced by writing beams A_1 and A_3. An important parameter of the interaction is the diffraction efficiency of beam A_2 which can defined as $\eta = |A_4/A_2|^2 = I_4/I_2$. For equal writing beam intensities ($I_1 = I_3$) the diffraction efficiency can be shown to be given by the simple expression

$$\eta(t) = \exp(-\alpha L)\sin^2[\beta \cdot U(t) \cdot L'] \tag{52}$$

where $U(t) = \tau(t) = \int^t A_1 \cdot A_3^* \, dt' = \int^t I_1(t') \, dt'$ (in this case of equal writing beams) is the incident optical fluence that writes the thermal grating, and $L' = (1 - \exp(-\alpha L))/\alpha$ is an effective length of the grating. This effective length relates to the attenuation of the writing beams and hence the reduction in the induced grating with depth of the medium. It is noted that for small absorption L' is approximately equal to L (the physical length of the medium) and for high absorption L' tends to the absorption depth of the medium $1/\alpha$. Since the thermal parameter β is proportional to α, the argument of the sine is maximized for large values of αL; however, the premultiplying factor $\exp(-\alpha L)$ in Eq. (52) limits the maximum diffraction efficiency that can be achieved in a highly absorbing medium to a value of $\exp(-\alpha L)$. When the writing beams are not equal, Eq. (52) is only approximately valid because of the change in the relative phase of beams A_1 and A_3 due to the difference in the last terms in Eqs. (50a) and (50c).

For small values of diffraction efficiency, the sine term in Eq. (52) is approximately equal to its argument and if consideration of the spatial form of the beams is made ($A_i = A_i(r)$) the diffracted component $A_4(r)$ is given by

$$A_4(r, t) = ib\left[\int^t (A_1(r) \cdot A_3^*(r))\, dt'\right] A_2(r, t) \tag{53}$$

where $b = (\beta/\alpha) \cdot \exp(-\alpha L/2) \cdot [1 - \exp(-\alpha L)]$. If beams A_1 and A_2 are counter-propagating plane waves, then diffracted beam A_4 is the spatial complex conjugate (or phase conjugate) of beam A_3. In this scheme, any distortions placed in the path of beam A_3 will be corrected by propagation of beam A_4 back through those same distortions. For high values of diffraction efficiency, the depletion of the beam A_2 leads to the sinusoidal (nonlinear) variation of the diffraction efficiency with writing beam strength, according to Eq. (52). This deviation from linearity will tend to reduce the quality of the phase conjugation at high writing beam strengths. It is also noted that the normal experimental situation in four-wave mixing is for the counterpropagating beams to have a Gaussian spatial intensity distribution. This effect will lead to a spatial apodization of the phase conjugate beam A_4 relative to input beam A_3. It is therefore necessary to consider these issues in an experimental system in which it is required to produce high-quality wavefront correction of beam distortion.

11.10 EXPERIMENTAL CHARACTERIZATION OF A THERMAL GRATING

A experimental system as shown schematically in Fig. 11.15 was used to characterize the diffraction efficiency and spatial properties of an optically induced thermal grating. The absorption cell investigated was a 2-mm-thick glass cell containing a solution of copper nitrate diluted in acetone. The absorbing solution was flowed through the cell in order to reduce cumulative heating effects. The concentration of the copper nitrate determined the absorption coefficient of the cell. Acetone was chosen as the host because of its high dn/dT ($= -8.95 \times 10^{-4}$ K^{-1} [37]). In experiments, the concentration was chosen to give a solution transmission of approximately 80% (corresponding to an absorption coefficient of $\alpha = 1.1$ cm^{-1}) and, together with the reflection losses of the uncoated cell windows, leads to a total cell transmission of 74%.

The optical beams (A_1–A_3) were derived from a single-longitudinal-mode (SLM), Q-switched Nd:YAG laser system operating at a wavelength of 1064 nm and producing pulses of $\sim$ 16-ns duration (FWHM) with a TEM$_{00}$ spatial mode output at a repetition rate of 10 Hz. An optical isolator was placed between the laser system

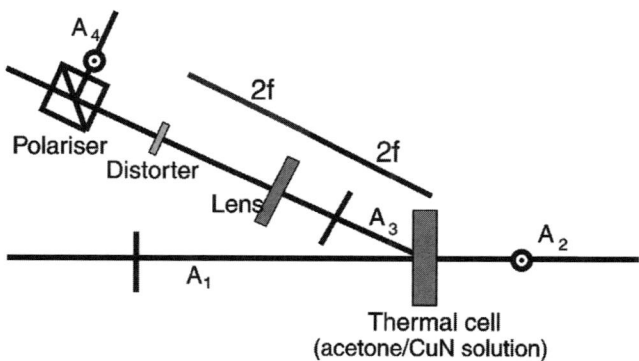

Figure 11.15. Experimental system to study the diffraction efficiency and spatial properties of an optically induced thermal grating.

and experimental setup to prevent optical feedback. The beams A_1 and A_2 were aligned to counterpropagate with respect to each other, and beam A_3 was incident at an angle $(2\theta) \sim 50$ mrad. The diffracted beam A_4 was monitored from a polarizer placed in the path of beam A_3. The beams A_1 and A_3 were linear p-polarized, and beam A_2 (and hence diffracted component A_4) was linear s-polarized. Temporal signals were monitored with a fast vacuum photodiode on a digital oscilloscope (time resolution ~ 1 ns), and spatial signals were monitored on a CCD camera with a beam profiling computer system. In order to study the effects of phase conjugation and distortion compensation from the thermal cell, we could incorporate into the path of beam A_3 a combination of a phase distortion (a glass slide etched in hydrofluoric acid) and an imaging lens.

11.10.1 Time dynamics and diffraction efficiency results

To study the time dynamics of the formation of the thermal grating and the diffraction efficiency of a probe beam from the grating, the experimental four-wave mixing system was operated without phase plate or lens in the path of beam 3. The energy of pulses incident on the absorbing solution were $E_1 = 23$ mJ, $E_2 = 4.7$ mJ, and $E_3 = 34$ mJ, where the Fresnel loss of the input window on the energy in beams 1 and 3 have been incorporated. The beams were all Gaussian in spatial shape (but with a slightly ellipticity) and mean diameter $d = 2w$ (at $1/e^2$ intensity points) was approximately 4.3 mm, giving an average beam fluence $(U = E/(\pi \cdot w^2))$ for each input beam of $U_1 = 0.16$ J/cm^2, $U_2 = 0.03$ J/cm^2 and $U_3 = 0.23$ J/cm^2. The temporal profile of the diffracted signal $(I_4(t))$ and the input probe beam $(I_2(t))$ are shown in Fig. 11.16 together with the instantaneous diffraction efficiency

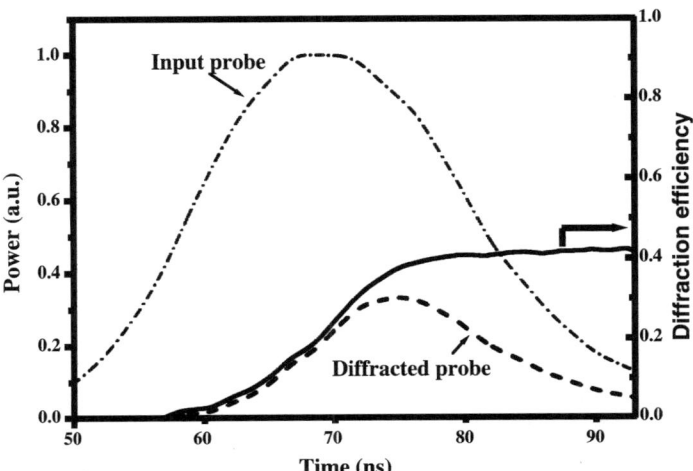

Figure 11.16. The temporal profile of the diffracted signal $(P_4(t))$ and the input probe beam $(P_2(t))$, together with the instantaneous diffraction efficiency $(\eta(t) = P_4(t)/P_2(t))$.

($\eta(t) = I_4(t)/I_2(t)$). It can be seen that the diffraction efficiency grows throughout the writing pulses, reaching a maximum at the end of the pulse interaction of 42%. This compares to a maximum possible diffraction efficiency equal to the transmission of the cell of 74%.

Figure 11.17 shows the corresponding numerical solution of the coupled wave equations (50a)–(50d) for the experimental input beam fluences and the thermal parameters for acetone [37] $\kappa = 0.2$ W/m·K, $\rho = 900$ kg/m^3, $c_p = 2180$ J/kg·K, $dn/dT = -8.95 \times 10^{-4}$ K^{-1} and the absorption of the cell $\alpha = 1.1 \times 10^2$ m^{-1}, $L = 2 \times 10^{-3}$ m, giving the thermal parameter $\beta = -0.3$ m/J and effective length $L' = 1.8 \times 10^{-3}$ m. Figure 11.17 shows a good match in the trends with the experimental results of Fig. 11.16. There is, however, a discrepancy between the maximum diffraction efficiency predicted by the numerical simulation (62%) and the experiment (42%). This discrepancy is not surprising considering the assumptions in the simulation (plane-wave analysis and no losses). If correction is made for air–glass surface reflections from the cell, then the experimental result increases to a diffraction efficiency of 45%. The spatial overlap of the spatially Gaussian beams (and hence spatial variation of the diffraction efficiency) could readily account for the remainder of the lower experimental diffraction compared to the plane wave theory.

11.10.2 Spatial issues and phase conjugation results

The spatial forms of the beams involved in the thermal grating diffraction experiments were monitored on a CCD camera and a beam profiling computer system. Figures 11.18a show a spatial profile of the grating writing beam A_3 (which is similar to the other writing beam A_1 and probing beam A_2); the diffracted beam,

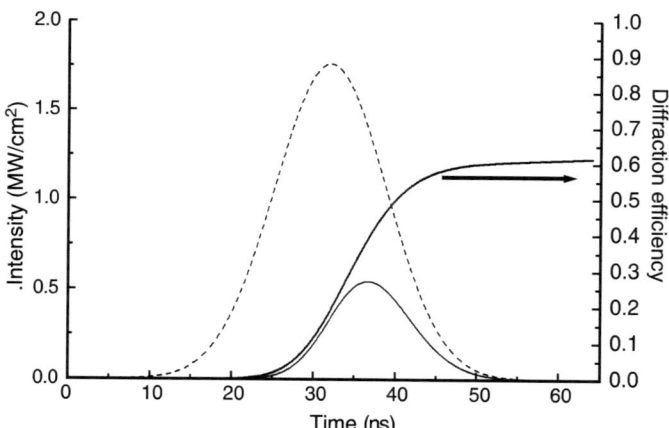

Figure 11.17. Numerical simulation of diffraction efficiency from a thermal grating under the same conditions as the experimental system.

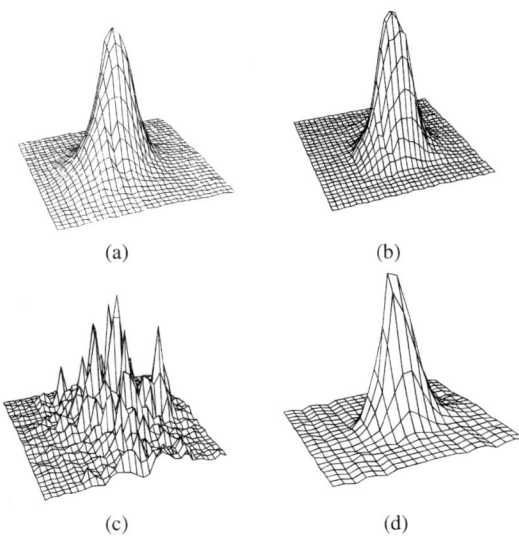

Figure 11.18. Spatial profiles of (a) the undistorted writing beam, (b) the diffracted beam, (c) the distorted writing beam, and (d) the (corrected) diffracted beam having retraversed the distorter.

A_4, is shown in Fig. 11.18b. The writing beams had a Gaussian spatial form with a slight ellipticity, and the diffracted beam is also Gaussian in shape but is somewhat smaller in size than the input beam. This reduction in beam size of the diffracted beam compared to the probe beam is expected from the spatial apodization produced by the finite spatial distribution of the thermal diffraction grating written by the Gaussian writing beams. To test the distortion correcting ability of the thermal hologram, the phase distortor and imaging lens (to ensure beam overlap in the thermal cell) was placed in the path of the writing beam to produce the resultant distorted beam A_3 as shown in Fig. 11.18c. Figure. 11.18d shows the diffracted beam having retraversed the distortor. The phase distortions are seen to be well-corrected and demonstrate the ability of the thermal grating "hologram" for aberration correction.

11.11 EXPERIMENTAL OPERATION OF A SELF-ADAPTIVE LOOP RESONATOR USING A THERMAL GRATING "HOLOGRAM"

In the previous sections, we have discussed the basic theory behind the thermal grating and performed characterizing experiments on a optically induced thermal grating formed in an absorbing solution; results have been discussed and related to the general theory including comparison with numerical modeling based on a plane-wave analysis. It has been shown that the thermal grating can have very high diffraction efficiency, and we have shown its capability in a FWM geometry for phase conjugation. In this section, an optically induced thermal grating is incorporated in an Nd:YAG system to form an adaptive laser resonator and test its efficiency and ability to correct for phase distortions in this configuration.

11.11.1 Experimental adaptive laser system

The adaptive laser resonator was constructed as shown schematically in Fig. 11.19. The resonator is based on an injected self-intersecting nonlinear loop configuration incorporating the thermal cell as the nonlinear medium. The laser gain is provided by a dual-rod Nd:YAG amplifier module with rods 115 mm in length and 6 mm in diameter and pumped by a single flashlamp. The loop also incorporates a nonreciprocal transmission element (NRTE) comprising a 45° Faraday rotator and half-wave retardation plate between a pair of polarizers. The angle of the waveplate can set the NRTE to have different transmission in the forward (injected) direction (T_+) compared to the backward direction (T_-). The transmission thermal grating formed by the self-intersection of the injection beam (A_1 and A_3) in the thermal cell acts as a diffraction element to form a ring cavity. The absorbing thermal cell was the same as described previously, using the acetone and copper nitrate solution with a 74% transmission factor. The principle of operation of the ring cavity is that it can produce oscillation of a lasing mode (A_2 and A_4) in the backward direction when the net loop gain ($G_T T_- \eta$) is greater than unity, where G_T is the total (saturated) gain of the amplifiers and η is the diffraction efficiency of the thermal grating. By setting the NRTE to have low forward transmission, it is possible to create an optimum combination of grating diffraction efficiency and low saturation of the amplifiers by the injection beam, while setting the backward transmission to near unity ensures

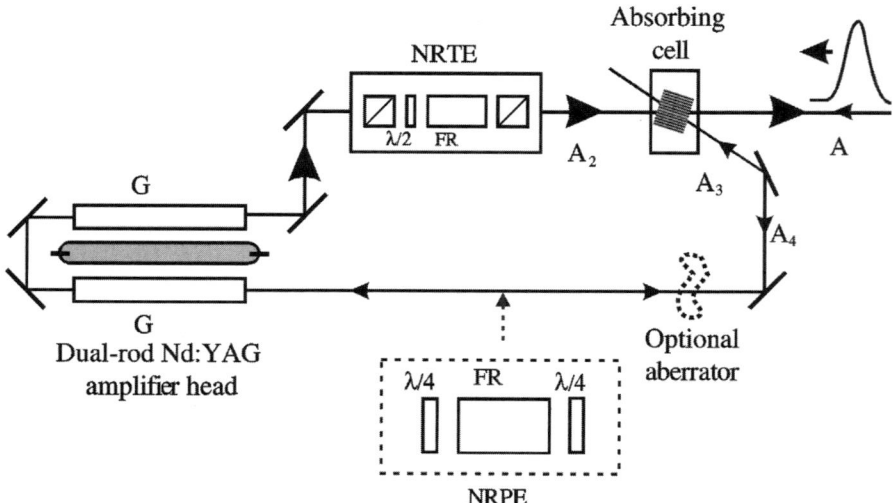

Figure 11.19. Self-adaptive loop resonator using a thermal grating hologram.

that the system can operate with minimum losses. The injection beam A_1 was derived from a Q-switched Nd:YAG laser system and had a TEM$_{00}$ input beam size (radius at $1/e^2$ intensity) of approximately 2 mm at the thermal cell.

11.11.2 Experimental results of adaptive resonator

Figure 11.20 shows the variation of the output energy (E_2) of the of the adaptive laser resonator as a function of input energy (E_1) for a small-signal gain in each amplifier of 60. Also shown in Fig. 11.20 is the ratio of output to input energy that we define as the reflectivity ($R = E_2/E_1$) of the system. The threshold energy for oscillation was ~ 1 mJ. A maximum output energy of 230 mJ was obtained for an input energy of approximately 10 mJ, corresponding to a reflectivity of 23 (2300%). A maximum value of reflectivity of 34 (3400%) was obtained at input energy of approximately 4 mJ. Because the total stored energy in the laser rods is approximately 1.1 J, the maximum output energy corresponds to an energy extraction efficiency of 20%. It should be noted that the input beam mode size underfills the laser rods, so it is also interesting to define the extraction efficiency of the energy stored in the rods within the input beam mode volume. This gives an efficiency value of 45%.

It is interesting to compare the efficiency of this thermal grating adaptive resonator with a previous study of an equivalent gain-grating adaptive resonator with one of the Nd:YAG amplifier rods used to form the gain grating element [21]. Running at similar gain levels, the gain-grating laser system produced an output energy of up to 330 mJ and corresponds to 43% greater system efficiency. It is

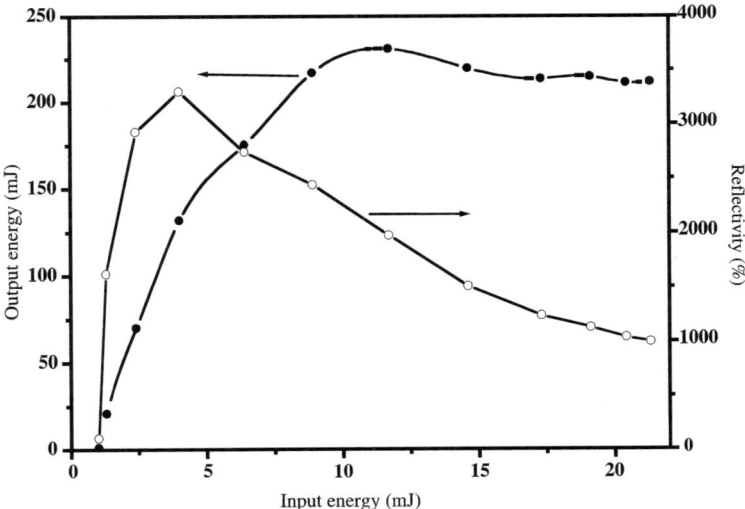

Figure 11.20. Energy output of the adaptive thermal grating resonator as a function of the injected energy. The reflectivity of the system refers to the ratio of the output to input energy.

necessary, however, to note that the thermal grating system had greater losses than the gain-grating system. Losses arise from the NRTE that had a backward transmission (T_-) somewhat less than 80% (principally due to the Glan-air polarizer losses), and the thermal absorption cell has a transmission of 74%. These loss elements are most severe because they are the final elements after the amplifiers and before the output of the laser system. It implies an energy before these elements of almost 400 mJ (or 35% energy extraction efficiency and 80% efficiency within the lasing mode volume). The NRTE in the gain-grating system is not so severe since it is intermediate to the two amplifier rods and hence the transmission loss is not at the high power output side of the cavity. This discussion suggests that a more efficient version of this cavity might be to place the final amplifier next to the thermal cell, although this would mean that this amplifier is more saturated by the input beam before it has been attenuated by the forward pass of the NRTE. While the loss of the NRTE can, in principle, be reduced to a negligible level, the loss due to the absorption of the solution in the thermal cell is inherent to the system since it is necessary for production of the dynamic nonlinear grating mechanism. A lower absorption in the cell can be used then in this case, but the threshold input energy would be increased since the induced grating diffraction efficiency is dependent on the absorption of the cell [see Eq. (52)].

At maximum output energy, the output has a pulse duration of 13 ns (FWHM) with a smooth temporal form and occurs with a delay of approximately 50 ns after the incident writing pulse. At lower input energy the output pulse occurs with a longer delay, and near threshold it showed pronounced temporal modulation corresponding to beating of two adjacent cavity modes.

The ability of the adaptive laser system for spatial correction was investigated by placing the phase distortor in the loop system as shown in Fig. 11.19. Figure 11.21 shows the spatial profiles of the input beam A_1, the aberrated loop beam A_3, and the adaptive cavity output beam A_2. The output is observed to have a substantially Gaussian spatial output and demonstrates good correction of the loop phase distortions.

The spectral properties of the adaptive resonator was investigated by monitoring the incident and output radiation with a Fabry–Perot interferometer. The Fabry–Perot had a free spectral range of 600 MHz and a finesse of about 7, corresponding to a spectral resolution of approximately 85 MHz. The results shown in Fig. 11.22 are scans through the Fabry–Perot ring pattern. The scan shows that the output radiation is single longitudinal mode and has a spectral width limited by the spectral resolution of the interferometer. It is observed that the oscillation frequency, however, is not the same as the injection frequency. This nondegenerate behavior has been predicted in other publications of this type of adaptive resonator [21, 36] and is expected in this system due to the phase shift of $\pi/2$ on diffraction from the thermally induced refractive index grating [36]. This grating induced phase shift makes the loop nonresonant with the incident frequency and causes the cavity oscillation to detune in frequency from the input to bring the oscillation to resonance. A similar effect was observed in a self-adaptive resonator employing a gain grating [21] in which case the grating phase shift is π. To compensate for the

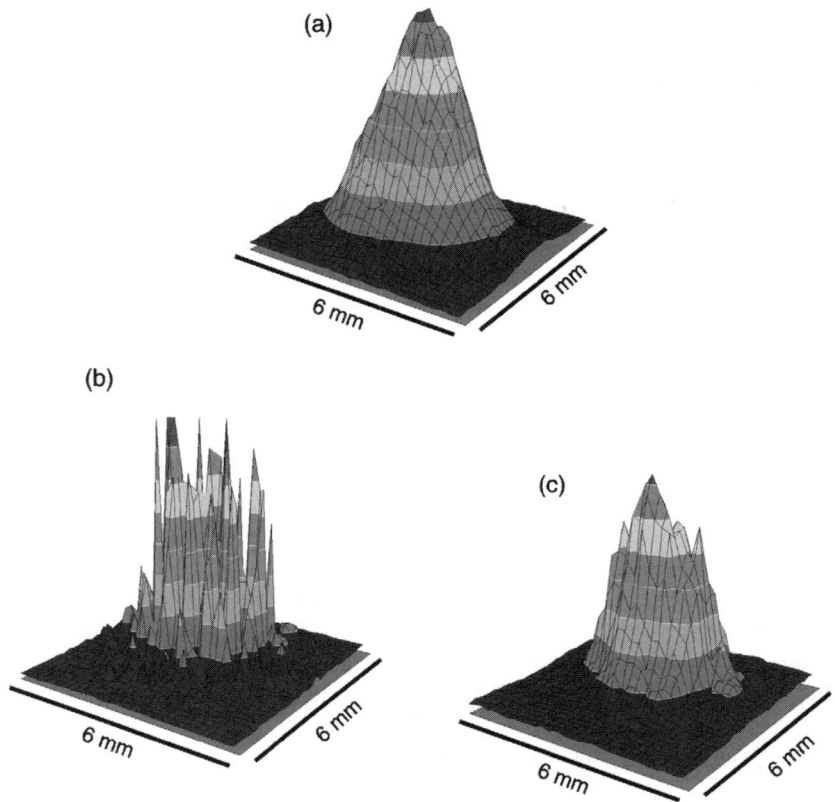

Figure 11.21. Spatial output of the adaptive thermal grating resonator. (a) Input beam, A_1; (b) aberrated loop beam, A_3; and (c) the adaptive laser output.

refractive index phase shift, a nonreciprocal phase element (NRPE) was incorporated into the loop as indicated in Fig. 11.19. This device consisted of a 45° Faraday rotator placed between a pair of quarter-wave retardation plates. The first quarter-wave plate is set to convert the linearly polarized input loop beam to circular polarized light which on passing through the Faraday rotator emerges still circularly polarized but with an added (or subtracted) 45° phase shift depending on the handedness of the circular polarization. The light then passes through the second waveplate, which is set to return the radiation back to its original linear polarization state but with the additional 45° phase advance (or lag) due to the Faraday effect. When light oscillates in the backward direction the NRPE produces the same effect but, due to the nonreciprocity of the Faraday effect, imparts a 45° phase lag (or advance) that is opposite to the forward case. The net effect is a differential phase shift of 90° or $\pi/2$ between the backward and forward paths. It is thereby possible to set the NRPE phase shift to offset the phase shift induced by the refractive index grating. Figure 11.22 shows the spectral output with the NRPE inserted. The output

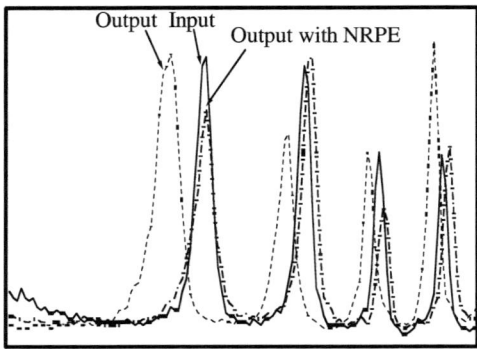

Figure 11.22. The spectral output of the adaptive thermal grating resonator. With a nonreciprocal 90° phase shift in the cavity, the output is degenerate with the input beam; without the shifter, the cavity is not resonant at the injection frequency and the output frequency is shifted.

frequency is now seen to coincide quite closely with the input frequency and to corroborate the validity of the above discussion. This measurement also clearly confirms why the NRPE device was necessary in a self-starting version of this system in earlier work [36] since the forward and backward modes are coupled and must be degenerate in frequency for cavity oscillation.

It is noted that the thermal cell offers considerable simplicity compared to using a saturable gain grating nonlinearity induced in a high-power laser amplifier and would allow the adaptive laser resonator technique to be used in a wider class of laser systems than currently possible. A problem for long-pulse or continuous-wave operation is that the strong thermal lensing induced by the absorbing thermal lens may limit the operational power range of adaptive lasers for this type of nonlinearity.

REFERENCES

1. R. A. Fisher (ed.), *Optical Phase Conjugation*, Academic Press, New York (1983).
2. A. Tomita, Phase conjugation using phase conjugation of a Nd:YAG laser, *Appl. Phys. Lett.* **34**, 463 (1979).
3. J. Reintjes and L. J. Palumbo, Phase conjugation in saturable amplifiers by degenerate four wave mixing, *IEEE J. Quantum Electron.* **18**, 1934 (1982).
4. R. A. Fisher and B. J. Feldman, On-resonant phase-conjugation reflection and amplification at 10.6 μm in inverted CO_2, *Opt. Lett.* **4**, 140 (1979).
5. P. Soan, A. D. Case, M. J. Damzen, and M. H. R. Hutchinson, High reflectivity four-wave mixing by saturable gain in Rhodamine 6G dye, *Opt. Lett.* **17**, 781 (1992).
6. G. J. Crofts, R. P. M. Green, and M. J. Damzen, Investigation of multipass geometries for efficient four-wave mixing in Nd:YAG, *Opt. Lett.* **17**, 920 (1992).

7. M. J. Damzen, R. P. M. Green, and G. J. Crofts, High-reflectivity four-wave mixing by gain saturation of nanosecond and microsecond radiation in Nd:YAG, *Opt. Lett.* **19**, 1331 (1992).
8. A. A. Denisov, O. L. Kulikov, and N. F. Pilipetskii, Pulsed CO_2 laser with a self-pumped mirror based on four-wave mixing in the laser active medium, *Sov. J. Quantum Electron.* **19**, 429 (1989).
9. M. J. Damzen, R. P. M. Green, and G. J. Crofts, Reflectivity and oscillation conditions of a gain medium in a self-conjugating loop geometry, *Opt. Lett.* **19**, 34 (1994).
10. K. S. Syed, R. P. M. Green, G. J. Crofts, and M. J. Damzen, Transient modelling of pulsed phase conjugation experiments in a saturable Nd:YAG amplifier, *Opt. Commun.* **112**, 175 (1994).
11. R. P. M. Green, G. J. Crofts, and M. J. Damzen, Novel method for double phase conjugation in gain media, *Opt. Commun.* **124**, 488 (1996).
12. D. Udaiyan, R. P. M. Green, D. H. Kim, and M. J. Damzen, Double-pumped phase conjugation in inverted Nd:YAG, *J. Opt. Soc. B* **13**, 1766 (1996).
13. I. M. Bel'dyugin, M. V. Zolotarev, S. E. Kireev, and A. I. Odintsov, Copper vapour laser with a self-pumped wavefront-reversing mirror, *Sov. J. Quantum Electron.* **16**, 535 (1986).
14. M. J. Damzen, R. P. M. Green, and K. S. Syed, Self-adaptive solid-state laser oscillator formed by dynamic gain-grating holograms, *Opt. Lett.* **20**, 1704 (1995).
15. R. P. M. Green, D. Udaiyan, G. J. Crofts, D. H. Kim, and M. J. Damzen, Holographic laser oscillator which adaptively corrects for polarization and phase distortions, *Phys. Rev. Lett.* **77**, 3533 (1996).
16. M. J. Damzen, R. P. M. Green, and G. J. Crofts, Reflectivity and oscillation conditions of a gain medium in a self-conjugating loop geometry, *Opt. Lett.* **19**, 34 (1994).
17. P. A. Routledge and T. A. King, Phase conjugation in the gain saturation of a flashlamp pumped dye laser, *Opt. Commun.* **62**, 357 (1987).
18. A. C. Cefalas and T. A. King, Phase conjugation by four wave mixing in an ArF excimer amplifier, *Opt. Commun.* **51**, 105 (1984).
19. A. Brignon, G. Feugnet, J.-P. Huignard, and J.-P. Pocholle, Efficient degenerate four-wave mixing in diode-pumped microchip Nd:YVO4 amplifier, *Opt. Lett.* **20**, 548 (1995).
20. F. V. Bunkin, V. V. Savranskii, and G. A. Shafeev, Resonant wavefront reversal in an active medium containing copper vapour, *Sov. J. Quantum Electron.* **11**, 810 (1981).
21. O. Wittler, D. Udaiyan, G. J. Crofts, K. S. Syed, and M. J. Damzen, Characterisation of a distortion corrected Nd:YAG laser with a self-conjugating loop geometry, *IEEE J. Quantum Electron.* **35**, 656 (1999).
22. M. J. Damzen, R. P. M. Green, and G. J. Crofts, Spatial characteristics of a laser oscillator formed by optically-written holographic gain-grating, *Opt. Commun.* **110**, 152 (1994).
23. O. Svelto, *Principles of Lasers*, 3rd ed., Plenum, New York, (1989), p. 291.
24. M. Cronin-Golomb, B. Fischer, J. O. White, and A. Yariv, Theory and applications of four-wave mixing in photorefractive media, *IEEE J. Qunatum Electron.* **QE-20**, 12 (1984).
25. S. Weiss, S. Sternklar, and B. Fischer, Double phase conjugate mirror: Analysis, demonstration, and applications, *Opt. Lett.* **12**, 114 (1987).

26. I. M. Bel'dyugin et al., Solid-state lasers with self-pumped phase conjugate mirrors in an active medium, *Sov. J. Quantum Electron.* **19**, 740 (1989).
27. A. Minassian, G. J. Crofts, and M. J. Damzen, Self-starting Ti:sapphire holographic laser oscillator, *Opt. Lett.* **22**, 697 (1997).
28. I. M. Bel'dyugin, M. V. Zolotarev, S. E. Kireev, and A. I. Odintsov, Copper vapour laser with a self-pumped wavefront-reversing mirror, *Sov. J. Quantum Electron.* **16**, 535 (1986).
29. R. W. Boyd, *Nonlinear Optics*, Academic Press, Boston (1992).
30. R. P. M. Green, S. Camacho-Lopez, and M. J. Damzen, Experimental investigation of vector phase conjugation in Nd^{3+}:YAG', *Opt. Lett.* **21**, 1214 (1996).
31. K. S. Syed, G. J. Crofts, R. P. M. Green, and M. J. Damzen, Vectorial phase conjugation via four-wave mixing in isotropic saturable-gain media, *J. Opt. Soc. Am. B* **14**, 2067 (1997).
32. A. Brignon and J.-P. Huignard, Continuous-wave operation of saturable-gain degenerate four-wave mixing in a $Nd:YVO_4$ amplifier, *Opt. Lett.* **20**, 2096 (1995).
33. S. Mailis, J. Hendricks, D. P. Shepherd, A. C. Tropper, N. Moore, R. W. Eason, G. J. Crofts, M. Trew, and M. J. Damzen, High phase conjugate reflectivity (>800%) obtained by degenerate four-wave mixing in a continuous-wave diode-side-pumped $Nd:YVO_4$ amplifier, *Opt. Lett.* **24**, 972 (1999).
34. M. Trew, G. J. Crofts, M. J. Damzen, J. Hendricks, S. Mailis, D. P. Shepherd, A. C. Tropper, and R. W. Eason, Multiwatt continuous-wave adaptive laser resonator, *Opt. Lett.* **25**, 1346 (2000).
35. A. A. Betin and M. S. Mangir, *CLEO Conference*, OSA Technical Digest Series, Vol. 9, Washington, D.C., (1996), p. 448.
36. S. Camacho-Lopez and M. J. Damzen, Self-starting Nd:YAG holographic laser oscillator with a thermal grating, *Opt. Lett.* **24**, 753 (1999).
37. *Handbook of Chemistry and Physics*, 70th ed., CRC Press, Boca Raton, FL (1989–1990).

INDEX

ABCD matrix, 70–74
Aberration compensation via phase conjugation, *see also* Polarization aberration correction
principle, 1–2
with Brillouin mirrors, 99–103, 162–166, 213–214
with gain gratings, 389–390
with liquid crystal mirrors, 360–363
with photorefractive mirrors, 274–275, 288–289
with thermal gratings, 398–399, 402–403
Adaptive laser resonators, *see* Laser resonators, self-starting adaptive lasers

$BaTiO_3$, 11, 257, 306
$BaTiO_3$: Co, 291–292
$BaTiO_3$: Rh, 257–284, 289–290, 292, 307, 315
oxydized crystals, 264–267
Beam cleanup, 291–292
Beam combining, 184–197
Beam quality, *see also* Fidelity of phase conjugation
definition, 109
measurement, 29–30, 116–117, 128, 150–152
Bragg condition, 20
Brillouin
effect, *see* Stimulated Brillouin scattering
enhanced Four-wave mixing, *see* Four-wave mixing, Brillouin-enhanced
gain coefficient, 24, 207–209
linewidth, 30–31, 207–209
mirrors, *see* Stimulated Brillouin scattering, phase conjugate mirrors
pulse compression, *see* Pulse compression by SBS

χ, *see* Susceptibility
Capillaries, *see* Liquid waveguide
"Cat" configuration, *see* Self-pumped phase conjugation, in photorefractive materials, internal loop geometry
CCl_4, 33, 160, 238, 242
CO_2 lasers, 369

Coherence length, 180–182
Coherent imaging, *see* Laser illumination applications
Conjugation, *see* Phase conjugation
Cr^{4+}-doped crystals, 12–13
Critical energy, 238–239
CS_2, 10, 33, 47–48, 54–56, 111, 137–142

Damage threshold, 50, 118, 207
Debye wave numbers, 261
Degeneracy circle, 279–282
Degenerate Four-Wave Mixing, *see* Four-Wave Mixing
Diffraction efficiency, *see* Gain, gratings; Thermal gratings
DLAP, 205–216
Doppler effect, 21
Double phase conjugation, 387–390

Electrostrictive coefficient, 23
Emission cross section, *see* Stimulated emission cross section
Energy extraction, 161–162, 171–172, 215–218
"Energy in the bucket" technique, 29–30, 150–152. *See also* Fidelity of phase conjugation
External loop geometry, *see* Self-pumped phase conjugation, in photorefractive materials, ring geometry

Fiber phase conjugators, 49–54, 117–137
characteristics, 117–122
continuous-wave operation, 49
damage, 50, 118
gradient index, 50–51
long pulse operation, 51–52
phase conjugate reflectivity, 50–56, 121–122
step index, 49–52, 78, 125
tapered fibers, 52–54
threshold, 118
Fidelity of phase conjugation, 28–30, 149–154, 158, 176–179, 279–283. *See also* Beam quality

407

Fourier decomposition, 380–382
Four-level laser system, 371–372
Four-wave mixing
  Brillouin-enhanced, 63, 82–84
  by free carrier generation, 11–12
  by thermal effect, 13–14, 393–399
  degenerate, 5
  in gain media, 12, 369–382
  in liquid crystal, 13
  in semiconductor amplifier, 305–306
  Kerr effect, 10
  nearly degenerate, 5
  photorefractive, 11
  principles, 6–8
Free carriers in semiconductors, see Four-wave mixing, by free carrier generation
Freon-113, 37–39, 47–49, 238
Frequency doubling, see Second harmonic generation
Frequency shift in SBS, 21

Gain
  coefficient, see Laser gain coefficient; Photorefractive, gain
  gratings, 371–382
  holograms, see Gain, gratings
  saturation, 368, 374, 378
Gas, see Stimulated Brillouin scattering, materials, gas
Glass for SBS, see Silica glass for SBS
Grating
  gain, see Gain, gratings
  intracavity, 314–318
  period, 5

Heat diffusion equation, 393
High-resolution speckle imaging, see Laser illumination applications

Internal loop geometry, see Self-pumped phase conjugation, in photorefractive materials, internal loop geometry
Intracavity phase conjugation, see Laser resonators; Phase conjugate lasers

KDP, 205. See also Second harmonic generation
Kerr effect, 10
$KNbO_3$, 294, 306

Laser diode arrays with phase conjugate feedback, 304–325
  divergence angle, 303–304, 309–310
  spatial modes, 303
  spectral characteristics, 312–313
  stability, 318–323
  temporal coherence, 313–314
Laser gain coefficient, 368
Laser illumination applications, 168–169, 180, 187
Laser resonators, 2–3, see also Q-switch, Mode locking
  self-starting adaptive lasers
    with gain grating, 390–392
    with liquid crystal, 351–363
  with Brillouin mirror, 64–105
    laser architectures, 64–70
    longitudinal modes, 84–99
    stability range, 70–78, 101–102
    theory, 70–78
  with thermal grating hologram, 399–404
LAP, 205–216
Lifetime, upper-state, 371–372
Light-induced absorption, 260–262
Liquid crystal, 13, 292–293, 331–336, 345–363
Liquids, see Stimulated Brillouin scattering, materials, liquids
Liquid waveguides, 54–56
Long pulse laser system, 168–183, 361–364, 404
Long range coherent radar, see Laser illumination applications
Loop resonators, see Ring resonators

M^2 value, see Beam quality
Master oscillator power amplifier, 2–3
  with a SBS phase conjugate mirror
    architectures, 110–112, 169–171, 215
    design rules, 114–116
  with a photorefractive phase conjugate mirror, 285–291
Materials for SBS, see Stimulated Brillouin scattering, materials
Maxwell's wave equation, see Wave equation
Mode locking, 90–93
Molecular reorientation, 13

n_2, see Nonlinear, refractive index
N_2, 32, 151–152, 172
Nd:Glass, 154–155
Nd:YAG, 12, 78, 112–113, 368
Nd:YALO, 78, 101, 113–114, 125
Nd:YLF, 157–158, 186
Nd:YVO_4, 12, 369
Nematic liquid crystal, see Liquid crystal
Nonreciprocal transmission element, 382–383, 400–404
Nonlinear
  materials, 10–14
  polarization, 23, 368

refractive index, 10, 393

Optical breakdown, 160–161
Optically-addressed light valve, 292–293
Optical phase conjugation, *see* Phase conjugation
Orientational scattering, 350

Parasitic
  grating, 274–276
  oscillations, 274
Phase conjugate lasers, 2–3. *See also* Laser resonators; Master oscillator power amplifier
Phase conjugate reflectivity, 5, 7
  via FWM in gain media, 369, 376–377
  via photorefractive effect, 273, 276–277
  via SBS in fiber phase conjugators, 50–56, 121–122
  via SBS in gas, 37–38, 47, 152
  via SBS in liquids, 37–38, 47, 160, 248–249
  via SBS in solid materials, 209–213, 217–219
  via thermal gratings, 401
Phase conjugate mirrors, *see* Phase conjugation
Phase conjugation, *see also* Four-wave mixing; Stimulated Brillouin scattering, Phase conjugate mirrors; Self-pumped phase conjugation
  choice of materials, 14, 370
  principle, 1–3
Phase locking, 184–197
Phonon
  decay rate, 24
  lifetime, 24
Photoexcitation cross sections, 259–260, 263
Photolithography, lasers for, 154, 241
Photorefractive, *see also* Three charge state model
  effect, 11
  gain, 262
  nanosecond regime, 267–271, 274, 276, 286–291
  phase conjugator, *see* Self-pumped phase conjugation, in photorefractive materials
  response time, 270–271, 278–279, 370
  two-beam coupling, 262, 269–271, 291–292, 370
Piezoelectric mirror, *see* Vibrating mirror
Pointing stability, 179–180, 196–197
Polarization aberration correction, 112, 285, *see also* Vector phase conjugation
Population inversion, 368–369
Pulse compression by SBS
  long cell configuration, 241
  materials for pulse compression, 237–239
  principles, 224–227

  tapered waveguide, 241
  theory, 227–233
  two-cell configuration, 226–227, 230–233, 242
  UV operation, 244–246
Pulse compression by SRS, 251–252
Purity of liquid SBS media, 160–161

Q-switch, 123–124, 129–130
  via Stimulated Brillouin scattering, 78–82, 103–104
  via gain gratings, 390–391

Recombination rates, 259–260, 263
Reflectivity, *see* Phase conjugate reflectivity
Refractive index grating, 20–22, 332–333, 351
Relay imaging, 155–156, 159, 179
Resonators, *see* Laser resonators
Rhodium doped barium titanate, *see* $BaTiO_3$: Rh
Ring resonators, 123–124
  with a phase conjugate mirror, *see* Laser resonators, self-starting adaptive lasers
  with Brillouin mirror, 68–70
Ring self-pumped phase conjugator, *see* Self-pumped phase conjugation, in photorefractive materials, ring geometry
Roof cut, 274

Saturable absorption, 12
Saturable amplifiers, 12, 368, 374, 378
Saturation
  fluence, 372
  intensity, 368
Second harmonic generation, 167–168, 177–179, 194–196, 323–325
Self-pumped phase conjugation
  by joint stimulated Rayleigh scatterings, 331–364
  by stimulated Brillouin scattering, *see* Stimulated Brillouin scattering
  by thermal gratings, 399–402
  in gain media, 382–387
  in photorefractive materials
    internal loop geometry, 272–274, 305–306
    response time, 278–279, 287–288
    ring geometry, 274–285, 331
    threshold, 274–276
  in semiconductor amplifier, 305–306
  principles, 8–10
Self-starting resonators, *see* Laser resonators, self-starting adaptive lasers
SF_6, 37–40, 47, 78, 82
$SiCl_4$, 33, 248–250
Silica glass for SBS, 217–220
$Sn_2P_2S_6$, 294

Space-charge field, 261, 277–278
Speckle-inhomogeneous beams, 342–345
Stability range, *see* Laser resonators
Stimulated Brillouin scattering, 10, 292, 370–371
  amplifier-oscillator configuration, 158–159. *See also* Pulse compression by SBS, two-cell configuration
  in fibers, *see* Fiber phase conjugators
  in laser resonators, *see* Laser resonators, with Brillouin mirror
  long pulse operation, 172–176
  loop geometry, 172–174, 188–191
  materials
    gas, 32, 45–47, 237. *See also* SF_6; N_2
    liquids, 33, 45–47, 238. *See also* CS_2; CCl_4; $SiCl_4$
    solids, 33. *See also* DLAP; LAP; Silica glass for SBS
  phase conjugate mirrors, 25–30, 43–56
  principles, 19–22
  pulse compression, *see* Pulse compression by SBS
  theory, 22–25, 31–38
  threshold, 30–31, 82, 118, 172–175, 212–213
  3D model, 35–38
  2D model, 31–35
  waveguides, 38–43. *See also* Fiber phase conjugators
Stimulated emission cross section, 372
Stimulated Raman scattering, *see* Pulse compression by SRS
Stimulated Rayleigh scattering, 331
Susceptibility, 4, 368

Tapered fibers, 52–54. *See also* Pulse compression by SBS, tapered waveguide
Temporal fidelity, 159–160
Thermal
  excitation coefficients, 259–260, 263
  gratings, 13–14, 393–398
  lensing, 115, 215–216, 285–286, 360
  scattering, 345–350
Three charge state model, 259–264, 278
Threshold of SBS, *see* Stimulated Brillouin scattering, threshold
Ti : Sapphire, 12
Two-beam coupling, *see* Photorefractive, two-beam coupling
Two-wave mixing, *see* Photorefractive, two-beam coupling

UV operation of SBS, *see* Pulse compression by SBS, UV operation

Vector phase conjugation, 391–392
Vibrating mirror, 274–276

Wave equation, 23, 368

X ray generation, *see* Photolithography, lasers for

Zigzag amplifier, 154–155